Proceedings of

Coastal Dynamics 2009

Impacts of Human Activities on Dynamic Coastal Processes

Proceedings of
Coastal Dynamics 2009

Impacts of Human Activities on Dynamic Coastal Processes

Tokyo, Japan   7 – 11 September 2009

Edited by

Masaru Mizuguchi

Chuo University, Japan

Shinji Sato

The University of Tokyo, Japan

World Scientific

NEW JERSEY · LONDON · SINGAPORE · BEIJING · SHANGHAI · HONG KONG · TAIPEI · CHENNAI

Published by

World Scientific Publishing Co. Pte. Ltd.
5 Toh Tuck Link, Singapore 596224
USA office: 27 Warren Street, Suite 401-402, Hackensack, NJ 07601
UK office: 57 Shelton Street, Covent Garden, London WC2H 9HE

British Library Cataloguing-in-Publication Data
A catalogue record for this book is available from the British Library.

PROCEEDINGS OF COASTAL DYNAMICS 2009
Impacts of Human Activities on Dynamic Coastal Processes
(With CD-ROM)

ISBN-13 978-981-4282-46-8 (pbk)
ISBN-10 981-4282-46-4 (pbk)

Printed in Singapore by World Scientific Printers

Foreword

Coastal area is the place where the human activities have been historically very high and the protection of the coastal area has always been an important issue. In the present industrial era, the human activities have reached a stage, where the anthropogenic effects on the natural process of coastal zone development can not be ignored, particularly in the developed countries. Some human (economical) activities may not only introduce the construction of the structures which significantly interfere with the natural development of the coastal zone but also may bring some foreign materials to the sea directly affecting the natural life in the coastal zone. It is needed for our society to improve our quality of life that we are able to predict (and evaluate) the change of coastal area in both natural and anthropogenic senses and to find the best solution when any problem arises. To this end, it is obvious that good knowledge and understandings of the dynamics played in the coastal zone are indispensable.

Many researches are being conducted to study the dynamics of the coastal system world wide. This conference Coastal Dynamics 2009, which is the sixth conference in the series following Coastal Dynamics 2005 held in April 2005 in Barcelona, Spain, brings together the field and laboratory experimentalists, theoreticians and modelers conducting research on the dynamics of the coastal systems. Their approaches include theoretical formulations, field and laboratory observations, and numerical simulations.

The main theme of Coastal Dynamics 2009 is "Impacts of human activity on dynamic coastal processes". Example topics within this theme may be listed as in the following.

Coastal hydrodynamics

Coastal sediment transport

Coastal morph-dynamics

Coastal hydro-meteorology (waves, currents, tides, tsunami, storm surge, etc.)

Coastal environmental problems and marine ecology

Observational techniques and results (field and laboratory)

Numerical simulation (models and results)

Miscellaneous coastal dynamics

In the Book of Abstract are printed the accepted abstracts together with three full keynote lectures, and the attached CD contains the corresponding full papers presented at the conference. Many abstracts (234 in total) were submitted to be reviewed by the members of the International Steering Board with help of LOC and 168 papers were accepted. Final number to be presented in the conference and to be found in BoA and CD is 142, excluding the keynotes.

Finally the International Steering Board of Costal Dynamics conference and the Local Organizing Committee would like to express sincere thanks to all those contributed to Coastal Dynamics 2009. We also acknowledge the support of Coastal Engineering Committee, JSCE, and the financial aid by Kajima Science Foundation.

Masaru MIZUGUCHI, Chair of LOC for Coastal Dynamics 2009 and member of Coastal Dynamics Conference International Steering Board

Shinji SATO, Executive chair of LOC for Coastal Dynamics 2009

CONTENTS

Keynotes

Coastal Dynamics 2009
Keynote 1

BEACH EROSION ARISING FROM ANTHROPOGENIC FACTORS

Takaaki Uda[1]

Abstract

The causes of beach erosion in Japan are classified into four main types: (a) obstruction of longshore sand transport, (b) beach changes associated with the formation of wave-shelter zones, (c) decreased fluvial sediment supply, and (d) offshore sand mining or dredging. Of these, examples associated with the formation of wave-shelter zones and decreased fluvial sediment supply were studied, typified by the changes in beaches at Otsu and Shimobara fishing ports, Takahama Port, and along the Enshu-nada coast caused by decreased fluvial sediment supply from the Tenryu River. Such erosion is closely related to land management and the "sector-by-sector" administrative system in Japan. The prevention of beach erosion thus requires both scientific research and improved coastal management through enhancing the public consensus.

Key words: beach erosion, wave-sheltering structure, fluvial sediment supply, land management

1. Introduction

In recent years, beach erosion has become a serious problem in Japan and natural sandy beaches have been rapidly disappearing, resulting in an artificial coastline excessively protected by seawalls and concrete armour units. When visiting a fondly remembered coast after a long interval, we find that much of the coast has been covered with concrete blocks and that the coast we remember has gone. Coastal areas suffering severe beach erosion are not limited only to particular regions but can be found throughout Japan. When investigating the issue of beach erosion, we soon arrive at a very important conclusion: almost all beach erosion originates from anthropogenic factors as a consequence that Japan has extensively altered its land during the last 20 to 30 years. In Japan, coastal work is carried out by four sectors. Port construction is carried out by the Fisheries Agency and the Port Department. Other coastal works are carried out by the River Department, and protective works involving planting coastal forests and the protection of farmland are overseen by the Ministry of Agriculture. These works have been carried out independently within their respective areas of jurisdiction. For example, when a long port breakwater is extended, sand is accumulated in the wave-shelter zone of the breakwater owing to longshore sand transport, resulting in beach erosion in the surrounding area. The construction work at the port is carried out by the port or fishing port authority, with little connection to the works aiming to protect the surrounding coastline. Overall measures to solve problems are difficult to adopt because of the sector-by-sector administrative system. In this study, the factors that hinder the solution of coastal problems are investigated from a fundamental viewpoint through the examination of several coastal areas as examples.

[1]Dr. Eng., Executive Director, Public Works Research Center, 1-6-4 Taito, Taito-ku, Tokyo 110-0016, Japan.
uda@pwrc.or.jp

2. Types of Beach Erosion

Beach erosion is caused by an imbalance in the sediment budget of a coast, with the exception of ground subsidence associated with the pumping of excess groundwater or a change in the earth's crust. Seasonal variations in beaches due to the occurrence of storms and calm waves are commonly observed, but the long-term stability of a beach is governed by longshore sand transport (Komar, 1976). When sand supplied by rivers and sea cliffs decreases compared with the longshore sand transport of a coast, the inevitable result is beach erosion. Similarly, artificially removing coastal sediment by dredging or mining results in beach erosion on neighboring coasts. The causes of beach erosion in Japan are classified into four main types: (a) obstruction of longshore sand transport, (b) beach changes associated with the formation of wave-shelter zones, (c) decreased fluvial sediment supply, and (d) offshore sand mining or dredging.

When a breakwater, jetty, or groin is extended offshore on a coast with predominant longshore sand transport, part or all of the longshore sand transport is obstructed, causing erosion downcoast and accretion upcoast. On a coast where waves are incident from the direction normal to the shoreline, longshore sand transport is induced from outside to inside the wave-shelter zone near a large port breakwater, resulting in erosion outside the wave-shelter zone and accretion inside. The dredging of sand deposited behind an oblique breakwater soon induces longshore sand transport from outside to inside the wave-shelter zone, resulting in erosion in the adjacent area.

Since many Japanese rivers originate from mountains with a high sediment yield, a large amount of sediment is transported to coasts, resulting in the formation of fluvial lowlands. The shoreline near the river mouth advances as sediment is deposited. However, 2,532 large dams have been constructed in Japan, resulting in the decrease in sediment supply to the coasts and the recession of the coastline. A sea cliff functions similarly to a river as a source of coastal sediment. The collapse of a cliff composed of unconsolidated layers mainly due to wave action produces a large amount of sediment, which is transported to nearby coasts. The application of measures against cliff erosion has the same effect on surrounding coasts as the decreased fluvial sediment supply from rivers.

Sand mining in river channels was prohibited in Japan in 1967 owing to excessive riverbed degradation. Sand mining of the seabed has been extensive, however, and still continues at many places offshore, particularly in western Japan. The effect of offshore mining may be extensive, depending on wave conditions, even though sand is not removed directly from the shoreline. Sand movement due to longshore sand transport occurs at depths less than the depth of closure, h_c, which is roughly equal to 10 m on well-exposed beaches. Since sand is continually being exchanged at this depth, the removal of sand from a depth less than h_c leads to the same results as sand mining near the shoreline. In this study, beach erosion triggered by the extension of a port breakwater and beach erosion due to the decrease in fluvial sediment supply from rivers caused by excess riverbed mining and the construction of dams are investigated, taking Otsu fishing port in Ibaraki Prefecture, facing the Pacific Ocean, Shimobara fishing port in Chiba Prefecture, Wada Port in Fukui Prefecture, and the Tenryu River and the Enshu-nada coast in Shizuoka Prefecture, as shown in Fig. 1, as typical examples.

3. Beach Erosion Associated with Formation of Wave-Shelter Zone upon Construction of Offshore Breakwater

3.1. *Otsu fishing port*

Typical beach changes associated with the formation of a wave-shelter zone were triggered by the construction of a port breakwater near Otsu fishing port, which faces the Pacific Ocean (Fig. 1).

Figure 2(a) shows an aerial photograph of the area taken in October 1975. In 1975, because the port breakwater was constructed into the wave-shelter zone of Otsu Point, the influence of this port breakwater on the coastline south of the fishing port was negligible, and a sandy beach of 50 m width extended continuously. By April 1992, the fishing port breakwater had been extended (Fig. 2(b)). An offshore breakwater of 750 m length was built parallel to the old offshore breakwater at a distance of 250 m from the shoreline. Since the new offshore breakwater protruded out of the wave-shelter zone formed naturally by Otsu Point, a large wave-shelter zone was formed west of the offshore breakwater, significantly accelerating the advance of the shoreline. On the coast south of the fishing port, the shoreline greatly retreated and it was protected heavily by concrete armour units (Fig. 3).

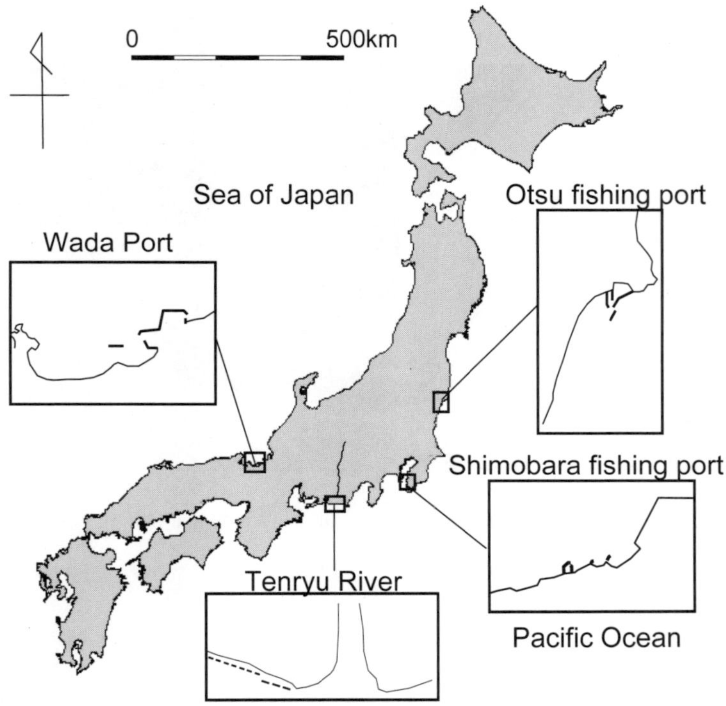

Figure 1. Location of study areas.

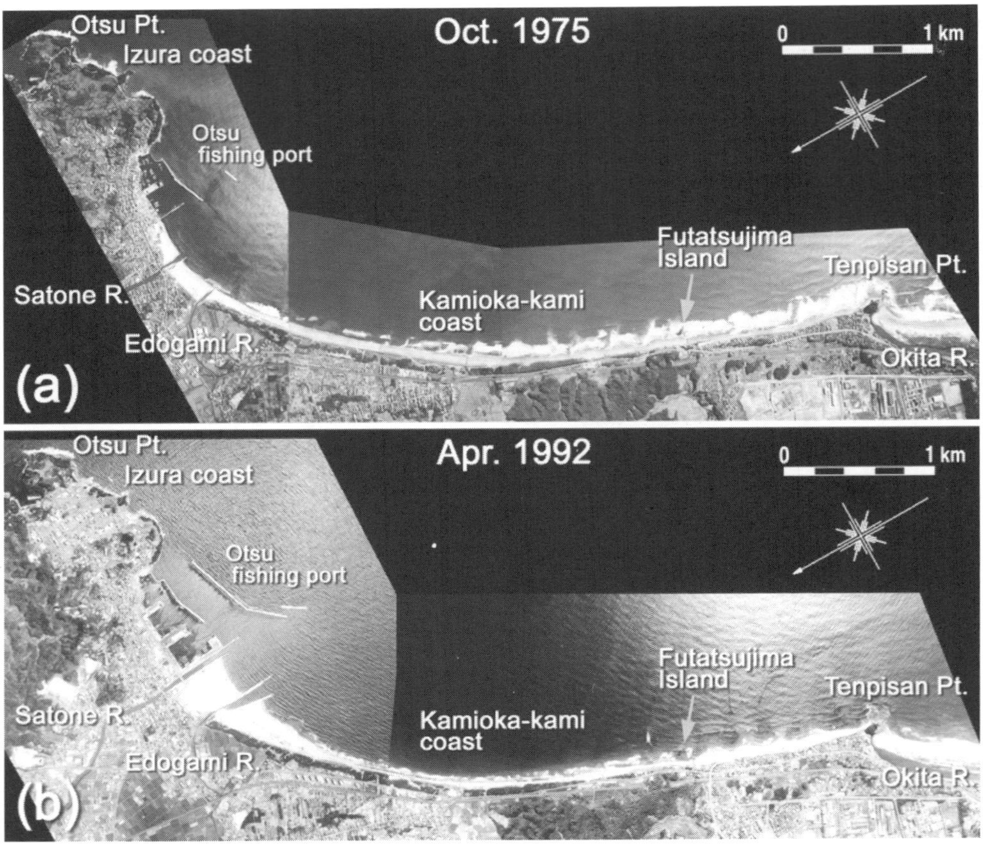

Figure 2. Aerial photographs of coastline around Otsu fishing port.

Figure 3. Shoreline of eroded coast taken on June 19, 2004.

3.2. *Shimobara fishing port*

The second example of beach erosion triggered by the construction of wave-sheltering structures is the coast surrounding Shimobara fishing port located in Tateyama Bay (Fig. 1). Wind waves from the north are mainly incident to the coast. The fishing port breakwater has been rapidly extended since 1997, and beach erosion has occurred because of the wave-sheltering effect of this breakwater. Figure 4 shows aerial photographs of the area between 1997 and 2007. Before 1997, there was a natural hooked shoreline 600 m long west of Shiomi fishing port. Shimobara fishing port was constructed along 500 m of the coastline and is bounded by a pile of small rocks and the breakwater of Shiomi fishing port at the eastern and western ends, respectively. By 1999, the port breakwater had been obliquely extended. Since then, shoreline recession has occurred from the eastern coastline. By 2002, extensive shoreline recession could be seen between the central part of the coast and the rocks at the eastern part as well as a major shoreline advance near the fishing port. As a result, the shoreline in the area adjacent to the fishing port advanced greatly, whereas the shoreline receded in the central and eastern parts of the coast. By 2007, the eastern breakwater had been extended and sand was trapped inside the fishing port, resulting in a shortage of sand on the nearby coast.

Figure 5 shows the shoreline configuration and its changes with reference to the configuration in 1997. Rocks in the eastern part acted as a solid boundary preventing longshore sand movement, and the shoreline retreated west of the rocks. The transition point between the advance and recession of the shoreline is located at x=300 m, and the shoreline greatly advanced toward the lee of the port breakwater west of this point. The maximum shoreline advance was 47 m and the maximum recession was 15 m between 1997 and 2002. Shoreline changes occurred monotonically with time after the breakwater was extended.

3.3. *Wada Port*

In general, the coastline of a pocket beach has a stable form, which depends on the distribution of incident wave height and the wave direction. When a breakwater is extended at an end of a pocket beach, the shoreline approaches another stable form in response to the newly formed wave field. There are many examples of beach changes caused by this mechanism in Japan. A typical example is the beach changes around Wada Port in Wakasa Bay in Fukui Prefecture (Fig. 1). The study area is a pocket beach with a length of 2 km.

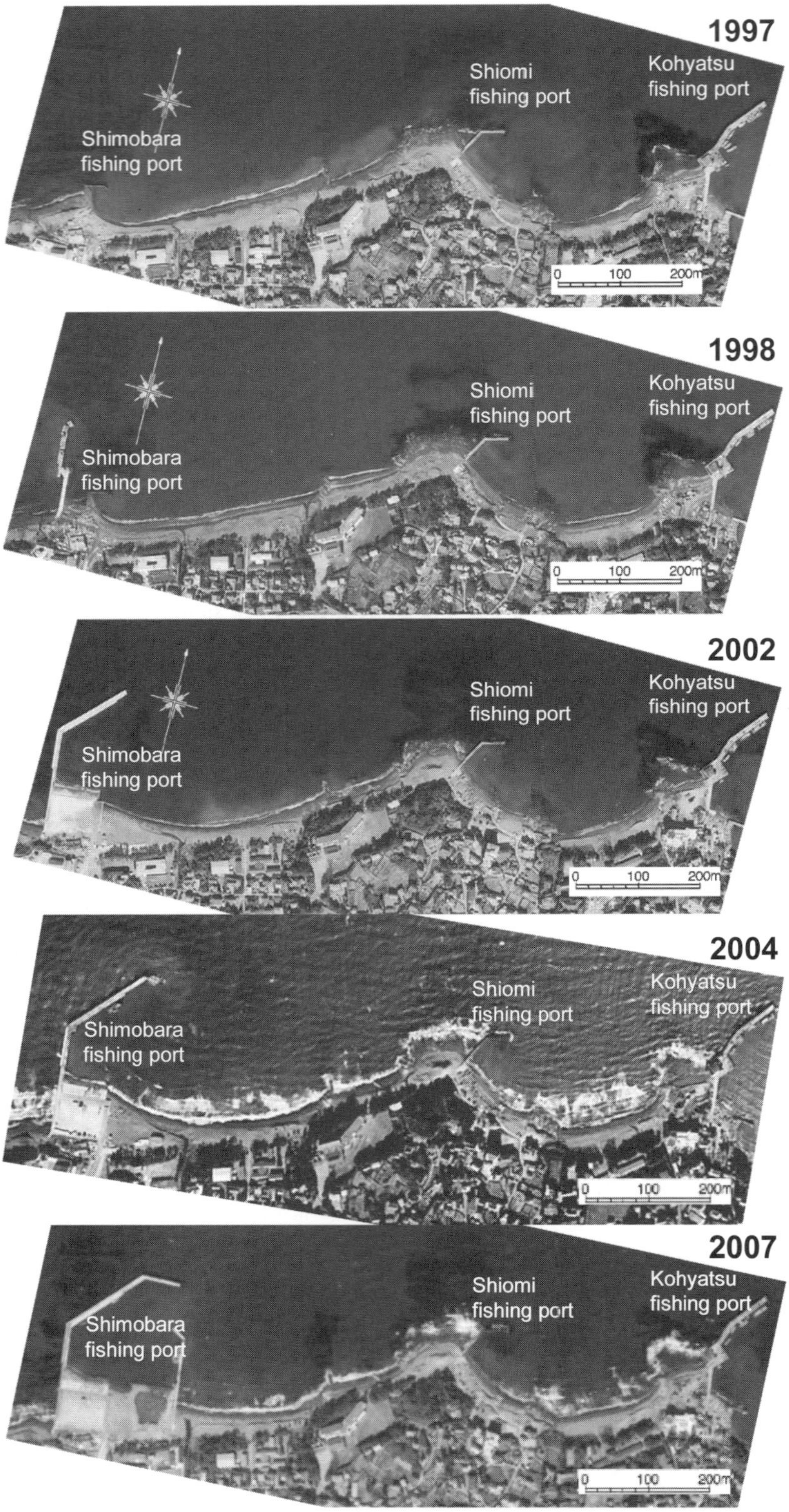

Figure 4. Aerial photographs of Shimobara fishing port between 1997 and 2007.

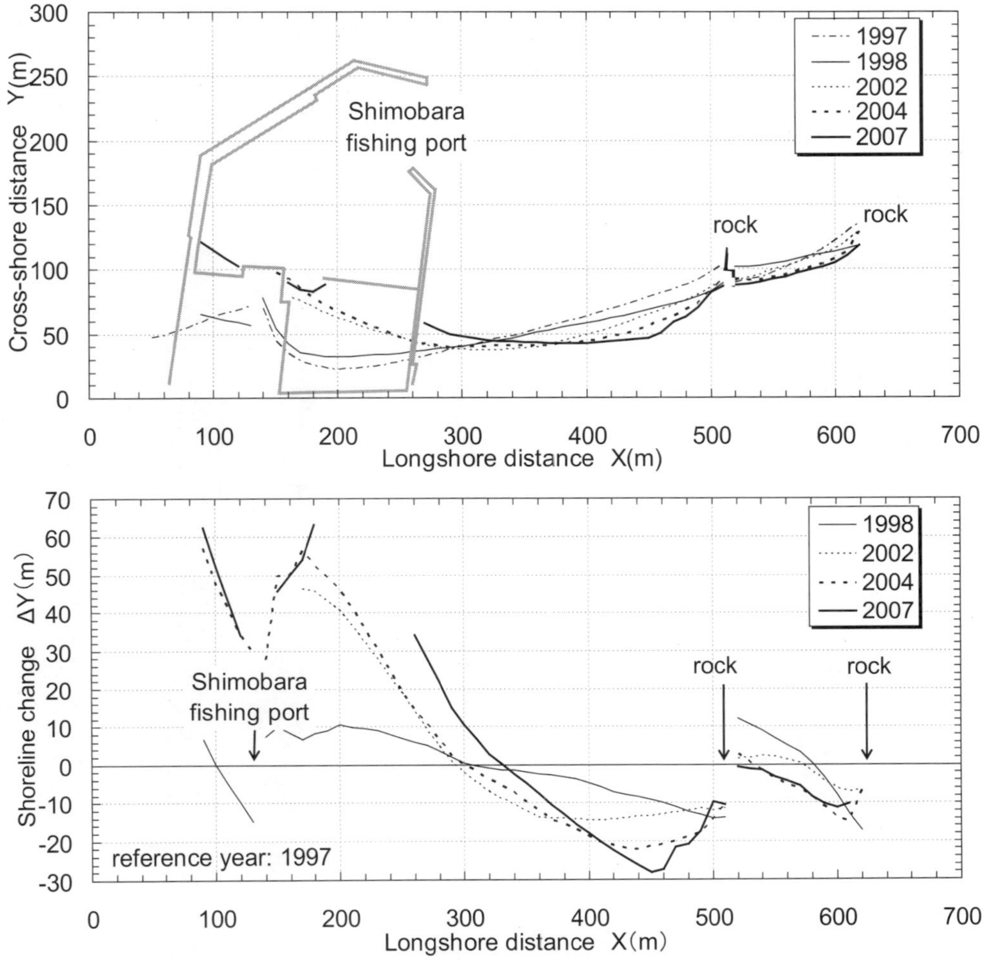

Figure 5. Shoreline configuration and changes.

Figure 6 shows aerial photographs taken between 1975 and 2004. The study area is bordered by the rocky headland of Shiroyama on the west side and the breakwater of Wada Port on the east side of the pocket beach (Fig. 6(a)). Hatsumi Island is located offshore of the coastline with a cuspate foreland behind the island. In 1981, a breakwater of Wada Port started to be extended eastward (Fig. 6(b)). By 1990, another port breakwater had been extended westward, crossing half the distance of 600 m between Hatsumi Island and Wada Port (Fig. 6(c)). As a result, a wave-shelter zone was formed behind the port breakwater and sand started to accumulate in the wave-shelter zone. The shoreline at the base of the breakwater advanced 60 m between 1981 and 1990, and sand deposition in the wave-shelter zone can be clearly observed. Because this sand was supplied from the western and central parts of the pocket beach, beach erosion occurred on the coast west of Hatsumi Island.

By 1995 the foreshore had greatly expanded in the adjacent area west of Wada Port, with a shoreline advance of up to 150 m (Fig. 6(d)). The shoreline behind Hatsumi Island had receded by 1990 owing to the construction of the port breakwater, and a detached breakwater with a length of 200 m was constructed behind the island as a measure against this beach erosion. This resulted in the enclosure of sand deposited between Hatsumi Island and Wada Port with no possibility of it returning to the west coast. Since this detached breakwater created another wave-shelter zone, beach erosion was further accelerated on the western part of the pocket beach. By 2001 the beach width had been markedly reduced in the central part of the pocket beach in contrast with the excessive accumulation of sand between the detached breakwater and Wada Port (Fig. 6(e)). In the wave-shelter zone of Wada Port, the shoreline width increased from 50 m in 1975 to 300 m in 2004, whereas the shoreline in the western part of the pocket beach uniformly retreated by about 30 m.

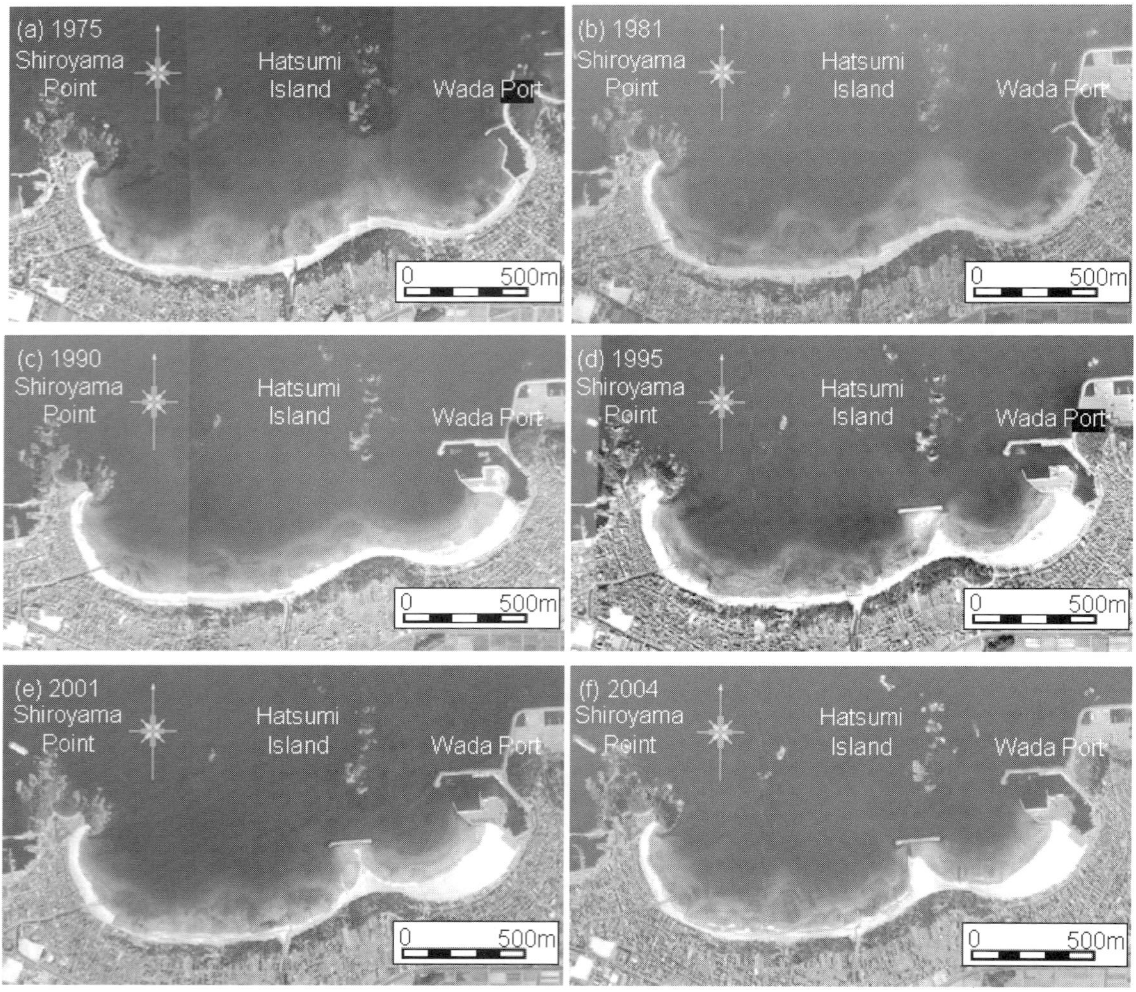

Figure 6. Aerial photographs of coastline around Wada Port.

On the basis of the aerial photographs, the change in shoreline configuration between 1975 and 2004 and the change in shoreline position with reference to that in 1975 were investigated (Figs. 7 and 8). It is clear that excessive sand accumulated east of the detached breakwater, whereas the beach width decreased to the west of the detached breakwater. The main cause of the beach changes is the extension of the breakwater of Wada Port, which formed a large wave-shelter zone behind it, with an additional effect resulting from the construction of the detached breakwater. Because a groin was extended behind the detached breakwater, longshore sand transport, which moved freely eastward or westward depending on the wave direction without such a facility, was completely disrupted, resulting in the formation of two isolated pocket beaches. Thus, typical beach erosion associated with the formation of a wave-shelter zone occurred around Wada Port; excessive sand accumulated in the eastern part of the pocket beach and beach erosion occurred in the western half. This example clearly shows that when a breakwater is extended at one side of a pocket beach, the entire beach is affected.

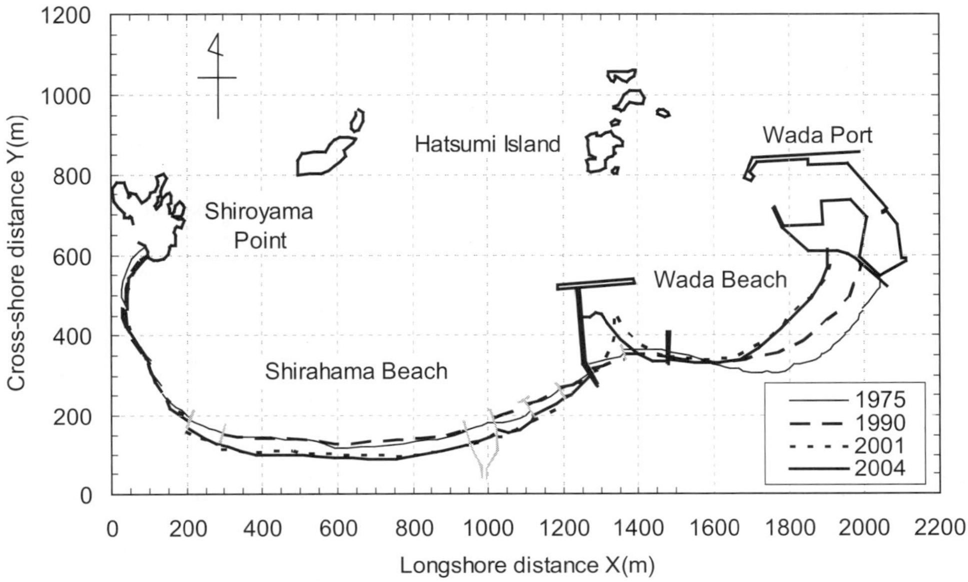

Figure 7. Shoreline configuration around Wada Port.

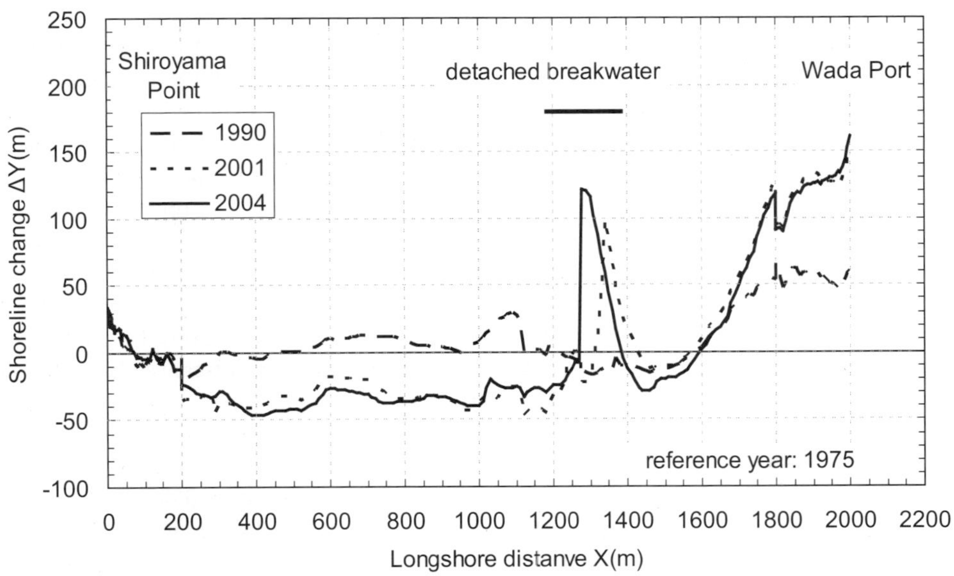

Figure 8. Shoreline changes around Wada Port.

4. Beach Erosion due to Decreased Fluvial Sediment Supply

The Enshu-nada coast, where the Tenryu River meets the sea, is given as an example of beach erosion around a river mouth triggered by decreased fluvial sand supply due to the construction of dams and excess riverbed mining. This river originates in the central Honshu mountains and flows downstream at a steep riverbed slope of 1/871 to the river mouth (Fig. 1). The riverbed consist of coarse gravel with a mean diameter of 14 mm. Riverbed excavation for sand mining was conducted extensively before 1968, and many dams were constructed, resulting in greatly decreased sediment discharge and beach erosion on coasts around the river mouth (Uda, 2007). The Tenryu River previously supplied a large amount of sediment to coasts, because the riverbed has a steep slope and the watershed is close to the median tectonic line of the Japanese archepelago, which yields much sediment. After World War II, however, several dams for electric power plants, such as the Sakuma Dam, were built, resulting in a decreased sediment supply to coasts and severe erosion. The river mouth terrace serving as a sand reservoir is now disappearing, and severe erosion is occurring near the shoreline.

Figure 9 shows sand accumulation in the reservoirs of major dams and sand mining in the Tenryu River watershed. On the Enshu-nada coast formed by sediment supplied by the Tenryu River, beach erosion has occurred near the river mouth owing to decreased fluvial sand supply caused by the accumulation of sand upstream in the many dams constructed in this watershed and sand mining of the riverbed for construction materials. The annual volume of sand mined since 1970 and the volume of sand accumulated upstream since 1956 are shown in the figure.

Given that sand mining had been conducted extensively before 1967, when the legal prohibition of sand mining went into effect, and that in the past sand volume of sand actually mined was generally twice as much as the permitted volume, the volume in Fig. 9 differs from the total volume of mined sand in all reaches of the Tenryu River. However, the total amount of sand mined since 1970 has reached at least 2.5×10^7 m^3. During this time, 1.25×10^8 m^3 of sand has been deposited upstream in dams. Therefore, a total of 1.5×10^8 m^3 of sand has been prevented from reaching the Enshu-nada coast.

Kawata and Uemoto (1998) showed that fine materials such as the wash load of rivers are ineffective for forming sandy beaches and only coarser materials are effective, which accounts for about 8% of the total sediment yield in the Tenryu River watershed. If we assume that the sand component related to beach erosion is 8% of the total volume of sand deposited in dams and 100% of the sand volume mined in the riverbed, then the cumulative volume of such sand reached 3.5×10^7 m^3 between 1956 and 2004, corresponding to an annual sand discharge of 7.0×10^5 m^3/yr. Torii et al. (2004) estimated that the annual long-term sand supply of this river is 8.0×10^5 m^3/yr from calculations using a model for predicting one-dimensional riverbed changes. In comparing this sand supply, approximately 90% of the annual sand supply of the river has been artificially removed.

A 110 km stretch of the Enshu-nada coast facing the Pacific Ocean extends on both sides of the mouth of the Tenryu River from Omaezaki Point in the east to Irago Point in the west (Fig. 1). Part of this coast was formed as a river delta from abundant sand from the Tenryu River, but after the construction of dams in the upper basin, sand discharge markedly decreased, resulting in the shoreline recession. Beach erosion around the river mouth due to reduced sediment discharge from the river can be clearly observed in aerial photographs of the river mouth (Fig. 10), as shown in detail by the alignment of five transects at 200 m intervals from the left bank of the river and the shoreline in 1962. Figure 11 shows longitudinal profiles along four transects between 1984 and 2001. At No. 218, a terrace at the river mouth in 1984 had almost disappeared by 2004. The seabed depth at the outer edge of the terrace increased by a maximum of about 5 m. The depth of closure, where the longitudinal profile change diminishes, was 14 m at No. 216 and 15 m at No. 218. Both severe shoreline recession and the almost complete disappearance of the terrace, which was a reservoir of sand for longshore sand transport, thus occurred simultaneously at the mouth of the Tenryu River. To determine the volume of sand removed from the terrace, we selected a rectangular area 1.6 km long alongshore and

1 km long cross-shore for examination (Fig. 10), and calculated the change in volume (Fig. 12) using bathymetric survey data. The rate of decrease in the total sand volume was very high between 1986 and 1993, reaching 4.0×10^5 m^3/yr. After 1993, the rate of decrease declined but remained at 1.5×10^5 m^3/yr. The total decrease in volume was 4.6×10^6 m^3 between 1984 and 2004, corresponding to a mean rate of decrease of 2.3×10^5 m^3/yr.

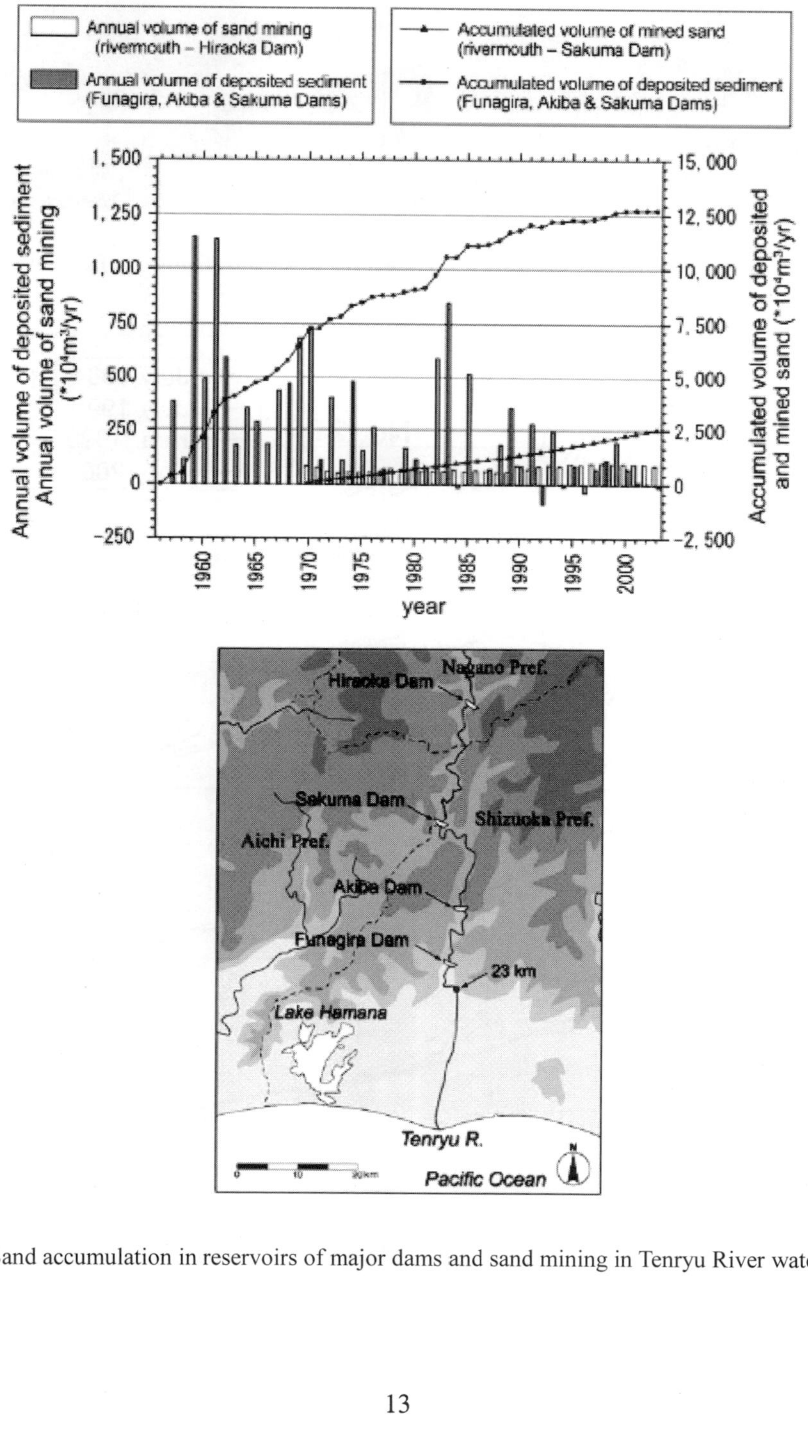

Figure 9. Sand accumulation in reservoirs of major dams and sand mining in Tenryu River watershed.

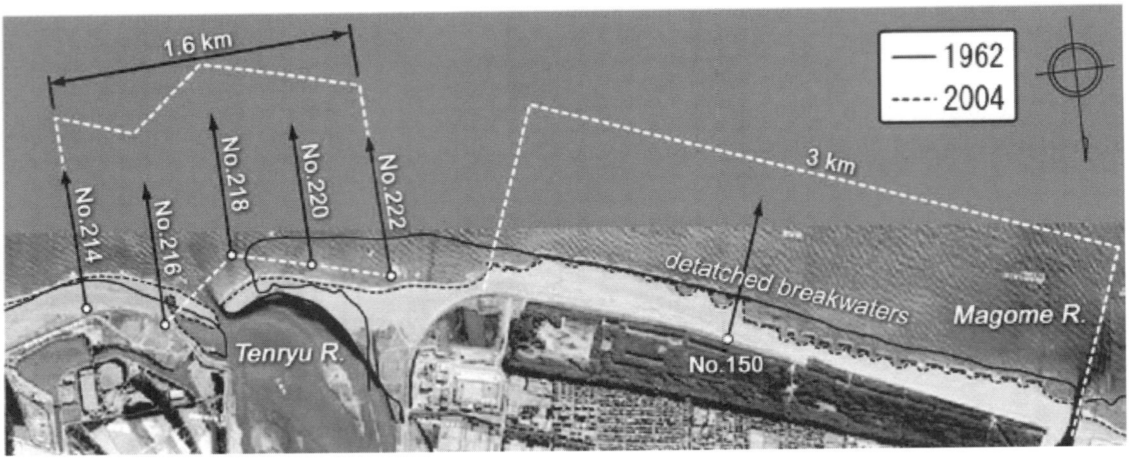

Figure 10. Shoreline recession at mouth of Tenryu River and alignment of survey lines.

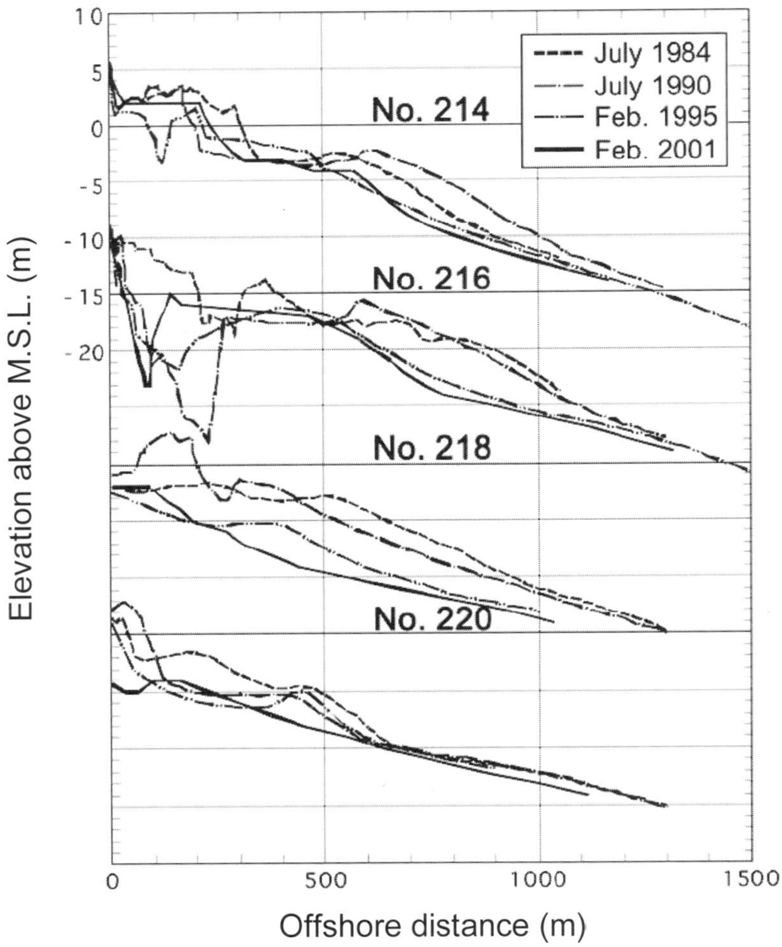

Figure 11. Shoreline recession at mouth of Tenryu River and alignment of survey lines.

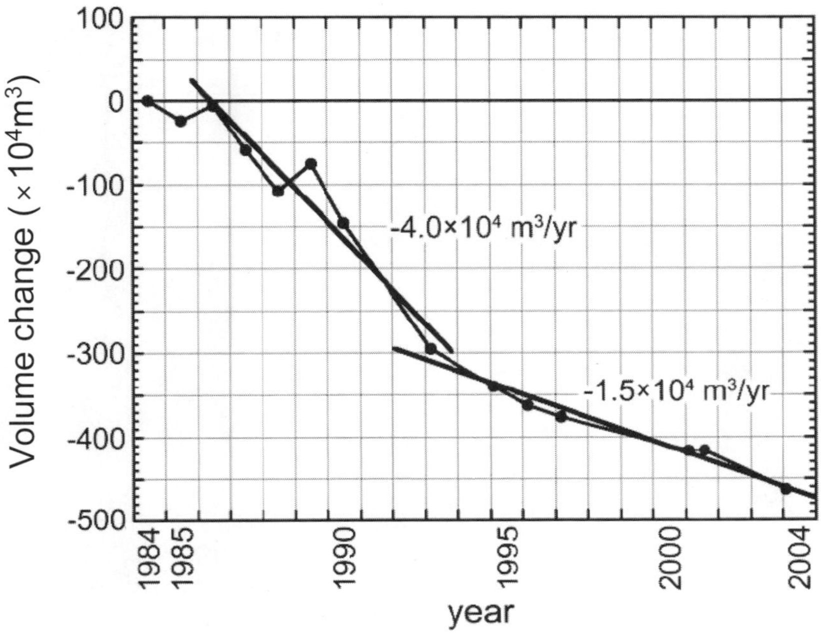

Figure 12. Changes in sand volume offshore of the river mouth.

The 500-m-terrace offshore of the Tenryu River mouth in 1984 had entirely disappeared by 2004, increasing the seabed depth by 5 m at the outer edge of the terrace. The shoreline zone is thus being severely eroded in addition to the offshore zone including the terrace at the river mouth. Although measures against beach erosion have been taken locally, the erosion of the entire long stretch of shoreline protruding as a river delta owing to the interruption of longshore sand supply is feared.

As measures against severe beach erosion around the mouth of the Tenryu River, 22 detached breakwaters have been constructed on the Hamamatsu-goto coast west of the river mouth. Figure 13 shows an oblique aerial photograph of this coast taken in January 2006. Tombolos have developed markedly behind the six detached breakwaters next to the mouth, and these breakwaters protrude significantly from the shoreline of the river mouth bar. These structures, built along the previous river delta shoreline before erosion, were left as a result of severe shoreline recession. At present, fluvial sand only flows toward the Ryuyo coast located east of the river mouth, and the flow of sand toward the west coast has become hindered by the protruding detached breakwaters, which, although effective in shore protection, simultaneously block the continuous westward movement of sand, destroying the overall balance of longshore sand transport. Fluvial sand supply from the Tenryu River has markedly decreased to the present level, and even though sand supply from the river may by artificially increased in the future by sand bypassing, it may be difficult for such sand to be transported by longshore sand transport because of the blockage caused by these structures (Fig. 13), resulting in downcoast erosion. The basic issue is not the function of individual detached breakwaters, but how overall beach changes under these conditions with longshore sand transport are realized, and how continuous longshore sand transport can be maintained.

Figure 13. Oblique aerial photograph of Hamamatsu-goto coast taken in January 2006.

5. Legal Issues Related to Beach Erosion

We consider the case of Otsu fishing port. The coastal management areas around this fishing port are schematically drawn in Fig. 14. One breakwater of the fishing port is assumed to extend at the corner of the coastline. In this case, the management area of the coast is subdivided into the fishing port or port area, and the natural shore protection zone outside of the port covered by the Seacoast Law under the jurisdiction of the River Bureau of the Ministry of Land, Transport and Infrastructure. Landward of the shore protection zone is the coastal forest area planted to prevent wind-blown sand, which is under the jurisdiction of the Forestry Agency of the Ministry of Agriculture, Forestry and Fisheries (Uda et al. 2005).

When a port breakwater is extended, longshore sand transport is induced from the outside to the inside of the wave-shelter zone as a result of the wave-sheltering effect of the breakwater on the sandy beach. The occurrence of this phenomenon can be easily predicted theoretically. The fundamental problem lies not in this physical process, but in the management of the coastal land. The construction of fishing ports, the management of coastal forests, and shore protection works are independently carried out by the Fishery Bureau, the Agricultural and Forestry Bureau, and the Civil Engineering Bureau, respectively, of the prefectural government in Japan. Each work is planned so as to achieve the maximum economic rationality in the short term without consideration of the degradation of the environment in the surrounding area; this is local optimization.

For instance, sand deposited in the navigation channel of Otsu fishing port was regarded as an obstacle and dumped in the offshore zone or used as a material for land reclamation by one management office, as schematically shown in Fig. 14. At the southern coast where sand was removed by longshore sand transport, the beach slope became very steep, and a seawall, concrete armour units, and detached breakwaters were installed as countermeasures against beach erosion by a different coastal management office. The optimal adjustment of the flow of sand, which is a physical problem that can be easily analyzed, through cooperative work among various management offices was troublesome and has been postponed. Regarding the issues common to several agencies, there is no common principle for the sound maintenance of national land, which must be realized to solve the problems at a fundamental level. Thus, fundamental measures were difficult to adopt and stopgap measures such as installing a seawall with concrete armour units were selected instead, which rapidly produced an artificial coastline.

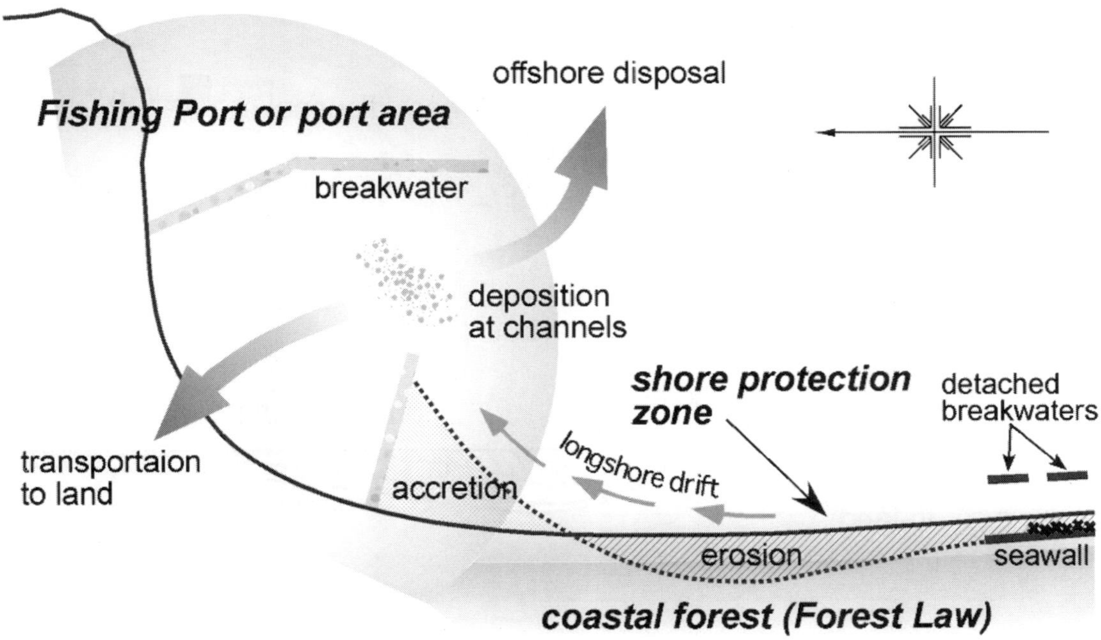

Figure 14. Coastal management areas and related laws in Japan.

6. Discussion

When an offshore breakwater is constructed on a straight coast, forming a wave-shelter zone behind the breakwater, longshore sand transport from outside to inside the wave-shelter zone is induced, resulting in erosion outside the wave-shelter zone and accretion inside. The stable shoreline configuration in this case can be practically predicted using the method of Hsu and Evans (1989), for example. Nevertheless, this is exactly what occurred between 1998 and 2007 at Shimobara fishing port, and erosion intensified in the area surrounding the fishing port as a result of sand movement from outside to inside the wave-shelter zone. The natural coast, which had long been used for swimming, was altered to an artificial coastline, losing its value for recreational use. Although the natural shoreline changes are generally reversible, as seasonal beach rotation (Short and Masselink, 1999), the shoreline changes associated with the formation of wave-shelter zone caused by the construction of port breakwater described in this study were irreversible, the same as the case occurred on the Morroccan coasts (Chaibi and Sedrati, 2009).

Another issue exists in port maintenance. Consider a general case where the wave-shelter zone formed by an offshore breakwater is used as a navigation channel or for anchorage (Fig. 15). Dredging is conducted to maintain a sufficient channel depth, and dredged materials have been disposed of offshore or transported to land on the basis of standards designated by the Fishing Port Act or Port Law. When an offshore breakwater or a jetty preventing sand deposition inside the port is shallower than the depth of closure, h_c, of sand movement by waves, sand is again transported by longshore sand transport inside the port after dredging owing to the wave-sheltering effect of the breakwater, meaning that successive dredging decreases the volume of sand on nearby beaches. Dredging conducted on the basis of general standards, with dredged sand disposed of independently by individual management authorities, results in beach erosion on surrounding coasts (Serizawa et al., 2005).

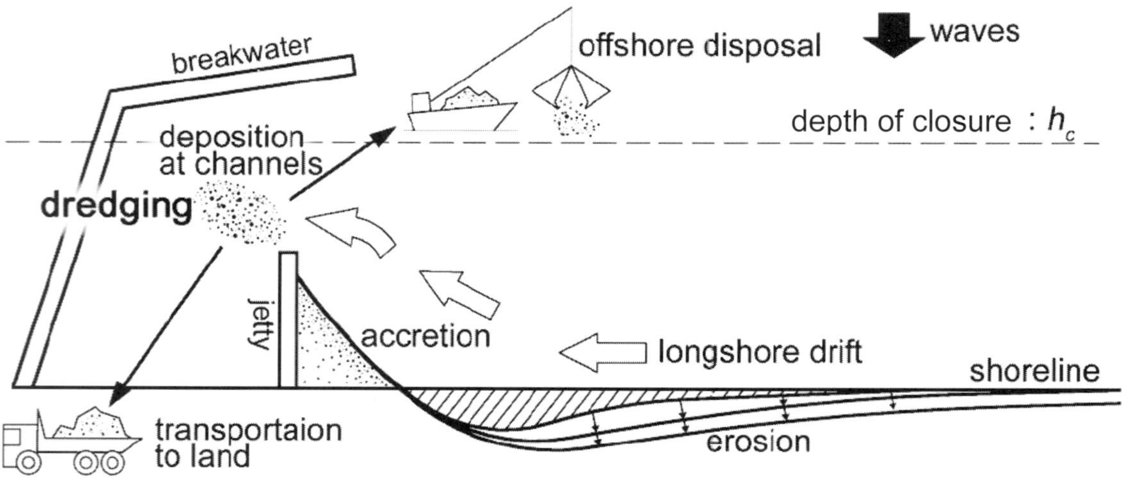

Figure 15. Sand deposition inside port and erosion triggered by dredging.

The numbers of fishing ports and commercial ports in Japan are 2,931 and 1,079, respectively, at present. This means that, on average, there is one port per 8.5 km nationwide, since the total length of coastline in Japan is 34,000 km. Many of the ports that are constructed on sandy beaches have the same problems as those discussed in this paper. Therefore, beach erosion and accretion associated with the formation of wave-shelter zone as mentioned above can be seen elsewhere in Japan. Furthermore, measures have been taken only after erosion has occurred; it is rare for preventive measures to be taken beforehand. Thus, natural beaches in Japan have been disappearing rapidly; the natural sandy beach is becoming an 'endangered geomorphology' in Japan.

At present, the quantitative prediction of 3-D beach changes and the change in grain size associated with beach nourishment is possible using advanced numerical simulations, such as the contour-line-change model (Serizawa et al., 2003), the BG model (Serizawa et al., 2006) and others. New measures based on comprehensive sediment management must be taken instead of local optimization using hard structures at sites of erosion (Yasumoto et al., 2007). For example, sand back-passing is required to prevent coastal deterioration.

7. Future Perspective

Coastal problems in Japan are not simply due to problems in coastal engineering but are mainly a result of the social system in Japan, including the legal system. It is very difficult to change this situation through only administrative efforts because it is deeply related to the mindset past administrations. Through the collaboration of specialists, engineers, and NPOs, a flexible social system that can be modified efficiently must be developed. For this purpose, the mutual exchange of information must be enhanced. Instead of each coastal management office narrowly pursuing its set of responsibilities, the long-term planning of coastal works including set-back of the coastline, if necessary, is strongly required, as well as a necessity of recycling sand dredged from inside ports as a material for beach nourishment.

References

Chaibi, M. and Sedrati, M., 2009. Coastal erosion induced by human activities: the case of two embayed beaches on the Moroccan coast, *Jour. Coastal Res., SI 56 (Proc. 10th ICS)*, Portugal: 1184-1188.

Hsu, J. R. C. and Evans, C., 1989. Parabolic bay shapes and applications, *Proc. Inter. Civ. Engnrs.*, Part 2, Vol. 87: 557-570.

Kawata, Y. and Uemoto, M., 1998. Coastal processes of Tenryu River and Enshu-nada coast, *Annual Coastal Eng. Jour.*, Vol. 45: 616-620. (in Japanese)

Komar, P. D., 1976. *Beach processes and sedimentation*, Prentice-Hall, Inc.

Serizawa, M., Uda, T., San-nami, T., Furuike K. and Kumada, T., 2003. Improvement of contour line change model in terms of stabilization mechanism of longitudinal profile, *Coastal Sediments '03*: 1-15.

Serizawa, M., Uda, T., Kumada, T., San-nami T. and Furuike, K., 2005. Beach erosion caused by dredging of navigation channels at ports and river mouth, *Proc. 14th Biennial Coastal Zone Conf.*, New Orleans, Louisiana: 1-5.

Serizawa, M., Uda, T., San-nami T. and Furuike, K., 2006. Three-dimensional model for predicting beach changes based on Bagnold's concept, *Proc. 30th ICCE*: 3155-3167.

Short, A. D. and Masselink, G., 1999. *Embayed and structurally controlled beaches, In: Short, A. D. (ed.), Handbook of Beach and Shoreface Morphodynamics*, John Wiley & Sons, Chichester: 230-250.

Torii, K., Sato, S., Uda, T. and Okayasu, T., 2004. Regional sediment management based on sediment budget for graded sediments - a case study of Tenryu watershed and Enshu-nada coast, *Proc. 29th ICCE*: 3110-3122.

Uda, T., San-nami, T., Serizawa M. and Furuike, K., 2005. Beach erosion in Japan as a structural problem, *Proc. 14th Biennial Coastal Zone Conf.*, New Orleans, Louisiana: 1-5.

Uda, T., 2007. Beach erosion arising from artificial land modification, *Jour. Disaster Res.*, Vol. 2, No. 1: 29-36.

Yasumoto, Y., Uda, T., Matsubara, Y. and Hirano, G., 2007. Beach erosion along Tottori coast and comprehensive sediment management, *Jour. Coastal Res., SI 50 (Proc. 9th ICS)*, Gold Coast, Australia: 82-87.

Coastal Dynamics 2009
Keynote 2

HOW DO BEDFORMS RESPOND TO CHANGING CONDITIONS? A STUDY USING ABSTRACTED MODELS

David Huntley[1] and Giovanni Coco[2]

Abstract

Field, laboratory and modelling studies generally agree that the rate at which bedforms develop in response to changing conditions depends upon the pre-existing state of the bed but there is currently no consensus about the nature of this dependence. This study uses a simple abstracted model to investigate the way a field of ripples responds to a step change in forcing conditions. The results show that the influence of pre-existing bedforms is significant but subtle. Large pre-existing bedforms retard the growth of new bedforms but defects in the old bedform pattern increase the rate of development even above the growth rate from a flat bed. Wavelet analysis correctly identifies the sites of new ripple growth at defects. We suggest that defects trigger new bedform growth because they are regions of relatively flat bed, with upstream perturbations. Prediction of non-equilibrium bedforms will require information on defect densities.

Key words: bedforms, ripples, morphodynamics, numerical modelling, wavelets, bedform growth rate

1. Introduction

A bed of mobile sediment subject to the action of a flow field is rarely planar and is much more likely to exhibit quasi-regular undulations, or bedforms, at a range of spatial and temporal scales. These bedforms are observed in a variety of settings including aeolian (e.g. Anderson, 1990), arctic (e.g. Kosugi et al., 1991) and underwater environments (e.g. Blondeaux, 2001). In the marine environment, for example, long-crested bedforms can range from small scale ripples, with wavelengths of the order of ten centimetres, up to sand ridges with scales of many kilometres. Other types of bedform can also develop. For example, close to the shoreline a rich variety of quasi-regular bedforms are typically found, from beach cusps with scales of tens of metres to surf-zone bars with scales of hundreds of metres. Recent work has also shown that, similar to the riverine environments, a seabed of mixed grain sizes can be formed into quasi-regular patterns of coarser and finer sediments, so-called sorted bedforms, where undulations in the sediment size rather than bed relief are the dominant characteristic. All of these bedforms are of considerable importance for coastal marine science since they change the roughness of the seabed in ways that influence not only the movement of sediments but also the flow of water above them. Understanding the processes responsible for moulding these bedforms and predicting their occurrence are therefore major challenges for coastal marine science.

[1] Marine Institute, University of Plymouth, Plymouth, Devon, UK. dhuntley@plymouth.ac.uk
[2] National Institute of Water and Atmosphere Research, Hamilton, New Zealand. g.coco@niwa.co.nz

Prediction of the rate of bedform development from a plane bed has been the subject of many studies (see Blondeaux, 2001 for a review). However in the field bedforms rarely develop from planar beds. Generally the seabed consists of pre-existing bedforms created by earlier wave or current conditions and a key question therefore is: how do bedforms adjust or reform in response to new conditions? More specifically, do previous bedforms change the rate of development of new bedforms compared to starting from a planar bed, and if so what features of pre-existing bedforms influence the growth of new bedforms? As the next section shows, current answers to these questions are ambiguous and even contradictory.

Abstracted modelling of morphology (eg Werner and Fink, 1993; Murray, 2003; Huntley et al., 2007) has arisen from the realisation that bedforms are created by non-linear feedback between the flow field and the seabed response. Abstracted models are designed to encapsulate only the basic, essential physical processes in a highly simplified form but usually contain positive feedbacks so that small perturbations can grow and eventually self-organize into a regular pattern, and negative feedbacks which limit growth of perturbations. Such models are proving remarkably successful in simulating and even predicting characteristics of coastal morphology (eg Huntley et al 2007; Coco and Murray, 2007). In this paper we plan to demonstrate that they can not only simulate bedform development in realistic ways but also have a role in investigating processes and identifying key issues which can direct future field and theoretical investigations. In particular we show how a simple abstracted model for ripples can be used in a systematic study of the evolution of long-crested bedforms as they change in response to a change in conditions.

Section 2 outlines the current understanding of bedform response to changing conditions, particularly emphasising the role of pre-existing bedforms. Section 3 describes in more detail the nature of abstracted modelling and the particular model used in this paper. Section 4 then presents the results of model simulations which are designed to determine what characteristics of pre-existing bedforms (hereafter called 'old' bedforms) influence the formation of new bedforms. The implications and conclusions are contained in section 5.

2. Observations of bedform response

Accurate and long term measurement of evolving bedforms in coastal and nearshore environments has become possible recently through developments in remote sensing techniques. In the nearshore zone, video camera stations provide time series showing the evolution of sand bar and shoreline features within ranges of the order of a kilometre (Lippmann and Holman, 1983, van Enckevort et al., 2003; Ruessink et al., 2007). For smaller scale features such as ripples and megaripples video systems can also provide insight (Clarke and Werner, 2003; Becker et al., 2007) although acoustic techniques such as sector scanning sonar and ripple profilers provide similar information (Hay and Wilson, 1994; Bell and Thorne, 1997; Traykovski et al., 1999; Hanes et al., 2001; Austin et al., 2007). One clear conclusion from these observations is that bedforms are frequently out of equilibrium with the prevailing conditions. Limited skill of bedform predictors (e.g. Grassmeijer and Kleinhans 2004; Williams et al., 2007) is likely to result from lack of knowledge of temporal evolution of morphology in response to changes in hydrodynamics. For example, Osborne and Vincent (1993) found poor agreement between observed and predicted ripple geometry and attributed the failure at least in part to slow bedform response to rapidly changing tidal conditions. Austin et al. (2007) provide a clear demonstration of the asymmetry of ripple geometry over a rising and falling tide level on a beach; ripples increase in scale progressively over a rising tide but remain essentially constant over the ebbing tide. They suggest that the ripples become too large to be changed by the waning energy conditions and hence display what they call 'arrested development', only being finally wiped out to begin a new cycle by the energetic flows in the swash zone. Traykovski et al. (1999) found similar hysteresis effects for ripples generated over a growing and waning storm; again the observed ripples grew quickly as the storm developed but the large ripples which eventually formed decayed much more slowly in the dying stages of the storm. Thus these observations provide good evidence that the timescale of ripple development is an important factor in predicting their geometry. However, although such observations can be used to indicate approaches to prediction which take account of these hysteresis effects (eg Trakovski, 2007; Austin et al., 2007), the uncontrolled nature of the environmental variability in the field makes it difficult to quantify growth rates in response to particular changes or to study processes responsible for growth of new bedforms from old. Laboratory studies under controlled conditions have therefore been used to try to elucidate these factors, with most studies focussing on small scale ripples.

Unfortunately the laboratory studies to date have failed to provide a clear consensus on how old bedforms influence the development of new bedforms. Davis et al. (2004) used a laboratory flume to study the development of small ripples, with typical heights of 1cm, in fine sand under oscillatory flows. They find that the transition from old to new ripples can be described by a logistic equation based only upon the 'energies' (wavenumber- averaged along-flow spectra of the bed) in the initial and final ripple fields and an exponential rate constant which depends only on the skin friction Shields parameter, with no difference in value whether starting from flat bed or from a field of old ripples. Thus they suggest that growth rates are larger for larger differences in initial and final ripples heights but are insensitive to the relative wavelengths of the old and new ripples. However there was considerable scatter in their results and unfortunately, although their test region was many wavelengths wide in the direction perpendicular to the longitudinal (flow) direction, they did not report on the three-dimensional geometries of their ripple fields. Smith and Sleath (2005) carried out a very large number of runs with oscillatory flows over a narrow test bed in a laboratory flume, studying ripple growth from flat bed and from old ripples. They interpret their observations in terms of two mechanisms. In some cases, new ripples were found to grow from Fourier components of the initial bed which had the same wavelength as the final ripple field. In other cases, generally when the change in wavelength from old to new was small and typically increasing, the ripples were found to 'slide' progressively from the old to the new geometry with ripple height growing or even remaining constant. By fitting their observed growth rates to exponential forms they conclude that, at least for the first case, the number of cycles required for the new ripples to reach 80% of their final height is a function, generally close to linear, of the ratio of initial to final wavelengths. However they admit that their results show considerable scatter which makes it difficult to reach firm conclusions. A third set of laboratory studies, by Doucette and O'Donoghue (2006), was carried out at full prototype scale using both second-order Stokes waves and irregular asymmetric waves. They find that ripples change their geometry in response to changed conditions by 'sliding' and 'merging' when growing in scale and by 'splitting' and 'merging' when reducing in scale. However they find that the time for new ripples to reach equilibrium, though strongly dependent on sediment transport rates (ie mobility number), is 'largely independent' of the initial bed. They conclude that the <u>rate</u> of new ripple evolution must be greater for a pre-existing ripple bed because an initial bed that is very different needs to evolve faster to reach equilibrium at the same time as evolution from a flat bed. Unfortunately the test sections used by both Smith and Sleath (2005) and Doucette and O'Donoghue (2006) had similar widths to the ripple wavelengths so were unable to develop significant three-dimensional variability.

Thus these three laboratory studies reach very different, even contradictory, conclusions about the influence of old bedforms on developing new bedforms in response to changed conditions. The degree to which old bedforms effect the time taken for new bedforms to approach equilibrium ranges from a negligible effect (Doucette and Donoghue, 2006) to an effect which depends upon either the relative energies or amplitudes of old and new ripples (Davis et al, 2004) or on their relative wavelengths (Smith and Sleath, 2005).

Bedform response has also been investigated using a range of theoretical and modelling approaches applied to a wide range of bedform types, though there have been few systematic studies of the influence of initial bedforms. Several studies have shown that small perturbations (order of mms to cms) of initial bathymetry influence the details of the geometry of emergent bedforms, such as the locations of crests, but have a relatively small effect on final bedform length scales compared to sensitivity to forcing conditions. Examples include Smit et al. (2008), who used a 2DH numerical model to study the development of nearshore crescentic bars from an initially longshore-uniform double sandbar system, and Caballeria et al. (2002) who used a similar numerical model for nearshore crescentic sand bars. Calvete et al. (2007) studied the development of rip channel systems on beaches using a linear stability model and found that changes in the initial cross-shore profile (order of cms to meters) can affect the emerging channel spacing as much as offshore wave conditions.

Some model studies have investigated the influence of isolated features in the initial topography. Landry and Werner (1994) used an abstracted model for ripples to show that isolated small ridges or troughs in an otherwise flat initial bed trigger rapid growth of ripples downwind, thus influencing the overall growth rate of emerging ripples but not their length scale. However the processes were not quantified or investigated in detail. Coco et al. (2000) investigated beach cusps using an abstracted model and found that isolated features on an otherwise plane initial beach slope were wiped out by emerging cusps. In this case the size and shape of the features affected the growth rate of the new pattern but not the final spacing.

The influence of pre-existing bedforms of different scales has been considered in a few studies. When presented with an initial field of beach cusps, the model of Coco et al. (2000) showed that the response was dependent on the relative wavelengths of the initial and expected emerging cusps; initial cusps with wavelengths within approximately a factor of two of the expected new cusps remained effectively unchanged whilst those with larger or smaller wavelengths were destroyed and replaced by new cusps. With respect to long-crested bedforms, Werner and Kocurek (1999) described a model where developing bedform spacing depends on time and pattern defects (defined as the ends of crest lines). As shown earlier, a dependence on relative wavelengths was also found for ripples by Smith and Sleath (2005). Marieu et al. (2008) use their detailed process model of wave vortex ripples to study the evolution of quasi-regular ripples from an initial randomly rippled bed. They suggest that evolution to the new ripple field takes place in three stages; organisation, growth and saturation. Development from a flat bed requires only the last two of these stages. In the organisation stage, the initial bed adapts to the new hydrodynamics by breaking down into small ripples, accompanied by a general decrease in 'bed energy' (ie variance of the bed elevation). They comment that the organisation stage can be slow compared to the growth stage and depends strongly on the bed history, though they do not quantify these comments. A systematic investigation of the influence of initial bedforms was presented by Huntley et al. (2008) using a numerical model initially developed to simulate the formation of sorted bedforms (Coco et al., 2007a). By running a set of initial forcing conditions for different lengths of time they created initial bedforms with different densities of defects (crest terminations, mergings and eyes); longer runs with constant forcing developed bedforms with fewer defects. By using these different initial bedform fields as starting conditions for runs with different forcing conditions they found that initial patterns with lower defect densities (defined as the number of defects per unit crest length) cause longer delays in new bedform development with, in the limit, an initial pattern with no defects delaying new bedform development indefinitely. They propose that this delay occurs during a first phase of new bedform growth when new bedforms are growing slowly by 'radiating' out from the sites of defects. Figure 1 shows some stills from a numerical simulation of sorted bedform development for one of the runs described in Huntley et al. (2008) to illustrate the key role played by defects during this phase of growth. This phase is followed by a second phase of growth when the old bedforms have been reduced in amplitude below that of the new and in this phase new bedforms develop across the entire domain at a rate which is independent of initial defect density. However the overall time taken for the new bedforms to reach equilibrium was found to decrease linearly with increasing defect density, showing the dominant influence of the delay during the first phase of growth. As found by Marieu et al. (2008), the first stage was associated with a decrease in overall elevation variance as the initial bedform field was broken down. The conclusion was that the three-dimensional shape of the initial bed, and in particular the presence of defects, has a major controlling influence on the rate at which new bedforms can evolve.

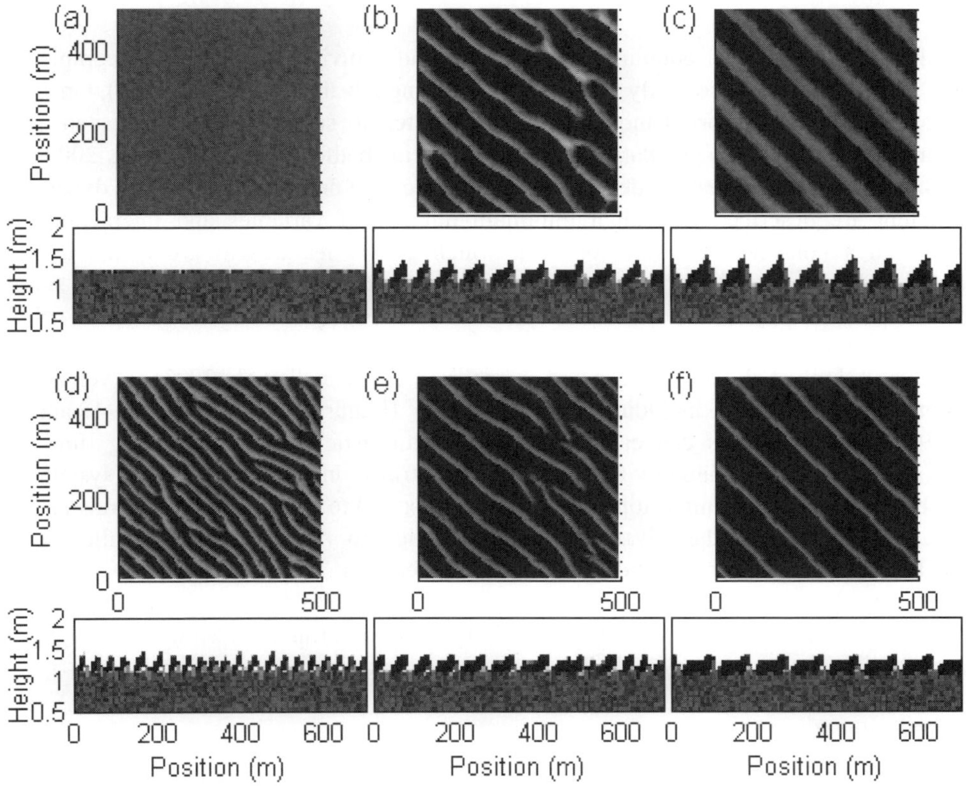

Figure 1. Development of sorted bedforms from numerical simulations. Top row of images shows initial conditions: (a) planar bed with random differences in grain size composition at each cell; (b) and (c) sorted bedforms developing under H=3 m wave forcing. Notice in (c) no defect is present. Bottom row of images shows development of sorted bedforms after 100 days of H = 2 m forcing from the above initial seabed conditions: (d) shows development of a bedform field; (e) shows little changes (all close to the defects) from the initial configuration and (f) shows no change at all (meaning the initial seabed configuration is forcing the system not to change from its initial configuration). Darker colours represent finer sand mixtures, lighter shades of gray represent coarser sand mixtures. Rectangular inserts show a vertical cross-section of the domain along the line from (0,0) to (500,500). Periodic boundary conditions are imposed in the numerical simulations.

Overall, this survey presents a rather confused picture of the influence of intial conditions on the evolution of new bedforms in response to changing conditions. It is clear that pre-existing bedforms can influence on the rate of development of new bedforms. However, the specific characteristics of the pre-existing bed which are responsible for influencing the response and their relative importance are effectively unknown and predictions of response rates are therefore likely to be substantially in error even if the initial state of the bed is known. We have therefore chosen to address this issue using a highly simplified abstracted model for ripples. The model allows us to isolate and vary specific characteristics of the initial conditions and thus identify their individual effects, with the aim of determining the key factors influencing bedform evolution.

3. Abstracted modelling

3.1 Introduction

Modelling the highly non-linear coupling between flow and evolving morphology is complex and a reductionist approach is not necessarily best suited to making advances (Werner, 2003). On the other hand, abstracted (or "exploratory") modelling is well suited to such a non-linear system and has proved remarkably successful in simulating a variety of marine bedforms (Huntley et al., 2007).

An abstracted model may be defined as one in which the processes driving the dynamics of a physical system are described through a minimum number of variables and interactions which encapsulate both a non-linear positive feedback which allows features to grow, and a negative feedback which limits the growth of such features. Abstracted models, using highly simplified descriptions of basic physical processes, are able to reproduce a wide range of observed morphodynamic features such as beach cusps (Werner and Fink, 1993; Coco et al., 2000), large scale coastal features (Ashton et al., 2001), wave and current ripples (Pannell et al., 2003) as well as a range of other non-marine bedforms including aeolian ripples (Landry and Werner, 1994) and dunes (Werner, 1995). These models can easily incorporate stochastic forcing, providing time-domain simulations of developing morphology and showing the intrinsic instability of coastal systems. They can readily deal with arbitrary initial topography and can extend to finite amplitude forms and quasi-equilibrium. They also have the advantage of being quick to run and therefore allow multiple simulations for the study of the statistical behaviour of the features.

The aim of this paper is to use a simple model for ripples to demonstrate that abstracted models can be used not only to simulate the development of bedforms but also to throw light on the processes controlling the response of bedforms to changing conditions.

3.2 Ripple model

The ripple model used here is based on the model of Landry and Werner (1994). A hexagonal close packed array of grains is oriented so that adjacent grains can lie on an angle of 30°, an appropriate angle of repose, relative to each other. Grains are chosen at random from the model domain and moved a distance determined by a prescribed rule. Grains at the departure point and the arrival point are then avalanched to the angle of repose. The process is then repeated many times to provide a model for grain movement under flow. Landry and Werner (1994) used several different rules for grain movement but here we use only a simple rule where the distance a grain moves is a constant, related to the flow velocity, multiplied by a small percentage (in our simulations we use 5%) of the starting height above the mean grain level. This rule simulates in a highly simplified manner the extra exposure to the flow experienced by grains at higher elevation and provides positive feedback through creating convergence of grain transport at ripple crests as they grow. Runs of this abstracted model show that regular ripples develop from this simple rule (Landry and Werner, 1994; Pannell et al., 2003). The model domain is a region 1000x1000 grain radii in area with periodic boundary conditions at each edge. Grain movement is confined to one direction except for the avalanching process, which occurs in the direction of the steepest local slope. We do not claim that our model reproduces accurately the features of particular natural ripples. It simulates unidirectional flow but does not simulate the 'reptation' process characteristic of aeolian ripples (Landry and Werner, 1994) and does not include the vortex shedding effects typical of larger natural ripples. For this reason we make no attempt to relate the number of grain movements to real time durations or the mean movement of individual grains to particular flow magnitudes. Our justification is that we want to use the simplest possible model to investigate the primary processes responsible for bedform change without the complication of multiple effects. Clearly the application of our findings to particular bedforms will need to be tested using more specific models.

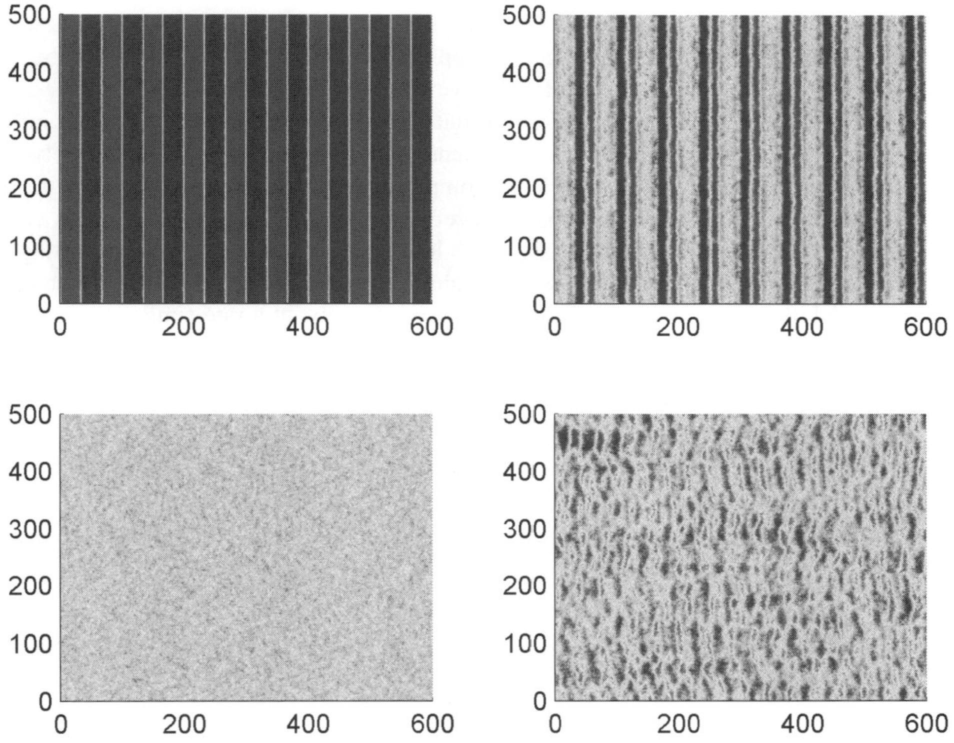

Figure 2. Development of new ripples from regular old ripples (top) and flat bed (bottom). The left-hand figures show the starting domain and the right hand figures show the domain after 8 x 10^6 grain movements. The influence of old ripples in suppressing the development of new ripples is clear.

4. Modelling response to changing conditions

4.1 Introduction

In this section we describe the use of our ripple model to investigate the influence of pre-existing old bedforms on the development of new bedforms in response to a step change in forcing conditions. Key questions are: does the presence of old bedforms influence the development of new bedforms? If it does, what influence does it have on rates of change? And what characteristics of old bedforms are most important in influencing response to new conditions?

4.2 Influence of pre-existing regular bedforms

We have used the ripple model to simulate an initial domain of entirely regular sinusoidal ripples and run the model for conditions which, from flat bed, produce ripples of a smaller wavelength. Figure 2 shows an example, comparing ripple development after a similar length of time from flat bed and from the pre-existing regular ripples. Clearly the old ripples are severely hampering the development of new ripples.

To quantify this effect we have used a spectral technique (Lyons et al., 2002; Davis et al., 2004; Huntley et al., 2008) to separate the elevation variances at the old and new bedform wavelengths. Figure 3 shows results for different amplitudes of initial ripples, one twice the other, and includes the rate of growth from a flat bed for comparison. This figure shows that in all cases the new ripples eventually saturate to the same elevation, but also demonstrates clearly that the influence of the pre-existing ripples in slowing the development of new ripples extends effectively until this saturation is reached. Of particular interest is the comparison with the decreasing amplitude of the old ripples, also shown in this figure. The variance levels where the old and new variances cross in the two cases are remarkably similar, differing by much less than a percent, strongly suggesting that new ripple growth is heavily suppressed until the old ripples are reduced to an amplitude matching the new ripples, after which the new ripples can grow much more rapidly, at a rate similar to that from a flat bed.

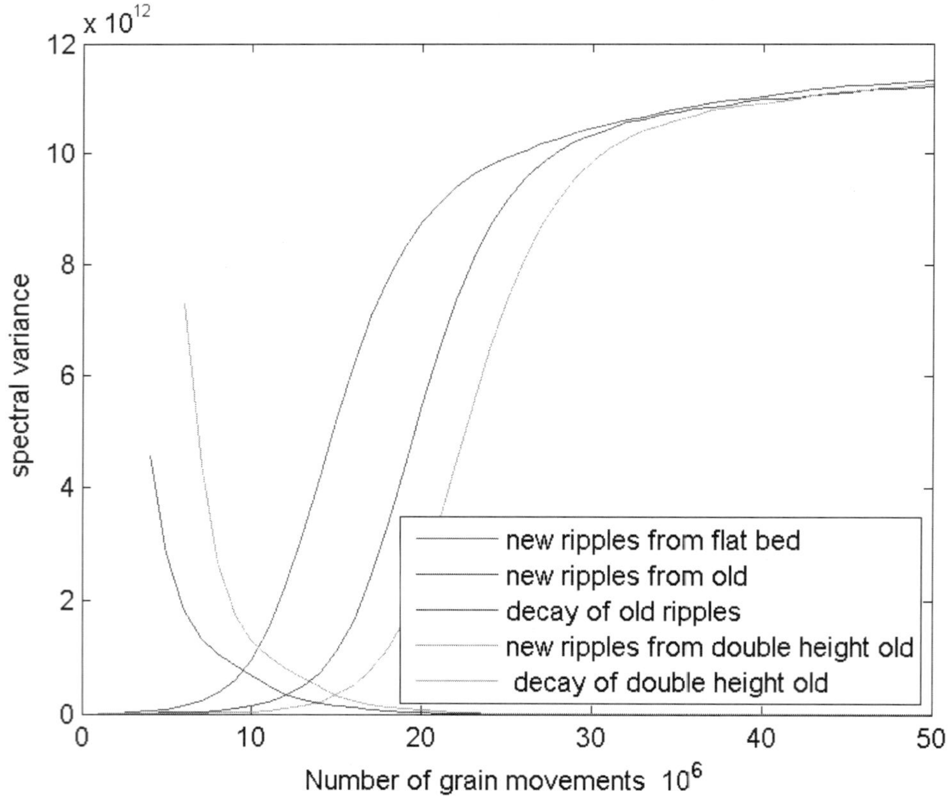

Figure 3. Development of new ripples from different initial (old) regular bedforms. Also shown are curves for the decay of the initial bedform variances. Note that the larger initial ripples cause a larger delay in the development of new ripples.

As we have seen, a similar but more extreme influence was also found for the sorted bedform model of Huntley et al. (2008) where an initial regular field of sorted bedforms was found effectively to entirely suppress the development of bedforms of a new wavelength (Figure 1). The grain size pattern of the old bedforms was modified to some extent by the new conditions but there was no development of the expected new wavelength.

This strong influence of the size of pre-existing bedforms is at least qualitatively consistent with the observations of 'arrested development' of new ripples by Austin et al. (2007) and the hysteresis observed by Traykovski et al. (1999). It does however appear to be at odds with the conclusion of Davis et al (2004); our modelling results suggest that it is the magnitude of the variance ('energy' in Davis et al., 2004) of the old bedforms, not the difference in the energy levels between the old and new bedforms that determines the evolution to the new bedform configuration. Reasons for this disagreement are unclear but may be related to the small scale of their laboratory ripples and the limited range of wavelengths they studied. Our model results may also be inconsistent with the suggestion of Doucette and Donoghue (2006) that the time taken for new ripples to reach saturation height is independent of shape of the initial bed The extent to which these inconsistencies reflect the simplified nature of our model is unclear and can only be resolved by further modelling work using more realistic models and initial bedforms.

4.3 The influence of pattern defects

Real bedforms are not, of course, regular and infinitely long crested. As illustrated in figure 4, they contain 'defects' in the form of crest termination, crest mergers or joins and 'eyes' (localised regions of double crests). To study the influence of more realistic old bedform fields we need to include defects.

Figure 4. Aeolian ripples (courtesy of C. Pilditch)

Schematic defects have been added to the initial bedform field in our ripple model. These schematic defects are simulated by multiplying the amplitude of a regular ripple field by a factor f defined by equation 1:

$$f = 0.5*(1-\cos(\pi\, r/w)) \qquad 1$$

where r is the radial distance from a specified central location for the defect and w is a specified defect width. Different defect densities have been simulated by adding defects at randomly chosen positions, all on ripple crests.

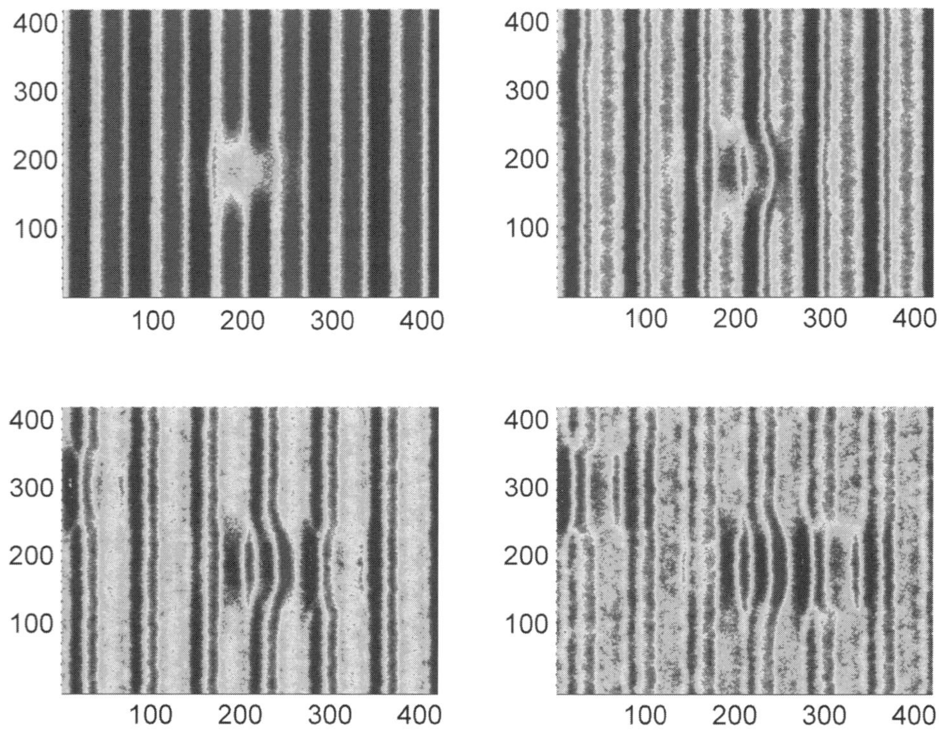

Figure 5. Development of new ripples in a domain with simulated defects. A segment of the domain containing one defect is shown. From top left to bottom right the figures show the segment after 1, 3, 5 and 7 x 10^6 grain movements. The changes to the top left of the final picture are the result of diffusion from another defect.

Figure 5 shows that the 'radiation' effect proposed from the sorted bedform runs (Huntley et al., 2008) is also present for ripples; new ripples develop rapidly at the centres of the defects and diffuse outwards radially and longitudinally during a similar first phase of growth. A similar second phase of growth is also evident at a later stage (not shown in figure 5), where the old ripple pattern has been broken down and the new pattern grows throughout the domain.

In figure 6 the influence of defect density is again clearly seen with initial growth rates increasing approximately linearly with the number of defects in the domain, in agreement with the sorted bedform runs. Interestingly, except for the lowest defect densities, this initial growth rate is faster than from a flat bed, perhaps because ripple crests upstream of the defect provide a 'trigger' for the development of new ripples within the defect. However, ultimately growth from a flat bed overtakes growth from previous ripples, reflecting the observation that rapid more general growth beyond the influence of defects has to await the destruction of the old bedforms. Nevertheless, the influence of defects is still felt right up to the time to reach equilibrium.

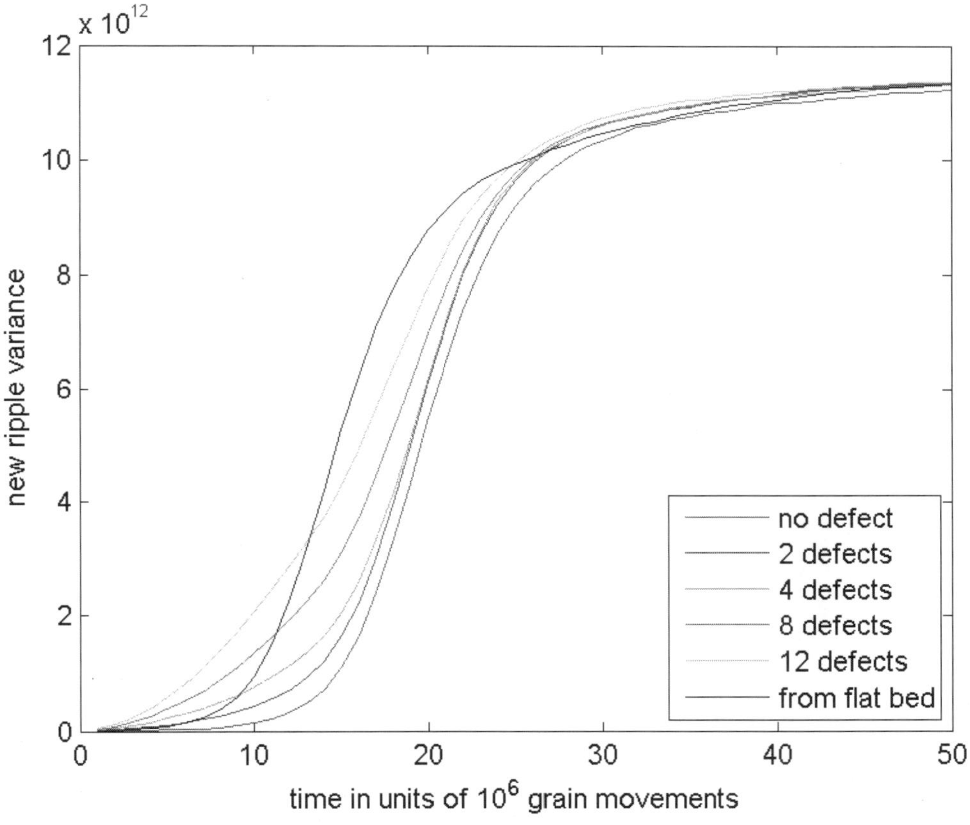

Figure 6. Growth rates of new ripples starting from initial regular ripples with different numbers of defects within the model domain (defect densities, defined as number of defects per m, are 5 times the number of defects). Initial growth rates are proportional to the number of defects but are eventually overtaken by the growth from a flat bed. Note that the line for 2 defects increases faster than expected after about 12 x 10^6 grain movements and eventually merges with the curve for four defects. This behaviour undoubtedly reflects the fact that the two defects used in this simulation were just touching each other, whilst the other runs used separated defects. This illustrates the significant, but relatively small, influence of the details of the layout of defects.

An obvious question leading from these observations is: what characteristics of defects make them so important? One possibility, suggested by the work of Smith and Sleath (2005), is that they contain within them Fourier components at the wavenumber of the emerging bedforms. Fourier analysis, however, suffers from the disadvantage that it measures Fourier components across the whole domain and is therefore inappropriate for studying localised defects. We have therefore applied a wavelet analysis technique (eg Torrence and Campo, 1998; Gorman and Hicks, 2005; Ruessink et al., 2007) using a Morlet wavelet to investigate the distribution of wavelet power for wavenumbers near the emergent ripple wavelength. Figure 7 shows an example for the ripple simulations and demonstrates that the wavelet technique provides a useful objective means of identifying the locations of defects. As might be expected, depending on the scale of interest, the strongest signals can be associated with the sides of the defects, where the slopes of the seabed are most abrupt (Figure 7, top-right panel) or to the centre of the defect (Figure 7, bottom-left panel). For large spatial scales, the strongest signal might be associated with the overall pattern and such a signal might overwhelm any signature related to the defect (Figure 7, bottom-right panel).

However on closer study it becomes clear that for some scales (for example the right hand panels of Figure 7) wavelets are strongest around the steep edges of defects, whereas the new ripples seem to start at the centre of defects and radiate outwards towards the edges. Thus the locations of fastest growth of the new ripples do not coincide with the locations of maximum signal for wavelets of the appropriate wavelength. The presence of Fourier components does not therefore appear to be the key characteristic of defects defining their influence of new bedform development. This observation is supported by further runs using initial regular bedforms with triangular rather than sinusoidal profiles and thus including a much richer range of Fourier components throughout the domain. These runs show growth rates of new ripples which are not greatly different from those for old sinusoidal bedforms.

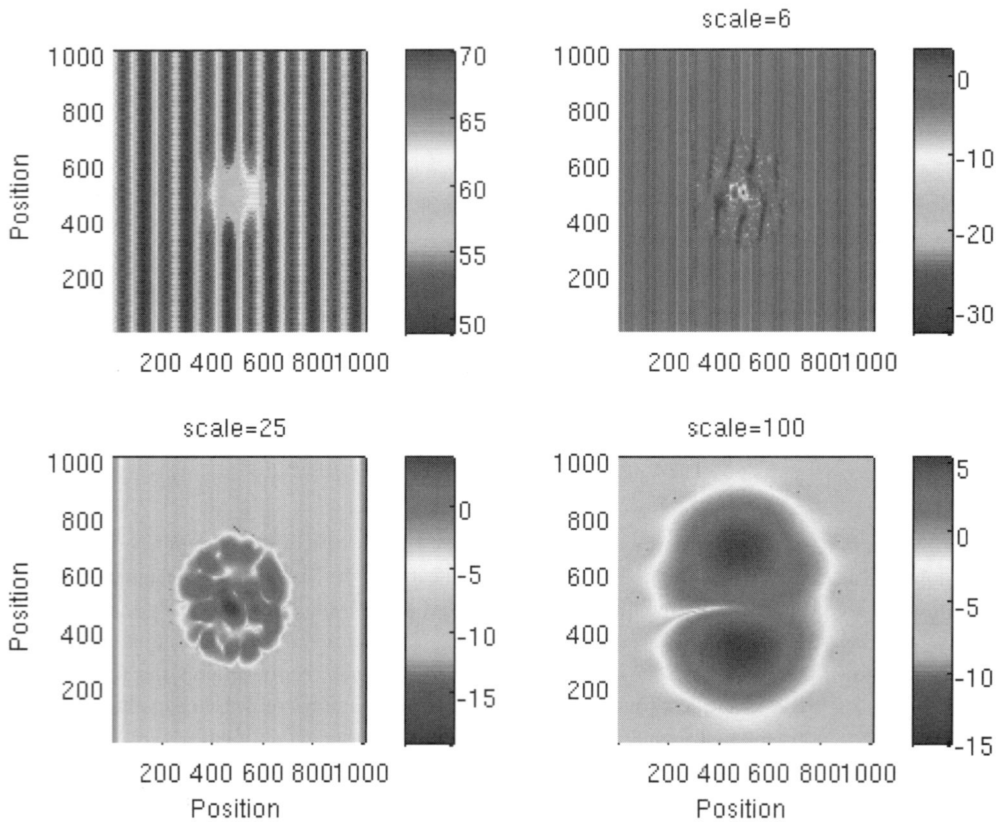

Figure 7. Example of wavelet analysis showing the variance associated with a defect. (a) ripple morphology with a spacing of 166 and one defect located approximately at the centre of the domain. Logarithmic wavelet spectra integrated over all directional sectors (to cover 360 degrees) for wavelet scales equal to (b) 6, (c) 25 and (d) 100 units. Notice the defect is not positioned exactly at the centre of the domain giving rise to a non-symmetric distribution of variance.

Another possibility is that the key feature of defects is the low amplitude of the old bedforms in their vicinity. This suggestion can explain why the new ripples develop first in the centre of a defect, where the amplitude of the old bedforms is zero, and over time 'radiate' outwards through regions of increasing old bedform height. It is consistent with the observations described in section 4.2 above which confirm that the rate of growth of new ripples is strongly influenced by the height of the initial old bedforms. It is also consistent with the rule for grain movement, which is a function of the absolute height of grains above the mean level rather than the frequency content of the modelled domain. For this non-linear system Fourier components may not be relevant. Further work is continuing to investigate this simple but potentially powerful idea.

5. Discussion

The results described in this paper are at this stage preliminary and, in the nature of abstracted models, deal with highly simplified versions of the physical processes responsible for bedform evolution. Nevertheless abstracted models have been shown to reproduce to a remarkable extent the behaviour of real morphodynamic systems (eg Huntley et al., 2007). The fact that the many of the results described using our simple ripple model are similar to those found for a much more physically realistic sorted bedform model (Huntley et al., 2008) lends weight to suggestion that they reflect real generic processes controlling the response of bedforms to changing conditions. The abstracted modelling approach does therefore appear to offer considerable benefits in investigating bedform behaviour.

The implications of this work are that the initial state of the seabed has a profound influence on the rate at which the bed responds to new conditions, but that this influence is subtle and will be very hard to predict in real situations. The growth rate of new bedforms depends upon the size of pre-existing bedforms, with larger bedforms causing increasing delays in the development of new bedforms, and upon the density of defects in the initial bedform pattern, with larger densities allowing more rapid responses to new conditions. The overall response rate can therefore be either larger or smaller than from a flat bed depending on the balance of these factors and the length of time the new conditions prevail. Prediction of response rates therefore depends on either detailed measurements or on prediction of the initial seabed morphology. Under conditions where the system is 'reset' from time to time, as for example in the beach setting investigated by Austin et al. (2007), it may be possible to predict the evolution of bedforms with some accuracy. There is also some evidence that defect density depends on the duration of specific forcing conditions, with defect densities progressively reducing over time (Werner and Kocurek, 1999; Ewing et al., 2006; Coco et al., 2007a), so knowledge of previous forcing conditions might assist with predictions. However in most cases our observations suggest that accurate prediction depends upon much more knowledge of the initial state of the bed than will generally be available.

Our results also suggest that previous attempts at predicting response rates are inadequate. Equilibrium approaches are clearly inadequate though, depending on the conditions, they could overestimate or underestimate rates and might even fortuitously provide an accurate prediction under a narrow range of conditions when these factors balance. The hysteresis approach is also limited since a simple time delay will not simulate the observed behaviour. None of the approaches proposed from the laboratory measurements reviewed in section 2 is consistent with our results although a systematic study that addresses the role of defects is still lacking.

In summary we conclude that the response of bedforms to changing conditions is strongly dependent on the pre-existing state of the seabed and that both the amplitude and the density of defects in the initial bedform pattern have strong influences. Our tentative conclusion is that defects are important because they are sites where the initial bedform pattern has its smallest amplitude so that they are the places where new bedforms can grow most rapidly. Away from defects, new bedforms cannot grow rapidly until initial bedforms are reduced to a smaller amplitude than the emerging bedforms. We are continuing to investigate this hypothesis.

Acknowledgments

GC funded by the (New Zealand) Foundation for Research, Science and Technology. Thanks to Richard Gorman for providing the code for the 2d-wavelet analysis.

References

Anderson, R.S., 1990, Eolian ripples as examples of self-organization in geomorphological systems, *Earth-Science Reviews*, 29: 77-96.

Ashton, A., Murray, A.B., and Arnault, O., 2001. Formation of coastlines features by large-scale instabilities induced by high-angle waves, *Nature*, 414: 296-300.

Austin M. J., G. Masselink, T. J. O'Hare, P. E. Russell (2007), Relaxation time effects of wave ripples on tidal beaches, *Geophysical Research Letters*, 34, L16606, doi:10.1029/2007GL030696.

Becker, J. M., Y. L. Firing, J. Aucan, R. Holman, M. Merrifield, and G. Pawlak (2007), Video-based observations of nearshore sand ripples and ripple migration, *Journal of Geophysical Research*, 112, C01007,

Bell, P.S., and Thorne, P.D., 1997. Measurements of sea bed ripple evolution in an estuarine environment using a high resolution acoustic sand ripple profiling system, *Oceans '97 MTS/IEEE Conference*, Halifax, Nova Scotia.

Blondeaux, P., 2001, Mechanics of coastal forms, *Annual Review of Fluid Mechanics*, 33, 339-370.

Calvete, D., Coco, G., Falqués, A., Dodd, N., 2007, (Un)predictability in rip channel formation, *Geophysical Research Letters*, 34, L05605, doi:10.1029/2006GL028162.

Clarke L. B., and B. T. Werner, 2003, Synoptic imaging of nearshore bathymetric patterns, *Journal of Geophysical Research*, 108 (C1), 3005, doi:10.1029/2000JC000780.

Coco, G., Huntley, D.A., and O'Hare, T.J., 2000, Investigation of a self-organisation model for beach cusp formation and development, *Journal of Geophysical Research*, 105(C9), 21991-22002.

Coco, G., Werner, B.T., Burnet, T., and Elgar, S. (2004) The role of tides in beach cusp formation, *Journal of Geophysical Research*, 109, CO4011, doi: 10.1029/2003JC002154.

Coco, G., and Murray, A.B., 2007, Patterns in the sand: from forcing templates to self-organization, *Geomorphology*, 91(3-4), 271-290.

Coco, G., Murray, A.B., Green, M.O., 2007a, Sorted bedforms as self-organizing patterns. Part 1: model development, *Journal of Geophysical Research*, 112, F03015, doi:10.1029/2006JF000665.

Coco, G., Murray, A.B., Green, M.O., Hume, T., Thieler, R., 2007b, Sorted bedforms as self-organizing patterns. Part 2: complex forcing scenarios, *Journal of Geophysical Research*, 112, F03016, doi:10.1029/2006JF000666.

Davis J. P., D. J. Walker, M. Townsend, I. R. Young, 2004, Wave-formed sediment ripples: Transient analysis of ripple spectral development, *Journal of Geophysical Research*, 109, C07020, doi:10.1029/2004JC002307.

Doucette, J. S., and O'Donoghue, T., 2006, Response of sand ripples to changes in oscillatory flow, *Sedimentology*, 53: 581–596.

Ewing, R.C., Kocurek, G., and Lake, L.W., 2006, Pattern analysis of dune-field paramters, *Earth Surface Processes and Landforms*, 31: 1176-1191.

Gorman, R.M., and Hicks, D.M., 2005. Directional wavelet analysis of inhomogeneity in the surface wave field from aerial laser scanning data, *Journal of Physical Oceanography*, 35: 949-963.

Hanes, D.M., Alymov, V., Chang, Y.S and Jette, C., 2001. Wave-formed sand ripples at Duck, North Carolina. *Journal of Geophysical Research,* 106(C10), 22,575-22,592.

Hay, A.E. and Wilson, D.J., 1994, Rotary sidescan images of nearshore bedform evolution during a storm, *Marine Geology*, 119(1-2): 57-65.

Huntley, D.A., Coco, G., and O'Hare, T.J., 2007. Abstracted modelling as a tool for understanding and predicting coastal morphodynamics, *Journal of Coastal Research*, SI39, 21-27.

Huntley, D.A., Coco, G., Bryan, K.R., and Murray, A.B., 2008, Influence of "defects" on sorted bedform dynamics, *Geophysical Research Letters*, 35, L02601, doi:10.1029/2007GL030512.

Kosugi, K., Nishimura, K. and Maneno, N., 1992, Snow ripples and their contribution to the mass transport in drifting snow, *Boundary-Layer Meteorology*, 59: 59-66.

Landry, W., and Werner, B.T., 1994, Computer simulations of self-organized wind ripple patterns, *Physica D*, 77: 238-260.

Lippmann, T.C., and Holman, R.A., 1989, Quantification of sand bar morphology: a video technique based on wave dissipation, *Journal of Geophysical Research* , 94(C1), 995-1011.

Lyons, A. P., W. L. J. Fox, T. Hasiotis, and E. Pouliquen, 2002. Characterization of the two-dimensional roughness of wave-rippled sea floors using digital photogrammetry, IEEE J. Oceanic Eng., 27: 515– 524.

Murray, A.B., 2003. Contrasting the goals, strategies and predictions associated with simplified numerical models and detailed simulations. In Prediction in Geomorphology, P.R. Wilcock and R.M. Iverson Editors, Geophysical Monograph Series, American Geophysical Union, Washington, DC, USA, 151-165.

Murray, A.B., and Thieler, E.R., 2004, A new hypothesis and exploratory model for the formation of large-scale inner-shelf sediment sorting and "rippled scour depressions", *Continental Shelf Research*, 24: 295-315.

Osborne, P.D. and Vincent, C.E., 1993. Dynamics of large and small scale bedforms on a macrotidal shoreface under shoaling and breaking waves. *Marine Geology,* 115, 207-226.

Pannell, M.A., O'Hare, T.J. and Huntley, D.A., 2003. Modelling sand ripple development by self organisation in unsteady flows. *Proceedings of the 28th International Conference on Coastal Engineering* (Cardiff, Wales). J. McKee Smith (ed.). World Scientific, New Jersey. pp. 2837-2849.

Ruessink, B.G., Coco, G., Ranasinghe, R., and Turner, I.L., 2007, Coupled and noncoupled behavior of three-dimensional morphological patterns in a double sandbar system, *Journal of Geophysical Research,* 112(C7), 10.1029/2006JC003799.

Smith, D., and Sleath, J.F.A., 2005, Transient ripples in oscillatory flows, *Continental Shelf Research*, 25(4), 485-501

Torrence, C., and Compo, G.P., 1998, A practical guide to wavelet analysis, Bulletin of the American Meteorological Society, 61-77.

Traykovski, P., Hay, A.E., Irish, J.D., and Lynch, J.F., 1999, Geometry, migration, and evolution of wave orbital ripples at LEO-15, *Journal of Geophysical Research*, 104, C1, 1505-1524.

Traykovski P., 2007. Observations of wave orbital scale ripples and a nonequilibrium time-dependent model, *Journal of Geophysical Research* , 112, C06026, doi:10.1029/2006JC003811.

Van Enckevort, I.M.J., B.G. Ruessink, G. Coco, K. Suzuki, I.L. Turner, N.G. Plant, and R.A. Holman, 2003, Observations of nearshore crescentic sandbars, *Journal of Geophysical Research* , 109(C06028), doi: 10.1029/2003JC002214.

Werner, B.T., 1995. Eolian dunes: Computer simulations and attractor interpretation, *Geology*, 23,:1107-1110.

Werner, B.T., and Fink, T.M., 1993, Beach cusps as self-organised patterns, *Science*, 260: 968-971.

Werner, B.T., and Kocurek, G., 1999. Bedform spacing from defect dynamics, *Geology*, 27(8): 727-730.

Werner, B.T., 2003. Modeling landforms as self-organized, hierarchical dynamical systems. In Prediction in Geomorphology, P.R. Wilcock and R.M. Iverson Editors, Geophysical Monograph Series, American Geophysical Union, Washington, DC, USA, 133-150.

Coastal Dynamics 2009
Keynote 3

NEARSHORE REMOTE SENSING

Rob Holman[1]

Abstract

The performance of most operational nearshore models is less limited by the quality of model physic or numerics than by the availability of input data for initial and boundary conditions. For example, circulation and sediment transport are very sensitive to the details of bathymetry, a data field that is notoriously hard to measure well and changes rapidly. Remote sensing approaches offer an increasingly attractive solution. While active sensors like radar and LIDAR have some advantages and will be discussed, this paper will focus more on optical remote sensing through the Argus program. Starting with a discussion of the required capabilities of an ideal sampling system, the paper will discuss the technical capabilities and future directions of optical remote sensing of the nearshore. Sampling gaps will be identified and possible solutions discussed.

Key words: remote sensing, hydrodynamics, sediment transport, morphodynamics, Argus

1. Introduction

The goal of nearshore processes research is to develop a predictive understanding of the dynamics of waves as they approach a beach over a shoaling bathymetry, understand the resulting nearshore flows, and predict the response of the bathymetry through sediment transport to those waves and currents. The usual embodiment of this understanding will be numerical models that allow prediction of all variables over some region of interest. Significant research investment has lead to great improvements in these models and acceptable accuracy in many fluid properties can be achieved, particularly in simplified situations of 1DH bathymetry and oblique waves [*Feddersen and Guza*, 2003; *Ruessink, et al.*, 2001]. Yet model performance weakens considerably for conditions of 2DH bathymetry and near normal incidence, when 2D circulation becomes significant [*Wilson, et al.*, in review-a]. Moreover, sensitivity of model output to inaccuracies in time-varying input variables is not well understood, although Wilson et al [in review-b] show that even daily surveys may not be sufficient (in that model-data discrepancies of mean currents measured between times of surveys can be effectively used to estimate intra-survey bathymetric changes).

There is strong reason to believe that the main limitations to model performance many not be due to weakness in the represented physics or numerics but instead in limitations in our ability to supply accurate input data in the form of initial and boundary conditions, particularly accurate bathymetry data. Traditionally, input wave and current data have been collected using in-situ instrumentation at a few fixed points or in an array, while survey data are collected occasionally using optical or GPS methods and a mobile platform like a jet-ski of the CRAB. As anyone who has sought funding for these efforts knows, these approaches are expensive and logistically difficult, so sampling is usually limited to short (O(month)) experiments or is sparse in time (bathymetry surveys). These approaches are not compatible with sustained sampling needed for observing systems, for example.

[1]College of Oceanic and Atmospheric Sciences, 104 Ocean Admin. Building, Oregon State University, Corvallis, Oregon, USA, 97331-5503, Holman@coas.oregonstate.edu

In this paper we will discuss alternatives that can be provided through the use of nearshore remote sensing (RS). A number of methods are available including active sensors such as radar and LIDAR that provide their own illumination and passive sensors like cameras that sense the reflection and scattering of ambient electromagnetic energy (light and infrared). Remote sensors can make measurements with high spatial resolution over large areas (in contrast to the point sampling of in situ sensors) and over extended sampling periods. Because they are removed from the water they are much more survivable and, in the event of need, more easily serviced. In contrast, most remote sensors return image data that must be mapped to world locations and whose relationship to geophysical variables such as sea surface elevation is not well known.

We will describe some of the characteristics of an ideal nearshore sampling system, then describe the characteristics of the most common remote sensors. The focus will then shift to optical methods and a number of applications discussed as examples of the power of these systems.

2. An Idealized Nearshore Sampling System

Sampling the full range of time and space scales in the nearshore is challenging. There is significant spatial variability from turbulent to regional scales, particularly associated with strong cross-shore gradients from wave breaking over sand bars or alongshore gradients near rip channels. In time, signals vary on the scales of turbulence, waves, storms, seasons and interannual. Thus, a sampling system needs high resolution in time and space, but also dense coverage and large span. Data collection systems should be unmanned and should return the data automatically to a central facility for analysis. Both instrumentation and data collection protocols should be robust and able to recover from the variety of problems that inevitably arise over multi-year sampling.

These are characteristics of many remote-sensing systems, although only for the case of ground-based systems that can sample persistently with no tending (as opposed to overhead RS, requiring an aircraft or equivalent). In addition, data will only be meaningful if mapped to world locations and converted to equivalent geophysical variables.

2.1. *Active Sensors*

Remote sensing systems can be classed by their electromagnetic (EM) frequency band, their imaging approach and their illumination mode (active or passive). Because EM waves are transverse, polarization can also be exploited. Non-EM approaches such as acoustics will not be considered.

Active sensors like radar and LIDAR learn about their imaging environment from the characteristics of reflected pulses of EM radiation that were emitted from the sensor and scattered from the ocean surface. Because they don't rely on ambient radiation they work at night and under poor visibility (all weather) and because they know the emit time of each pulse, the distant to any object can be found via time of flight.

2.1.1. *Radar*

Radar spans EM wavelengths from 1mm to 100m but most commercial systems are X-band with 3cm wavelengths, much longer than visible light (0.5 micron). Because of this, a radar 'camera' would need to be prohibitively large, roughly one hundred thousand times the size of an optical camera chip. Thus radar imaging is done in a scan mode, essentially rotating the aim point of a single pixel with only modest horizontal resolution (dictated by the length of the long axis of the antenna). Poor vertical resolution (the short axis of the antenna) is not a hindrance for imaging a flattish sea surface since delay time easily translates to range to achieve effective horizontal mapping.

The physics by which radar pulses reflect from the ocean for the typical low graze angles (nearly horizontal views) is only roughly known. The primary accepted mechanism is Bragg reflection, whereby scattering from a surface that has roughness at half the radar wavelength is self-reinforcing and can be strong. This corresponds to the roughly centimeter scale of wind-generated capillary waves. Thus, glassy seas on windless days will back-reflect no energy. Additionally, the bubbles of breaking waves provide strong reflections known as sea spikes that is not primarily Bragg related. Thus, radar does not directly see ocean waves but instead sees modulations of scattering associated with waves (either front face breaking or steepened capillary waves at the flow convergences on the front faces of waves). Radar emits completely polarized radiation with either a horizontal or vertical orientation and is usually tuned to receive the same polarization. Scattering processes are different for either orientation, so the choice can reinforce or squelch ocean wave signals.

The all-weather, day-night robust nature of radars makes them attractive for long-term measurements and can compensate for disadvantages including the above complicated imaging physics (they don't actually see waves) and the fact that spatial sampling in non-synchronous (the radar sweeps through the domain ever second or two so samples to the left and right are not synchronous and must be interpolated prior to cross-spectral analyses such as the determination of wave direction. Nevertheless, excellent data can be collected continuously over a large domain. Figure 1 should an example 'snapshot' of approaching waves from an X-band radar at Duck, NC.

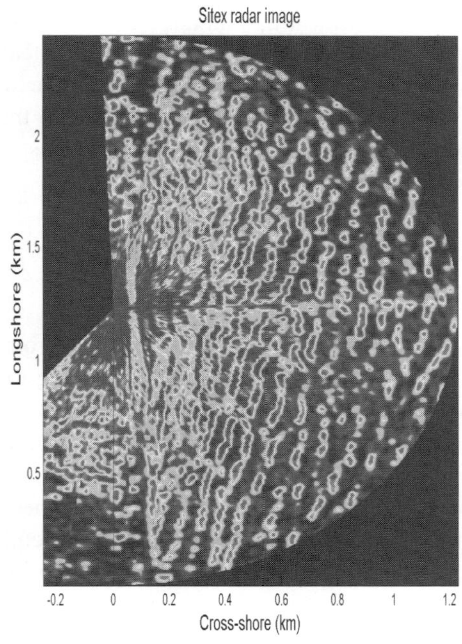

Figure 1. Example X-band radar image from the Sitex experiment at Duck, NC. The shoreline runs vertically near 0.1 km cross-shore distance. The line to the right is the FRF pier. Waves approaching from the southeast are clear visible and can be tracked as the shoal and refract toward shore. Radar 'frames' are collected every 1.33s allowing a variety of cross-spectral methods. Image courtesy of M. Haller, Oregon State University.

With correct spatial mapping and compensation through interpolation for the rotational sampling lag, radar provides good data in space and time of the nearshore wave field, similar to optical collections but with reduced resolution and better day-night capability. Thus, many of the same analyses can be carried out on either instrument, for instance the inversion of measured wave celerity data to yield bathymetry. For efficiency these approaches by which remote sensing data can be converted to geophysical data will be described in the Argus sampling section below.

Radar can be run in more sophisticated modes such as Doppler radar that detects velocities of scatterers from their radar Doppler shift (common in weather radar) or interferometric radars that can detect changes that are smaller than a radar wavelength from interference patterns between two sets of returns. However, these methods remain experimental in the nearshore.

2.1.2. *LIDAR*

LIDAR (Light Detection and Ranging) is an active optical sensor that scans a nearshore region with a laser and records the strength of the returned pulse. Like radar, time of flight between pulse emission and detection is translated into distance from the sensor for object location. Unlike radar, the time history of each pulse is not recorded. Instead only the time to the first dominant return (assumed to be the ocean surface or sandy bottom) is usually recorded. Both the surface and bottom can be detected to find water depth, often using lasers of different colors to maximize each signal. Also unlike radar, LIDAR scans in both horizontal directions, covering a region in a pattern at a typical rate of 5,000 shots per second, each with a roughly 1m^2 footprint.

The great strength of LIDAR is the potential for the rapid acquisition of a huge number of accurate (15 cm rmse) 3D locations of either the ocean surface, subaerial topography or, for clear water, sub-aqueous bathymetry. However, even this acquisition rate is small compared the demands of nearshore sampling, so choices must be made between spatial and temporal coverage and resolution. Typically LIDAR are mounted on a mobile platform like an airplane in order to sample over large areas. The loss of time series information is not critical for relatively fixed features like beach topography, but leaves a confused picture of rapidly changing features like ocean waves. It would be feasible and attractive to experiment with compromise sampling strategies that scan a more limited region (O(100m) box) rapidly enough to properly sample ocean waves. This would require instrument mounting on a tower or other fixed platform.

2.2. *Passive Sensors*

In contrast to active sensors, passive sensors rely on ambient electromagnetic radiation like sunlight to illuminate a region of interest. Visible band cameras, mimicking the successful performance of the human eye, have been is use for centuries. They are typically high resolution, large field of view, use robust technology and are cheap. More recently, cameras have become available to image different parts of the infrared band. The infrared band of EM waves is divided into a number of sub-bands all of which are longer wavelength that visible light, so cameras are necessarily lower resolution. At the mid and long wavelengths, a significant portion of the energy is emitted from objects like the ocean due to blackbody radiation. These sensors act more like temperature sensors and can provide rich images of unexpected temperature structure due to turbulence or current structure (e.g. Figure 2). Because these are usually imaging cameras and can sample at video rates (30 Hz), they can provide the same space and time sampling as visible band cameras, albeit at lower resolution. Thus, available analysis methods appropriate to nearshore processes work can be the same as for visible band cameras and will be described below. However, it should be noted that thermal structures visible in infrared imagery often have no visible counterpart so are only exploitable in the infrared.

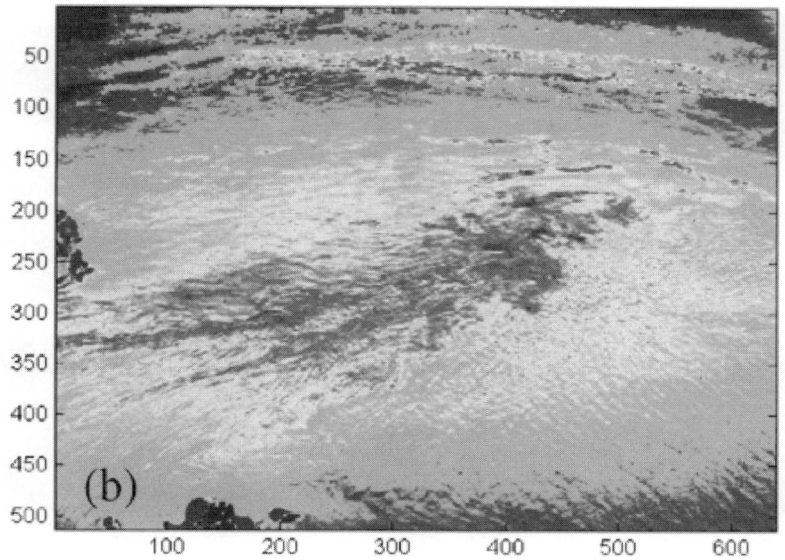

Figure 2. Example infrared image from the COHSTREX 2005 experiment, courtesy of UW APL. Warm colors indicate warmer surface temperatures. The main feature is a tidal jet moving to the right through a gap in a groin. Waves are also visible at several length scales. (see http://cohstrex.apl.washington.edu/2005Infrared.html).

2.3. *Argus Optical Sampling*

Visible band cameras come in a wide range of models with varying resolution, quality and price. For the remaining discussion, we will focus on the current generation of Argus cameras. The Argus Program is an ongoing research program begun by the Coastal Imaging Lab (CIL) at Oregon State University but with active stations now covering a large number of sites around the world. An Argus Station is a computer-controlled collection of cameras that are programmed to carry out a number of different types of data collection and to return the data to central sites for subsequent analysis.

A full discussion of the technical specifications of Argus Stations is contained in Holman and Stanley [2007] and will only be abstracted here. A typical station consists of five FireWire digital cameras mounted on a high vantage point such as a lighthouse or tower, with each camera aimed to cover a portion of the 180° field of view. Cameras typically operate at 2 Hz and acquire frames of 1280x960 pixels with a typical ground footprint of 10 to 100cm depending on range. Images are multi-spectral with RGB pixels arranged in a Bayer pattern on the sensing chip.

Unlike radar or LIDAR, the transformation between image (pixel) and world coordinates requires accurate knowledge of intrinsic calibration parameters of the camera as well as of the camera aim directions (Euler angles; [*Holland, et al.*, 1997]). Typically this requires automated recognition of the image location of known ground features. Absolute accuracy to within one pixel is readily achievable. Bookkeeping of the many calibration and collection details of an Argus Station requires a robust database and toolbox of acquisition and analysis algorithms [*Holman and Stanley*, 2007]. For example, figure 3 shows a rectified snapshot from the Argus Station at the Field Research Facility in Duck, NC. The image is a composite from five cameras mapped into the local coordinate system and shows how, with proper photogrammetry and database support, even five cameras can be treated as part of a single composite system.

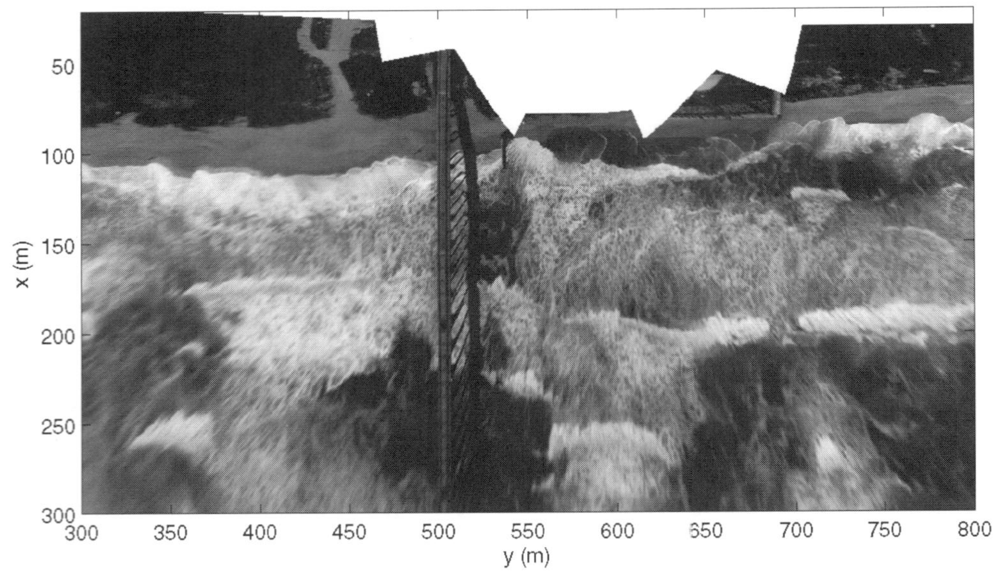

Figure 3. Example rectified composite snapshot from the Duck FRF Argus Station for April 17, 2009, 1700 GMT. The image seamlessly joins remapped images from five separate cameras, for example merging different views of the pier at y=505m from several cameras. Such a capability allows accurate spatial sampling over a large domain.

Because Argus Stations are remote and designed for long-term collections with no human maintenance, the volume of collected data cannot exceed that which can be returned through the available communication lines (ideally internet). Data are typically collected on an hourly basis and consist of two general collection forms, image data and time series data. It is not taxing to return a single image every hour. However, it is nearly impossible to return a time series of 2048 such frames from each of five cameras and collected at 2 Hz through anything but a specialized communications network. Thus, for analyses that require time series analysis such as bathymetry or wave spectral estimation, care is taken to collect and return data not from every pixel in the image but only from those pixels that are needed for the specific analysis. Called pixel instruments, this collection style is analogous to deploying arrays of in-situ instruments. Typically, dense spatial sampling of a nearshore region of interest can be achieved using less than 1% of image pixels.

The physics by which light is reflected or scattered from the ocean surface determines how we see waves and how we must convert between optical signals and the geophysical wave signals they represent. In general, we see the ocean surface due to light that reflects from that surface. The source of this light can be the actual solar disk (in which case the reflected light is called glitter) or the blue light of the skydome. However, the reflection coefficient for normally-incident light is only 2% so a great deal of light enters the water, some of which is scattered back up by ocean particulates in what is called upwelling radiance. To learn about the ocean surface (waves) we wish to maximize the reflected signal (coming from the ocean surface) compared to the upwelling radiance. This naturally occurs at lower graze angles (the reflection coefficients rise quickly within 15° of the horizon) or we can use polarizers set to maximize the passage of horizontally polarized reflected light (polarizers are usually set to reduce reflection and glare, so this is the opposite orientation from normal).

We see non-breaking waves for two reasons. First, the reflection coefficient increases with the angle of incidence, becoming large near tangential viewing, so we see slope-dependent effects. Second, the brightness of the skydome, the source of most reflected light, increases toward the horizon. Since reflection from the ocean typically is specular (angle of incidence equals the angle of reflection), the front and rear faces of an approaching wave will reflect to different parts of the skydome so will have different brightness. We can exploit this variation to 'see' the waves and carry out a number of analyses based on these visible signals.

Light is also strongly scattered from bubbles associated with wave breaking. This is due to multiple reflections from the many air-water interfaces of a patch of bubbles and total reflection can reach 50%. Unlike specular reflection, scattering from bubbles is Lambertian (has no dependence on viewing angle).

3. Example Applications

Nearshore remote sensing has been used to make a wide variety of useful measurements in the nearshore both of the fluid domain (waves and currents) and of the bathymetric response (bathymetry and morphology). One of the earliest applications of Argus was measuring the morphology of submerged sand bars, so the following discussion will start with these image products before building to more complex but powerful time-domain-based measurements.

3.1 Image-based Products

Everyone is familiar with the traditional oblique (unrectified) snapshot. While providing a good impression of the day's conditions, it has little scientific value principly because it is just one snapshot of a stochastic wave field. However, other derived images can be quite useful and can include better statistical representation of the nearshore domain. The most useful product is the 10-minute time-exposure image (for example Figure 4). By averaging over the time scale of not just individual waves but also over wave groups, we can produce a product that outlines areas of prefered wave breaking. Lippmann and Holman [1989] showed that the bands of brightness in these images corresponded to the dissipation of a random wave field over submerged sand bars. Other more detailed following studies have supported this basic idea, but with improvements in understanding [*Aarninkhof*, 1996; *van Enckevort and Ruessink*, 2001]. The longest continuous record of daily time exposure images started in 1986 at Duck, NC, and continues to this day. These long sequences offer unparalled records of natural sand bar variability that have completely changed our views of how sand bar and even rip currrent systems work [*Alexander and Holman*, 2004; *Holman, et al.*, 2006; *Lippmann and Holman*, 1990].

Time exposure images also show enhanced breaking at the shoreline (Figure 4). Since the width of the subaerial beach is a key variable for coastal zone management, a number of studies have investigated the accuracy of video-based shoreline measurements [*Plant, et al.*, 2007; *Plant and Holman*, 1997]. By measuring the variation of shoreline position with tide level variations, Plant and Holman [1997] demonstrated the ability to measure the foreshore beach slope, an important dynamical variable.

The relationship between optical signals and wave dissipation inspired Aarninkhof et al [2005] to consider whether this relationship could be inverted. Given time exposure images of the patterns of wave breaking, was it possible to find the underlying bathymetry for which the theoretical dissipation pattern best matched the observed time exposure pattern. Using an assimalitive approach they developed a method that yields excellent bathymetry estimates over extended records. This approach is currently being incorporated into the Beach Wizard program.

Figure 4. Example ten-minute time exposure image of the wave breaking patterns at Duck, NC, on March 10, 2005. White bands, areas of persistent wave breaking, correspond to the shallow waters of submerged sand bars. Part of a crescentic bar in the distance becomes very complex closer to the camera, including a number of cross-shore oriented features that are persistent hour to hour and have been shown to have bathymetric origin [*Konicki and Holman*, 2000]. The white band at the shoreline corresponds to the shore break and can be exploited to automatically find the changing shoreline position.

Stereo analysis is also a powerful method of finding 3D locations of observed features like the sub-aerial beach topography. The method requires two or more cameras to be separated by an appreciable baseline to provide converging views. For nearshore applications, it is convenient for the separation to be vertical, i.e. with cameras located at the top and middle of a tall tower. The method can sometimes be limited on a clean smooth beach by a lack of identifiable features for which corresponding locations can be found. Holland and Holman [1997] solved this problem by exploiting the runup edge as an easily identifiable feature whose 3D location can be estimated in a series of image pairs to accurately map the foreshore beach profile and document changes on rapid time scales.

3.2 Time Series-based Analyses

While image data can be very useful and images are easily retrievable with low bandwidth communications, the real power of optical sampling comes from time series data (sometimes called motion imagery) that allow samping of the corrresponding time and space scales of optical data. The heart of these methods is the pixel time series, a time series of varying optical intensity at a single pixel location (translate-able to a specific world location using normal photogrammetric principles [*Holland, et al.*, 1997]. Lippmann and Holman [1991] were among the first to show that optical time series were highly coherent with in-situ sensors over the incident wave band, so could be used as cheap proxies for in-situ instruments. Because it is trivial to collect time series from an array of pixels rather than just one pixel, powerful array processing methods become available. For instance, it has been shown that wave angle can be estimated using arrays of a small number of pixel time series with configurations and analysis methods similar to those used with arrays of in-situ sensors [*Herbers and*

Guza, 1990; *Lippmann and Holman*, 1991].

Wave runup is a powerful variable that is equivalent to a sea surface elevation time series but projected onto the beach face (so horizontally spread). Since runup is by definition an antinode of standing waves on the beach, it is a good tool for studying infragravity waves, avoiding the nodal patterns associated with offshore in-situ sampling [*Holman*, 1981]. Optical methods are ideal for measuring runup and avoid the complications of physical runup wires. Optical runup signals can be digitized from pixel time series of a cross-shore array of contiguous pixels (e.g. Figure 5). Such an array is called a time stack and shows not only the time varying water's edge (the runup) but also the speed (inverse slope in the time-space plot) of the advancing bore fronts, the water particle velocities as seen by foam streak slopes and the interactions among different bores. Chickadel and Holman [2003] used a comparable method but with the spatial axis oriented in the alongshore direction to find longshore current velocities from foam streaks in time stacks.

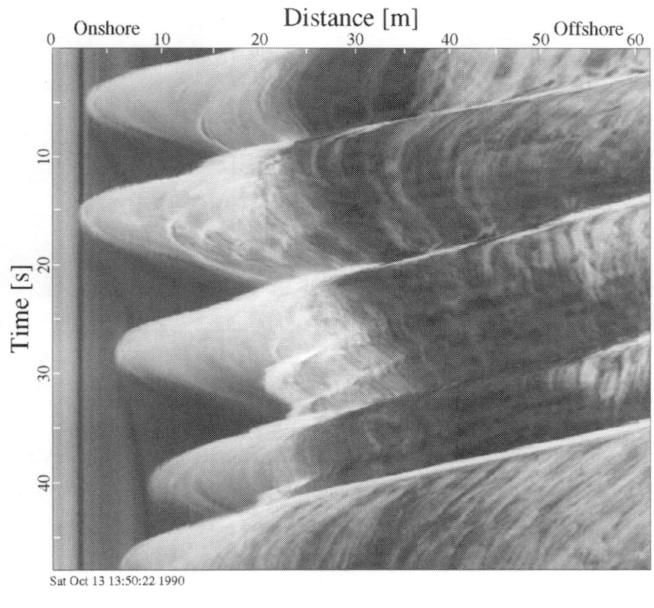

Figure 5. Example time stack showing wave nearshore bores and swash. The x and y-axes correspond to cross-shore distance and time, respectively (so the slope of any feature is equivalent to inverse speed). Foam streaks show the surface water velocity of the waves while the speed of the incoming bores can be found from the slope of the approaching bright bore fronts.

Perhaps the most exciting application of time-domain nearshore remote sensing is the estimation of bathymetry based the observed celerity of waves. It has long been known that the speed of wave propagation depends on the hyperbolic tangent of the non-dimensional depth, a dependence that simplifies to a square root dependence in shallow water. By collecting a spatial array of pixel time series and exploiting the phase relationships between locations, wavelengths can be found and converted to equivalent depths. This problem has been studied in various forms since WWII but has yielded ever more robust results as sampling and signal processing technologies have improved. Single snapshots have given way to motion imagery analysis (although snapshots are often much more readily available for strategically important regions and may be exploited using innovative new methods [*Splinter and Holman*, in press]). This method is applicable to radar or optical methods.

However, good signal processing practices are needed to subdue the high frequency noise that occurs in both sensors (in radar due to salt and pepper noise and in optics due to the dominance of short period chop in imagery). The problem is best done in frequency space, isolating the most coherent waves of the incident band and modeling the full cross-spectral matrix over a region of interest that is big enough to not be dominated by chop but small enough to include only coherent motions. Such a set of algorithms is being explored as part of the Beach Wizard program that seeks to predict bathymetry and fluid properties using all-source remote sensing data. An initial celerity-based component of this work is described in Plant et al [2008] and is found to provide excellent results (e.g. Figure 6).

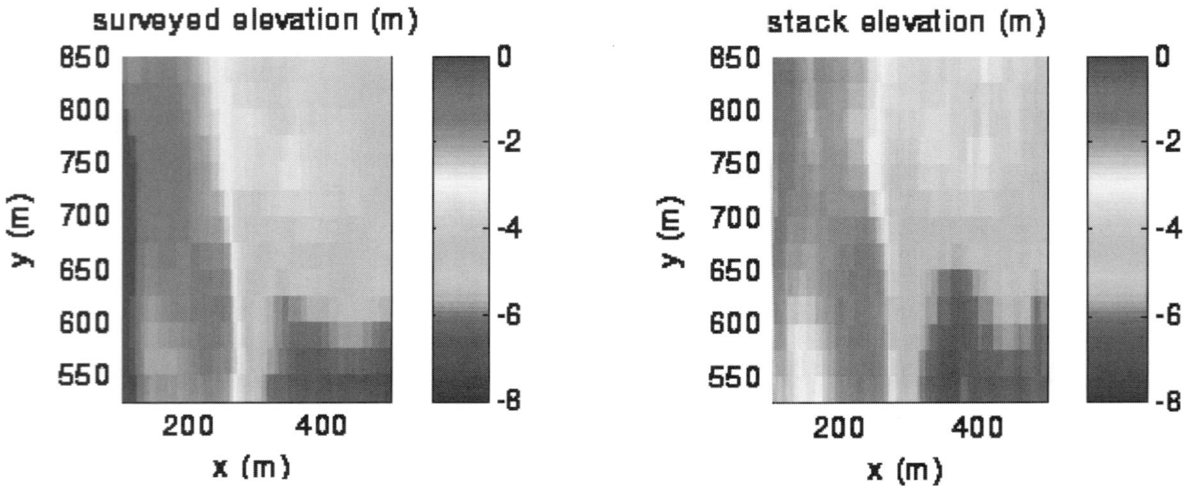

Figure 6. Comparison of a surveyed nearshore bathymetry (left) with one estimated using celerity-based algorithms. Results from this example are excellent and can be estimated for each daylight hour. Because confidence intervals accompany each estimate, these can be merged objectively using a Kalman filter to yield an evolving bathymetry map.

4. Weaknesses and Future Work

Nearshore remote sensing offers a low-cost, logistically simple approach to solving many nearshore monitoring and measurement problems. However there are some quantities that cannot easily be measured and there are other limitations inherent in these approaches.

The main limitation is the inability of most remote sensing methods to estimate wave heights or, more specifically time series of sea surface elevation. This is because images are two-dimensional so do not contain sufficient information to find 3D quantities without further information. Several possible solutions have been or are being investigated. Some specialized radar sensors measure Doppler shift as well as the strength of backscatter. From Doppler shift, the velocity of the scatterers can be found. These should in turn correspond to the orbital velocity of the wave motion, something that can be directly related to wave height for each wave. However, several technical difficulties will need to be surmounted. Radar preferentially samples the crests of waves since troughs are in the radar shadow of closer waves. So Doppler results are an unknown partial average of the full wave cycle. Radar backscatter is also dominated by sea spikes of breaking waves that may be traveling at speeds different from the underlying incident wave. Clarification of these issues will require further research.

Optical systems can exploit the variation of the reflection coefficient with angle of incidence to try to estimate wave heights, but this requires knowledge also of the skydome variability with vertical angle, something that likely will have to be measured. One solution may be to use and optical polarimeter to exploit the polarimetric characteristics of light reflected from the ocean surface. The process of reflection introduces polarization with a preferred axis tangential to the reflecting surface [*Collett*, 1993]. Thus images of the azimuth of polarized light from an imaging polarimeter can be transformed into sea surface slope maps from which sea surface elevations can be found.

A further complication is based on the fact that the mechanisms by which we see breaking versus non-breaking waves are quite different (breakers are dominated by white crests). This is a particular problem at the edge of the surf zone where waves vary between breaking and not (coherence becomes very weak in these circumstances).

Wave breaking is the largest signal in both radar and optical images, typically an order of magnitude larger than non-breaking variability. While this would seem to be a nasty noise problem it actually offers an opportunity. If wave breaking can be quantitatively related to a geophysical variable like wave dissipation (as initial tests find to be roughly true [*Killian*, 2007]) then optics and radar can provide a direct measurement of this important variable. Dissipation, in turn, is directly related to radiation stress gradients, the prime forcing for nearshore flow. In situ direct measurement of dissipation is extremely hard and would be localized to an instrument. Time-evolving maps of dissipation from remote sensing tools would be a wonderful addition to nearshore observing systems (and is being investigated in the Surf Zone Optics program).

Finally, most research on remote sensing has exploited just one of the potential sensors. Radar and the various flavors of passive sensors each see the ocean differently due to the different scattering mechanisms. Thus, each may provide complementary rather than redundant information. The potential for multi-sensor fusion methods needs much more investigation.

5. Summary

The nearshore is a complex domain with variability at a wide range of space and time scales that are difficult to sample with sparse in-situ instrumentation. It is also a harsh environment that takes a toll on exposed equipment. Many, but not all, of the sampling needs can be met instead using remote sensing methods that can provide good spatial resolution and span and good temporal resolution. Because the equipment can be protected from the elements, data can be collected for long periods at low costs.

Different sensors are available, each with different strengths and each 'seeing' the world through different physical processes. While many remote sensing imagers are thought of as providing only a qualitative sense of the nearshore, they can equally well be treated as instruments whose data can be calibrated, geolocated and quantified using the same signal processing algorithms as are used for in-situ sensors. For example bathymetry, a key and usually limiting variable in any numerical model, is notoriously hard and expensive to measure using survey methods but could become a relatively robust remote sensing product if appropriate signal processing methods are used.

As we move toward an observing world where regular predictions of the nearshore domain are needed, remotes sensing will become increasingly important.

Acknowledgements

The Argus Program that is the main focus of discussion has developed over the years with contributions from too many individuals to mention (although many are referenced). However, it is important to mention John Stanley, the glue that holds the program together and makes it actually work. I want to anonymously thank all of you who have been part of the Argus world over the years, contributing to the science as well as the excitement of our work. Funds to support this work have come from ONR through both the Coastal Geosciences Program (award #N00014-07-1-0490) and the SECNAV/CNO Chair Program (award # N00014-03-1-0973) and through the USGS (Coop #06ERAG0028).

References

Aarninkhof, S. G. J. (1996), Quantification of bar bathymetry from video observations, 91 pp, Delft University, Delft, The Netherlands.

Aarninkhof, S. G. J., B. G. Ruessink, and J. A. Roelvink (2005), Nearshore subtidal bathymetry from time-exposure video images, *Journal of Geophysical Research*, DOI: 10.1029/2004JC002791.

Alexander, P. S., and R. A. Holman (2004), Quantitative analysis of nearshore morphological variability based on video imaging, *Marine Geology*, *208*, 101-111.

Chickadel, C. C., R. A. Holman, and M. F. Freilich (2003), An optical technique for the measurement of longshore currents, *Journal of Geophysical Research*, *108*, 3364.

Collett, E. (1993), *Polaried light: fundamentals and applications*, 581 pp., Marcel Dekker Inc, Rochester, New York.

Feddersen, F., and R. T. Guza (2003), Observations of nearshore circulation: Alongshore uniformity, *Journal of Geophysical Research*, *108*, 3006.

Herbers, T. H. C., and R. T. Guza (1990), Estimation of directional wave spectra from multicomponent observations, *Journal of Physical Oceanography*, *20*, 1703-1724.

Holland, K. T., and R. A. Holman (1997), Video estimation of foreshore topography using trinocular stereo, *Journal of Coastal Research*, *13*.

Holland, K. T., R. A. Holman, T. C. Lippmann, J. Stanley, and N. Plant (1997), Practical use of video imagery in nearshore oceanographic field studies, *IEEE Journal of Ocean Engineering*, *22*.

Holman, R. A. (1981), Infragravity energy in the surf zone, *Journal of Geophysical Research*, *86*, 6442-6450.

Holman, R. A., and J. Stanley (2007), The history and technical capabilities of Argus, *Coastal Engineering*, *54*, 477-491.

Holman, R. A., G. Symonds, E. B. Thornton, and R. Ranashinghe (2006), Rip spacing and persistence on an embayed beach, *Journal of Geophysical Research*, *111*.

Killian, J. R. (2007), Remote sensing of radiation stress gradients from optical imagery, 97 pp, Oregon State University, Corvallis, OR.

Konicki, K., and R. A. Holman (2000), The statistics and kinematics of transverse sand bars on an open coast, *Marine Geology*, *169*, 69-101.

Lippmann, T. C., and R. A. Holman (1989), Quantification of sand bar morphology: A video technique based on wave dissipation, *Journal of Geophysical Research*, *94*, 995-1011.

Lippmann, T. C., and R. A. Holman (1990), The spatial and temporal variability of sand bar morphology, *Journal of Geophysical Research*, *95*, 11,575-511,590.

Lippmann, T. C., and R. A. Holman (1991), Phase speed and angle of breaking waves measured with video techniques, in *Coastal Sediments, '91*, edited by N. Kraus, pp. 542-556, ASCE, New York.

Plant, N. G., S. G. J. Aarninkhof, I. L. Turner, and K. S. Kingston (2007), The performance of shoreline detection models applied to video imagery, *Journal of Coastal Research*, *23*, 658-670.

Plant, N. G., K. T. Holland, and M. Haller (2008), Ocean wavenumber estimation from wave-resolving time series imagery, *IEEE Transactions on Geosciences and Remote Sensing*, *46*, 2644-2658.

Plant, N. G., and R. A. Holman (1997), Intertidal beach profile estimation using video images, *Marine Geology*, *140*, 1-24.

Ruessink, B. G., J. R. Miles, F. Feddersen, R. T. Guza, and S. Elgar (2001), Modeling the alongshore current on barred beaches, *J. Geophys. Res.*, *106*, 22451-22464.

Splinter, K. D., and R. A. Holman (in press), A method for determining bathymetry gradients from single frame images of nearshore waves, *IEEE Transactions on Geosciences and Remote Sensing*.

van Enckevort, I. M. J., and B. G. Ruessink (2001), Effect of hydrodynamics and bathymetry of video estimates of nearshore sand bar position, *Journal of Geophysical Research*, *106*, 16,969 - 916,979.

Wilson, G. W., H. T. Özkan-Haller, and R. A. Holman (in review-a), Alongshore nonuniform dynamics in the surf zone: 2. application to field data, *Journal of Geophysical Research*.

Wilson, G. W., H. T. Özkan-Haller, and R. A. Holman (in review-b), Data assimilation and bathymetric inversion in a 2DH surf zone model, *Journal of Geophysical Research*.

Papers

Coastal Dynamics 2009
Paper No. 1

REAL-TIME WAVE PREDICTION USING HOURLY ANALYZED ATMOSPHERIC GPV

Tracey H. Tom[1], Hajime Mase[2] and Tomohiro Yasuda[3]

1. Introduction

This study developed a real-time wave prediction system by using SWAN (Simulating WAves Nearshore) together with the data of hourly analyzed atmospheric GPV (Grid Point Value). Validation of this system was carried out by comparing predictions with observations by NOWPHAS (Nationwide Ocean Wave information network for Ports and HArbourS) wave gauges during the month of December, 2007. The predicted wave heights and periods corresponded well to observed values. This system provides current and near future ocean condition estimations for locations where observational wave data is unavailable.

2. Wave Model

Wave characteristics are calculated using SWAN, a third-generation wave model developed by Delft University of Technology. The wave prediction system consists of an outer domain covering most of the Northern Pacific Ocean (latitude 5° N to 55° N, and a longitude 120°E to 175°E) with a spatial resolution of 0.25° and a nested inner domain covering the seas surrounding Japan (latitude 24°N to 47°N, and a longitude 126°E to 149°E) with a spatial resolution of 0.02°. Boundary wave spectra input for the inner domain is provided by the outer domain and initial condition setup is provided by the wave field data saved from the previous calculation cycle. The outer and inner domain uses bathymetry with spatial resolutions of 10 minutes and 2 minutes respectively.

3. Meteorological Input Data

The system utilizes two types of meteorological 10 meter wind data grids. Global Forecast System (GFS) 10 meter wind data with a spatial resolution of 0.5 degrees is used on the outer domain and Per Hour GPV Atmospheric Analysis (GPV-QMA) 10 meter wind data, with a spatial resolution of roughly 0.05 degrees, is used on the inner domain. The GFS data is received every six hours and contains an analysis along with 180 hours of forecast. The GPV-QMA is received half past every hour, and covers an area from latitude 22.4N to 47.6N and longitude 120E to 150E with a spatial resolution of five kilometers. The GPV-QMA includes 17 levels of wind and temperature analysis and is assimilated with Wind Profiler Network and Data Acquisition System (WINDAS) Doppler radar data and Automated Meteorological Data Acquisition System (AMEDAS) surface data.

4. Process Flow

The entire process cycle from start to finish executed once every hour as new GPV-QMA analysis data is received (half past the hour). There are four steps in the entire process cycle: (1) Process outer domain forecast for step (2) boundary condition input, (2) Process inner domain hindcast, (3) Process outer domain forecast for step (4) boundary condition input, and (4) Process inner domain for forecast. Each step has two types of output: Initial condition for next processing cycle (next hour), and statistical ocean parameters. Wind input for the inner domain hindcast consists of the previous and current hour GPV-QMA 10m wind data whereas the wind input for the inner domain only consist of the current hour GPV-QMA 10m wind data set to constant. The upper figure in Figure 1 shows wind input used for both hindcast and forecast and the lower figure shows the processing cycles.

[1]Ocean Research Division, Surflegend, Inc., Fujisawa, Kanagawa, 251-0046, Japan, tracey@surflegend.co.jp
[2]Disaster Prev. Res. Inst., Kyoto Univ., Uji, Kyoto, 611-0011, Japan, mase@engan.mbox.media.kyoto-u.ac.jp
[3]Disaster Prev. Res. Inst., Kyoto Univ., Uji, Kyoto, 611-0011, Japan, yasuda@engan.mbox.media.kyoto-u.ac.jp

5. Verification

Real-time testing was performed from December 1st, 2007 through 31st. The processing cycle was performed once every hour upon arrival of the GPV-QMA data from the Japan Meteorological Agency (JMA). Predicted results were compared with the observed ones at seven NOWPAS wave gauge points, situated in locations open to the ocean. Figure 2 shows the comparison of predicted significant wave heights and mean wave periods with observed ones at Wajima for the entire test period.

6. Conclusions

Evaluation showed overall good performance in prediction accuracy and correlation coefficient of significant wave heights. Mean wave periods are rather scattered for the observations; however, the tendency is good. Further testing is needed during the Japanese typhoon season to show the systems' performance during high wind conditions. This study has shown that this forecast system can show the current state of the ocean in real-time. Real-time analysis and short-term forecast can work in conjunction with existing observational data systems to offer a better real-time picture of the oceans state. Potentially any point in the computational domain can serve as a virtual buoy where buoy data is unavailable. Archived hindcast allow quicker evaluation of events that occur along the coast and seas surrounding Japan, at lower overall cost.

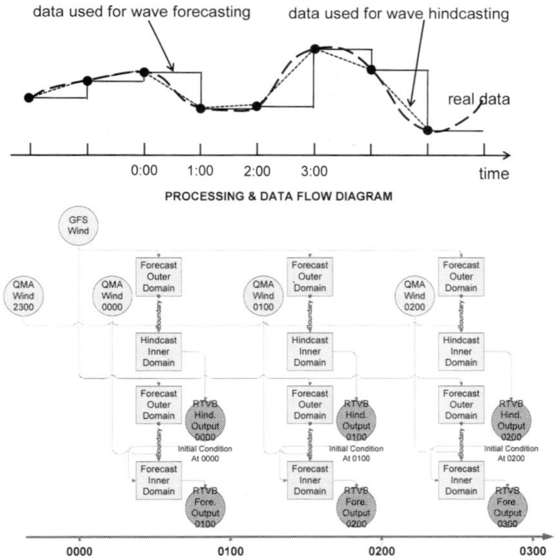

Figure 1 (Upper) Sig. Wave height and (Lower) Mean Wave Period comparison with observational values

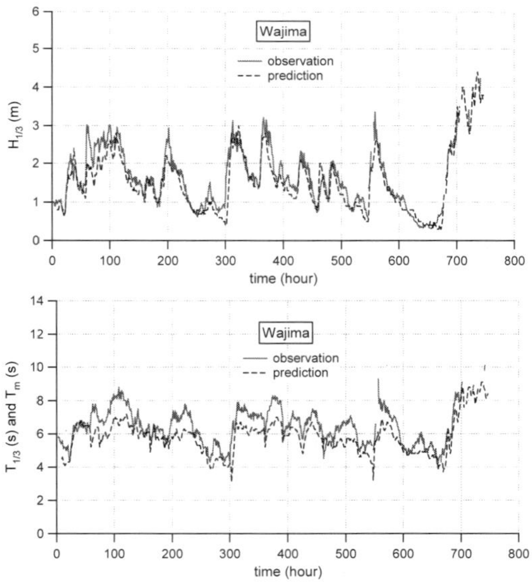

Figure 2 (Upper) Sig. Wave height and (Lower) Mean Wave Period comparison with observational values

References

Mase, H., S. Katsui, T. Yasuda, and T. H. Tom, 2005. Verification of GFS-WRF-SWAN Wave Prediction System By Three Seasons' Comparison. *Annual Journal of Coastal Engineering,* 52(1): 181-185.

Tom, T. H., H. Mase, S. Katsui, T. Yasuda, and K. Ogawa, 2006. Analysis of High Seas Generated by Hurricane Katrina. *Annual Journal of Coastal Engineering,* 53(1): 421-425.

Lalbeharry, R., Behrens, A., Guenther, H., and Wilson, L., 2004. An evaluation of wave model performances with linear and nonlinear dissipation source terms in Lake Erie, *Proc .8th Int. Workshop on Wave Hindcasting and Forecasting,* Hawaii, USA.

Booij, N., L.H. Holthuijsen and R.C. Ris, 1996: The "SWAN" wave model for shallow water, *Proc. 25th Int. Conf. Coastal Engng.*, Orlando, USA., 668-676

Coastal Dynamics 2009
Paper No. 2

MULTISCALE SIMULATIONS USING UNSTRUCTURED MESH SWAN MODEL FOR WAVE HINDCASTING IN THE DUTCH WADDEN SEA

Marcel Zijlema[1]

1. Introduction

The Dutch Wadden Sea in the northern part of the Netherlands is a tidal inlet system that is partly sheltered from the North Sea by barrier islands. Waves entering a tidal inlet from the open sea and subsequently propagating over tidal shoals vary substantially in length scales. In addition, the Wadden Sea is a particularly challenging area due to its complex bathymetry and the occurrence of many physical processes, such as nonlinear wave-wave interaction, wave breaking, local wave generation, and wave-current interaction. All these processes can be characterised as spectral evolution of the wind waves and swells. Several numerical models, known as spectral wave models, have been developed for wave prediction. One of them is SWAN (Booij et al., 1999) which is the most widely applied wave model in coastal engineering that is available for both research and consultancy studies (www.swan.tudelft.nl).

Although the quality of the wave height prediction in a tidal inlet is quite good, the prediction of the wave periods and of the spectra is generally poor. In Groeneweg et al. (2008), the recently obtained measured data have been used in extensive hindcasts of the tidal inlet of the island Ameland in the Wadden Sea for a variety of storms. These hindcasts provided useful insight in the performance of SWAN and in the relevant processes in the Wadden Sea. For instance, the low-frequency wave energy over the ebb-tidal delta appears to be significantly underpredicted.

The associated SWAN simulations were performed on two curvi-linear grids. The first is a coarse grid covering a large part of the Wadden Sea. The second grid is a detailed one covering the tidal inlet of Ameland and was obtained from refining a part of the larger grid. The typical resolution in this inlet is 30 to 50 m. Offshore wave boundary conditions for the fine grid were obtained from the coarse grid computation. This so-called nesting approach is a traditional practise to resolve both large scales in coastal waters and small scales in surf zones.

2. Use of unstructured grids within a terascale environment

Although nesting is very common for spectral wave models, it does not always result in optimality since these models only reflect the numerical techniques and computational resources that used to be available. Moreover, it is essential to simultaneously simulate all scales, since small- and large-scale processes may experience interactions. The use of an unstructured grid seems a natural next step. Recently, a new version of SWAN utilizing unstructured meshes has been developed at the Delft University of Technology (Zijlema, 2008). Using an MPI parallelization, this model also works efficiently for high-resolution applications and thus greatly improves the accuracy of the model.

In this project, we have set up an optimal and highly resolved triangular mesh of the Dutch Wadden Sea to capture local bathymetry and to model physical processes in an accurate and efficient manner. Resolution is down to 15 to 20 m in the tidal inlet of Ameland and in the shallow part behind this island.

The unstructured mesh SWAN model is ported to an IBM Power6 cluster (max. 60 Teraflops), Huygens, a new Dutch national supercomputer at the supercomputing center SARA at Amsterdam. Typically, the simulations with the obtained unstructured grid for the Wadden Sea were run on 128

[1]Environmental Fluid Mechanics Section, Faculty of Civil Engineering and Geosciences, Delft University of Technology, P.O. Box 5048, 2600 GA Delft, The Netherlands. m.zijlema@tudelft.nl

processors and the execution took no more than 5 minutes. The nesting approach employing generally non-optimal curvi-linear grids, as described by Groeneweg et al. (2008), would take a few hours to run it on a simple PC.

3. Hindcast of storm events

A number of storm instants was selected covering a wide range of physical conditions, mostly around the peak of the storms. In this study, we use the state-of-the-art formulations and default parameterizations for the source terms as given by Booij et al. (1999), except for two terms: 1) the triad interaction is omitted and 2) a modified setting of whitecapping proposed by Rogers et al. (2003) is adopted (the so-called n2.0 setting). As an example, some results in terms of 1D wave spectra at some buoy locations of a simulated storm event occuring on January 2, 2005 are shown in Figure 1. For comparison, the results of Groeneweg et al. (2008) are included and are obtained with the physical settings of Van der Westhuysen et al. (2007).

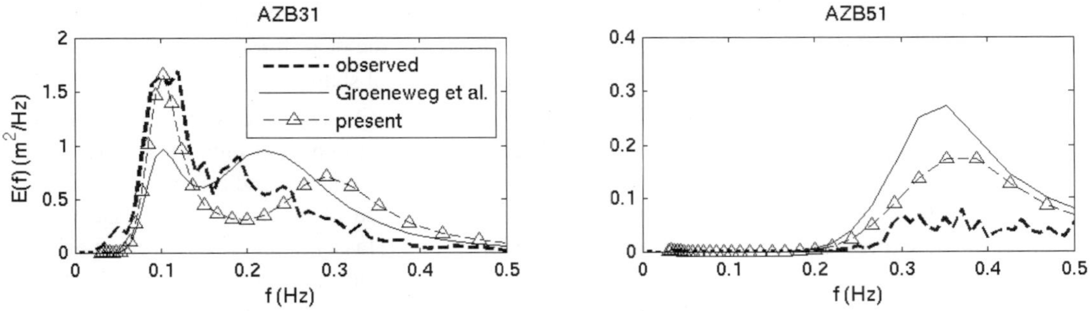

Figure 1. Comparison of simulated and measured spectra at some buoys in the tidal inlet of Ameland

Our experiences to date with simulating storm events in the Dutch Wadden Sea indicate that improvements in wave modeling, including optimization of the SWAN source term settings, and in domain definition and mesh resolution, significantly increases the accuracy and reliability of SWAN. Moreover, high-performance computing with unstructured meshes opens the possibility to simulating storms at unprecedented fidelity and gives rise to a number of modeling challenges.

Acknowledgements

This work was partly supported by ONR Grant N00014-06-1-0256. This work was also sponsored by the National Computing Facilities Foundation (NCF) for the use of supercomputer facilities, with financial support from the Netherlands Organization for Scientific Research (NWO).

References

Booij, N., Ris, R.C. and Holthuijsen, L.H., 1999. A third-generation wave model for coastal regions, 1. Model description and validation. *J. Geophys. Res.,* 104(C4): 7649-7666.

Groeneweg, J., Van der Westhuysen, A., Van Vledder, G., Jacobse, S., Lansen, J. and Van Dongeren, A., 2008. Wave modelling in a tidal inlet: performance of SWAN in the Wadden Sea. *Proc. 31th Int. Conf. on Coast. Engng.,* ASCE.

Rogers, W.E., Hwang, P.A. and Wang, D.W., 2003. Investigation of wave growth and decay in the SWAN model: three regional-scale applications. *J. Phys. Oceanogr.,* 33: 366-389.

Van der Westhuysen, A.J., Zijlema, M. and Battjes, J.A., 2007. Nonlinear saturation-based whitecapping dissipation in SWAN for deep and shallow water. *Coast. Eng.,* 54: 151-170.

Zijlema, M., 2008. Parallel, unstructured mesh implementation for SWAN. *Proc. 31th Int. Conf. on Coast. Engng.,* ASCE.

Coastal Dynamics 2009
Paper No. 3

A DISTRIBUTED COLINEAR TRIAD APPROXIMATION IN SWAN

N. Booij[1], L.H. Holthuijsen[2] and M. Benit[3]

1. Introduction

Third-generation spectral wave models have now been used for nearly two decades to predict wave conditions in coastal waters. Even in the surf zone the performance is good in terms of significant wave height but mixed in terms of mean wave period and spectra. The reason seems to be that triad wave-wave interactions are not, or only poorly, represented in these models. The present study addresses a parameterization of these interactions that is expected to improve the model results.

2. The parameterization

When random waves, with a unimodal spectrum, enter the surf zone, triad wave-wave interactions initially generate secondary peaks in the spectrum, at multiples of the peak frequency. After this initial stage, the interactions force the spectral tail into a smooth, universal shape ($k^{-4/3}$ for $kd < 1$ where k is wave number and d is water depth and $k^{-5/2}$ for $kd > 1$). Other observations and also computations with extended Boussinesq models indicate that this smoothing occurs for a wide variety of other spectral shapes. To simulate this behaviour, we formulate our parameterization of the triad interactions in terms of a source term of the spectral action balance of the waves. It is essentially a reduction of the expression for resonant quadruplet interactions (Hasselmann, 1968) from which we remove one component. For triad interactions, the resonance conditions are strictly met only by collinear, non-dispersive waves (i.e. in extremely shallow water), but before waves enter such shallow water, *near*-resonance occurs. We therefore consider collinear waves under quasi-resonance conditions ($k_{1+2} = k_1 + k_2$) and find in terms of the action density $N(k,\theta)$:

$$S_{nl3}^{N}(k_1,\theta) = \int N(k_{1+2},\theta)\left[\beta_a N(k_2,\theta) - \beta_b N(k_1,\theta)\right] dk_2$$

in which k is wave number, θ is spectral direction and β_a and β_b are unknown interaction coefficients. We infer expressions for β_a and β_b from the observed shape of the spectral tail and from a dimension analysis, and from scaling information in the Lumped Triad Approximation (LTA) of Eldeberky (1996). As we consider all triads (per spectral direction), we call our parameterization the Distributed Collinear Triad Approximation (DCTA).

3. Observations and model

To calibrate one unknown coefficient in the expressions for the interaction coefficients β_a and β_b and to verify the parameterization, we use published observations in the laboratory and in the field under a variety of conditions (approximately 60 cases). The laboratory observations were made in one-dimensional flumes with random, unidirectional waves propagating at normal incidence towards a gentle beach, with and without an offshore bar (e.g., Smith 2004). The field observations were made of random, short-crested waves approaching an open ocean coast (the FRF near Duck, North Carolina, USA, Birkemeier et al., 1997). The wave model that we use is the SWAN wave model (Booij et al.,

---

[1]) Digital Hydraulics, Willemsplein 535, 3016 DR, Rotterdam, the Netherlands, n.booij@digitalhydraulics.com
[2]) Faculty of Civil Engineering and Geosciences, Delft University of Technology, Stevinweg 1, 2628 CN, Delft, the Netherlands, l.h.holthuijsen@citg.tudelft.nl
3) Faculty of Civil Engineering and Geosciences, Delft University of Technology, Stevinweg 1, 2628 CN, Delft, the Netherlands, m.p.benit@student.tudelft.nl

1999). This model accounts for all relevant processes affecting waves in coastal waters: shoaling, refraction, wave-current interactions, generation by wind, quadruplet wave-wave interactions, depth-induced breaking, bottom friction and triad wave-wave interactions. For the latter, the above LTA is normally used, with the result that only second harmonics in the surf zone are generated and no subsequent smoothing of the spectral tail occurs. For the present study we replace the LTA with the DCTA, the coefficient of which we calibrate using half the number of observations. The remaining observations are used to verify the calibrated model and to show any improvement over the LTA.

4. Preliminary results

Below we show for a *unimodal* incident spectrum, a comparison of some preliminary computational results with observations, using the DCTA and the LTA (note the scaling of the variance density $E(f)$). It is obvious that the DCTA reproduces both the generation of all (transient) higher harmonics at multiples of the peak frequency and the subsequent smoothing of the spectral tail (continuing shoreward in the flume; not shown here). It is equally obvious that the LTA produces only (persistent) second harmonics that are too large.

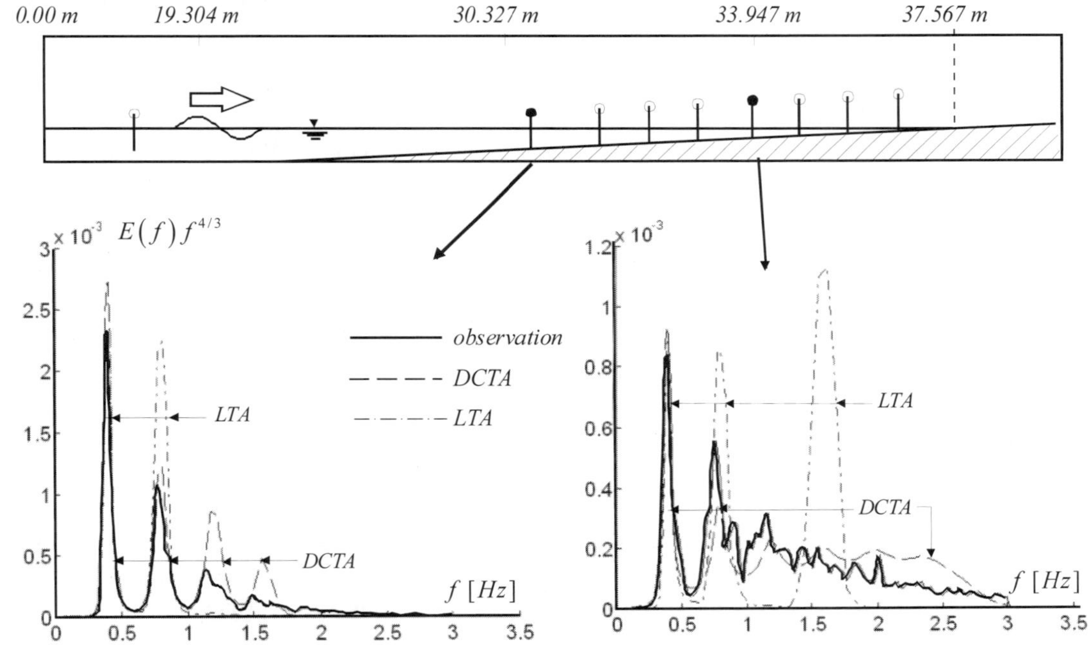

The observed spectra (Smith, 2004) and computed spectra (DCTA and LTA) at the two indicated locations in a flume with waves propagating from left to right over a flat beach with 1:300 slope.

References

Birkemeier, W.A., et al., 1997. 1990 DELILAH Nearshore Experiment: Summary Report. Technical Report CHL-97-24, US Army Corps of Engineers Waterways Experiment Station, 213 p.

Booij, N., et al., 1999. A third-generation wave model for coastal regions, Part I, Model description and validation, *J.Geoph.Research*, 104, C4, 7649-7666

Eldeberky, Y., 1996. Nonlinear transformation of wave spectra in the nearshore zone, Ph.D. thesis, Delft University of Technology

Hasselmann, K., 1968. Weak-interaction theory of ocean waves, *Basic Developments in Fluid Dynamics*, ed. M. Holt, Academic Press, N.Y., London, 2, 117-182

Smith, J. McKee, 2004. Shallow-water spectral shapes, *Proc. 29th Coastal Engineering Conference*, 206 - 217

Coastal Dynamics 2009
Paper No. 4

IMPROVED MPS METHODS FOR WAVE IMPACT CALCULATIONS

Abbas Khayyer[1] and Hitoshi Gotoh[2]

The key role of the so-called wave impact pressure in design of coastal structures has been underlined by both theoretical (e.g. Cooker and Peregrine, 1994) and experimental (e.g. Hattori et al., 1994) studies. Therefore, development of reliable design tools for calculation of wave impact pressure would be essential.

Recently particle methods or the Lagrangian gridless methods have been applied in a variety of engineering applications including the free-surface hydrodynamic flows. The MPS (Moving Particle Semi-implicit) method is a capable and simple particle method developed by Koshizuka and Oka (1996) for simulation of incompressible free-surface flows, particularly those accompanied by large deformations and fragmentations. Despite its simplicity and wide range of applicability, the MPS method has a few drawbacks including non-conservation of momentum (Khayyer and Gotoh, 2008) and artificial pressure fluctuation (Gotoh et al., 2005). By focusing on the momentum conservation properties of original MPS formulations, a new pressure gradient term which conserves both linear and angular momentum is derived. The MPS method modified by the new pressure gradient term has been given the name CMPS (Corrected MPS; Khayyer and Gotoh, 2008). To resolve the problem of pressure fluctuation, the derivation of the Poisson Pressure Equation (PPE) in MPS method is revisited. It is found that the original PPE in MPS method has been derived on the basis of an assumption which is not fully valid as a result of the existence of numerical errors due to particle-based discretization. By deriving a higher order source term for the PPE, a less-fluctuating and more-accurate pressure field is obtained by the CMPS-HS (CMPS with a Higher-order Source term) method.

The refined calculation of wave impact pressure by improved MPS methods is shown through the simulation of a dam break with impact (Hu and Kashiwagi, 2004) and a flip-through impact (Hattori et al., 1994). **Fig. 1** shows the snapshots of water particle together with pressure field during a dam break with impact at the impact instant and at $t = 0.750$ s. The figure shows the step-by-step enhanced calculation of pressure field by improved MPS methods. **Fig. 2** shows the snapshots by MPS and improved MPS methods during a flip-through at the time of the maximum recorded pressure at P_3 ($t^* = 200$; $t^* = t\,C_s/H_F$, t = calculation time, C_s = sound speed, H_F = maximum wave height) and at $t^* = 500$. The improvements are evident in terms of both pressure field and free-surface profile. **Fig. 3** depicts the time variation of pressure at measuring point P_3. From this figure, the CMPS-HS method has provided an acceptable calculation of wave impact pressure during a flip-through impact.

References

Cooker, M.J., Peregrine, D.H., 1992. Wave impact pressures and its effect upon bodies lying on the sea bed. *Coastal Engineering* 18 (3-4), 205-229.

Gotoh, H., Ikari, H., Memita, T. and Sakai, T., 2005. Lagrangian particle method for simulation of wave overtopping on a vertical seawall, *Coast. Eng. J.*, 47(2 & 3), 157-181.

Hattori, M., Arami, A., Yui, T., 1994. Wave impact pressure on vertical walls under breaking waves of various types. *Coastal Engineering*, 22 (1-2), 79–114.

Hu, C.H., Kashiwagi, M., 2004. A CIP Method for Numerical Simulations of Violent Free Surface Flows, *Journal of Marine Science and Technology* 9 (4), 143-157.

Khayyer, A. and Gotoh, H., 2008. Development of CMPS method for accurate water-surface tracking in breaking waves, *Coast. Eng. J.*, 50(2), 179-207.

[1,2] Dept of Urban and Environmental Eng., Kyoto University, Katsura Campus, Nishikyo-ku, Kyoto 615-8540, Japan. [1] khayyer@wfee.mbox.media.kyoto-u.ac.jp ; [2] gotoh@mbox.kudpc.kyoto-u.ac.jp

Koshizuka, S. and Oka, Y., 1996. Moving particle semi-implicit method for fragmentation of incompressible fluid. *Nuclear Science and Engineering*, 123, 421-434.

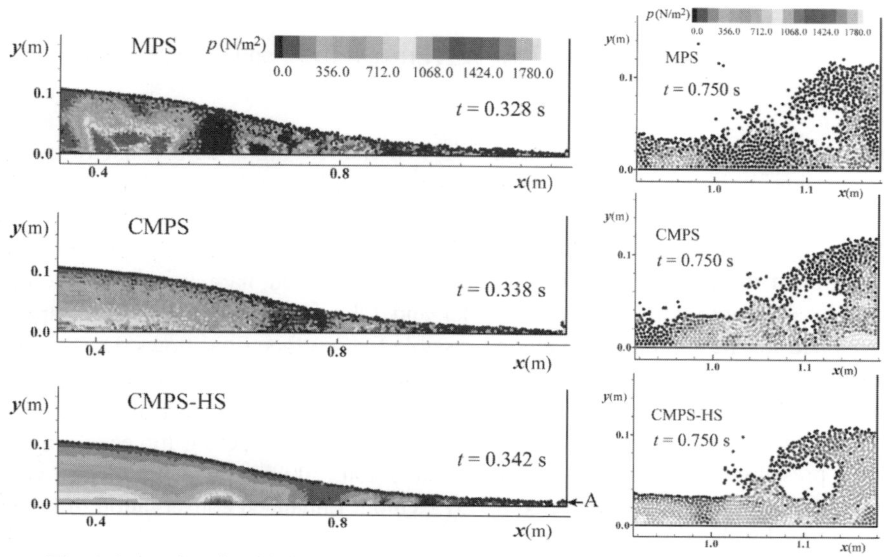

Fig. 1 A dam break with impact – Snapshots by MPS and Improved MPS methods

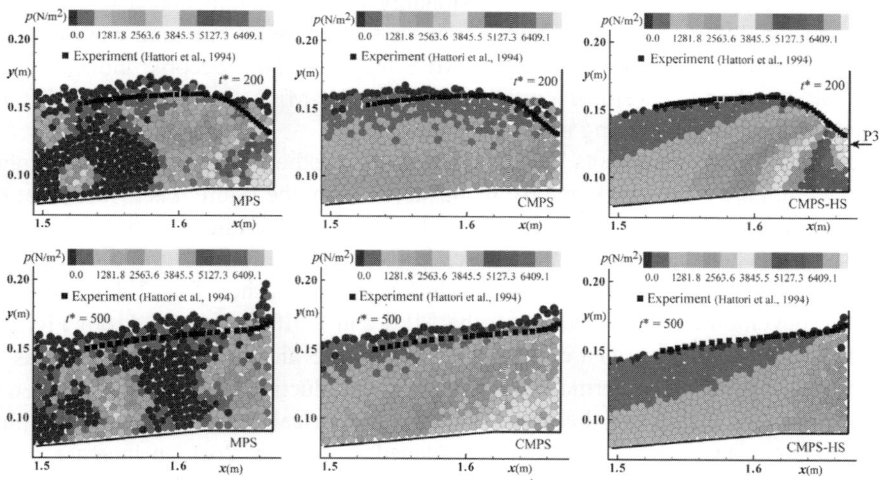

Fig. 2 A flip-through impact – Snapshots by MPS and Improved MPS methods

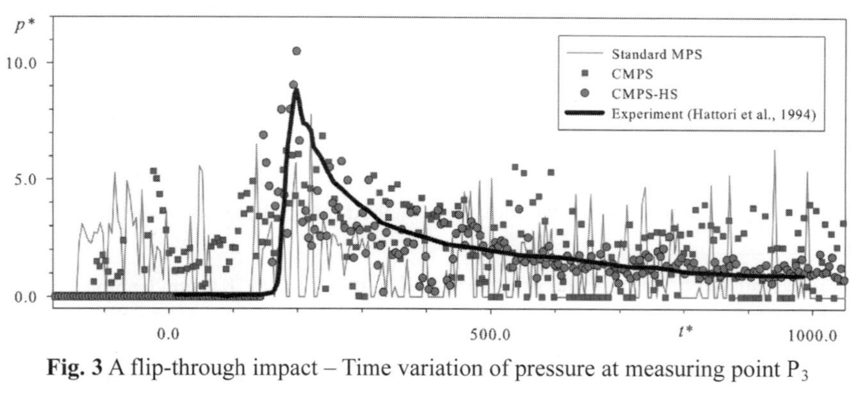

Fig. 3 A flip-through impact – Time variation of pressure at measuring point P_3

Coastal Dynamics 2009
Paper No. 5

3D-CMPS METHOD FOR IMPROVEMENT OF WATER SURFACE TRACKING IN BREAKING WAVES

Hitoshi Gotoh[1], Abbas Khayyer[2], Hiroyuki Ikari[3] and Chiemi Hori[4]

Accurate simulation of a free-surface fluid flow is a challenging issue because of the presence of an arbitrary moving interface. The difficulty in simulation is further increased when the free-surface experiences large deformations accompanied by fluid fragmentations as in case of Breaking Waves.

A relatively new class of numerical methods which provides a substantial potential for the simulation of violent fluid flows is the Lagrangian gridless method, namely, the particle method. The Moving Particle Semi-implicit (MPS; Koshizuka and Oka, 1996) is one of the capable particle methods developed for the simulation of incompressible free-surface fluid flows. Despite its capability, the MPS method has a few drawbacks including artificial pressure fluctuation (Gotoh et al., 2005a) and non-conservation of momentum (Khayyer and Gotoh, 2008a). By focusing on the momentum conservation properties of the MPS approximations for internal interacting forces and by deriving an anti-symmetric pressure gradient model, Khayyer and Gotoh (2008) proposed a revised version of the MPS method, namely, the CMPS (Corrected MPS) method. Enhanced conservation of both linear and angular momentum by the CMPS method resulted in refined simulation of 2-D plunging breaking waves and resultant splash-up (Khayyer and Gotoh, 2008). However, an important issue in the simulation of wave breaking is the three-dimensionality of the phenomenon. Even when generated over a two-dimensional topography, the fluid motions associated with breaking waves are essentially three-dimensional. Hence, development of 3D-CMPS method would become indispensable. In this paper, a 3D-CMPS method is developed on the basis of the 3D-MPS (Gotoh et al., 2005b) method for improvement of water surface tracking in breaking waves.

Fig. 1 shows two typical snapshots by MPS and CMPS method illustrating the development of a plunging jet in a plunging wave breaking on a plane slope. From the figure it is evident that the CMPS method has provided a more integrated plunging jet with less unphysical particle dispersiveness as seen in the MPS snapshot. In addition, the x-direction velocity ($=u$) field in the CMPS snapshot is smoother and more regularly distributed than that in the MPS snapshot. **Fig. 2** depicts the impact of plunging jet and development of splash-up by the MPS and CMPS methods. From **Fig. 2(a-b)**, the CMPS method has resulted in a clearer plunging jet and the air chamber beneath it as well as a smoother free-surface profile. Furthermore, an improved reproduction of splash-up has been portrayed by the CMPS method as the curl of the splash-up, the geometrical characteristics of the plunging jet and the air chamber beneath it are all better reproduced. From **Fig. 2(c)**, an increase in the spatial resolution by refinement of particle size (from $d_0 = 2$mm to $d_0 = 1$mm) has resulted in a further improved reproduction of plunging jet and resultant splash-up by the CMPS method.

References

Gotoh, H., Ikari, H., Memita, T. and Sakai, T., 2005a. Lagrangian particle method for simulation of wave overtopping on a vertical seawall, *Coast. Eng. J.*, 47(2 & 3), 157-181.

Gotoh, H., Ikari, H. and Sakai, T., 2005b. Development of numerical wave 3D MPS method, Proc. Waves 2005, Madrid, Spain.

Khayyer, A. and Gotoh, H., 2008. Development of CMPS method for accurate water-surface tracking in breaking

[1,2,4] Department of Urban and Environmental Engineering, Kyoto University, Katsura Campus, Nishikyo-ku, Kyoto, 615-8540, Japan. [1] gotoh@mbox.kudpc.kyoto-u.ac.jp , [2] khayyer@wfee.mbox.media.kyoto-u.ac.jp , [4] c.hori@kt2.ecs.kyoto-u.ac.jp

[3] Researcher, NEWJEC Inc., Honjo-Higashi, Kita-ku, Osaka 531-0074, Japan. ikarihr@newjec.co.jp

waves, *Coast. Eng. J.*, 50(2), 179-207.
Koshizuka, S. and Oka, Y., 1996. Moving particle semi-implicit method for fragmentation of incompressible fluid. *Nuclear Science and Engineering*, 123, 421-434.

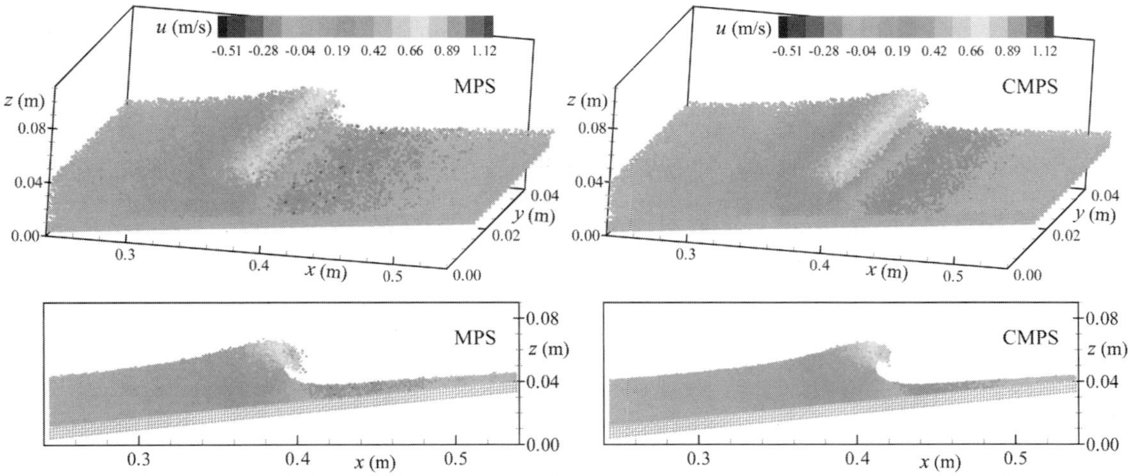

Fig. 1 Typical snapshots by MPS and CMPS methods illustrating the development of a plunging breaking wave

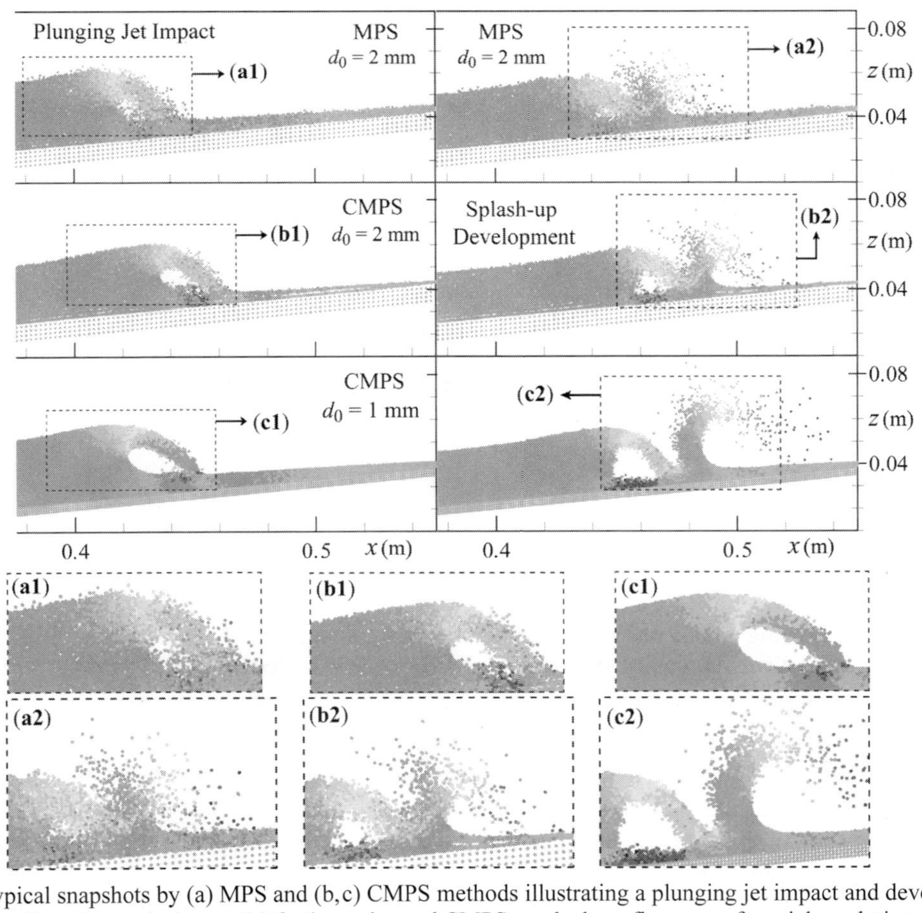

Fig. 2 Typical snapshots by (a) MPS and (b,c) CMPS methods illustrating a plunging jet impact and development of resultant splash-up – (c) further enhanced CMPS results by refinement of spatial resolution

Coastal Dynamics 2009
Paper No. 6

SPH METHOD FOR SIMULATION OF WAVE BREAKING AND OVERTOPPING WITH EXPERIMENTAL VALIDATION

Zhigang,bai [1] and Jingjin,zhao [2]

One of the most typical phenomena on which the water-surface-tracking model plays an important role is wave breaking. This paper describes a SPH method on the simulation of breaking and overtopping of water waves over sloping and vertical structures in a numerical wave tank. As one of the major Lagrangian solvers of the Navier-Stokes equation, SPH offers a variety of advantages to handle large deformations of the free surface with an easy and high accurate way, particularly suiting to those cases where there is wave breaking, splashing or flow separation.

Employing the SPH particle approach, the governing mass and momentum equations for a turbulent flow are presented in the following Lagrangian form:

$$\frac{D\rho}{Dt} = -\rho\nabla \cdot u \tag{1}$$

$$\frac{D\vec{v}}{Dt} = -\frac{1}{\rho}\vec{\nabla}P + \vec{g} + \vec{\Theta} \tag{2}$$

Where $\vec{\Theta}$ is an empirical term representing the effects of viscosity.

Several improvements that we have implemented are presented here to handle turbulence, the fluid viscosity and density, and different boundaries are used.

On one hand, The sub-particle scale (SPS) turbulence model is proved to be capable of adequately predicting the turbulence levels under the breaking waves following a phase averaging approach. Sub-particle scaling (SPS) for a compressible fluid requires a special averaging methodology: we use a flat-top spatial filter to the governing equations.

$$\frac{D\overline{\rho}}{Dt} = -\overline{\rho}\nabla \cdot \tilde{u} \tag{3}$$

$$\frac{D\tilde{u}}{Dt} = -\frac{1}{\overline{\rho}}\nabla\overline{p} + g + \frac{1}{\overline{\rho}}(\nabla\overline{\rho v_0}\nabla)\tilde{u} + \frac{1}{\overline{\rho}}\nabla\tau^* \tag{4}$$

where: $\overline{\bullet}$ denotes an arbitrary spatial filtering, $\tilde{u} = \overline{\rho u}/\overline{\rho}$, τ^* is the sub-particle scale (SPS) stress tensor with elements.

$$\tau_{ij}^* = \overline{\rho}\left(2C_s^2\Delta^2\left|\overline{S}\right|\widetilde{S_{ij}} - \frac{2}{3}\widetilde{S_{kk}}\delta_{ij}\right) - \frac{2}{3}\overline{\rho}C_I\Delta^2\delta_{ij} \tag{5}$$

We take $C_I = 0.00066$, $C_s=0.12$, Δ is the initial particle spacing.

On the other hand, three boundary conditions (Dynamic, Repulsive, Periodic open Boundary) are introduced for treatment of free surfaces and boundary conditions. The incident wave is produced by moving a numerical wave paddle located at the offshore boundary. It is also consisted of wall particles but they move following a prescribed motion and thus generate the desired wave series.

Different types of wave breaking on a different slope are simulated with SPH method, a sequence of 2-D results are shown. Fig. 1 shows that a plunging breaking wave on a beach of slope 15° with an offshore wave height of 0.09 m and a period of 1.4 s. The wave is created by a paddle and breaks as a weakly plunging breaker. Throughout the process of breaking, several important features are

[1]Department of Port Eng., School of Civil Eng., Tianjin University, 92 Weijin Rd., Nankai district,Tianjin 300072, People's Republic of China. Zhigangbai@tju.edu.cn

[2]Department of Port Eng., School of Civil Eng., Tianjin University, 92 Weijin Rd., Nankai district,Tianjin 300072, People's Republic of China. Zhaojj@sohu.com

visible: the formation of the plunger, the initiation of the vorticity, the splash-up of the plunger, the reverse breaking that occurs, and the creation of a downburst by the reverse breaker pounding the vortex generated by the initial breaking into the bottom. In Fig. 2, a spilling breaking wave is shown on a beach of slope 4° with an offshore wave height of 0.09 m and a period of 1.4 s. overtopping of water waves over a vertical structure is shown in Fig. 3 and the wave overtopping volume over a vertical wall is estimated.

Accordingly, the project is now in its validation phase where the numerical code is taken through a series of validation tests. The experiment is conducted in a two-dimensional wave flume at Tianjin University in China, which is 35.0 m long, 0.5 m wide and 0.8 m high, the slope of wave flume can be adjusted automatically. The wave breaking on a slope is observed and the wave overtopping volume over a vertical wall is compared with the numerical results. The calculation results of SPH method are validated against the experimental data found in the literatures and good agreement is observed. Through comparisons with the results of hydraulic experiments and conventional free-surface flow analysis, it demonstrates the conservation properties of SPH and its ability to deal efficiently with violent nonlinear free-surface flows associated with water breaking and overtopping, and even wave-structure interaction.

References

Khayyer A., Gotoh H., Shao S.D., 2008. Corrected Incompressible SPH method for accurate water-surface tracking in breaking waves. *Coastal Engineering* 55 236–250

Dalrymple R.A., Rogers B.D., 2006. Numerical modeling of water waves with the SPH method. *Coastal Engineering* 53 141 – 147

Shao, S.D., Ji, C., Graham, D.I., Reeve, D.E., James, P.W., Chadwick, A.J.,2006. Simulation of wave overtopping by an incompressible SPH model. *Coastal Engineering Journal*, 53 (9), 723–735.

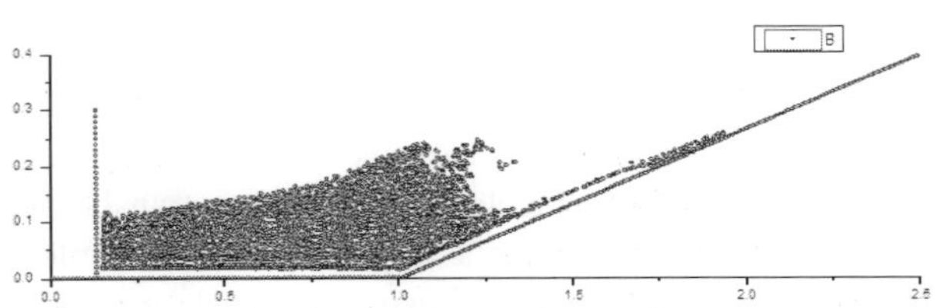

Figure 1. plunging wave breaking

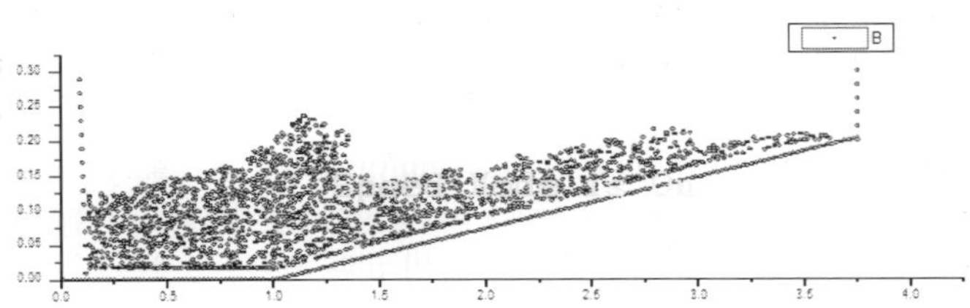

Figure 2. spilling wave breaking

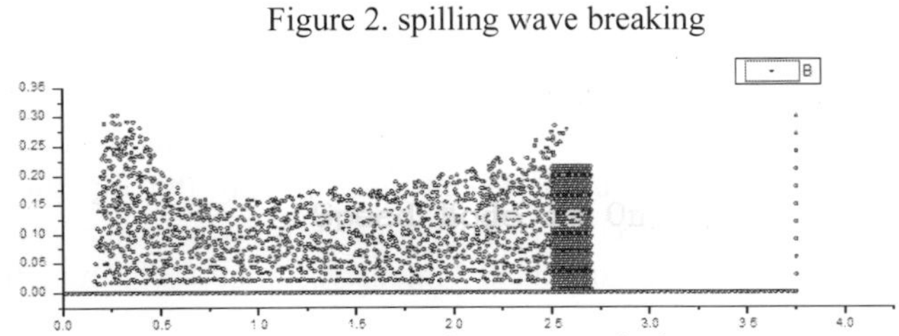

Figure 3. wave overtopping over vertical structure

Coastal Dynamics 2009
Paper No. 7

USING A HYBRID PARTICLE METHOD TO MODEL COASTAL BLUFF COLLAPSE DURING EXTREME EVENTS

Johan Vandamme[1]; Dr Qingping Zou[1]; Prof Dominic Reeve[1].

1. Introduction

Coastal bluff erosion is currently increasing in prominence amongst both the media and the general population. However, these long-term models treat coastal erosion as a probabilistic function, extrapolating data from past years and decades to predict future erosion. Not only does this cost accuracy in the long term models, but it also restricts the end user from postulating the effects of cliff stability works, sea-level rise or any site changes. Therefore, a reliable method to model both the wave hydrodynamics and the stability and potential movement of coastal cliffs is necessary to improve the understanding of extreme waves and coastal cliffs interactions. This understanding will help engineers and coastal planners to design defensive works for storm attacks.

2. Model

A coastal bluff erosion model has been developed using a hybrid approach, combining Smoothed Particle Hydrodynamics (SPH) to model wave run-up, breaking, and consequent erosion, and a similar particle based method to model the bluff stability. SPH was developed by Dalrymple (2006), Rogers (2004) and Monaghan (2000) amongst others, proving both robust and accurate when dealing with breaking waves. Examples of this are well documented within the literature, and an example of SPH wave breaking can be seen in Figure 1. The SPH model is based on the one developed by Zou (2007), and simulates erosion under breaking waves.

The erosion data is then fed into a separate module that has computes the stability of the coastal bluff under the changes of erosion. The stability method is based on the multiple wedge slope stability analysis presented by McCombie (2008), which is a displacement stepping method, computing the overall stability of an arbitrary slip surface on a slope. This allows translational, rotational and hybrid failure surfaces to have their stability analyzed in a way that offers a simple comparison between the many types of failure surfaces. One advantage of using a particle method for this is that no mesh generation or lattice straining takes place. This causes inaccuracies in situations of large relative displacement. Using a particle method eliminates this, and allows episodic failures to occur continuously without the need for time consuming mesh regeneration subroutines.

Contrary to other methods of slope stability, the multiple wedge analysis does not rely on guessed assumptions about the interaction of interwedge boundary forces, but rather only requires the knowledge of the soils effective mobilization under displacement of φ'_{mob}.

3. Results

Although the bluff stability module of this hybrid method can be run in isolation to observe the critical slip surfaces and the effect of engineering works or rainfall, the useful results of this method run beyond this. Figure 2 shows an initial cliff profile showing the slip surfaces that are deemed critical for a simple case, a homogeneous slope with a low hydrostatic water table. The hybrid method of this model allows the computationally intensive process of simulating the erosion at the

[1]Centre for Coastal Dynamics and Engineering, School of Engineering, University of Plymouth, UK.
Fax: +44 (0)1752 232638. E-mail: johan.vandamme@plymouth.ac.uk

base of the cliff to be run in separation to the stability module. This means that having simulated the likely erosion on a certain site, the data can be used to monitor the likely bluff failure for a variety of geotechnical situations, allowing for a more complete picture of the likely failure mechanisms.

In the full paper, the mechanics and profiles of cliff collapse under attack of extreme storms are explored in greater detail, and the SPH hybrid model data are evaluated for the accuracy of both single and multiple soil types

4. Figures

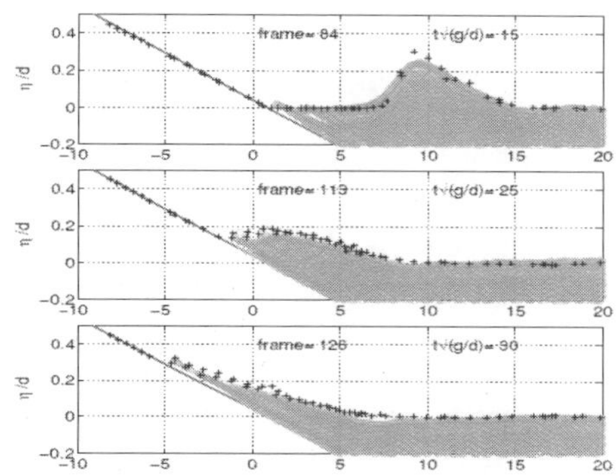

Figure 1. SPH simulation of run-up of a breaking wave over a steep slope. Green open circles, SPH model; blue star symbols, experiment. (From Zou and Dalrymple (2006))

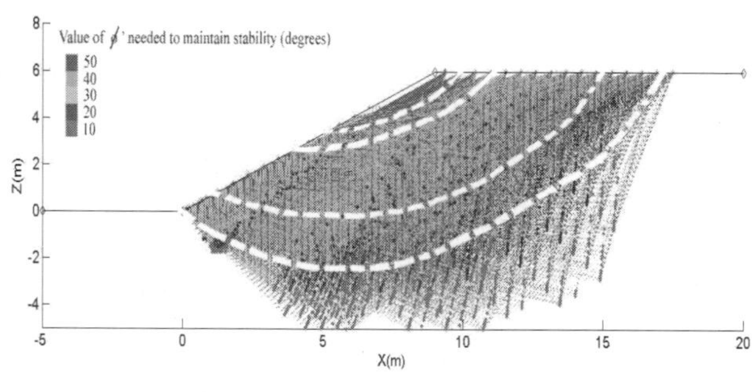

Figure 2. GeoSPH stability analysis of critical slip surfaces of a dry homogenous soil. Dashed lines to ease distinction between colour areas that refer to the friction angle mobilized by the slip surface in order to maintain stability

Figure 3. Example of coastal bluff erosion at Happisburgh (July 2006)

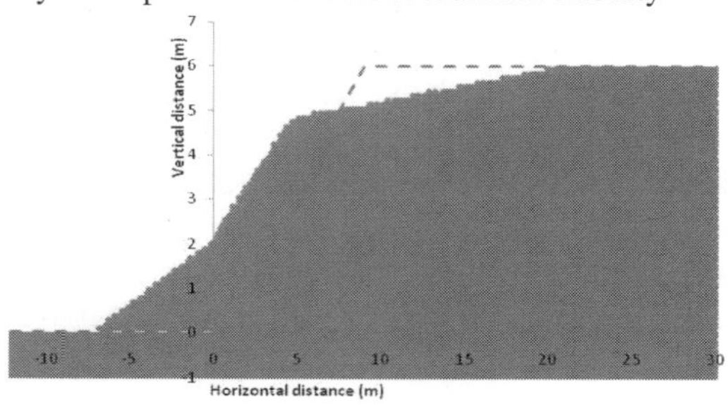

Figure 4. The equilibrium position of the cliff slope, after multiple failures. The dashed line shows the original position.

References

Dalrymple, R. A. and Rogers, B. D. 2006. Numerical modeling of water waves with the SPH method. *Coastal Engineering*, 53, p141-147

McCombie, P. 2009. Displacement based multi. wedge slope analysis. C*omputers and geotechnics*, 36, p332-341

Monaghan, J. J. 2000. SPH without a tensile instability. *Journal of Computational Physics*, 159, p290-311

Rogers, B. D. and Dalrymple, R. A. 2004. SPH modeling of breaking waves. In *Proc. 29th International Conference Coastal Engineering*. Lisbon. p1147-1157.

Zou, S. and R.A. Dalrymple, Sediment suspension simulation under oscillatory flow by SPH-SPS method. Coastal Engineering Conference, 2006. ASCE American Society of Civil Engineers

Zou, S. 2007. Coastal Sediment Transport Simulation by SPH. PhD Thesis. Johns Hopkins University.

Coastal Dynamics 2009
Paper No. 8

BOUND WAVE RELEASE INDUCED BY SHORT WAVE BREAKING – TRUE OR FALSE?

Tom E. Baldock[1]

1. Introduction

Wave groups approaching the shore force associated long waves (Longuet-Higgins and Stewart, 1962), commonly termed bound long waves. It is widely assumed that the bound long waves are released during short wave breaking, subsequently propagating to the shore as a free wave, where they are reflected (see reviews in Baldock and Huntley, 2002; Janssen et al., 2003). The emphasis here is on the short wave breaking process being the trigger. Statements asserting this release are either unattributed with no citation of convincing evidence, or loosely attributed to Longuet-Higgins and Stewart (1962). However, the author is unaware of convincing evidence of such release of bound long waves as a result of short wave breaking, while there appears to be strong evidence to the contrary. Furthermore, in the author's view, the statements of Longuet-Higgins and Stewart (1962) do not mention bound wave release on short wave breaking, and actually state the opposite. In particular, the author's interpretation is that Longuet-Higgins and Stewart (1962) suggest that the bound wave will decay in amplitude following short wave breaking. This is in agreement with a number of the author's data sets (Baldock and O'Hare, 2004) and some data from other recent data sets (Battjes et al., 2004; Dong et al., 2009). The full paper will present examples of these data and consider if the data and statements of Longuet-Higgins and Stewart (1962) are consistent with the notion of bound wave release upon short wave breaking. A brief overview is given below.

2. Data

Baldock and O'Hare (2004) presented a number of data sets that illustrated the strong dissipation of incident bound waves once short wave breaking commenced. This is inconsistent with bound wave release, which suggests that the incident long wave amplitude should remain constant or increase further as a result of shoaling. Two examples are shown in figures 1 and 2, respectively. In both cases, the amplitude decreases immediately after short wave breaking. As noted by Battjes et al. (2004), it is possible that the bound waves break in very shallow water. However, in order for this to explain the data illustrated, this would require that breaking occurred when the long wave height wave height to water depth ratio, $\gamma_L=H_L/d$, was about equal to 0.2, instead of the usual value of $\gamma\approx0.8\text{-}1$. Battjes et al. (2004) present data that show a similar dissipation of the incident long waves after short wave breaking, and again the dissipation commences at a location where $\gamma_L\approx0.1\text{-}0.3$. Similarly, data from Dong et al. (2009) show that dissipation of the long waves only occurs if the short waves break. Baldock and O'Hare (2004) proposed that these data are consistent with the idea that the bound wave is not released at short wave breaking unless the *short* wave groups are in shallow water at the breakpoint, and that this condition is not met in many cases, particularly for steep waves. In this case, the bound wave could remain a forced wave inside the breakpoint, and reduce in amplitude along with the reduced radiation stress forcing.

3. Interpretation of Longuet-Higgins and Stewart (1962)

The general interpretation of Longuet-Higgins and Stewart (1962) is usually of the form:
'... the incident bound waves are not destroyed in the surf zone, but are released as free-wave motions... ....after reflection at the shore, the long waves travel back offshore as leaky waves or remain refractively trapped to the coast as edge waves...'. However, the statements in Longuet-

[1]Division of Civil Eng., University of Queensland, St Lucia, QLD 4072, Australia. t.baldock@uq.edu.au

Higgins and Stewart (1962) do not say this. In the author's opinion, the statements in Longuet-Higgins and Stewart (1962) imply either

i) reflection of the long waves occurs before short wave breaking.

ii) if reflection of the long wave occurs after short wave breaking, then the amplitude of the forced wave reduces within the surf zone (until it is reflected).

The latter point implies that the bound wave amplitude is determined by the local forcing that continues after short wave breaking, and therefore that the bound wave is not released as a free wave during breaking. This does not exclude release and reflection of the bound wave further shoreward.

There does not appear to be convincing evidence of the first statement from experiments of long waves propagating over bars, but the latter statement is in agreement with the data described above and that shown in figures 1 and 2, where the bound harmonics start to dissipate energy after short wave breaking and when $\gamma_L<0.15$. Some analytical and numerical model results also appear to support this conclusion, although it is not entirely clear if the forced wave is released or not in those models. It is however, inconsistent with the conventional view that bound waves are released upon short wave breaking, and subsequently propagate to the shore as free waves. The distinction is important, since the amplitudes of the forced wave or released long wave inside the surf zone are important in determining the total long wave motion. Therefore, the location where the long wave is released is important.

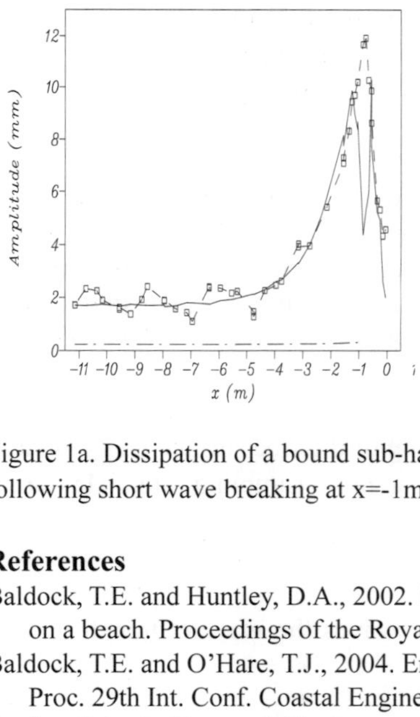

Figure 1a. Dissipation of a bound sub-harmonic following short wave breaking at x=-1m.

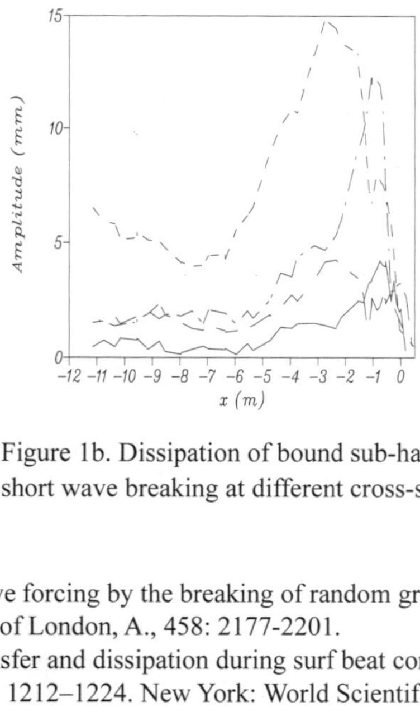

Figure 1b. Dissipation of bound sub-harmonics following short wave breaking at different cross-shore locations

References

Baldock, T.E. and Huntley, D.A., 2002. Long-wave forcing by the breaking of random gravity waves on a beach. Proceedings of the Royal Society of London, A., 458: 2177-2201.

Baldock, T.E. and O'Hare, T.J., 2004. Energy transfer and dissipation during surf beat conditions Proc. 29th Int. Conf. Coastal Engineering. pp. 1212–1224. New York: World Scientific.

Battjes, J.A., Bakkenes, H.J., Janssen, T.T. and van Dongeren, A.R., 2004. Shoaling of subharmonic gravity waves. Journal of Geophysical Research-Oceans, 109(C2).

Dong, G. et al., 2009. Experimental study of long wave generation on sloping bottoms. Coastal Engineering, 56(1): 82-89.

Janssen, T.T., Battjes, J.A. and van Dongeren, A.R., 2003. Long waves induced by short-wave groups over a sloping bottom. Journal of Geophysical Research-Oceans, 108(C8).

Longuet-Higgins, M.S. and Stewart, R.W., 1962. Radiation stress and mass transport in gravity waves, with application to `surf beats'. Journal of Fluid Mechanics, 13: 481-504.

Coastal Dynamics 2009
Paper No. 9

MECHANISMS OF FORMATION AND DYNAMICS OF SECONDARY WAVES IN COASTAL ZONE

Sergey Kuznetsov[1] and Yana Saprykina[1]

1. Introduction

The waves formed as separate peaks on a surface of an initial wave are the secondary waves. For example, such secondary waves are formed in a coastal zone at propagation of storm waves or swell above submerged bar, reefs, trenches, breakwaters and other underwater structures sharply changing water depth. The dispersion mechanism of formation of such secondary waves is described in detail by many researchers (Kuznetsov, Speranskii, 1986; Beji, Battjes, 1993; Masselink, 1998). The presence of secondary waves, that looks as additional independent waves, reduce the average period of waves and is important for taking into account at calculation and modeling of any processes in a coastal zone. Formation of secondary waves can suppress or at least detain wave breaking and wave energy dissipation up to the coastal line (Senechal, Bonneton, Dupuis, 2002). Unfortunately, despite of numerous field and laboratory experiments the nature of secondary waves is investigated insufficiently. In particular, not clearly, is usual the occurrence of secondary waves on any natural coast? Whether all regimes of waves form the secondary waves at their transformation above a submerged bars and trenches? Whether the mechanism of formation of observed secondary waves is identical for all conditions or there are some different features? The purpose of this investigation - detailed studying of mechanisms and conditions of formation of secondary waves on the base of the data of field and laboratory experiments, and also by using of numerical modeling.

2. Experiments

Laboratory experiments were carried out in the flume of Institute of Hydroengineering of the Polish Academy of Sciences (Gdansk) in 2005 and in wave tank of the Research center "Sea coast" (Sochi, Russia) in 2004. Formation of secondary waves during transformation of initially monochromatic and bichromatic waves above the summered bar was studied. Free surface elevation was registered using capacity and resistance gauges with sampling frequency 200 Hz. Duration of records was from 60 s up to 4 min.

For analysis of natural waves the data of field experiments "Novomikhailovka-2002" and "Shkorpilovtsy-2007" (Black sea) were used. Synchronous recording of the free surface elevation at 7 or 15 points of coastal zone was performed by resistant type wire gauges at sampling rate 5 Hz during one hour.

For the modeling the Boussinesq type equation with improved dispersion characteristics were used (Madsen, Sorensen, 1993).

For demonstration of frequency and phase structure of secondary waves the wavelet transform and mutual spectral analysis of wave chronograms were applied.

3. Results

Because the most part of coastal zone belong to so-called intermediate water depth, where the waves are transformed due to both near resonant triad interactions and dispersion, the basic mechanism of formation of secondary waves in coastal zone is nonlinear-dispersive.

There are two stages of secondary wave formation. During the first stage, at the moments when the energy transfer from the first harmonic to the second one is dominated, the amplitude of second

---

[1]P.P. Shirshov Institute of Oceanology of Russian Academy of Sciences, Nakhimovskii prospect, 36, Moscow, 117997,Russia. e-mail: saprykina@ocean.ru, kuznetsov@ocean.ru

harmonic increases and the first and second harmonics propagate with the almost the same phase speed due to the opposite influence of nonlinearity and dispersion. At the second stage, during backward energy transfer from the second harmonic to the first one, the nonlinearity and dispersion both tend to reduce the phase speed of second harmonic and the second harmonic becomes to lag behind the first harmonic that looks as separation of secondary waves from the primary ones. This mechanism exists independently from the presence of underwater bars or trenches, but they are able to affect on secondary waves formation due to the depth changing, that increases or decreases the intensity both of nonlinear processes and dispersion. So formation of secondary waves depends on a stage of wave transformation defined by length of wave run. Therefore, the increasing of a dispersion caused by peculiarities of a bottom relief (bars, trenches, etc.) not always leads to formation of visible secondary waves. The cases of absence of secondary waves behind the underwater bars and their existing above uniformly increasing bottom are revealed in our field experiments.

It was found that the amplitude of the secondary waves depends on parameter of nonlinearity of waves. The Ursell number cannot be used as criterion of at secondary waves formation because does not take in account a stage of wave transformation.

In the natural irregular waves, where the spectrum peak of waves is wider than in the laboratory, different individual waves are at different stages of transformation, and a secondary waves are the result of an interference of many frequency components, but the main features of the processes demonstrate the same mechanism of secondary waves formation.

The numerical modeling was used to confirm the results of physical modeling and field experiments and to demonstrate of relative contribution of nonlinearity and dispersion.

Acknowledgements

This work has been supported of the Russian Foundation for Basic research, project 08-05-00648.

References

Beji, S, Battjes, J.A., 1993. Experimental investigation of wave propagation over a bar. *Coastal Engineering*, 19: 151-162.

Senechal, N, Bonneton, P., Dupuis H., 2002. Field experiment on secondary wave generation on a barred beach and the consequent evolution of energy dissipation on a beach face. *Coastal Engineering*, 46: 233-247.

Kuznetsov S.Yu., Speranskii N.S., 1986. About the dispersive nature of secondary waves in the surf zone. *Oceanology*, 31(3): 423-426 (In Russian).

Masselink G., 1998. Field investigation of wave propagation over a bar and the consequent generation of secondary waves. *Coastal Engineering*, 33: 1-9.

Madsen, P.A., Sørensen, O.R., 1993. Bound Waves and Triad Interactions in Shallow Water, *Journal of Ocean Engineering*, 20(4): 359-388.

Coastal Dynamics 2009
Paper No. 10

RESONANT EVOLUTION OF LONG PERIOD WAVES BY PERIODICALLY FLUCTUATING WIND VELOCITY

Takashi Izumiya[1] Satoshi Watanabe[2] and Kunihiko Ishibashi[1]

1. Introduction

Long period waves or surf beat are low frequency motions associated with higher frequency surface gravity waves and have a significant importance for nearshore processes. The bound long waves have been investigated by several researchers through theoretical analysis (e.g., Longuet-Higgins and Stewart 1962, Ottesen-Hansen et al. 1980) and/or field observations (e.g., Huntley et al.1981). However, the precise generation mechanisms of free long period waves are not well understood. This study investigates the spectral characteristics of low frequency fluctuations of natural wind velocity and atmospheric pressure, and the resonant evolution of long period waves by low frequency fluctuations of winds in a laboratory experiment.

2. Objectives

The overall objectives of the study are (1) to measure natural wind velocity and atmospheric pressure and estimate their frequency spectra, (2) to precisely measure the wind velocity and air pressure exerted on a water surface by a periodically fluctuating wind speed in a laboratory wind wave tank, (3) to investigate the effect of the time variations of wind velocity on wind waves and long period waves to find an alternative generation mechanism for long period waves.

3. Field and laboratory experiments

Field observations have been carried out to investigate the spectral properties of natural wind velocity and atmospheric pressure using an ultrasonic anemometer and precise pressure gauges. Laboratory experiments on the responses of wind waves under periodically fluctuating winds have also been conducted to measure the wind velocities and air pressures exerted on a water surface at several elevations. The periods of the low frequency fluctuations are set to be 0 s (constant wind speed), 3 s and 6 s and the mean wind speeds are about 5 m/s, 7 m/s, 9 m/s and 11 m/s. The frequency spectra of the fluctuations in wind velocity and air pressure were estimated as well as the cross spectra between wind waves and the fluctuating components of wind velocities and air pressures.

4. Results

The spectral densities of the fluctuating components of natural wind velocity and atmospheric pressure indicate a f^{-1} decay law as shown in Figure 1. This suggests that the low frequency components of wind velocity and pressure contain a significant amount of energy and could excite low frequency surface waves. Figures 2 and 3 show the time variations of wind speed, air pressure and water surface elevation for periods of 3 s and 6 s, respectively. The frequency spectra of the fluctuating components of wind speed, air pressure and water surface elevation are shown in Figure 4. The long period waves develop at the peak frequency of the low frequency fluctuations of wind speed and air pressure. The low-frequency components of wind velocity, air pressure and surface waves shown in Figure 5 indicate that the low-frequency fluctuating components of wind velocity change almost in phase with the long period waves, while those of the air pressure are almost completely out of phase with the wind waves. The quadrature spectrum, $Q_{p\eta}$ between water surface elevation and air pressure is also estimated in Figure 6, and shows a negative value of $2\pi f\, Q_{p\eta}$ at the low peak frequency. From these results, the long period waves appear to be generated by the periodically fluctuating components of wind velocity, and not by those of air pressure.

[1]Dept of Civil Eng. and Architecture, Niigata University, 8050, Ikarashi Nino-cho, Nishi-ku, Niigata 950-218, Japan. izm@eng.niigata-u.ac.jp
[2] Tokamachi District Promotion Bureau, 2-1 Tsuma-ari-Nishi, Tokamachi, Niigata 948-0037, Japan.

References

Huntley, D.A., Guza, R.T. and Thornton, E.B., 1981. Field observations of surf beat 1. Progressive edge waves. J. Geophys. Res. 86, 6451-6466.

Longuet-Higgins, M.S. and Stewart, R.W., 1962. Radiation stress and mass transport in gravity waves, with application to surf beat, J. Fluid Mech., 13, 481-504.

Ottesen-Hansen et al., 1980. Correct reproduction of group-induced long waves, Proc. of 17th ICCE, 784-800.

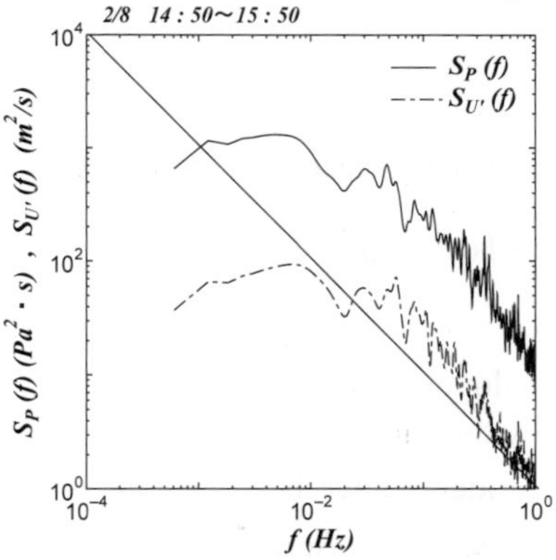

Figure 1. Spectra of fluctuating components of natural wind speed and atmospheric pressure.

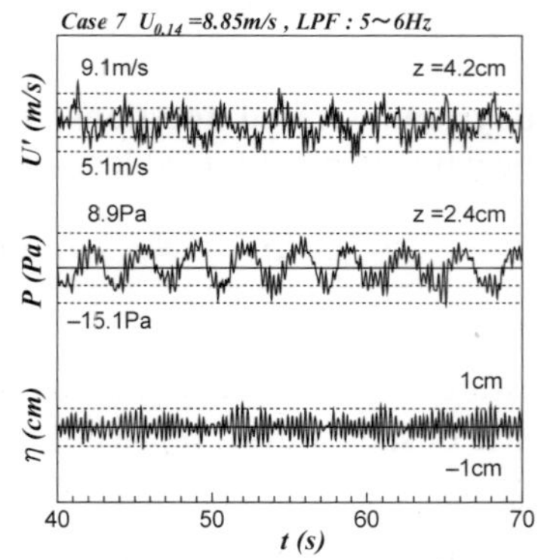

Figure 2. Time variations of wind speed, air pressure and water surface elevation.

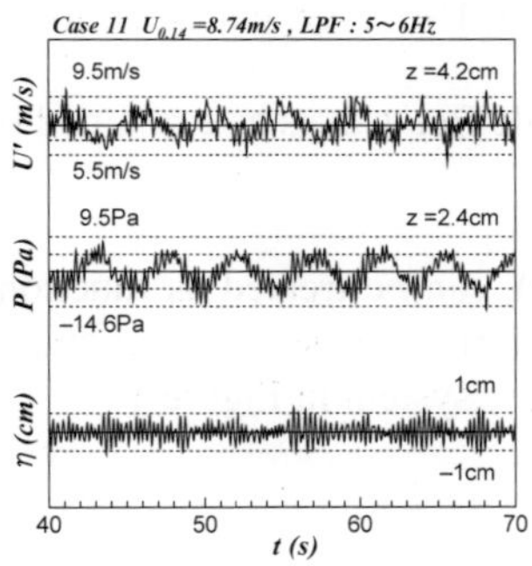

Figure 3. Time variations of wind speed, air pressure and water surface elevation.

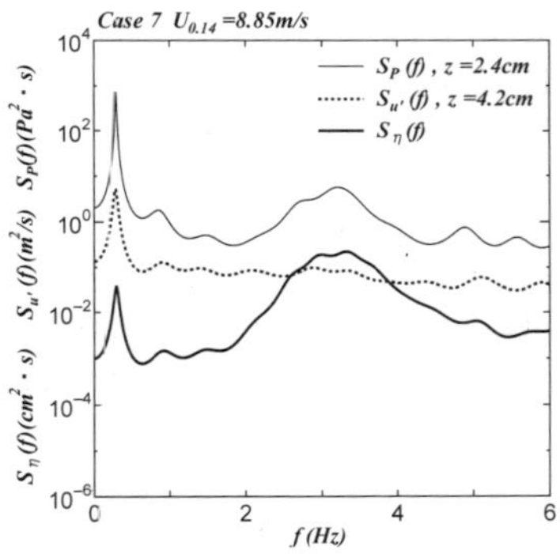

Figure 4. Spectra of wind speed, air pressure and water surface waves.

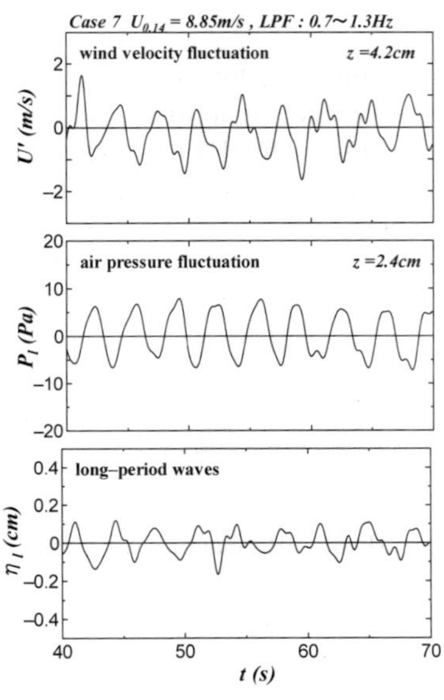

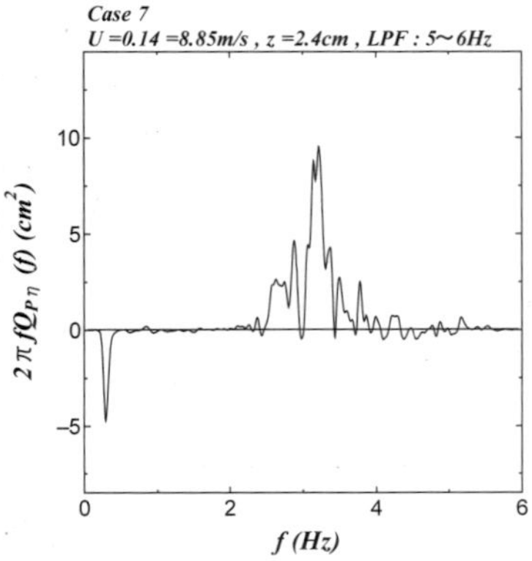

Figure 6. Quadrature-spectrum between water surface elevation and air pressure.

Figure 5. Long period components of wind velocity, air pressure and water surface elevation.

Coastal Dynamics 2009
Paper No. 11

ANALYTICAL DESCRIPTION OF TRANSIENT LONG WAVES LIKE STORM SURGES, SURF BEAT AND TSUNAMI

Peter Nielsen[1]

1. Yesterdays unanswered questions

Simple analytical treatment of forced long waves used to include only steady forced waves, e g,

$$\eta_{\text{forced}}(x,t) \;=\; A_{\text{forced}}\text{f}(x-Ut) \;=\; \frac{-\Delta_p/\rho g}{1-\dfrac{U^2}{gh}}\text{f}(x-Ut) \qquad (1)$$

for a low pressure system travelling at speed U, and an analogous solution for the long wave forced by a pulse of short waves, replace Δ_p by S_{xx}/h and U by the short wave group velocity c_g.

But what happens with (1) for $U > \sqrt{gh}$? - the alert students have been asking. Is all we get from fast moving low pressure systems really just a bit of low water? Surely not.

Problems also arise with the surf beat solution $\eta_{\text{forced}}(x,t) \;=\; \dfrac{-S_o/\rho gh}{1-c_g^{\,2}/gh}\text{f}(x-c_g t)$. That is, for a single pulse f( ), the forced wave has the shape of a bowl. However, experiments (Baldock 2006) show an И-shaped long wave.

More realistic solutions are obtained by including suitable free waves as explained below.

2. Today's new solutions

Waves which travel with constant form, e g, $\eta(x,t) = A\text{f}(x\text{-}ct)$, can be treated as function of just one variable. This enables analytical description of transient storm surges, surf beat and tsunami, in analogy with the treatment of the swinging up from rest of the system in Figure 1:

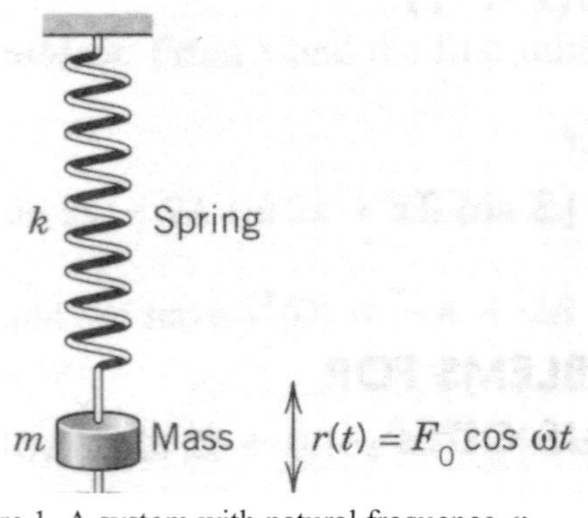

Figure 1. A system with natural frequence ω_o forced at ω. Starting from rest at $t = 0$.

$$\begin{aligned} y(t) \;&=\; y_{\text{forced}} + y_{\text{free}} \\ &=\; \frac{F_o/m}{\omega_o^2-\omega^2}\left[\cos\omega t - \cos\omega_o t\right] \end{aligned} \qquad (2)$$

i e, the complete solution is the forced oscillation and a free oscillation, which cancels it at time zero.

The analogous solutions for 1D long waves resulting from forcing which travels at speed U, consist of three waves:

$$\begin{aligned} \eta(x,t) \;&=\; \eta_{\text{forced}} + \eta_{\text{free}}^{\text{forward}} + \eta_{\text{free}}^{\text{backward}} \\ &=\; A_{\text{forced}}\text{f}(x-Ut) + A_{\text{free}}^{\text{forward}}\text{f}(x-\sqrt{gh}\,t) + A_{\text{free}}^{\text{backward}}\text{f}(x+\sqrt{gh}\,t) \end{aligned} \qquad (3)$$

---

[1] Department of Civil Engineering, The University of Queensland, Brisbane, Australia 4072. p.nielsen@uq.edu.au.

where A_{forced} is the steady forced wave amplitude, and the free wave amplitudes depend on the starting conditions, e g, flat surface and zero velocity everywhere for an initial quiescent state. Although (3) consists of waves which each propagate with constant form, the composite picture is highly transient as shown in Figure 2. The shown wave is a simple tsunami for which the forced wave is also analogous to (1).

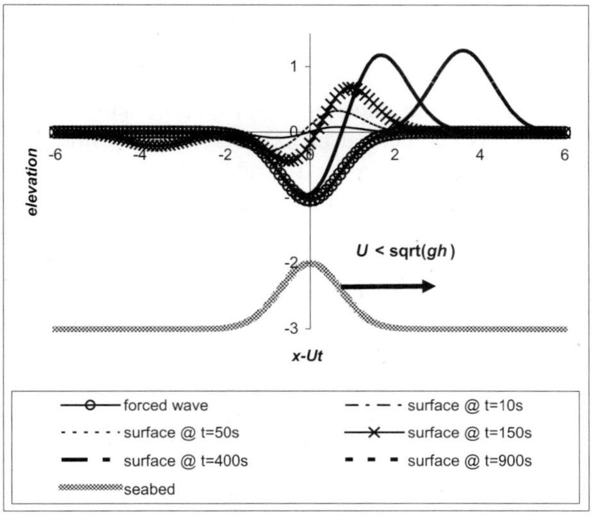

Figure 2. Tsunami driven by a travelling seabed hump, considered in the frame of reference which follows the hump. Initially the surface is flat. The tsunami grows as the free waves, travelling at absolute speeds $\pm\sqrt{gh}$, separate from the forced wave. By t=900s the free waves have just separated from the forced wave.

The scenario is analogous for a travelling low pressure system with the forced wave given by (1). The answer to the "low water only for $U > \sqrt{gh}$ conundrum" raised for that solution above is then illustrated by Figure 3.

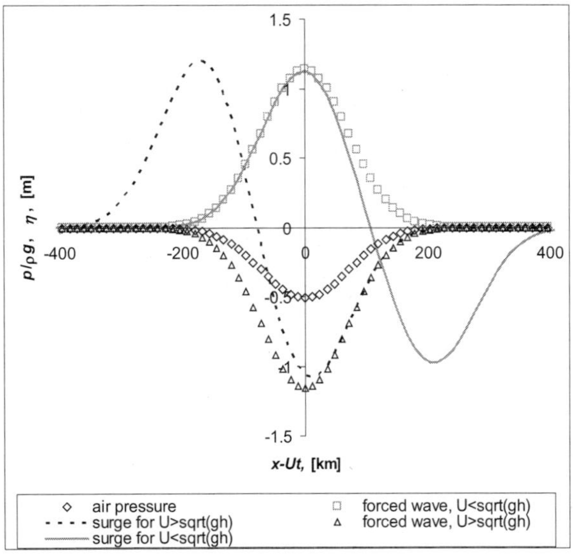

Figure 3. Surges corresponding to slow and fast low pressure systems considered in the frame of reference which follows the forcing. At the times considered, the backward free waves have left the picture to the far left. Hence each surge consists of a forced wave plus a forward free wave. For the fast system, the forward free wave trails the forcing since $U > \sqrt{gh}$. The complete landfall scenarios are then analogous for fast and slow systems despite (1) changing sign. That is: low water followed by dangerous high water. The N-shapes of the surges are analogous to the И-shaped surf beat measured by Baldock 2006.

References

Baldock T E (2006): Long wave generation by the shoaling and breaking of transient wave groups on a beach. *Proc Roy Soc Lond A*, *Vol 462*, pp 1853-1876.

Coastal Dynamics 2009
Paper No. 12

MODELLING OF WAVES & CURRENTS AROUND POROUS SUBMERGED BREAKWATERS

Roshan Suminda Ranasinghe[1], Shinji Sato[2] and Yoshimitsu Tajima[3]

1. Introduction

The Boussinesq-type model equations for studying waves and currents around coastal structures have made remarkable advance during the past few decades. However, contrast to the fast development in modeling of hydrodynamics around impermeable structures, that of porous bottom or porous structures have been very slow, inevitably due to the uncertainty on determination of few empirical coefficients associated with porous media. As submerged breakwaters result primarily in wave dissipation through wave breaking over the structure, bottom friction and turbulence in porous layer (most of the time submerged breakwaters are rubble mound porous structures), it is essential to include porous damping in modeling waves and currents around these structures.

Cruz et al. (1997), derived a set of 2DH Boussinesq equations over a porous bed of arbitrary thickness and tested their applicability on wave transformation over porous submerged breakwater with an opening for non-breaking waves. Hsiao at al. (2002) presented a fully non-linear Boussinesq-type of model, following the same approach as Nwogu (1993), to improve the dispersion properties inside the porous media. Very recently, Chen (2006) improved Hsiao et al. (2002) model by eliminating weak vertical vorticity assumption and retaining higher order porous damping terms in momentum equation. Though most of the models have been tested for non-breaking wave transformation over porous submerged breakwaters; a few studies investigating the evolution of waves and wave-induced nearshore currents around these structures.

The goal of the present study is to discuss the evolution of waves and wave-induced currents around submerged porous breakwaters and to develop a numerical model based on a modified version of Chen (2006) equations coupled with a transport equation to compute turbulent kinetic energy hence energy dissipation due to wave breaking.

2. Numerical Modeling

As mentioned previously, a modified version of Chen (2006) equations has been employed in simulating waves and currents over permeable submerged breakwaters, whereas Nwogu (1993) platform equations has been used in simulating those over impermeable submerged breakwaters for comparison.

Continuity equation

$$\eta_t + \nabla.[(h+\eta)u_\alpha] + n\nabla.(h_s u_\beta^s) + \left\{ \begin{aligned} &(h+\eta)\left[\left(z_\alpha^2/2 - h^2/6\right)\nabla(\nabla.u_\varepsilon) + \left(z_\alpha + h/2\right)\nabla\left[\nabla.(hu_\alpha) + n\nabla.(h_s u_\beta^s)\right]\right] \\ &+ nh_s\left[\left(z_\beta^2/2 - \left(h^2 + hh_b + h_b^2\right)/6\right)\nabla(\nabla.u_\beta^s) + \left(z_\beta + (h+h_b)/2\right)\nabla\left[\nabla.(h_s u_\beta^s)\right]\right] \end{aligned} \right\} = 0 \tag{1}$$

Momentum equation for clear water

$$u_{\alpha t} + (u_\alpha.\nabla)u_\alpha + \nabla\eta + \left(z_\alpha^2/2\right)\nabla(\nabla.u_{\alpha t}) + z_\alpha\nabla\left[\nabla.(hu_{\alpha t}) + n\nabla.(h_s u_{\beta t}^s)\right] = 0 \tag{2}$$

Momentum equation for porous medium

$$\begin{aligned} &u_{\beta t}^s + \nabla\eta + (u_\beta^s.\nabla)u_\beta^s + \left(z_\beta^2/2\right)\nabla(\nabla.u_{\beta t}^s) + z_\beta\nabla\left[\nabla.(h_b u_{\beta t}^s)\right] + \nabla\left(\left(h^2/2\right)\nabla.u_{\alpha t}\right) - \nabla\left[h\nabla.(hu_{\alpha t})\right] - \\ &n\nabla\left[h\nabla.(h_s u_{\beta t}^s)\right] - \nabla\left(\left(h^2/2\right)\nabla.u_{\beta t}^s\right) + \nabla\left[h\nabla.(h_b u_{\beta t}^s)\right] + R_\beta u_\beta^s + R_\beta\left(z_\beta^2/2\right)\nabla(\nabla.u_\beta^s) + R_\beta z_\beta\nabla\left[\nabla.(h_b u_\beta^s)\right] - \\ &\nabla\left(\left(R_\beta h^2/2\right)\nabla.u_\beta^s\right) + \nabla\left[R_\beta h\nabla.(h_b u_\beta^s)\right] + \nabla R_\beta\left\{\left(\left(h^2 + hh_b + h_b^2\right)/6\right)\nabla.u_\beta^s - \left((h+h_b)/2\right)\nabla.(h_b u_\beta^s)\right\} = 0 \end{aligned} \tag{3}$$

---

[1] Doctoral Student, Dept. of Civil Engineering, 7-3-1, Hongo, Bunkyo-ku, The University of Tokyo. 113-8656, Japan. sumindar@coastal.t.u-tokyo.ac.jp.
[2] Professor, ditto, sato@coastal.t.u-tokyo.ac.jp.
[3] Associate Professor, ditto, yoshitaji@coastal.t.u-tokyo.ac.jp.

Two additional terms have been introduced into momentum conservation equation for clear water to simulate energy dissipation due to wave breaking and bottom friction. A simple eddy viscosity-type of formulation is used to simulate the energy dissipation due to wave breaking and the rate of energy dissipation is expected to be governed by the magnitude of eddy viscosity, which is related to turbulent kinetic energy (computed using transport equation, Eq. 4) and turbulent length scale.

$$k_t + u_s.\nabla k = \sigma\nabla.\nabla(\nu_t k) + B\frac{l_t^2}{\sqrt{C_D}}\left[\left(\frac{\partial u}{\partial z}\right)^2 + \left(\frac{\partial v}{\partial z}\right)^2\right]_{z=\eta}^{3/2} - C_D\frac{k^{3/2}}{l_t} \tag{4}$$

In addition to that, we employ anisotropic eddy viscosity formulation following Tajima et al. (2007).

3. Results and Discussion

Figure 1 shows the transformation of wave height & mean water level over a porous submerged breakwater placed 2.0m offshore at 0.028cm submergence under normal incident regular waves ($H_0 = 0.041$ cm, T =1.0s). The model results seem to be in good agreement with the experimental data and comparison could be clearly made with the impermeable case (here we only present the 1DH model results for the simplicity and the 2DH model results will be discussed thoroughly in full paper).

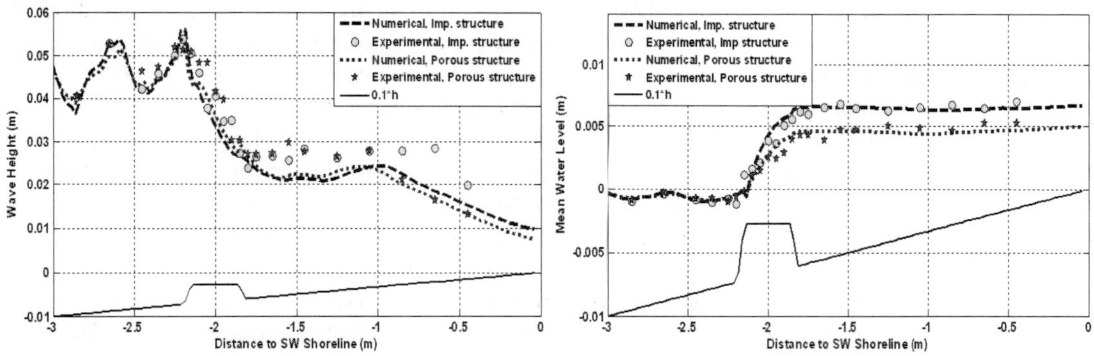

Figure1. Wave height distribution (left) and mean water level distribution (right) over a porous submerged breakwater

4. Summary

A numerical model has been developed based on Chen (2006), Boussinesq-type equations to simulate waves and wave-induced currents around porous submerged breakwaters coupling transport equation to introduce energy dissipation due to wave breaking. The model was verified with a new set of laboratory data in particular for 1DH conditions.

References

Nwogu, O.J. 1993. Alternative form of Boussinesq equations for nearshore wave propagation. *Journal of Waterways, Port, Coastal and Ocean Engineering*, ASCE, 119 (6), 618-638.

Tajima, Y., Sato, S., Shimozono, S., Isobe, M. 2007. Modelling of wave induced current around submerged detached breakwaters. *Proceedings, Coastal Structures '07,* ASCE, Venice, Italy, pp. 31-45.

Cruz, E.C., Isobe, M., Watanabe, A. 1997. Boussinesq equations for wave transformation on porous beds. *Coastal Engineering*, 30, 125-156.

Hsiao, S.C., Liu, P.L.F., Chen, Y. 2002. Non-linear water waves over a permeable bed. *Proceedings, Royal Society of London*, Series A 458, 1291-1322.

Chen, Q. 2006. Fully non-linear Boussinesq-type equations for waves and currents over porous beds. *Journal of Engineering Mechanics*, 132(2), 220-230.

Coastal Dynamics 2009
Paper No. 13

A MODEL OF WAVES AND CURRENTS AROUND COASTAL STRUCTURES

Pham Thanh Nam[1] and Magnus Larson[2]

1. Introduction

Structures are widely employed in coastal engineering projects. For example, detached breakwaters are used to reduce the incoming wave energy to a coastal stretch and protect it against erosion. Jetties are constructed at harbor entrances or estuary mouths to reduce channel infilling and inlet closure by longshore sediment transport. Thus, such structures are designed to directly affect the hydrodynamic and sediment transport processes, inducing erosion or accumulation of sediment. However, in order to achieve the desired response in the beach topography by placing coastal structures it is essential to determine the wave and current field around the structure.

The main objective of this study is to develop a robust numerical model to simulate waves and currents in the vicinity of coastal structures that can be applied with confidence in engineering studies. This model includes three submodels: (1) a multi-directional and frequency random wave transformation model; (2) a surface roller model; and (3) a model to compute currents and mean water elevation.

2. Methodology

The wave submodel employed calculates multi-directional and multi-frequency random wave transformation based on the energy balance equation according to Mase (2001). Wave shoaling, refraction, and diffraction is represented in the submodel as well as energy dissipation due to wave breaking. However, the energy dissipation term was modified following Dally *et al.* (1985) to obtain better agreement with measurements of the wave height in the surf zone.

The surface roller, associated with the wave breaking, was modeled following the approach by Dally and Brown (1995) and Larson and Kraus (2002). The energy flux in the alongshore direction was included in the energy balance equation for the roller. It improves the description of the forcing inside the surf zone, producing a better wave radiation stress field for the current model.

The depth-averaged two-dimensional submodel of nearshore currents employs the output from the wave and surface roller submodels. Wave-induced currents were derived using the momentum equations including the wave radiation stress and considering the effects of bottom friction. The mean water level was determined from the continuity equation.

3. Selected Results

Model validation was based on several unique, high-quality laboratory data sets from experiments with structures in the Large-Scale Sediment Transport Facility (LSTF) basin of the Coastal and Hydraulics Laboratory (CHL) in Vicksburg, U.S. (Gravens *et al.*, 2006; and Gravens and Wang, 2007). These data encompassed measured cross-shore variation in significant wave height, longshore current, and mean water level at many profile lines for different wave input and beach conditions. For the test cases from the LSTF basin with a detached breakwater, the calculation results on significant wave height of the modified Mase model (EBED) were better than those of the original model, producing very good agreement with the measurements (Fig. 1). The computed longshore current also agreed well with the measurements (Fig. 2). Although the calculated mean water level was somewhat overestimated compared to the measurements, the absolute error was small. Further validation of the model is underway with other LSTF data sets, including experiments on T-head

[1]Department of Water Resources Engineering, Lund University, P.O. Box 118, S-221 00 Lund, Sweden. Email: thanh_nam.pham@tvrl.lth.se
[2]Ditto. Email: magnus.larson@tvrl.lth.se

groins. The present model will form the basis for predicting morphological evolution in the nearshore with special focus on the topographic response to structures.

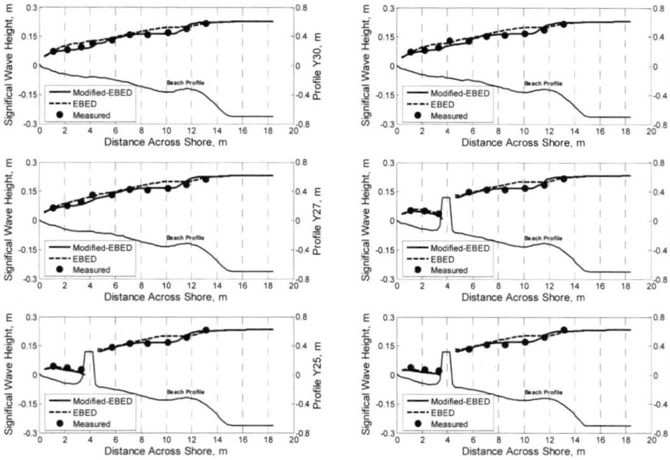

Figure 1. Comparison of calculated and measured significant wave height for LSTF Case T1C1

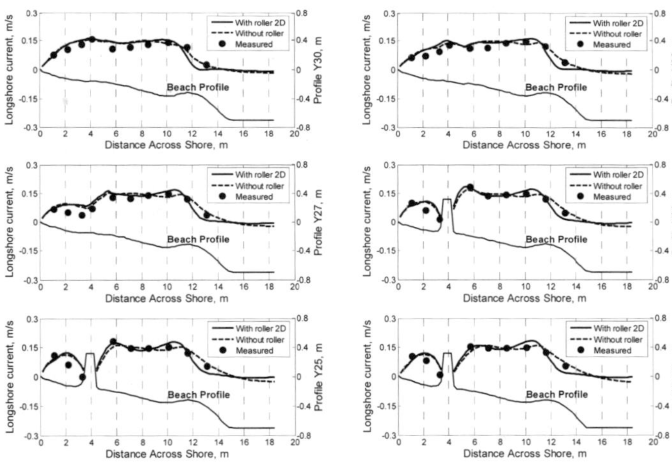

Figure 2. Comparison of calculated and measured longshore current for LSTF Case T1C1

References

Dally, W. R., Dean, R. G., and Dalrymple, R. A., 1985. Wave height variation across beaches of arbitrary profile. *J. Geophys. Res.*, 90(C6), 11917 – 11927.

Dally, W. R., and Brown, C. A., 1995. A modeling investigation of the breaking wave roller with application to cross-shore currents. *J. Geophys. Res.*, 100(C12), 24873 – 24883.

Gravens, M.B., and Wang, P., 2007. Data report: Laboratory testing of longshore sand transport by waves and currents; morphology change behind headland structures. *Technical Report*, ERDC/CHL TR-07-8, Coastal and Hydraulics Laboratory, US Army Engineer Research and Development Center, Vicksburg, MS.

Gravens, M.B., Wang, P., Kraus, N.C., and Hanson, H., 2006. Physical model investigation of morphology development at headland structures. *Proc., 30th Int. Conf. on Coastal Eng.*, ASCE, San Diego, 3617–3629.

Larson, M., and Kraus, N.C., 2002. NMLONG: Numerical model for simulating longshore current; Report 2: Wave-current interaction, roller modeling, and validation of model enhancements. *Technical Report ERDC/CHL TR-02-22*, US Army Engineer Research and Development Center, Vicksburg, MS.

Mase, H., 2001. Multi-directional random wave transformation model based on energy balance equation. *Coastal Engineering Journal*, 43(4), 317 – 337.

Coastal Dynamics 2009
Paper No. 14

ARTIFICIAL NEURAL NETWORK MODELLING OF WAVE-INDUCED CURRENT OVER SUBMERGED BREAKWATERS

Mojtaba Tajziehchi[1] and Amir Sharif Ahmadian[2]

1. Introduction

Recently, submerged breakwaters are being used in coastal management to protect shorelines with minimum environmental effects and spoiling coastal view. The hydrodynamic and morphological impacts of submerged breakwaters depend on breakwater submergence, wave climate, distance of structure to shoreline and crest width of the structure. The knowledge of submerged breakwaters effects on nearshore waves and currents leads to more accurate calculation of sediment transport and morphological evolution in the vicinity of the structure. Few literatures are available to predict wave-induced current over submerged breakwaters.

A new prediction method of wave-induced current over submerged breakwaters using coupling numerical model (Tajziehchi and Cox, 2006) and Artificial Neural Networks (ANNs) is proposed. Wave-induced flux over submerged breakwaters is a complex function of several parameters such as incident wave climate, submergence depth and breakwater crest width. One of the main advantages of applying the proposed ANNs is to predict wave-induced current over submerged breakwaters disregarding the variety of several effective parameters. On the other hand, ANNs models need an extensive set of corresponding data that may not be available. Therefore, a coupling numerical model and ANN was employed to address the two problems in predicting discharge over submerged breakwaters. The numerical model of Tajziehchi and Cox (2006) was employed to generate reliable training data. A new ANN model was designed and trained using the generated current data and the predicted mass flux were compared with the corresponding measured experimental data conducted by Cox and Tajziehchi (2005).

2. Numerical Model

The 2D depth averaged numerical modelling of wave-induced discharge and wave transmission over submerged breakwaters, proposed by Tajziehchi and Cox (2006), was employed to generate training data. Over 6000 data with different dimensionless values of submergence ratio (h_s/H_i), breakwater crest width ratio (B/L) and dimensionless incident wave height (H_i/gT^2) were generated using the calibrated/modified Delft3D numerical model. The model was validated using measured experimental data in 3m wide wave flume at Water Research Laboratory, Australia (Tajziehchi and Cox, 2005).

3. ANN Modeling

One of the main advantages of applying ANNs method in this research is to predict wave-induced current over submerged breakwaters disregarding the variety of several effective parameters. An ANN was designed and trained using current data generated with the calibrated Delft3D numerical model. In this study, a Resilient Back-propagation (Rprop) algorithm is used to train a multilayer feed forward network. A method called Early Stopping was used herein for improving generalization. Then, the hybrid model is tested and the architecture of neural network is optimized. A summary of final artificial neural network is presented in Table 1.Calculated discharge over submerged breakwater using the proposed model were compared with the measured experimental data collected by Cox and Tajziehchi (2005) in 3 m wide wave flume with variety of water depth, breakwater crest width and wave climate as well as data measured by Gourlay (1996) on a broad crested reef. Figure 1 shows a

[1]Dept of Civil Eng, Hormozgan University, P.O. Box 3995, Bandar Abbas, Iran, m.tajziehchi@hormozgan.ac.ir
[2]Dept of Civil Eng, Hormozgan University, P.O. Box 3995, Bandar Abbas, Iran, sharifahmadian@gmail.com

comparison of calculated discharge using the proposed coupled hybrid model and the numerical model of Delft3D with the measured discharge ratio, for B/L>2.5. Comparison of calculated and measured discharge over submerged breakwater reveals the accuracy of the trained ANN model (using artificial data) and its capability in predicting discharge over submerged breakwaters. More comparison and discussion of results will be provided in the full paper of the conference and it indicates that the proposed hybrid current model (HCM) predicts current over submerged reefs with higher accuracy in comparison with the numerical current model (NCM) of Delft3D for large variety ranges of dimensionless parameters.

Acknowledgement
The authors would like to thank Water Research Laboratory, University of NSW, Australia, for the instrumentation, facilities and support in setting up the physical model and Hormozgan University, Iran, for providing Delft3D software.

References

Cox, R.J. and Tajziehchi, M. 2005. 2D experimental modeling of hydrodynamic effects of submerged breakwaters, *Coastal Dynamics'05*, 12p. [CD-ROM].
Gourlay, M.R. 1996. Wave set-up on coral reefs. 1. Set-up and wave-generated flow on an idealized two dimensional horizontal reef. *Journal of Coastal Engineering*, 27: 161-193.
Tajziehchi, M., Cox, R. 2006. Wave transmission and Wave-induced current in Presence of Submerged Breakwaters. *Proc.30th Int. Conf. on Coastal Engineering*, ASCE: 4970-4982.

Table 1. Final architecture of Artificial Neural Network model

Architecture	Transfer function	Training algorithm	Generalization
$I_4H_7O_1$	Log sigmoid - Linear	Resilient Back-propagation (Rprop)	Early Stopping

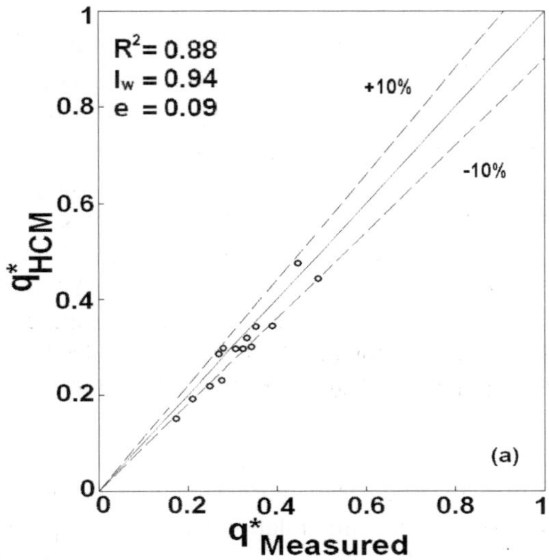

Figure 1. Comparison of measured and calculated dimensionless discharge for B/L>2.5

Coastal Dynamics 2009
Paper No. 15

MODEL-DATA COMPARISONS OF WAVE SHAPE CHANGES OVER LOW-CRESTED BREAKWATERS

Zhong Peng[1] and Qingping Zou[1]

1. Introduction

Understanding wave shape changes over Low-Crested Breakwaters (LCBs) is crucial to the assessment of the functionality and stability of coastal and flood defence schemes. In addition, it has long been recognized that wave asymmetries are directly related to sediment transport and subsequent changes in beach morphology. Previous studies (Elgar and Guza, 1985; Herbers et al., 2003) investigated the transformation of asymmetrical waves over a natural beach. On the other hand, investigations on LCBs, have mainly been concerned with wave transmission, reflection and wave spectral evolution (Van der Meer et al., 2005). Currently, there is a lack of knowledge on the transformation of asymmetrical waves over LCBs located on the beach. Laboratory data sets of wave surface elevations collected in the DELOS project and numerical models have been used to explore wave shape changes over LCBs.

2. Experiment

Two-dimensional wave transmission tests were conducted in the flume (24.0 m × 0.6 m × 0.8 m) at University of Cantabria, Spain. Two rubble mound LCBs with 0.25 m and 1.00 m crest width were tested. The two cross-sections were tested at three water levels giving three freeboards (R_c =-0.05, 0, and 0.05 m). The test programme included three incident wave heights (0.05 m, 0.1 m and 0.15 m) and three wave periods (1.6 s, 2.4 s and 3.2 s). Experimental set-up and test conditions were described in further detail in Kramer et al. (2005).

3. Numerical model

The RANS-VOF model developed by Lin and Liu (1998) is used to investigate the wave shape changes over LCBs. It consists of a Reynolds-Averaged Navier-Stokes solver, a k-ε turbulence model and a Volume of Fluid method to capture free surface. An internal wave-maker was used to generate target regular waves. A radiation boundary, composed of an open boundary and a sponge layer, is specified at the inlet boundary, while free-slip boundary condition is applied on the other boundaries. A zero-stress condition is applied at the free surface. Turbulent kinetic energy k and dissipation rate ε are assumed to be a function of distance from the solid surface and have zero gradient at the free surface. A total 995 cells are used in the x-direction with a minimum cell size of 0.01m, and in the y-direction 88 cells are used with a minimum cell size of 0.01m. Figure 1 shows the Sketch of the numerical setup. A return flow system, allowing the water to flow back to the flume right side of the source region, was used in the simulations. The duration of simulations is in excess of 150s.

4. Results

Figure 2 shows an example of wave skewness and asymmetry changes over a LCB. Our analysis demonstrates that wave skewness has small modulation when waves propagate on the incident side, but decays on the seaside slope of LCBs. On the crest and leeside slope of LCBs, wave skewness increases rapidly and attains the maximum value. On the transmission side, wave skewness decreases

[1] School of Engineering, University of Plymouth, Devon, PL4 8AA, UK. zhong.peng@plymouth.ac.uk

to similar values as those on the incident side. Wave asymmetry decreases gradually from around zero values to negative values on the incident side, attains a minimum value on the crest of LCBs, and then dramatically increases from negative values to positive values on the transmission side. It is evident from figure 2 that the minimum values of wave skewness and asymmetry occur on the seaside slope of LCBs, while the maximum values exist on the leeside slope of LCBs.

5. Discussions

Wave skewness and asymmetry are changed significantly by the presence of LCBs. Wave shape changes of LCBs with negative freeboard are similar to those over a sandbar beach shown by the field measurements and predictions in Herbers et al. (2003). Numerical results of wave skewness and asymmetry are in good agreement with measurements. In the full paper, further 2-D experimental irregular wave data will be analyzed to enhance our understanding of this problem. Based on the experimental and numerical results, empirical formulae of the cross-shore transformation of wave asymmetries over LCBs will be derived in terms of a wide range of wave conditions and structural geometry. Bispectral analysis will be used to explain wave shape changes over LCBs.

References

Elgar, S. L. and R. T. Guza, 1985. "Observations of bispectra of shoaling surface gravity waves." *Journal of Fluid Mechanics*, 161: 425-448.

Herbers, T. H. C., Orzech, M., Elgar, S. and Guza, R. T., 2003. "Shoaling transformation of wave frequency-directional spectra." *Journal of Geophysical Research-Oceans*, 108(C1): 3013.

Kramer, M., B. Zanuttigh, van der Meer, J. W. Vidal, C. and Gironella, F. X, 2005. "Laboratory experiments on low-crested breakwaters." *Coastal Engineering,* 52(10-11): 867-885.

Van der Meer, J. W., R. Briganti, Zanuttigh, B. and Wang, B.X, 2005. "Wave transmission and reflection at low-crested structures: Design formulae, oblique wave attack and spectral change." *Coastal Engineering,* 52(10-11): 915-929.

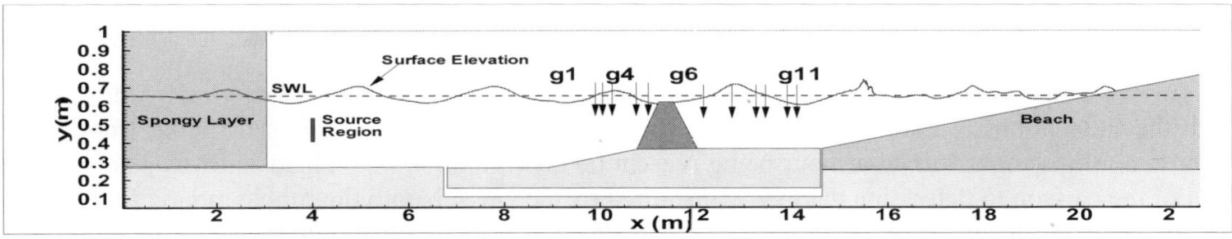

Figure 1 Sketch of the numerical setup of LCBs with narrow crest. Water depth h=0.4m. Solid line represents one of calculated surface elevations; dashed line stands still water level.

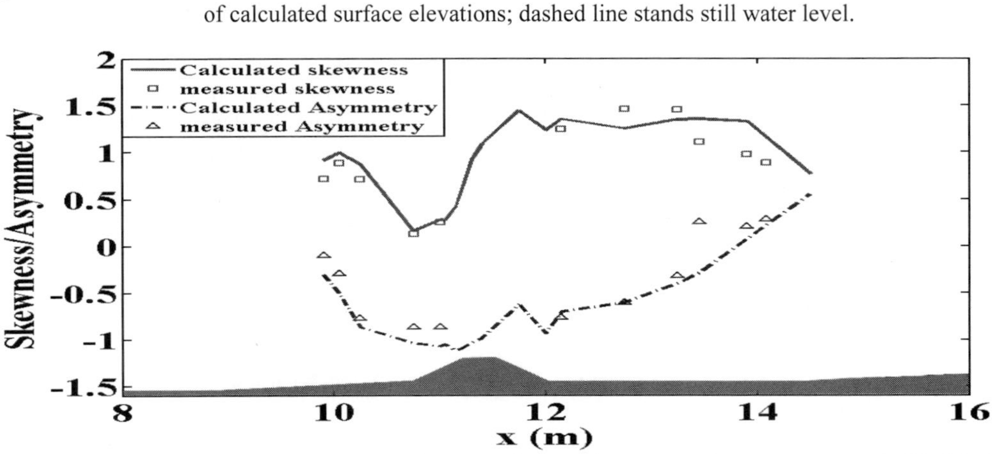

Figure 2 Transformations of calculated and measured wave skewness and asymmetry over LCB. (H_i=0.15m, T=2.4s and h=0.4m). Gray area represents the layout of LCBs and the beach.

Coastal Dynamics 2009
Paper No. 16

FAILURE MODE AGAINST IMPACT AND SOLITARY WAVES OF ARMOURED CAISSON BREAKWATERS PROTECTED BY A PARTIALLY FAILED ARMOUR LAYER

Miguel Esteban[1], Hiroshi Takagi[2] and Tomoya Shibayama[3]

1. Introduction

At present a number of formulas and methodologies exist to evaluate the effects of waves on caisson breakwaters subjected to either wind waves or tsunami attack. However, although formulas such as that introduced by Van der Meer (1988) can calculate the damage suffered by a rubble mound breakwater, no methodology exists to determine the effect that a partially destroyed rubble mound armour would have on the overall stability of a armoured caisson breakwater. This effect is know to affect current existing methodologies (Kim et al. (2005), Esteban and Shibayama (2008)), to calculate the failure of a caisson breakwater. Thus, the aim of the present study is to clarify the failure mechanism of a armoured caisson breakwater against a comprehensive range of wave types, including those produced by wind mechanisms (breaking and non-breaking) and those produced by a tsunami type event (bore-type, breaking and non-breaking).

2. Laboratory Experiments

Laboratory experiments were carried out with a variety of wave types at a 15.3m long x 0.6m wide x 0.55m deep wave flume located at the Hydraulics laboratory of Yokohama National University in Japan. To simulate wind-type waves regular waves were generated using a wave generator. These encompassed a wide range of wave periods and heights in order to differentiate between the effects of breaking and non-breaking waves. To simulate tsunami-type waves a reservoir of water was created on one side of the tank, which could be quickly opened using a pulley system to generate solitary waves. By varying the depth of water in the tank different types of waves could be simulated (bore-type, breaking and non-breaking). Fig. 1 shows the experimental set-up for the generation of solitary waves.

To investigate how different configurations of armour layers affect the load exerted by the caisson onto the foundations, a total of 4 different layouts of tetrapod armour were used (going from a full layer in configuration A to no armour being present on configuration D). A load cell was placed at the heel of the caisson to determine the load transmitted by the caisson onto the rubble mound foundation. Finally, all experiments were taped using a video camera to identify the failure mechanism.

3. Laboratory Results

The results of the laboratory experiments showed how the load exerted by the caisson breakwater onto the foundations greatly depends on the configuration of the armour layer in front of it and the type of wave. Fig. 2 shows for example the load time history for a breaking-type regular wave, for each of the tetrapod armour configurations. From this figure it can be seen how the greatest recorded loads are for a partially destroyed tetrapod layout (configuration B), which are much higher than the loads recorded for a non-armoured caisson breakwater. However, for the case of a solitary wave the depth of the water in front of the breakwater appears to be a crucial factor in determining the loads that are exerted on the foundation. For the experiments conducted at higher water depths the results are similar to those of the regular waves, although for those conducted at lower water depths configuration B does

[1]United Nations University Institute of Advanced Studies, 1-1-1 Minato Mirai, Nishi-ku, Yokohama 220-8802, Japan esteban@ias.unu.edu
[2]Penta-Ocean Construction Co., Japan. htaka@ynu.ac.jp
[3]Dept of Civil Eng., Yokohama National University, 79-5 Tokowadai, Hodogaya-ku, Yokohama 240-8501, Japan. tomo@ynu.ac.jp

not give the highest loading, as can be seen in Fig. 3.

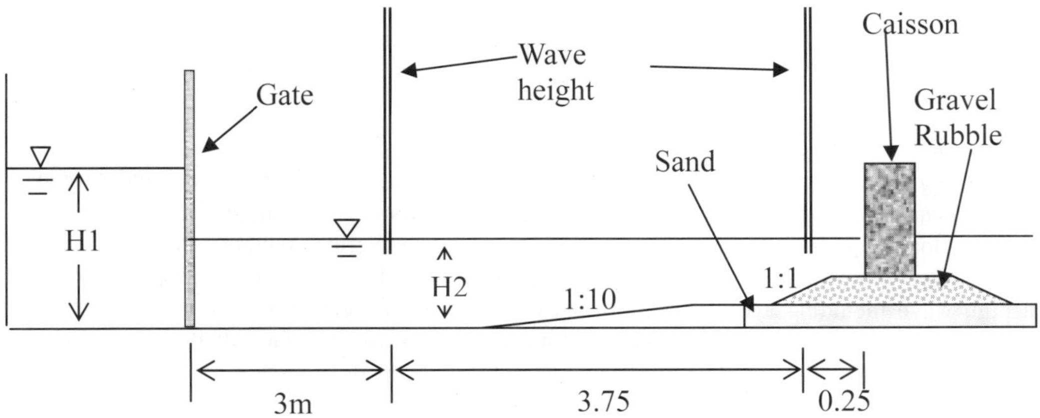

Fig.1 Experimental set-up for solitary wave generation (for configuration D, a Caisson without any armour in front of it)

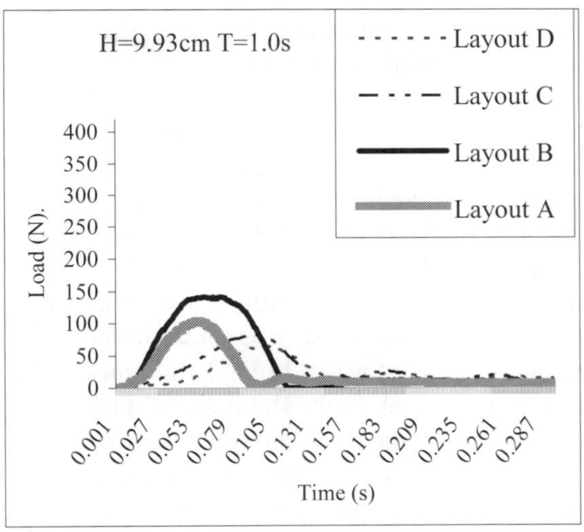

Fig. 2. Load profile of regular breaking wave with H=9.93cm ant T=1.0s

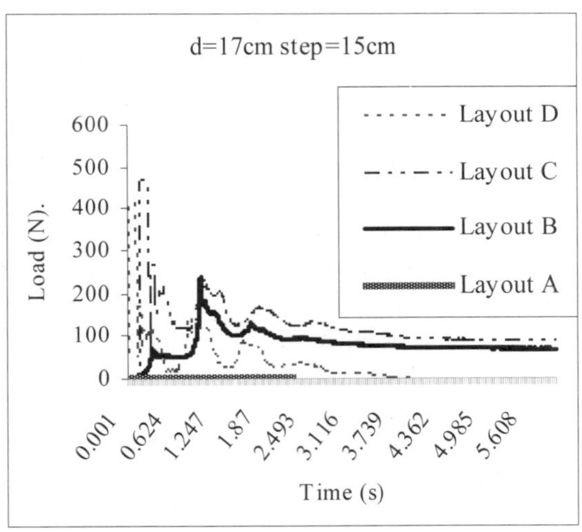

Fig. 3. Load profile of solitary wave in a water depth d=17cm

References

Van der Meer, J.W. (1987).Stability of breakwater armour layers – design formulae, *Coastal Engineering* 11: 219-239.

Kim, T.M., Yasuda, T., Mase, H. & Takayama, T. (2005). Computational Analysis of Caisson Sliding Distance Due to Typhoon Tokage, *Asian and Pacific Coasts Conference 2005*, pp 565-576.

Esteban M. and Shibayama T. (2008) Computational Estimation of Caisson Sliding and Tilting of Sus ami West Breakwater due to Typhoon Tokage, *33rd Annual Journal of Civil Engineering in the Ocean,* JSCE, Japan.

Coastal Dynamics 2009
Paper No. 17

INTERACTION MODEL OF MOORED BUOY AND BREAKING WAVE BASED ON LAGRANGIAN PARTICLE METHOD

Hiroyuki Ikari[1] and Hitoshi Gotoh[2]

1. Introduction

When a mooring chain is broken due to a high wave, a moored ship can be drifted and stranded. Because it can cause enormous damages to facilities in a port, it is important to predict the movement of a moored floating object driven by a high wave. Some numerical methods for this kind of prediction has been proposed, however, there is no model to simulate a floating object appropriately under a high wave including a fragmentation and coalescence of water like a breaking wave. The particle method is suitable for analysis under such a complicated water surface change. Therefore, in this study, a simulation model based on a particle method is developed.

2. Numerical Model

In this model, fluid and a buoy are tracked by the MPS method (Koshizuka et al., 1995). A buoy is constructed by connecting some particles and treated as a rigid body (Koshizuka et al., 1998). On the other hand, a mooring chain is calculated by a spring-mass model, however, in the 2-D field, the motion of water particle across a mooring chain is not allowed. Therefore, the force acting on a particle, or a mass point, is calculated indirectly by using the formula of drag force with allowing the overlap of water particle and mooring-chain particle.

3. Calculated Results

In a numerical wave flume with a horizontal bottom, a regular wave is generated. A buoy is under the one point relaxation mooring. A tensile force in the progress direction of wave is always imposed as a wind load of the buoy. The wave conditions are T=1.0 s and H=0.13 m in the case 1, and T=2.2 s and H=0.12 m in the case 2. Fig.1 shows snapshots. When a buoy is in the wave trough, a mooring chain is under relaxation, on the other hand, when a buoy is in the wave crest, it is under tension. The relaxation and tension of the chain are repeated with the progress of wave. In Fig.2, the track of the centroid of buoy is shown with comparing to a previous experimental result. The shape of track shows elliptical, of which the major axis is in the upper right direction. In the case 2 in which wave steepness is smaller, this ellipse is stretched. Both calculated results show good agreements with experimental results in the previous studies. In Fig.3, the mooring force acting on the buoy is shown time-dependently. In the calculated result, the mooring force becomes zero when a mooring chain is relaxed, however, as to a peak value of the mooring force, both calculated results show good agreements with experimental results.

References

Koshizuka, S., Tamako, H. and Oka, Y. 1995. A Particle Method for Incompressible Viscous Flow with Fluid Fragmentation. *Comp. Fluid Dyn. J.*, 4: 29-46.

Koshizuka, S., Nobe, A. and Oka, Y. 1998. Numerical Analysis of Breaking Waves Using the Moving Particle Semi-implicit Method. *Int. J. Numer. Meth. Fluids*, 26: 751-769.

[1]NEWJEC Inc., 2-3-20, Honjo-Higashi, Kita-ku, Osaka, 531-0074, Japan. ikarihr@newjec.co.jp
[2]Dept. of Urban and Environmental Eng., Kyoto Univ., Katsura Campas, Nishikyo-ku, Kyoto, 615-8540, Japan. gotoh@mbox.kudpc.kyoto-u.ac.jp

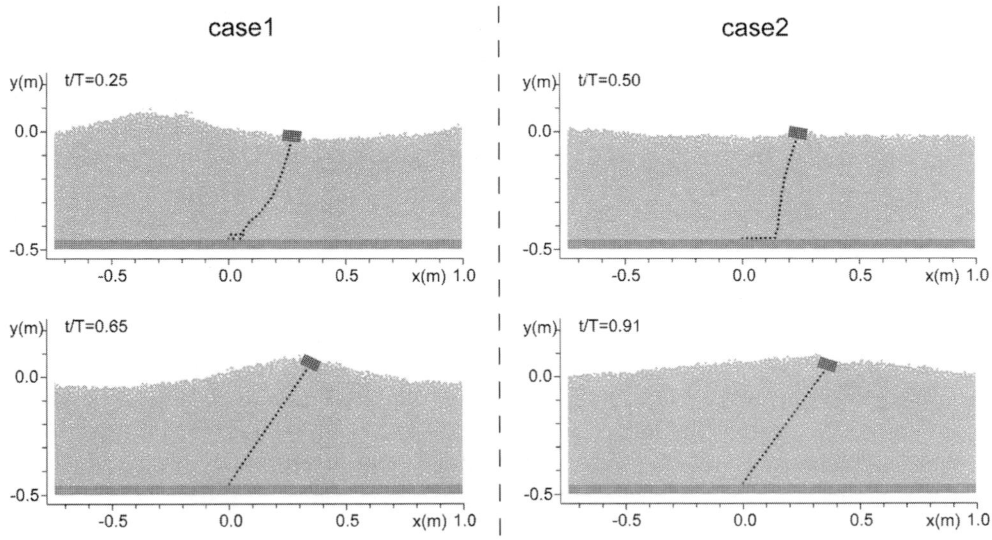

Figure 1. Snapshots

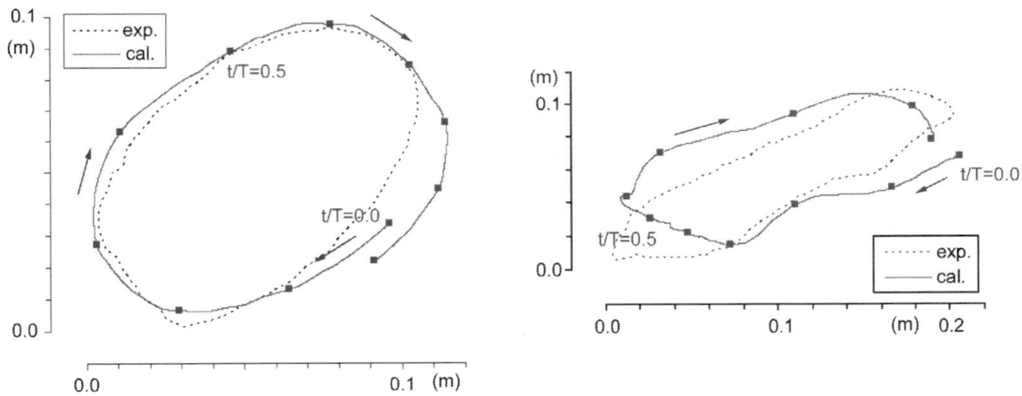

Figure 2. Track of centroid of buoy (left:case1, right:case2)

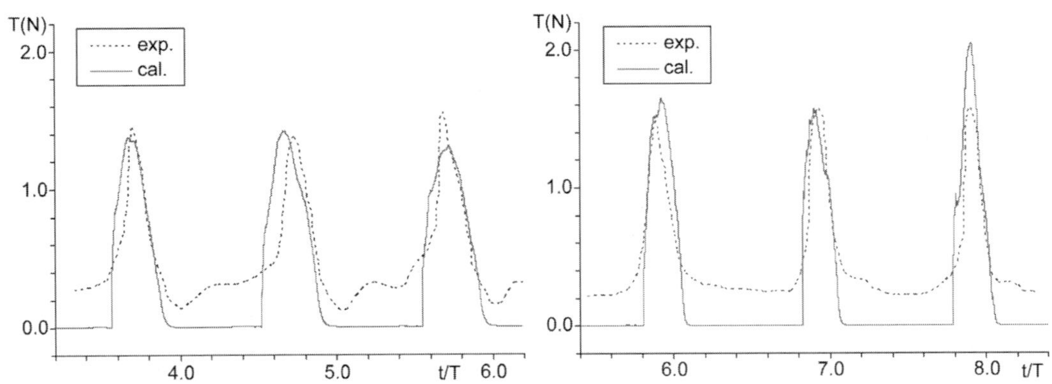

Figure 3. Mooring force (left:case1, right:case2)

Coastal Dynamics 2009
Paper No. 18

INVESTIGATIONS ON THE INFLUENCE OF A LOW-TIDE TERRACE USING THE BOUSSINESQ-TYPE WAVE MODEL TRITON

Sofia Caires[1] and Marcel R. A. van Gent[2]

1. Introduction

The profile of sandy bottoms can change in response to severe storms. Often measured bottom profiles (which in their vast majority are measured in the summer) are not representative of the winter bottom profile. These are, however, the bottom profiles used in physical modelling. It would be desirable to take the bathymetry uncertainty into account in the physical modelling. However, since that would in principle involve a lot of physical testing and the building of different nearshore bottom profiles, which is quite an expensive exercise, bathymetry uncertainty is never taken into account. Numerical wave models are in principle cheaper, but also less accurate.

In order assess the performance of the Deltares | Delft Hydraulics Boussinesq-type model, TRITON (Borsboom, 1998 and Borsboom et al, 2000), for modeling waves propagating over shallow foreshores towards sea defences a number of physical model tests have been executed in Deltares | Delft Hydraulics "Scheldt flume". The TRITON model was run using the same scale, approximately the same boundary conditions and the same monitoring locations as used in the Scheldt flume experiments. The TRITON and the Scheldt flume results were carefully compared using the same postprocessing and assuming the Scheldt flume results to represent the truth.

2. Experiment setup

The bottom profile considered is an idealization of the bottom profile in front of the Petten Sea defence (The Netherlands), since this foreshore is a characteristic shallow foreshore for many situations. Severe wave breaking occurs on this foreshore. Considerations regarding the schematization and modelling of this shallow foreshore are described in Van Gent (2001). Depth contours at this site are close to parallel to the coast and the angle of wave attack is mainly perpendicular. Such situation can therefore be accurately modelled in a wave flume and using 1-D numerical schematizations.

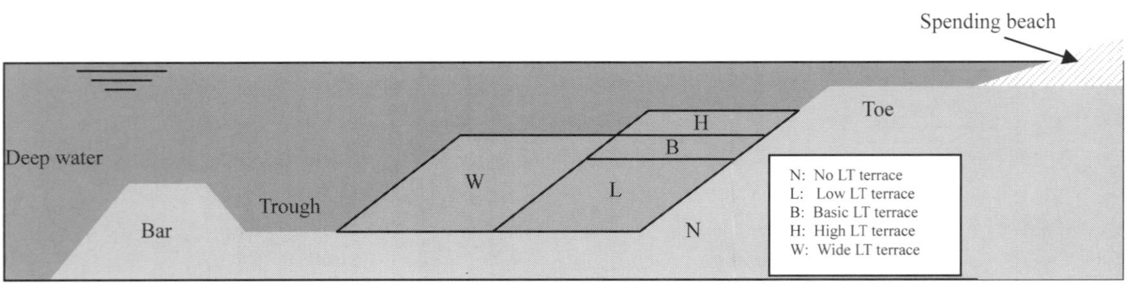

Figure 1. low-tide terrace geometries (not to scale for better viewing).

Previous studies (Van Gent and Giarrusso, 2005) have indicated that the wave loads are mainly

[1] Deltares | Delft Hydraulics, P.O. Box 177, 2600 MH Delft, The Netherlands. sofia.caires@deltares.nl
[2] Deltares | Delft Hydraulics, P.O. Box 177, 2600 MH Delft, The Netherlands. marcel.vangent@deltares.nl

influenced by the shape (depth and length) of the low-tide terrace. Therefore, several low-tide terrace configurations are taken into account in this study; more precisely, five different bottom profiles were tested, see Figure 1, under twelve hydraulic boundary conditions.

3. **Conclusions**

The comparisons between the physical model measurements and numerical model results show that TRITON can model the wave propagation quite accurately as long as highly non-linear effects do not play a significant role. Non-linear effects of $O(\varepsilon\mu)$, where μ is the water depth to wavelength ratio (the measure of the relative importance of linear dispersion) and ε the wave height to water depth ratio (the measure of the relative importance of nonlinear effects), are not fully accounted for by TRITON and it is shown that, even for relatively low values of $\varepsilon\mu$, the model underestimates the creation of low frequency wave energy. Unfortunately, the errors of TRITON cannot be characterized solely in terms of a non-dimensional number, since despite a positive relation between the relative bias and $\varepsilon\mu^2$, there is a strong dependence of the errors on the still water level. It was concluded that there is a strong dependence between the TRITON errors at the toe of the structure and ak^2 (where a is the wave amplitude and k the wave number) at the boundary.

The level of the low-tide terrace was found to influence the wave conditions at the toe of the structure as follows:

- As the level of the low-tide terrace increases there is more energy dissipation and the wave height decreases, decreasing the hydraulic load.
- On there other hand, as the waves dissipate low frequency energy is released and the spectral wave period increases, increasing the hydraulic load.

These results hence show that it is important to take the level of the low-tide terrace into account and that the most unfavourable bathymetry depends on to what extent processes on the structure depend on the wave height and period: a shallower foreshore may be more unfavourable than a deeper foreshore.

Acknowledgements

Dr. Jacco Groeneweg and Dr. Mart Borsboom of Deltares | Delft Hydraulics have actively collaborated in this research, providing valuable comments and advise. The present work is part of the CoMIBBS Joint Research Activity (JRA1) of the HYDRALAB-III project (project code 022441), funded by the EU Infrastructure Program and co-funded by Deltares | Delft Hydraulics.

References

Borsboom, M.J.A., 1998. *A Boussinesq-type wave model with improved linear and non-linear behaviour*. WL | Delft Hydraulics report H3202.51/X0231.40, October 1998, Delft, The Netherlands.

Borsboom, M.J.A., Doorn, N., Groeneweg, J. and Van Gent, M.R.A., 2000: A Boussinesq-type wave model that conserves both mass and momentum. *Proc. 27th Int. Conf. Coastal Eng.*, ASCE, Sydney, 148 – 161.

Van Gent, M.R.A., 2001: Wave run-up on dikes with shallow foreshores. *J. of Waterways, Port, Coastal and Ocean Eng.*, ASCE, Sept-Oct 2001.

Van Gent, M.R.A. and Giarrusso, C.C., 2005. Influence of foreshore mobility on wave boundary conditions. *Proc. Waves 2005*, ASCE.

Coastal Dynamics 2009
Paper No. 19

NUMERICAL ANALYSIS OF HARBOR OSCILLATION IN HARBORS OF VARIOUS SHAPES

Taro Kakinuma[1], Taishi Toyofuku[2], and Taisuke Inoue[3]

1. Introduction

Water oscillation in harbors of various shapes has been studied through linear theories, e.g. Hwang and Tuck (1970), hydraulic experiments, e.g. Horikawa et al. (1969), and numerical models, e.g. Derun et al. (2003). In this study, a numerical model based on the nonlinear shallow water equations is applied to study oscillation in harbors of different shapes including shapes of L, T, triangle, I with different width, I with a narrowed area, and so on, as well as I-type harbors where the static water depth is distributed. The harbor oscillation in a bay in Kagoshima, i.e., Urauchi Bay, the shape of which is similar to that of a T-type harbor, is also treated.

2. Numerical Calculation Method

The fundamental equations are the continuity and shallow water equations, which are solved using a finite difference method. The radiation condition is adopted on both incident-wave boundaries and boundaries in water areas, whereas boundaries between land and sea are assumed to be perfect-reflection walls. In each case, trains of regular waves are input on an offshore boundary, which is not a harbor mouth. The trains have the same wave height of 0.2 m but different wave periods. The incident wave direction is perpendicular to the cross sections of harbor mouths. The static water depth is 20 m in the flat-seabed cases, as well as at the harbor mouths and heads in sloping-seabed cases.

3. Numerical Results

Results of amplification factors in an I-type harbor were compared between the numerical model and a linear theory, from which the applicability of the present model was confirmed. The numerical calculation results of amplification factor R, which is defined by the ratio of response height over the incident wave height, i.e., 0.2 m, for harbors of several model shapes in the abstract.

(a) L-type harbors with different bending positions: Figure 1(a) shows I- and L-type harbors, where the harbor length ℓ is the same but their bending positions are different. Figure 2(a) shows the amplification factor R in the harbors shown in Figure 1(a), where $k\ell$ is the dimensionless wave number, i.e., $k\ell = 2\pi\ell/(T\sqrt{gh})$. The second modes show higher R as the bending position is nearer to the harbor mouth and R is larger than that in the I-type harbor when LA = 1,000 and 1,200 m.

(b) I-type harbors with a narrowed area: Figure 2(b) shows the factor R in the I-type harbors with a narrowed area shown in Figure 1(b). The first mode shows lower R at the head in Case (b-1), where the narrowed area is placed at the harbor mouth, than that in Case (b-2), whereas the second mode shows the contrary phenomenon. In Case (b-4) the second mode shows lower R at the head than that in Case (b-3), where the harbor width at the narrowed area is wider than that in Case (b-4).

(c) C-type harbor: Figure 2(c) shows the amplification factor R in the C-type harbor, where two I-type harbors are connected with a rectangular-section channel as shown in Figure 1(c). At the heads of the I-type harbors, the waves coming through two mouths are in opposite phase, such that the factor R becomes lower than that in the corresponding I-type harbor alone, which is shown in Figure 2(a). It should be noted that the phase difference between the waves through two mouths depends on

[1]Dept. of Ocean Civil Eng., Kagoshima University, 1-21-40 Korimoto, Kagoshima, Kagoshima 890-0065, Japan. taro@oce.kagoshima-u.ac.jp
[2]Ditto. gmjhh255@yahoo.co.jp
[3]Ditto. taisuke02092000@yahoo.co.jp

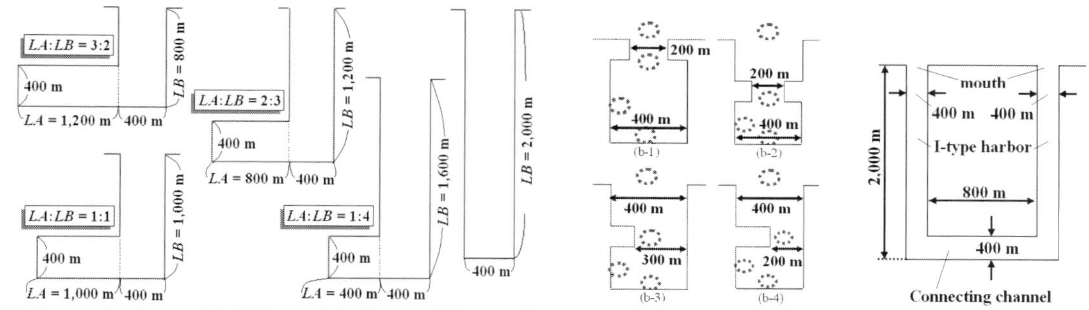

(a) L-type harbors with different bending positions (b) I-type harbors with a narrowed area (c) C-type harbor

Figure 1. Harbors of various shapes with flat bottom

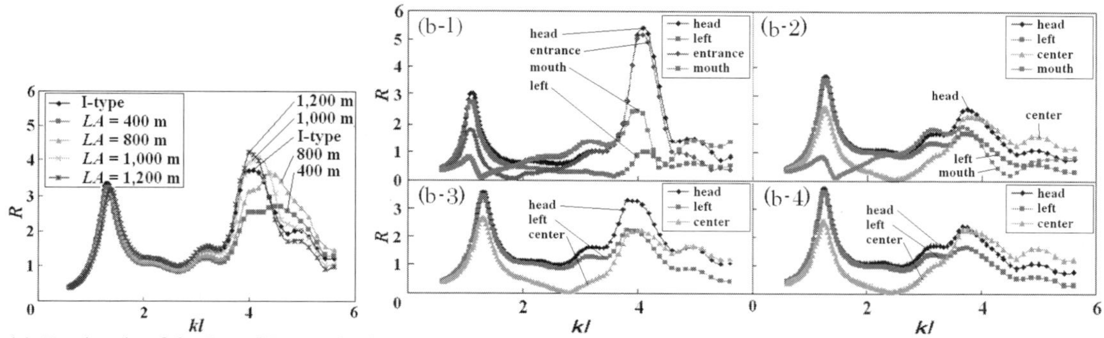

(a) R at heads of the I- and L-type harbors (b) R in the I-type harbors with a narrowed area

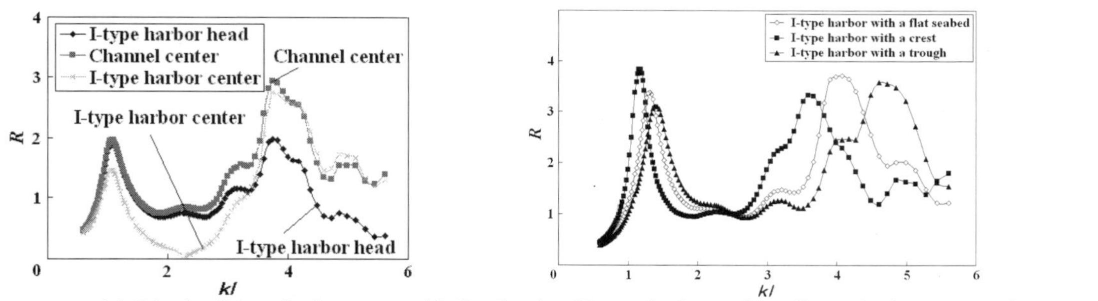

(c) R in the C-type harbor (d) R at heads of I-type harbors with a flat seabed, a crest, and a trough

Figure 2. Amplification factor R in the harbors shown in Figure 1 and the I-type harbors with a crest or trough

the position, resulting in rather high R at the centers of two I-type harbors and the connecting channel. *(d) I-type harbors with a seabed crest or trough*: Figure 2(d) shows R at the heads of I-type harbors with a seabed crest or trough, where the static water depth is 10.5 or 29.5 m at the center, respectively, and the harbor bottom is uniformly sloping towards the head and mouth. In the former case, both the first and second modes appear at the lower wave numbers than those in the I-type harbor with flat bottom and another peak appears clearly when $k\ell$ is equal to about 5.0. In the latter case, the wave numbers of the first and second modes become larger than those in the I-type harbor with flat bottom.

References

Derun, A. B., Kakinuma, T., and Isobe, M., 2003. A nonlinear numerical model of harbor oscillations and its application to responses of harbors of various shapes. *Proc. Coastal Engineering Journal*, JSCE, 50: 231-235. (in Japanese)

Horikawa, K., Shuto, N., and Nishimura, H., 1969. Characteristic oscillation of water in an L-shaped bay. *Coastal Engineering in Japan*, 12: 47-56.

Hwang, L-S. and Tuck, E. O., 1970. On the oscillations of harbours of arbitrary shape. *J. Fluid Mech.*, 42: 447-464.

Coastal Dynamics 2009
Paper No. 20

UNCERTAINTY OF NUMERICAL SIMULATIONS ON STORM SURGES BY USE OF BEST TRACK DATA OF TROPICAL CYCLONE

Hisamich Nobuoka[1]

1. Introduction

Storm surges must be one of the most serious impacts for coastal zone in the global warming. We have to know the spatial distribution of storm surges along whole of the coasts in order to estimate the impact together with sea-level rises for adaptations against the global warming especially in developing countries. Two major problems of the storm surge information are that some historical records of the surges only exist, and bias of winds and atmospheric pressures of future tropical cyclones estimated by global circulation model of meteorology have to be modified. A type of best track information of the cyclones is useful for calculation of storm surges, because the historical data has been stacked for half century or more and the biases can be modified comparatively easily. Accuracies of calculation of storm surge still depend on wind fields even by the best tracks data.

This study shows uncertainty of calculated storm surges due to the tropical cyclone. The nine cases of pressure and wind field of the cyclone Sidr (2007) are implemented as uncertain external force. The calculated results of the storm surges are compared with those results of a field survey.

2. Calculation method and cases

Shallow water equations with ADI method of finite differences were employed for calculating storm surges. This storm surge model calculated quite good results when high accuracy data of wind filed are provided. For example, the calculated storm surges of Hurricane Catrina in 2006 by use of reanalyzed wind data of NOAA were in good agreement with those surveyed results at New Orleans, Biloxy and so on. As the contribution of wave set-up in target area of Ganges delta is quite weak, this set-up is not included in this calculation.

Pressure and winds were calculated based on classic equations, Meyer's Equation and theoretical Equation. One of the functions of these equations is a radius at each time of each tropical cyclone. The radiuses are however unknown values, so three cases of radiuses based on the value of Kato's empirical equation (2005) were used in this calculation. Surface winds are empirically about 0.7 times weaker than geostrophic winds due to land/sea surface friction. On the other hand, observed winds in tropical cyclone were sometimes stronger than the normal geostrophic winds. These stronger winds are called 'Super Gradient Wind' (SGW). To express SGW, the coefficient is changed to over 1.0 instead of the coefficient for the friction. There are two types of the multiple value, those are proposed by Fujii and Mitsuta (1986), and Mitsuta and Fujii (1987). One is empirically constant value, and the other is variable according to the pressure at center of a cyclone. In this calculation, the 9 cases of pressure and wind calculation including SGW shown in Table 1 were used to calculate the storm surges of cyclone Sidr.

3. Simulated Results of Storm Surges

Comparing sea levels above astronomical tide of each simulation case of cyclone Sidr are shown in Figure 1. The blue, green and red line show the minimum (case-1), average in all cases and maximum (case-9) results of calculated storm surges. The bar shows the results of case-8 used variable type of SGW model, which are the closest to the heights by survey (Shibayama et al., 2008). The heights at

[1]Dept of Urban and Civil Eng., Ibaraki University, 4-12-1 Naka-Narusawa, Hitach, Ibaraki,316-8511, Japan. nobuoka@mx.ibaraki.ac.jp

Sothkhali, Naltona and Somboria are varied from 3.2 m to 7.6 m by different wind fields. Those points were close to center of the cyclone, where SGW works much. The radius of the tropical cyclone affected the heights much, only when SGW was taken into account.

In our knowledge, a lot of storm surges by tropical cyclones can be expressed by classical type of wind fields. Therefore, we can say winds of the cyclone are most uncertain factor in case best track data are used. On impact assessments of storm surges, not only a case of classical wind but also cases of variable type of SGW and constant type of SGW should be implemented for taking uncertain factor into consideration. The calculated results of spatial distribution of storm surge including SGW effects along the coast of Bengal bay in last 40 years are going to show in the proceeding.

References

Fujii, T. and mitsuta Y., 1986. Synthesis of a stochastic typhoon model and simulation of typhoon winds, *Annual disas*, *Prev. Res. Inst. Kyoto Univ.*, No. 29 B-1, pp.229-239 (in Japanese).

Kato, F. , 2005. No. 275, Technical note of NILIM, *National Institute for Land and Transport Managements*, *Japan,* 108p (in Japanese).

Mitsuta, Y. and Fujii, T., 1987. An analysis and synthesis of typhoon wind over japan, bull. *Disas, Prev. Res. Inst, Kyoto Univ*., Vol.37, Part4, No329, PP.169-185.

Shibayama, T. et al., 2008. Field survey of storm surge disaster due to cyclone sidr in bangladesh, A*nnual Journal of Coastal Engineering*, *JSCE*.

Table 1. Simulation Cases

Case	Radius of cyclone based on Kato's equation (2004)	Wind Type
CASE-1	-20%	Classic
CASE-2	0%	Classic
CASE-3	+20%	Classic
CASE-4	-20%	SGW (variable)
CASE-5	0%	SGW (variable)
CASE-6	+20%	SGW (variable)
CASE-7	-20%	SGW (constant)
CASE-8	0%	SGW (constant)
CASE-9	+20%	SGW (constant)

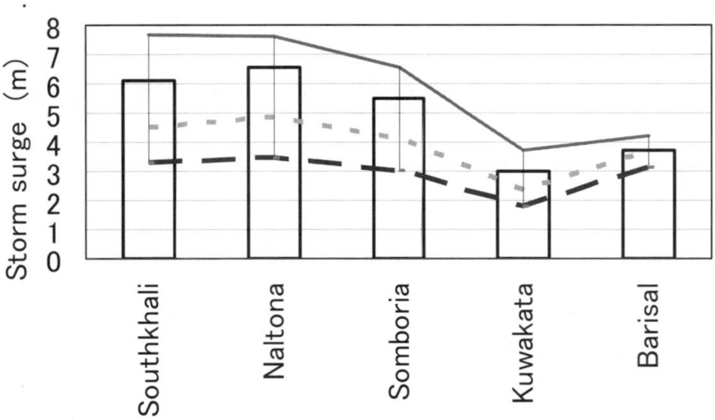

Figure 1. Comparing sea level above astronomical tide at 5 points.

Coastal Dynamics 2009
Paper No. 21

STORM SURGE HINDCAST IN TOSA BAY OF JAPAN USING A COUPLED MODEL OF SURGE, WAVE AND TIDE

Soo Youl Kim[1], Tomohiro Yasuda[2], Hajime Mase[3], Yoshiharu Matsumi[4]

1. Introduction

Recently an interest of storm surge modeling has been increased towards the inclusion of dynamic processes such as wave-current interaction including in bottom boundary layers and wind stress as a function of waves. In the present paper, the hindcast of storm surges occurred in the Tosa Bay in 1970 was performed by using a coupled model consisted of Surge, WAve and Tide, named SuWAT developed by Kim et al. (2007), and the effects of waves and tidal variations on surge heights were investigated from a viewpoint of the importance of coupling dynamic effect.

2. Numerical prediction model for storm surge (SuWAT)

A coupled model of Surge, WAve and Tide (SuWAT) was developed by Kim et al. (2007) in which the SuWAT code unifies numerical models of tide prediction, surge prediction and wave prediction by using the Message Passing Interface. This parallelization strategy enables the variability of geographic resolution in grid mapping. Several target locations can be nested from common larger (or, coarser) grid size domains. Such a coupled model with a multi-target nested grid system is new to parallelize and simultaneously calculate the interaction between surges, tides and waves.

3. Storm surge hindcast in the Tosa Bay, Japan

The SuWAT was implemented to the Tosa Bay in Japan to hindcast the storm surge due to Typhoon Anita in 1970. In the hindcast, six level grid sizes were employed as shown in Figure 1: 12,150 m in both x and y directions on the first domain; 4,050 m on the second; 1,350 m on the third; 450 m on the fourth; 150 m on the fifth; and 50 m on the sixth. The tide was imposed at open boundaries on only the first outer domain. Open boundary conditions on the finer grid for both of the tides and waves were interpolated from the coarser grid domain.

The coupling effects on storm surge were investigated in comparison with the results of various numerical conditions together with the observations as shown in Figure 2. From the results, it is seen that the water levels predicted from the traditional way using wind stress and atmospheric pressure are insufficient when comparing with the observations at all stations. However, when considering wave effects (wave set-up due to wave breaking) the predictions agree well with the observations; particularly, at the finer grid target locations (P5 ~ P8). The wave effect of set-up caused maximum water level rise up to 1.0 m at the most embedded region (a figure does not shown herein). The tidal effect on the surge generation was small, because the tidal range is relatively small as 1.0 m in the Tosa Bay.

4. Conclusions

[1]Department of Social Management, Tottori University, 4-101 Koyama-Minami, Tottori 680-8552, Japan. sooyoul.kim@sse.tottori-u.ac.jp
[2]Disaster Prevention Research Institute, Kyoto University, Gokasho, Uji, Kyoto 611-0011, Japan. yasuda@engan.mbox.media.kyoto-u.ac.jp
[3]Disaster Prevention Research Institute, Kyoto University, Gokasho, Uji, Kyoto 611-0011, Japan. hmase@engan.mbox.media.kyoto-u.ac.jp
[4]Department of Social Management, Tottori University, 4-101 Koyama-Minami, Tottori 680-8552, Japan. matsumi@sse.tottori-u.ac.jp

This study investigated the coupling effect on storm surge using a coupled model of surge, wave and tide (called SuWAT) with a multi-target nested grid system. Although the surge height could not be explained by the traditional storm surge model using wind and atmospheric pressure, the inclusion of wave effect (wave set-up) gave good predictions for storm surge height. It was also found that the effect of tidal variations is small and that the geographic resolution for the forecast of the water level rising under extreme wind events is important.

References

Kim, S.T., Yasuda, T., Mase, H., 2008. Effects of Tidal Variation on Storm Surges and Waves, *Applied Ocean Research,* (2nd revision).

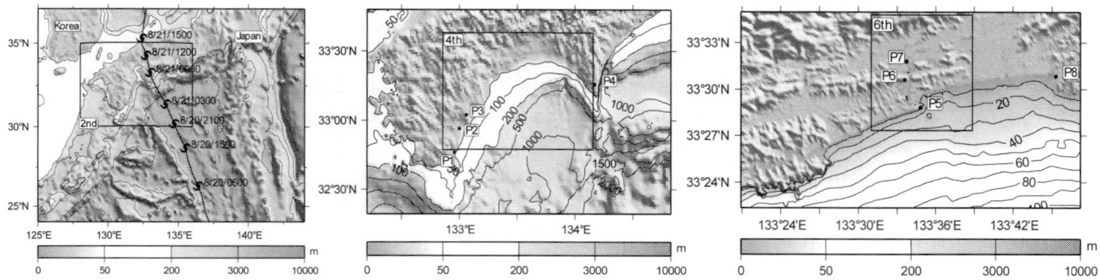

Figure 1. Geophysical areas of six levels, observation stations (P1~P8) and the best track of Typhoon Anita in 1970. The left panel indicates the most outer grid and the secondary embedded grid, the middle panel: the third and fourth grids, and the right panel: the fifth and the most inner grids in a multi-target nested grid system (Japan Standard Time (JST) is used).

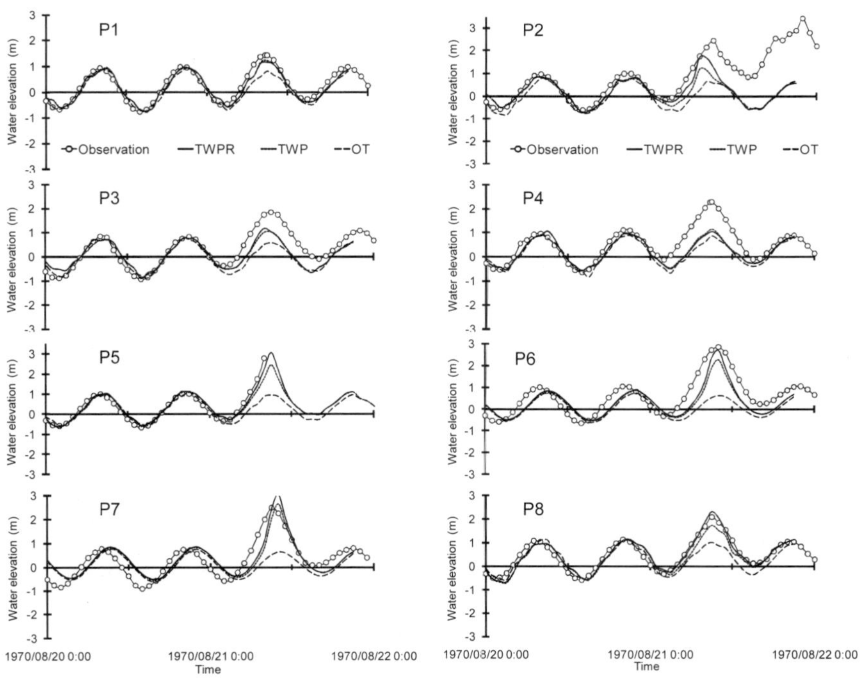

Figure 2. Comparisons of time series of observational water levels (○); TWPR: the computed total water levels considering tides, surges and wave induced force (—); TWP: the computed total water level of surges, waves and tides without the wave induced force (…); OT: the computed tide (---) (Time is JST).

Coastal Dynamics 2009
Paper No. 22

EFFECTS OF WAVE STRESS AND VERTIACAL MIXING ON STORM SURGE SIMULATIONS

Nobuhito Mori[1], Rie Takada[1], Hajime Mase[1], Tomohiro Yasuda[1] and Soonyul Kim[2]

1. Introduction

Major driving forces of storm surge are pressure depressions and wind stresses. Therefore, the depth integrated equations are used for storm surge simulations to estimate the anomaly sea level rise. However, there are other phenomena should be take into consideration. For example, wave induced radiation stresses increase the elevation in the surf zone, and distributions of eddy viscosity or turbulent diffusion also change velocity profiles in complex topography. In addition, recent observations are improved wind stress models in high wind region (U_{10}>30m/s, *i.e.* Jarosz, 2007). Therefore, it is important to verify the effects of each component on the storm surge simulations.

The main purpose of this study is estimations of wave induced setup, the improved drag coefficient for high wind speed region and vertical mixing model influences on the storm surge. The depth integrated two dimensional and three dimensional hydro static model including turbulence mixing and diffusion are used to predict the sea surface elevation in the storm surge. The numerical results are analyzed to compare the effects of physical components on the storm surge.

2. Numerical Implementation and Conditions

Two different models for storm surge simulations are used for simulations. One is the depth integrated long wave equation (denotes 2D model, hereafter) and another is the three dimensional primitive equation of Regional Ocean Modeling System (ROMS, denotes 3D model, hereafter). The 2D model can be coupled with SWAN model through the Message Passing Interface in parallel cluster. The 3D model was modified to include atmospheric pressure gradient effects for the storm surge. Two different models for drag coefficient are selected Honda-Mitsuyasu and modified Large and Pond formulae. The generic length scheme with Craig-Banner model is used for the vertical mixing due to turbulence of the 3D model. A series of numerical simulations were performed both a constant slope of 1/100 and a real topography, respectively. The influences of each component such as the modification of the drag coefficient formula, the wave-setup, the difference between the 2D and 3D models are verified by the constant slope case changing wind speeds. The numerical simulations for the storm surge hindcasts are performed for typhoon T7010. The typhoon T7010 induced more than 2.35 m high surge at Kochi, Japan in August 21th, 1970. Kochi faces the Pacific Ocean, directly, and the wind induced surge is relatively small than semi-closed coastal area.

3. Results and Discussion

Figure 1 shows comparison of effects of each component between 2D model, 2D model+wave, 3D model and empirical model for the constant slope test. The panel (b) indicates the elevation of 2D, 2D+SWAN, 3D, and the panel (c) indicates the wave height of 2D+wave case, respectively. The only 2D+wave simulation shows different behavior due to wave-induced setup in the surf zone, although the 3D simulation gives slightly higher elevation than the 2D simulation. Therefore, the wave induced setup is third major components for storm surge than other components.

Figure 2 shows comparison of elevations between observed data and numerical results for T7010 hindcasts. Since observation points were located near the coast, all of 2D+wave simulation gives higher elevation than the 2D simulation. The 2D+wave simulation shows better agreement than the 2D simulations at the most of points. Figure 3 shows spatial distributions of the elevation of 2D+wave simulations and wave induced elevation that is obtained by the difference between the 2D and the 2D+wave models. The wave setup increases the elevation near the coast that is more than 10% of total elevation. Therefore, the wave setup is important for the storm surge. The 3D model give different velocity distributions and bottom stresses but the influence of 3D model on the elevation is minor.

[1]Disaster Prevention Research Institute, Kyoto University, Gokasho, Uji, Kyoto 611-0011, Japan
[2]Dept of Civil Eng., Tottori University, Tottori, Japan.

The detail of results and discussion will be presented at the conference.

References

Jarosz, E. et al. (2007) Science, vol.315, pp.1707-1709.

Weisberg, R.H. and L. Zheng (2008) JGR, doi.10.1029/2008JC005115.

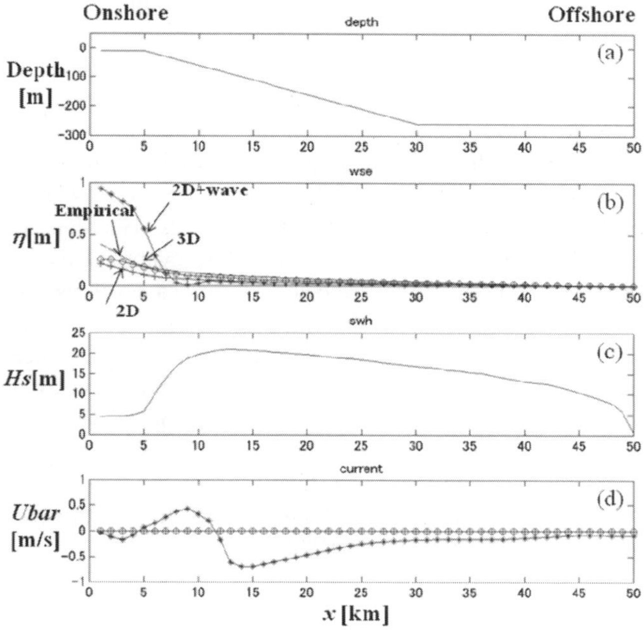

Figure 1. Comparison of numerical results between 2D model, 2D model+wave, 3D model and empirical model for constant slope test Of U_{10}=30m/s ((a): bottom topography, (b): elevation, (c): wave height, (d): depth integrated velocity)

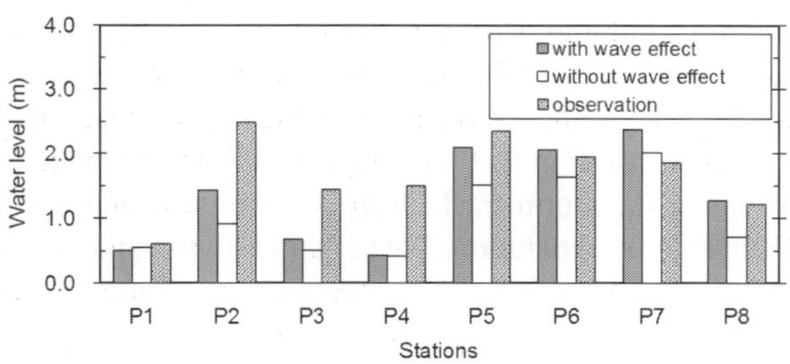

Figure 2. Comparison between observed sea surface elevations and numerical results at Kochi, Japan (P1-P8: observation station number; bars: 2D model, 2D-model+wave, observation from left to right)

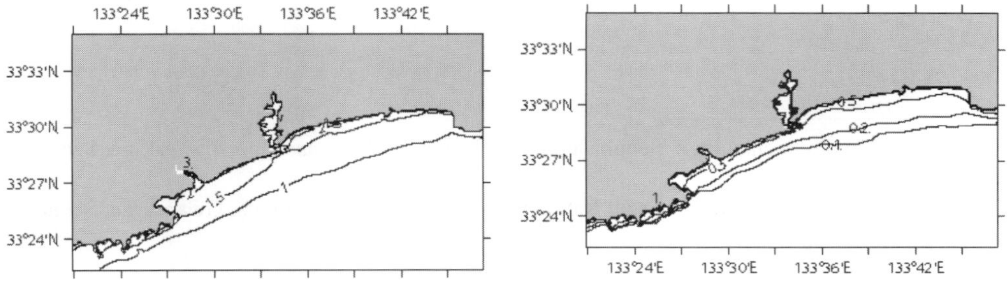

(a) Total elevation (b) Wave setup induced elevation

Figure 3. Planer views of surge by 2D+wave simulation relative to mean sea level at Kochi, Japan

Coastal Dynamics 2009
Paper No. 23

EFFECTS OF OCEAN ENVIRONMENTS UPON STORM SURGE AROUND KOREAN PENINSULA

Young Hyun Park[1] and Kyung-Duck Suh[2]

1. Introduction

Most of storm surge researches have been focused on the verification of developed models using observed data. In this study, we use the coupled numerical model which is composed of ADCIRC (Luettich et al., 1992; Westerink et al., 1992) and PBL hurricane model (Thompson and Cardone, 1996). The model has been successfully used in several researches (Heerden et al., 2007; Scheffner and Carson, 2001). We use it to understand the probable future events. This study is carried out to examine the effects of track of typhoons, tides, and long-term sea level rise upon the storm surge along the Korean peninsula. The details are described in section 2 of this paper. The Korean Peninsular is surrounded by unique ocean environments with quite different water depths and tides, so it is an interesting place for research in the world. Typhoon is the name of the tropical cyclone especially causing damage to the region of Southeast Asia. The Korean Peninsular experiences three typhoons per year on average, and the responses are different due to the environmental differences as mentioned above. The characteristics and track of virtually generated typhoon are based on typhoon Maemi which made landfall on September 12, 2003 in Korea. It was the record lowest pressure typhoon in Korea ever since measurement began in 1904.

2. Analysis of storm surge

2.1. Track of typhoon

The Yellow Sea is located between Korea and China and it is composed of a semi-enclosed continental shelf. Its average and the deepest water depth are 44 m and 103 m, respectively and spring tidal range is more than 7 m in Anhung, Korea. However, the East Sea is completely different and its average, the deepest water depth and tidal range are 1500 m, 4040 m and 0.3 m, respectively. The Southern Sea has about 2200 islands and its average and the deepest water depth are 100 m and 227 m, respectively. The coastline is so winding that the length is 8.8 times longer than the straight distance. Differences are apparent along the storm track in these three seas and they become extreme as the path is on the Korean Peninsula by separation. We set up eastward and westward moving four tracks and eight tracks shifted from typhoon Maemi, respectively in figure 1 and each distance between tracks is 30 minutes of latitude (about 45 km). The simulated storm surges are measured at four tidal stations located in east, southeast, southwest and west of Korea and the stations are Hupo (36°40′28″ N, 129°27′20″ E), Pusan (35°05′35″ N, 129°02′15″ E), Chujado (33°57′31″ N, 126°18′ 8″ E), Anhung (36°40′12″ N, 126°08′06″ E) in figure 1.

[1]Dept of Civil and Environmental Eng., Seoul National University, 599 Gwanangno, Gwanak-gu, Seoul 151-744, Korea. tomypark@snu.ac.kr

[2]Dept of Civil and Environmental Eng., Seoul National University, 599 Gwanangno, Gwanak-gu, Seoul 151-744, Korea. kdsuh@snu.ac.kr

2.2. Effects of tides

Dean and Dalrymple (2004) mentioned that Gulf of Mexico which is a shallow continental shelf showed higher storm surge comparing with Atlantic and Pacific shorelines and storm surge is related to water depth. Many storm surge researches were conducted without coupling with tidal simulations and then tides are simply added to the outputs. However, it may cause big differences especially in the Yellow Sea where its spring tide (9 m) reaches 20 % of the average water depth (44 m). The study is carried on spring tide and it is compared with the results simply added tide.

2.3. Sea level rise

Sea level rise by climate change is a big issue in coastal problem. Because the characteristics of future typhoons will be changed by sea level rise, the simulations are conducted at five sea level rises such as 1.6, 3.2, 6.4, 16.0 32.0 cm above current water level. When we consider that the sea level rise is 0.32 cm/year for the last 30 years around Korea, the results forecast changes in storm surge next 5 to 100 years.

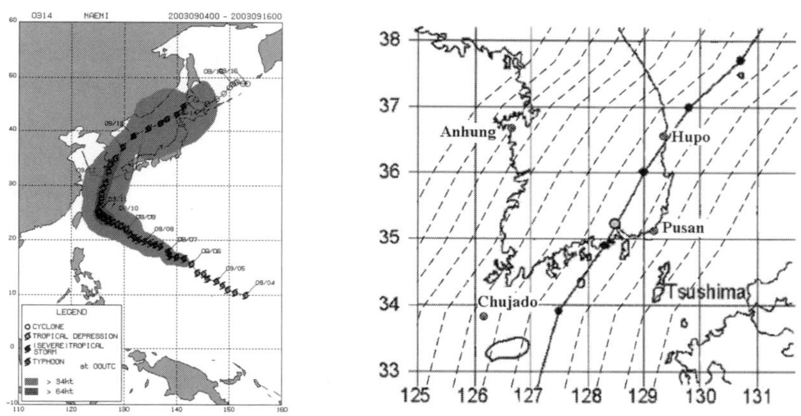

Figure 1. The track of typhoon Maemi (left), twelve virtual typhoons and locations of observations (right).

References

Dean, R.G. and Dalrymple, R.A., 2004. *Coastal processes: with engineering applications*, Cambridge University Press.

Heerden, I.L., Kemp, G.P. and Mashriqui, H., 2007. Use of the ADCIRC Storm Surge Model for Hurricane Katrina Surge Predictions and Levee Forensic Studies, *Proceedings of Sessions of Geo-Denver 2007*, Colorado, USA, 18–21.

Lee, J.C., Kwon, J.I., Park, K.S., and Jun, K.C., 2008. Calculations of storm surges, typhoon Maemi, *Journal of Korean Society of Coastal and Ocean Engineers,* 20 (1): 93-100.

Luettich, R.A., Westerink, J.J. and Scheffner, N.W., 1992. *ADCIRC: An Advanced Three-Dimensional Circulation Model For Shelves, Coasts, and Estuaries, Report 2, User's Manual for ADCIRC-2DDI. Technical Report DRP-92-6*, US Army Corps of Engineers.

Scheffner, N.W. and Carson, F.C., 2001. *Coast of south Carolina storm surge study. ERDC/CHL TR-01-11*, US Army Corps of Engineers.

Thompson, E.F. and Cardone, V.J., 1996. Practical modeling of hurricane surface wind fields, *Journal of Waterway, Port, Coastal and Ocean Engineering*, 122 (4): 195-205.

Westerink, J.J., Blain, C.A., Luettich, R.A. and Scheffner, N.W., 1992. *ADCIRC: An Advanced Three-Dimensional Circulation Model For Shelves, Coasts, and Estuaries, Report I, Theory and methodology of ADCIRC-2DDI and ADCIRC-3DL. Technical Report DRP-92-6*, US Army Corps of Engineers.

Coastal Dynamics 2009
Paper No. 24

ALONGSHORE VARIATION IN NONLINEARITIES IN SWASH HYDRODYNAMICS OVER BEACH CUSPS

K. R. Bryan[1] and Giovanni Coco[2]

1. Introduction

Cuspate topography provides a variation in the beachface slope, which can be used as a natural laboratory to study the influence of slope and downwash convergence on each swash cycle. The steeper-sloped cusp horns are likely to be characterized by smaller, faster swash cycles than the bays. The longer swash cycles in bays are more likely to collide with subsequent uprushes, and be influenced by increased downwash mass-flux caused by channeling. Here we use techniques for detecting non-linearities to determine the main controls on longshore swash variability.

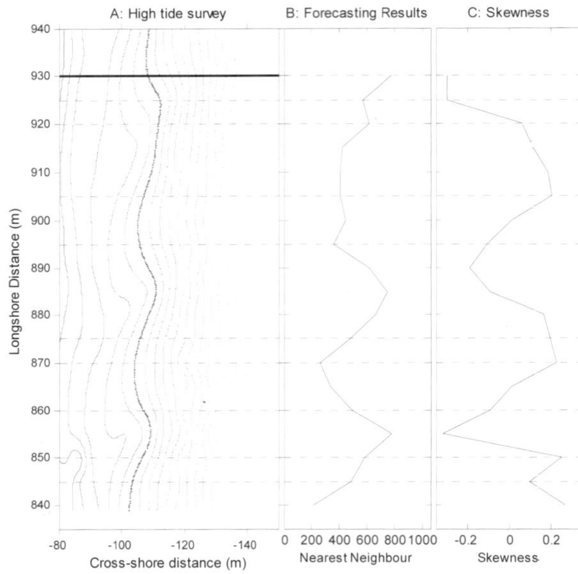

Figure 1. A: Surveyed intertidal beach morphology. The contours are at 0.25m separation. The contour with crosses marks the mean waterline +2 standard deviations. The dashed lines mark the video time stack locations. B: The optimum number of nearest neighbours, and, C, the skewness of the swash timeseries.

2. Methods

The swash timeseries were collected using video immediately prior to the DUCK94 field experiments in North Carolina and have been published previously (Ciriano *et al.*, 2005). Here we choose a timeseries which was collected on September 6th, at high tide, which was a day when the topography was strongly cusped (Figure 1A). The movement of the waterline along transects at 5m spacing (location shown as dashed lines in Figure 1A) was extracted from video footage, and corrected using beach face surveys to make 19 swash timeseries (sampling frequency: 2 Hz, duration: ~ 2hrs).

The two methods for measuring the degree of non-linearity in the swash timeseries were used. Firstly, weak non-linearities were measured using the skewness and asymmetry of the timeseries (e.g.

[1] Dept. of Earth and Ocean Sciences, Univ.of Waikato, Private Bag 3105, Hamilton, NZ. k.bryan@waikato.ac.nz
[2] National Inst. of Water and Atmospheric Res., P.O. Box 11-115, Hamilton, NZ. g.coco@niwa.co.nz

Elgar and Guza, 1985). The other technique is non-linear forecasting (e.g. Bryan and Coco, 2007). This method uses a simple model, such as an autoregressive model, to assess the difference between forecasts made using global versus local applications of the model. The model we use is:

$$x(t + d\Delta t) = a_o + \sum_{j=0}^{j=m-1} a_{j+1} x(t - j\Delta t) \tag{1}$$

where Δt is the time lag, a_{j+1} are coefficients and m is the length of the sequences used to make the forecast. In the global model, the whole timeseries is used to evaluate the coefficients a_{j+1}, and in the local model only segments which display similar local behaviour or are nearest neighbours are used. A linear system will behave globally (all swash cycles evolve similarly), where as a non-linear system is better forecast with a local model.

3. Results and Conclusions

The skewness showed an alongshore pattern, ranging from negative at the horns and positive at the bays (Figure 1C, Figure 2, top panel); the asymmetry did not have a distinct alongshore pattern. A negatively skewed result means that there are more points from high on the beach than low in the timeseries. The somewhat parabolic shape of the swash cycles gives a naturally negatively skewed signature. In the bay, interactions between the larger mass flux in the downrush and the following uprush can suppress the following uprush, leading to more points lower on the beach face than higher, or a positive skewness. The forecasting results also show horn/bay differences. Timeseries in the bay are better forecast using smaller numbers of nearest neighbours than timeseries in the horns (Figures 1B and Figure 2, Bottom panel). This means that there are local sequences of swash cycles within the timeseries that behave differently than the rest of the timeseries. This local behaviour most likely is another signature of the impact that channeling and swash interactions have on the nature of the timeseries. Ultimately, a 2-dimensional numerical model could be used to control for cusp geometry and isolate the effects of such processes on the timeseries.

Acknowledgements

Field data collection was led by B.T. Werner and S. Elgar and supported by the Office of Naval Research (USA). GC was supported by FRST (contract WRHC092).

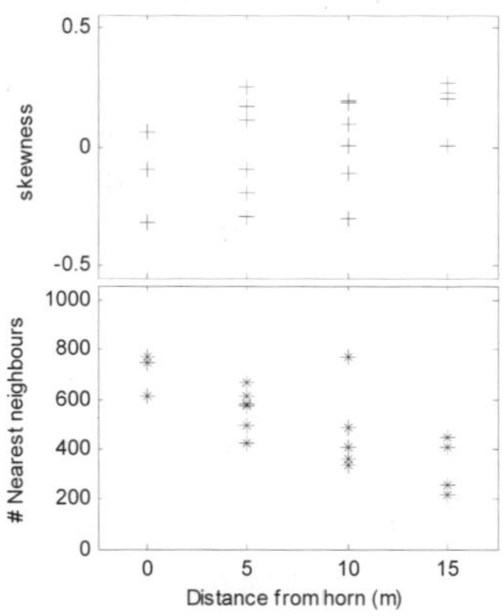

Figure 2. Skewness (Top Panel) and number of nearest neighbours (Bottom Panel) as a function of horn distance.

References

Bryan, K. R. Coco, G., 2007. Detecting nonlinearity in run-up on a natural beach, *Nonlinear Processes in Geophysics*, 14: 385–393.

Ciriano, Y., Coco, G., Bryan, K. R., Elgar, S., 2005. Field observations of swash zone infragravity motions and beach cusp evolution, *J. Geophys. Res*, 110, C02018, doi: 10.1029/2004JC002485.

Elgar, S., Guza, R. T., 1985. Observations of bispectral of shoaling surface gravity waves. *J. Fluid. Mech,*. 161:425–448.

Coastal Dynamics 2009
Paper No. 25

BED SHEAR STRESS IN BORE-GENERATED SWASH ON STEEP BEACHES

Gustaaf Kikkert[1], Dubravka Pokrajac[1], Tom O'Donoghue[1]

1. Introduction

Understanding bed shear stress is of vital importance for predicting sediment transport rates within the swash-zone. Swash models commonly predict bed shear stress using a simple quadratic resistance law with a constant friction factor. The approach is too simplistic given that swash flow is highly unsteady and non-uniform. Direct measurements have been obtained of swash bed shear stress in the field (Conley and Griffith, 2004) and direct (Barnes and Baldock, 2007) and indirect (Cowen et al. 2003, Hondebrink, 2007) measurements have been obtained in the laboratory. The combined dataset is still rather sparse however, and generally lacks detailed hydrodynamic measurements.

This paper reports new bed shear stress estimates obtained from detailed measurements of the hydrodynamics of bore-generated swash on steep beaches with different roughness. A number of approaches to determine the bed shear stress are explored in order to assess their practical viability and the shear stress results are used to investigate the relationship between the bed shear stress in the swash zone and the main flow parameters.

2. Experiments

The experiments were carried out using a swash rig built into a 20m long, 0.9m high and 0.45m wide, glass-sided flume, with a reservoir at one end (Kikkert et al. 2008). The reservoir is fronted by a gate which can be raised at high speed to produce a large plunging wave leading to a bore which propagates towards a beach. The experiment produces a single, well-controlled bore with similar height and velocity to full-scale wave-generated bores on steep beaches. For the present experiments, the reservoir was filled to a depth of 0.60m, which generates a bore approximately 0.25m high with velocity approximately 2.0m/s, producing a maximum runup of about 5m on a 1:10 rough, impermeable beach. Repeatability of the experiments is excellent, which means that measurements from many repeats of the same event can be used to obtain turbulence measurements by ensemble-averaging. Three impermeable beaches with different roughness have been tested using three sediment sizes: 1.5mm sand, 6mm gravel and 10mm gravel. Simultaneous velocity and depth measurements were recorded at six cross-shore locations using a combined Particle Image Velocimetry (PIV) and Laser-Induced Fluorescence (LIF) system for velocity and depth respectively. Individual experiments were repeated 50 times to determine ensemble-averaged velocities, depths and turbulent quantities.

3. Methods for Evaluating Bed Shear Stress

The bed shear was evaluated using the following four methods: *(i) By fitting the Log-law* through the lower end of the ensemble-averaged velocity profiles (see example in Fig. 1). *(ii) By extrapolating the Reynolds stress profile* to the bed. Throughout most of the swash event the boundary layer was not fully developed, so only the lower part of the Reynolds stress profile was used to obtain the bed shear stress (see example in Fig. 2). *(iii) From the turbulent kinetic energy dissipation rate.* Horizontal spatial velocity data were used to determine the second order structure function, D_{uu}, and estimate the dissipation rate ε using $D_{uu}(r)=c_u\varepsilon^{2/3}r^{2/3}$ where r is the distance between measurements and c_u the Kolmogorov constant with an experimentally determined value of 1.92 for steady, uniform flow. The dissipation rate was related to the shear velocity, u^*, using $\varepsilon=\kappa z(u^*)^3$, where κ is 0.4 and z the distance above the bed. This relationship is valid in the constant stress layer. The shear velocity was therefore

[1]School of Engineering, College of Physical Sciences, University of Aberdeen, Fraser Noble Building, King's College, Aberdeen, AB24 3UE, United Kingdom. g.a.kikkert@abdn.ac.uk

calculated as close as possible to the bed. *(iv) From the momentum balance*, by evaluating the individual terms within the Depth-Integrated Reynolds Averaged Navier Stokes equations. Stansby and Feng (2005) performed a similar momentum balance analysis for the surf zone.

The first three methods used for evaluating the bed shear stress are based on knowledge and principles that apply to steady uniform flow, while the momentum balance method only assumes two-dimensional flow.

4. Results

The paper will present a comparison of bed shear stress estimates obtained by all 4 methods applied to the detailed hydrodynamic measurements. The theoretical and practical issues associated with the methods, and the implications for the use of the methods for indirect bed shear stress measurement will be discussed.

Results will be presented showing the dependence of swash bed shear stress on cross-shore position, flow history, relative roughness and Reynolds number. These results can be used to guide the closure equations employed for bed shear stress in swash numerical models.

Acknowledgements

This work is part of a collaborative research project between the Universities of Aberdeen and Nottingham, funded by the UK's EPSRC (EP/E011330/1 and EP/E010407/1).

References

Barnes, M. and Baldock, T., 2007. Direct bed shear stress measurements in bore-driven swash and swash interactions. *Proceedings of Coastal Sediments 2007*, New Orleans, ASCE, pp. 1947-1960.

Conley, D. and Griffen, J., 2004. Direct measurements of bed stress under swash in the field. *Journal of Geophysical Research*, 109(C3), Art No C03050.

Cowen, E., Sou, I., Liu, P. and Raubenheimer, B., 2003. PIV measurements within a laboratory generated swash zone. *Journal of Engineering Mechanics*, 129: 1119-1129

Kikkert, G., O'Donoghue, T. and Pokrajac, D., 2008. Effects of roughness and permeability on swash hydrodynamics. *Proceedings of 31st International Conference on Coastal Engineering*, ASCE. In press

Hondebrink, L., 2007. Numerical and experimental study of bore-driven swash on impermeable slopes. *PhD Thesis*, University of Aberdeen, Aberdeen, United Kingdom.

Stansby, P. and Feng, T., 2005. Kinematics and depth-integrated terms in surf-zone waves from laboratory measurements. *Journal of Fluid Mechanics*, 529: 279-310.

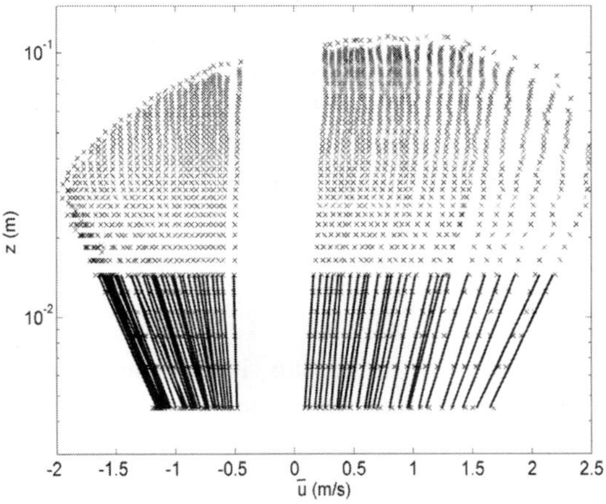

Figure 1. Example log-law fit to velocity profiles

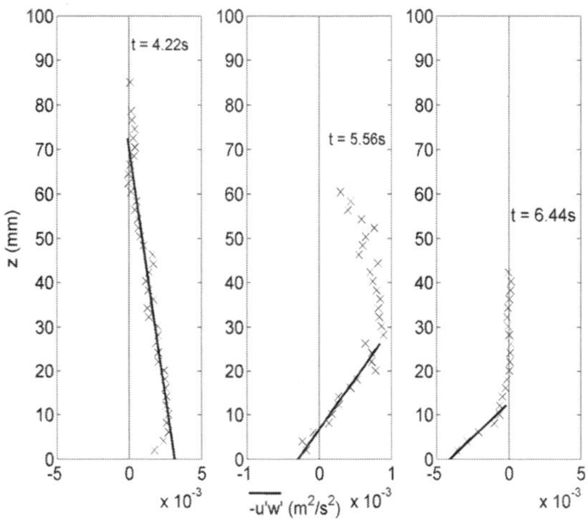

Figure 2. Example Reynolds stress profiles

Coastal Dynamics 2009
Paper No. 26

WAVES IN AND JUST OUTSIDE THR SWASH ZONE DURING A STORM

Masaru Mizuguchi[1]

1. Introduction

In laboratory experiments it is well-known that waves near the shoreline (or in the swash zone) behaves like a breaking bore with nearly-complete energy dissipation on one hand or like a standing wave loop with little dissipation on the other extreme. In the field the waves in the swash zone show the bore-like behaviour in most cases when the sea condition is calm and the measurement is possible. Here we report a trial to measure the waves near the shoreline during a storm. We also show some interesting features of the wave motion in and just outside the swash zone for the field storm waves.

2. Field observation

Field observation was made with twenty ultrasonic wave gages mounted on the Hasaki Pier of PARI. The gages covered around 105 m in the cross-shore direction to include the whole swash zone, as shown in Fig.1, where the bottom profiles during a storm period (Sept. 19 to 22, 2003), which will be analyzed later, are also plotted. Output data was sampled with frequency 5 Hz continuously from July 2002 to Feb. 2005.

The storm was brought by Typhoon 0315(15th in 2003) with maximum significant wave height of 6 m offshore (20m deep). We use about one hour data from 8:30 am to 9:30 am, Sept. 22 just before severe berm erosion started.

3. Data and its analysis

In Fig. 2 examples of the obtained data outside the swash zone are shown. Near the shoreline, at chs. 12 and14, broken short waves are riding on long waves and large short waves do correspond to peaks of long waves. In Fig. 3 obtained data in the swash zone are plotted in large scale together with swash front motion estimated from the measured fixed point data. Short waves contribute very little to the swash motion.

First cross spectra analysis is applied for wave data outside the swash zone as shown in Fig.4. Clearly long period waves are dominant in power with peak frequency 0.011Hz (90s) as seen in the bottom diagram and it is also clear that the long waves are of standing mode as the phase relation shows 0- π shift up to 0.022 Hz. Pure progressive waves with straight line of phase function are only observed in the frequency higher than 0.055 Hz, where very little power exists.

Next a typical individual swash motion is compared with Carrier-Greenspan (CG) solution and the ballistic motion models as shown in Fig. 5. For CG solution, wave amplitude and period are specified from the measured data. For the ballistic model, one with given initial velocity show much underestimated swash motion, indicating that acceleration of wave motion cannot be neglected. On the other hand one with given period overestimates the motion considerably, indicating non-breaking or no saturation of swash oscillation.

4. Conclusion

A clear demonstration is made that long waves, whose period is of order 100s, is dominant near the shoreline during a storm. It is also shown that the long waves are of standing mode in the cross-shore direction and the individual wave motion may be described by Carrier-Greenspan solution.

---

[1]Dept of Civil Eng., Chuo University, 1-23-27 Kasuga, Bunkyo-ku, Tokyo, Japan. nami@civil.chuo-u.ac.jp

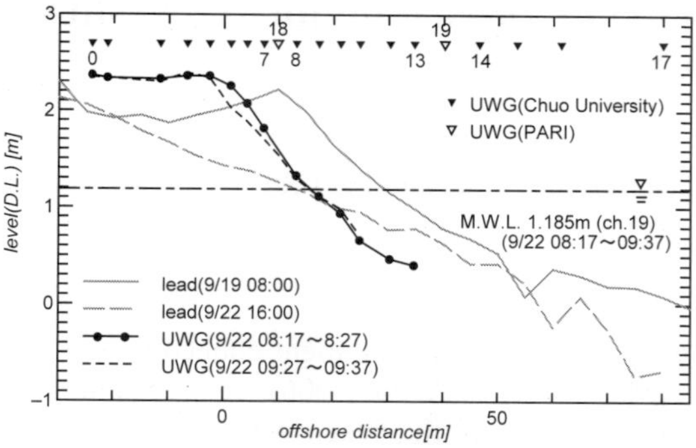

Fig.1

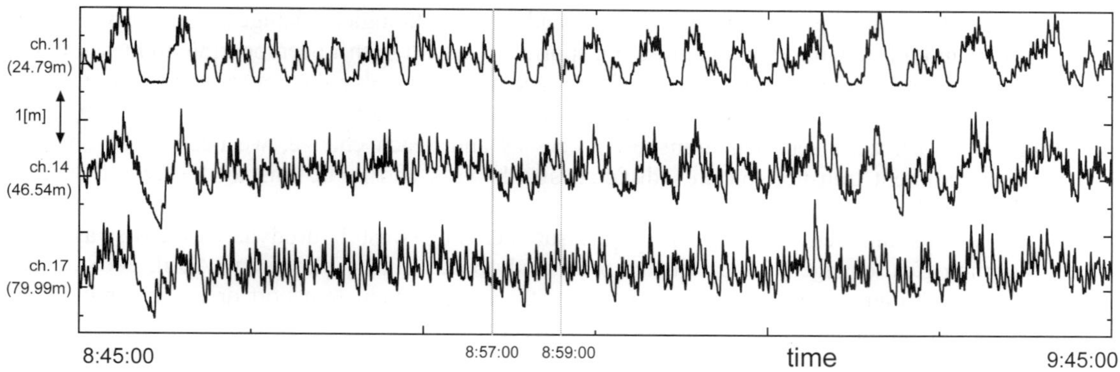

Fig.2

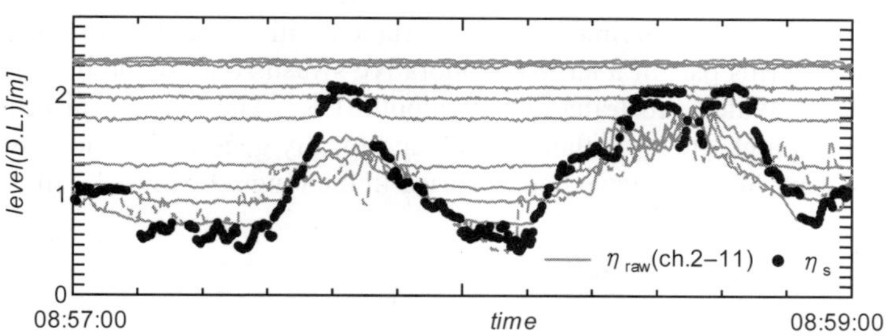

Fig.3

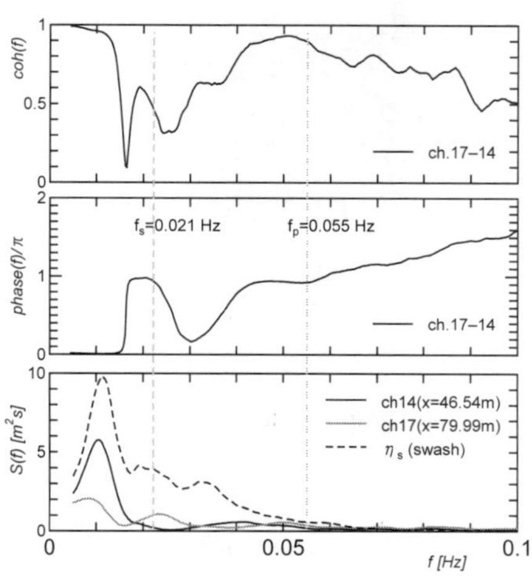

Fig.4

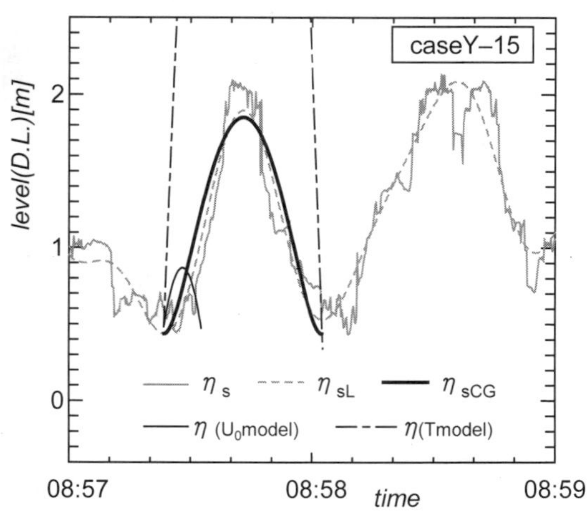

Fig.5

Coastal Dynamics 2009
Paper No. 27

FIELD MEASUREMENTS OF NET SEDIMENT FLUX FROM INDIVIDUAL SWASHES ON A SANDY BEACH

Chris E. Blenkinsopp[1], Ian L. Turner[1], Gerd Masselink[2] and Paul R. Russell[3]

1. Introduction

The swash zone is characterised by strong flows, high suspended sediment concentrations, large sediment fluxes and rapid morphological change (Masselink & Puleo, 2006). Large amounts of sediment (10's of kg/m beach width) are transported up and down the beach with each swash cycle, however the net transport per swash cycle which leads to morphological change is generally at least an order of magnitude smaller than the gross transport. To date, it hasn't been possible to measure the uprush and backwash transport rates with sufficient accuracy to resolve the net transport rate, mainly due to a lack of suitable field instrumentation. For progress to be made in our understanding and modelling of swash zone sediment transport processes and beach change, we need to be able to quantify net sediment transport on an individual swash basis and relate this to the hydrodynamic conditions.

To address this knowledge gap, extensive measurements of swash hydrodynamics and swash-by-swash morphological change were conducted on a sandy beach during March and April, 2008. This paper will present comprehensive measurements of the net sediment transport driven by individual swash events throughout the swash zone of a sandy beach.

2. Methodology

Field data were obtained at Le Truc Vert beach on the French Atlantic coast over 27 tidal cycles between 19th March and 4th April, 2008. The experiment was undertaken within the structure of the multi-institutional ECORS project and forms part of a large collaborative swash zone research project between the Universities of New South Wales, Australia and Plymouth, UK. The macrotidal beach (D_{50}=0.35mm) experiences a mixture of high-energy Atlantic swell and locally generated wind-waves and during the experiment, the significant offshore wave height varied in the range 0.9 m to 4.1 m.

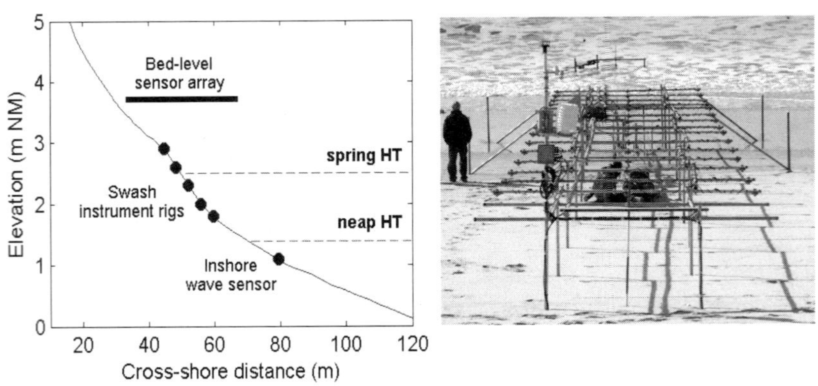

Figure 1. Beach profile at the start of the field experiment and photograph of the instrument rig.

---

[1] Water Research Laboratory, School of Civil and Environmental Engineering, University of New South Wales, Sydney, NSW 2052, Australia. c.blenkinsopp@unsw.edu.au, ian.turner@unsw.edu.au.

[2] School of Geography, University of Plymouth, Plymouth, PL4 8AA, UK. gerd.masselink@plymouth.ac.uk.

[3] School of Earth, Ocean and Environmental Sciences, University of Plymouth, paul.russell@plymouth.ac.uk.

High frequency bed-level data were recorded using an array of 45 ultrasonic bed-level sensors sampled at a rate of 4 Hz. The sensors were mounted on a large scaffolding rig approximately 1 m above the bed in three cross-shore lines with a spacing of 1.9 m (Figure 1). The bed-level sensors are described by Turner *et al.* (2008a) and make non-intrusive Eulerian measurements of the vertical distance to the closest target (bed-level when the bed is dry and swash height when the bed is wet) based on the time of flight of a reflected signal. Additional measurements of offshore and nearshore wave data, swash hydrodynamics and beach topography were made to support the bed-level data.

3. Results and Concluding Remarks

The array of ultrasonic sensors allows the bed-level change at multiple locations throughout the high tide swash zone to be monitored at high frequency. Figure 2(a) shows a time-series from a single sensor mounted approximately at the mid-swash position (x=57.6 m) during high tide. The values on the y-axis are the elevation of the upper surface (bed or swash) at the sensor location, the vertical spikes occur as a swash passes beneath the sensor, while between these events the elevation of the bed can be determined allowing the bed-level change due to individual swash events to be calculated. Turner *et al.* (2008b) noted that these swash-by-swash bed-level changes can be very large with magnitudes exceeding ±40mm, and as such there is the potential for rapid morphological change at the beachface. The large array of measurement points recorded at Le Truc Vert allows us to track individual swashes across the beachface and by integrating bed-level changes over a series of sensors we can compute net sediment flux throughout the swash zone per swash event. As an example, a frequency distribution of sediment flux past a single point (x=57.6 m) on the beachface throughout an entire high tide is presented in Figure 2(b). This figure shows that the sediment fluxes are remarkably normally distributed with a mean close to zero, ensuring that the beach remains stable. However very large individual fluxes of up to 200 kg/m were measured and these are of the same order of magnitude as the net flux over the whole high tide which was just +282 kg/m. These significant transport fluxes highlight the extremely dynamic state of equilibrum that persists across the beachface, where the net flux over time represents the fine balance between the positive and negative sediment fluxes caused by individual swashes. The relationship to the forcing swash hydrodynamics is under investigation.

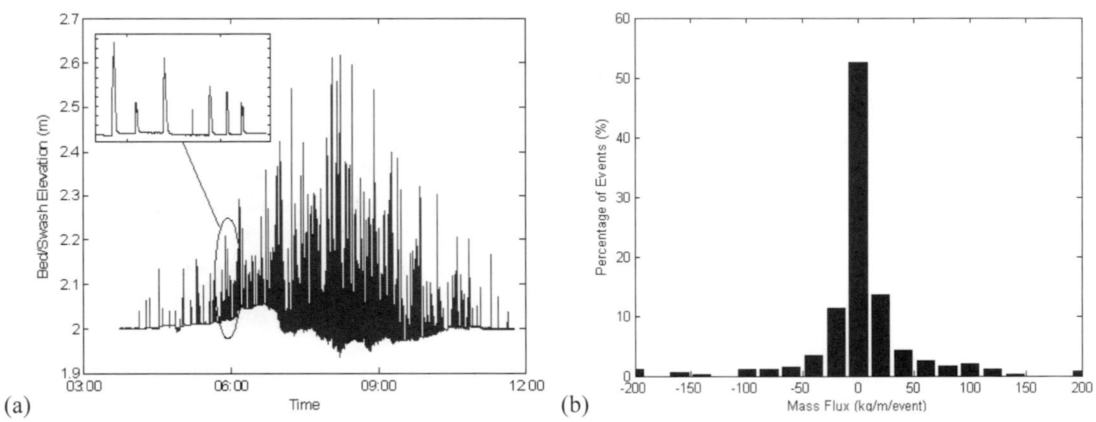

Figure 2. (a) Time-series of bed/swash elevation data, (b) frequency distribution of sediment fluxes at x=57.6 m.

References

Masselink, G. & Puleo, J., 2006. *Swash zone morphodynamics*. Continental Shelf Research, 26, 661–680.

Turner, I.L, Russell, P.E. & Butt, T., 2008a. *Measurement of wave-by-wave bed-levels in the swash zone.* Coastal Engineering, 55(12), 1237-1242

Turner, I.L., Masselink, G., Russell, P.E. & Blenkinsopp, C.E., 2008b. *Bed-level change over individual swash cycles on sand and gravel beaches*. Proceedings 31st International Conference on Coastal Engineering, P22.

Coastal Dynamics 2009
Paper No. 28

DYNAMIC CHANGES OF WAVES AND CURRENTS OVER THE COLLAPSING SANDBAR OF THE TENRYU RIVER MOUTH OBSERVED DURING TYPHOON T0704

Yoshimitsu Tajima[1], Haijiang Liu[1] and Shinji Sato[1]

1. Introduction

Process-based understandings on dynamic interactions among waves, currents and topography changes around the river mouth are essential to predict the movement of sediments discharged from the river. It is however difficult to directly observe such features by installing instruments since severe forces induced by flooding current and broken waves dramatically change the bed conditions.

Applying six field cameras, authors succeeded to capture the collapsing sandbar at the Tenryu River mouth during typhoon T0704. Using obtained images and other data, this study aims to explore the interactive characteristics of waves, currents and topography changes around the river mouth.

2. Field Observations and Analysis

As seen in Figure-1, six cameras were installed around the Tenryu River mouth and recorded snap-shot images at every 1.2 seconds. Modified rectification technique of Holland et al. (1997) was applied to transfer the obtained images to the top-view image and combined with the other images to yield the entire image of the sand bar of the river mouth (Figure-2). Drifting debris were traced in successive snap shot images during the storm and obtained tracks were transferred to the actual coordinates to yield surface water velocities. Figure-3 compares obtained velocity data with the other hydrodynamic data such as flow rate at Kashima, 25km upstream from the river mouth, surface water level at Kaketsuka(η_K), 3km upstream from the mouth, Ryuyo(η_R), just behind the sandbar of the river mouth and offshore(η_O) in front of the river mouth with a water depth of 40m.

3. Numerical Analysis

Numerical model was applied to represent the wave-current interaction fields during the storm. The entire model consists of: (i) a wave model based on the spectral wave action balance equations with broken wave and surface roller models(Tajima and Madsen, 2005); and (ii) a mean current model based on non-linear shallow water equations with forcing terms due to broken waves and rollers.

Referring to the time-series of the measured data (Figure-3), three computation cases A, B and C, were selected so that these cases cover characteristic conditions around the peak of flooding river flow. Case A is at 5:00 on July 15th when the surface water level at Ryuyo was at its peak, case B is at 13:00 when the river flow rate was at its peak; and case C is at 17:00 when offshore surface water level and river flow rate are nearly the same as case A. We also carried out sensitivity analysis of predicted mean current field with respect to different topographies, bottom roughness, and wave radiation forces.

4. Primary Results and Discussions

As seen in Figure-4, the numerical model reasonably represented the wave blocking at the river mouth. The model predicted south-eastward current patters around the river mouth induced by forces due to broken waves and surface rollers. This current pattern is consistent with the observed features of measured topography change in which most of the sand were transported to the eastward and clear sediment deposition was observed at the southeast area of the river mouth.

---

[1]Dept of Civil Eng., The University of Tokyo, Hongo 7-3-1, Bunkyo-ku, Tokyo 113-8656, Japan.
yoshitaji@coastal.t.u-tokyo.ac.jp

The gap of the surface water levels between η_R and η_O was larger in case A than the one in case C while the river flow rates were nearly the same in these two cases. The present numerical model reasonably explained this feature by applying different topography conditions in cases A and C. Referring to Figure 2, case A applied the "partially collapsed" sandbar topography obtained by interpolating topographies before and after the storm whereas case B and C simply applied the topography after the storm. Final paper discusses more details about the sensitivity analysis which clearly shows that the impact of topography change on the predicted mean current fields was predominant to the other factors such as bottom roughness and or wave-induced radiation stresses.

References

Holland, K.T., Holman, R.A., Lippmann, T.C., Stanley, J., 1997. Practical use of video imagery in nearshore oseanographic field dtudies, IEEE J. of Oceanic Eng., 22(1), pp. 81-92.

Tajima, Y. and Madsen O.S. 1996. Modeling near-shore waves, surface rollers, and undertow velocity profiles, J. of Waterway, Port, Coast. and Ocean Eng., 132(6), pp. 429-438.

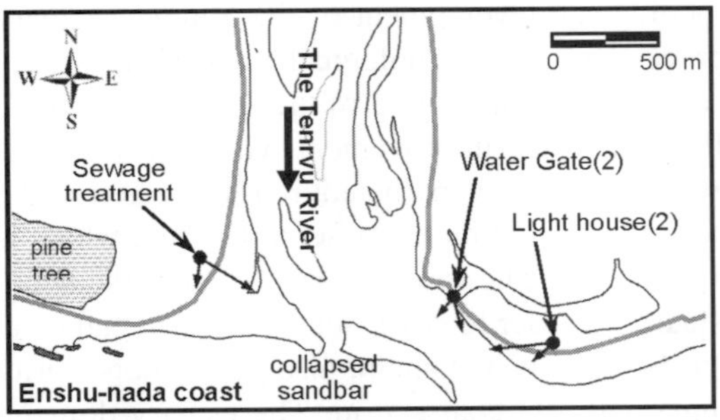

Figure 1 Overview of the Tenryu River mouth and angles of installed six cameras (arrows).

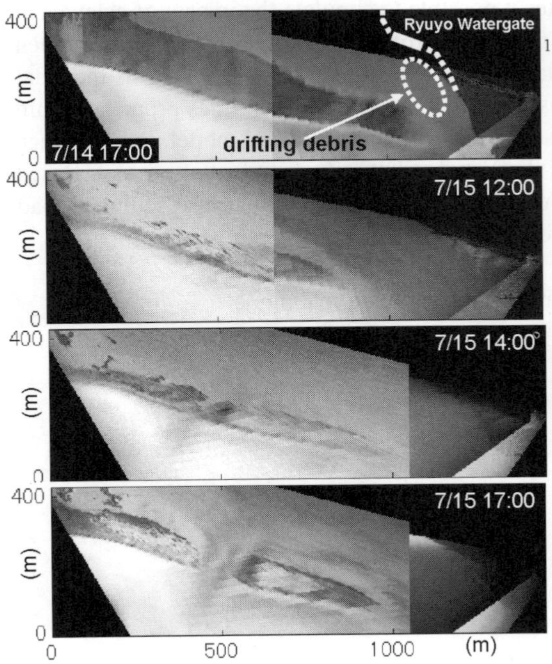

Figure 2 Rectified and combined top-view image of the Tenryu River mouth (before the storm).

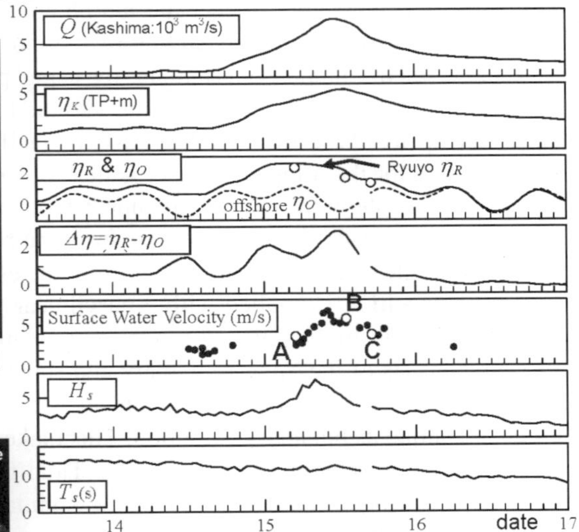

Figure 3 Measured flow rate at Kashima, surface water levels at Kaketsuka(η_K), Ryuyo(η_R) and offshore (η_O), surface water velocities at Ryuyo, and measured wave conditions at offshore. Blank circles are results of numerical computations.

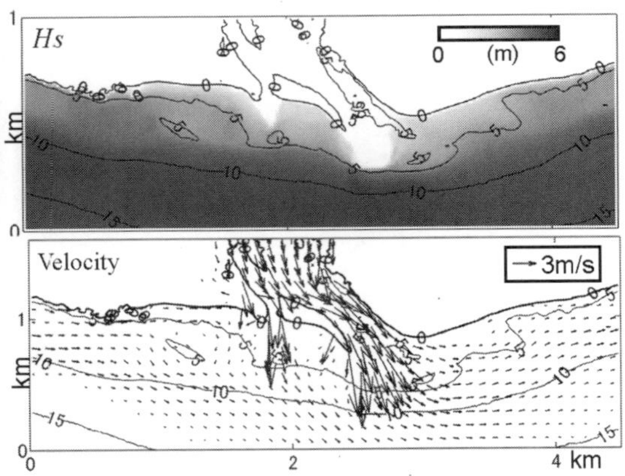

Figure 4 Predicted significant wave heights and depth-averaged mean current velocities (Case B).

Coastal Dynamics 2009
Paper No. 29

PROCESS-BASED MODELING FOR THE YANGTZE ESTUARY

Ao Chu[1,2], Z.B. Wang [1,3], H.J. de Vriend[1,3]

Abstract: The development of the Yangtze Estuary is impacted by the human interference in the estuary as well as in the river basin. River discharge to the Yangtze Estuary has been changed due to human interference in the Yangtze River Basin. Especially since the construction of the Three Gorges Dam, not only the total annual runoff is decreased, but also the seasonal variation of the discharge is changed. The changed flow regime has caused a significant decrease of sediment input from the Yangtze River to the estuary. These changes together with the large scale engineering works in the estuary itself will inevitably impact the morphological development of the estuary. The aim of our research is to improve our understanding of the morphological development of the Yangtze Estuary and the effect of the human interference by using process-based morphological model. This paper describes the first step of the model set-up, viz. the hydrodynamic modeling of the Estuary. A 2D/3D model is set up for the estuary on the basis of Delft3D. A major improvement with respect to the previous models for the Yangtze Estuary is that the present model covers the entire tidal region of the Yangtze River. As show in figure 1, the upper river boundary of the model is set at Datong, which is tidal limit of the estuary about 640 km far from the river mouth. The sea boundary is set at the depth about 50m with the distance about 160 km from the estuary. In order to cover such a large spatial area, domain decomposition is applied with 2 curvilinear grids. The river domain, also called upper grid, covers from Datong to Jiangyin (the tidal current limit) with 2D coarse grid. The estuary domain, lower grid, consists of the Yangtze Estuary, the Hangzhou Bay and a part of adjacent East China Sea. High resolution is used in the estuary domain with 2D or 3D grid. The observed daily discharge is specified at Datong as boundary condition. This makes it possible to simulate the impact of changes in the upstream, e.g. discharge regime changes. Tidal forcing is provided with tidal constitutes at open sea boundaries. Model results show that the simulated tidal constituents agree well with the tidal analysis results at observation stations. Co-tidal maps show that the semi-diurnal tides propagate from the southeast into the Yangtze Estuary. With the present calibrated and validated model, the influence of discharge regime changes in the upper stream on the Yangtze Estuary is studied at present. Based on this research the morphological evolution in the Yangtze Estuary under influence of sediment load reduction in the upper stream and other human interferences will be further investigated.

Key word: the Yangtze Estuary, Morphology, Hydrodynamics, Numerical model

[1] Delft University of Technology, P.O.Box5048, 2600GA Delft, the Netherlands, a.chu@tudelft.nl
[2] Hohai University, 1 Xikang Road, Nanjing, 210098, China
[3] Deltares, P.O. Box 177, 2600 MH Delft, The Netherlands

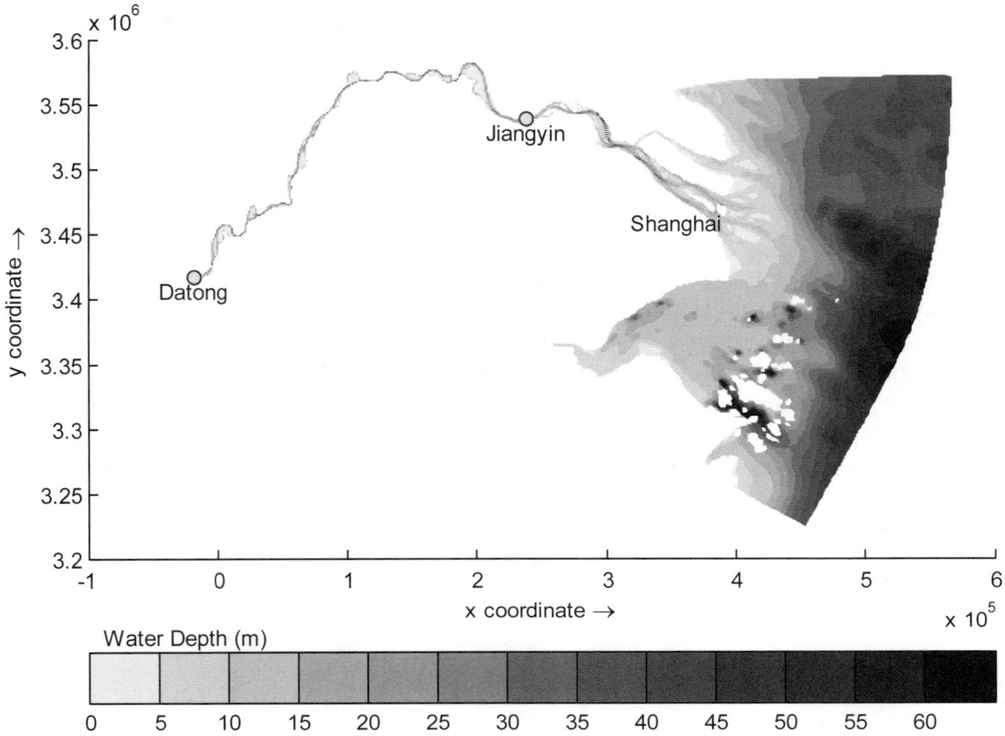

Figure 1 Sketch of domain for the Yangtze Estuary Model
River domain: Datong ~Jiangyin; Estuary domain: Jiangyin ~Sea

Conference info.
http://www.coastal.jp/cd09/index.php?Topic

Coastal Dynamics 2009
Paper No. 30

SEASONAL VARIATION OF RESIDUAL CURRENTS IN THE MEGHNA ESTUARY OF BANGLADESH

Mohammad Asad Hussain[1]

1. Introduction

During the recent period, coastal oceans and estuaries are the regions receiving a great deal of interest mainly due to the increased utilization of its resources. The Meghna Estuary is a 'coastal plain estuary' on the Bangladeshi coastline in the Bay of Bengal. These kinds of estuaries are normally difficult to describe in physical terms, because they are sensitive to the driving forces, which can be divided into bathymetry, oceanographic conditions outside the estuary (tides), hydrology of the adjacent watershed (river discharge) and meteorological conditions (e.g. wind) (Jacobsen *et. al.* 2002). Many scientific investigations have focused on storm surges in the Bay of Bengal but the erosion-accretion process in the estuary still remains as a challenge due to its complex hydrodynamic and morphological condition. The present research focused on the seasonal variation of residual flow during non-cyclonic conditions in the Meghna Estuary to enhance the present understanding of the hydrodynamic and morphological features of the region.

2. Methodology

A 3D numerical model which solves mass conservation equation and Navier-Stokes equations for the conservation of momentum was applied for the simulation of residual currents in the Meghna Estuary. The hydrodynamic model is an *f*-plane quasi-3D sigma-coordinate baroclinic circulation model including temperature and salinity (Sasaki and Isobe, 1996). In the present model, turbulence has been treated explicitly and equations for turbulent quantities such as turbulence velocity scale (or equivalently turbulence kinetic energy) and turbulence macroscale are solved following Mellor-Yamada's (1982) turbulence closure scheme (Hussain, 2006). The model was successfully applied to simulate the residual currents of Tokyo Bay (Hussain *et. al.*, 2006). Ten years of meteorological data was collected and analyzed to identify the general tendency of seasonal variation. Discharge and tide data at the Meghna Estuary was collected for a similar period to identify the seasonal variation and for model input. At first the present model was calibrated and validated using the existing data and records available from literature. Then different scenarios were generated to study and analyze the effects of wind stress, fresh water inflow, Coriolis force and tides into the estuarine system.

3. Results and Discussion

Wind data shows that its direction is mainly from NW and SW during pre-monsoon and monsoon peri ods and from SE and NE during post-monsoon and winter periods. Discharge during the monsoon peri od is as much as fifteen times greater than that during the dry season which has significant effect on sa linity (MWR, 1997).

Through numerical investigations it was observed that during the monsoon period a counter-clockwise circulation exists in the estuary with a northward flow in the Sandwip Channel and a southward flow in the Shahbazpur Channel, which is consistent with the results of Jacobsen *et. al.* (2002). Numerical experiments suggest that tidal current is the primary forcing factor behind the northward flow through the Sandwip Channel. During the monsoon period, residual current in the Northern Bay of Bengal ma y be occasionally dominated by strong South-west monsoon wind (average 4~5 m/s and maximum 20

[1]Institute of Water and Flood Management, Bangladesh University of Engineering and Technology, Dhaka 1000, Bangladesh asadh@iwfm.buet.ac.bd

~30 m/s) which increases water level in the estuary and may create back water effects in the Meghna e stuary. It was also observed that the residual circulation, to some extent, traps the river water inside th e Meghna Estuary and thus increases the residence time, which is one of the reasons for the relatively low salinity in the estuary even during the dry season. During the winter (dry) periods the counter cloc kwise circulation in the estuary exists but the circulation becomes weaker and it is not influenced by t he weak north-easterly (1.5~3 m/s) wind.

4. Conclusions

The forcing factors that govern the residual circulation in the Meghna Estuary were found to be river discharge and tidal current. During the monsoon season there is a relatively strong counter clockwise circulation in the estuary and as the river discharge through the estuary considerably reduces in the winter season, the circulation becomes weaker.

Acknowledgements

The cooperation by Water Resources Planning Organization (WARPO) and Bangladesh Meteorological Department (BMD) are gratefully acknowledged.

References

Hussain, M. A. (2006) Numerical Simulation of Residual Currents and Water Quality in Semi-Enclosed Bays, Ph.D. Thesis, Submitted to Department of Civil Engineering, The University of Tokyo, Japan.

Hussain, M.A., Isobe. M. and Koibuchi. Y. (2006), Numerical Simulation of Residual Currents in a Semi-Enclosed Bay, Proceedings of the International Conference on Coastal Engineering, ICCE-2006, San Diego, California, USA, September, 2006.

Jakobsen, F., Azam, M. H. and Kabir, M. M. (2002) Residual flow in the Meghna Estuary on the coastline of Bangladesh, Estuarine, Coastal and Shelf Science, 55, 587-597.

Mellor, G.L. and Yamada, T. (1982): Development of a turbulence closure model for geophysical fluid problems, Rev. of Geophysics and Space Physics. 20(4), pp 851-875.

MWR (1997) Meghna Estuary Study, Salinity distribution in the estuary. Ministry of Water Resources, Bangladesh Water Development Board, Technical Note MES-004. Report by DHV Consultants BV for DGIS/DANIDA and GOB, January 1997.

Sasaki, J. and Isobe, M. (1996): Development of a long-term predictive model of water quality in Tokyo Bay, Proc. Estuarine and Coastal Modeling, ASCE, pp. 564-580.

Coastal Dynamics 2009
Paper No. 31

MODELING OF LONG-TERM TIME SERIES OF WAVE AND WIND
- CASE STUDY AT JOETSU COAST -

Hajime Mase [1], Tomohiro Yasuda [2] and Nobuhito Mori[3]

1. Objectives

In the design of sand control facilities and in the assessment of beach erosion due to global climate change, long-term morphological changes should be estimated and evaluated. Wave and wind time series are important factors for the estimation of long-term shoreline change. This study proposes a method to model the long-term time series of wave and wind by using observation data, a wave prediction model and the Monte Calro method in a statistical sense.

2. Procedure

2.1 Observation data

Wave and wind data employed here are the observed data at the Naoetsu Port from 2001 to 2005, where is one of NOWPHAS (Nationwide Ocean Wave Information network for Ports and HarbourS) sites. Those data were used to know the probabilistic distribution of parameters representing stormy conditions. Figure 1 shows a sketch showing such parameters for storm waves: the peak significant wave height during a storm; the duration of storm; the time to the peak wave height; and repetition time of successive storms. The wind direction and speed were also summarized.

2.2 Estimation of spatial wave distribution

A wave prediction system, GFS-WRF-SWAN, was employed to determine the spatial distribution of waves. The wave prediction system consists of GFS (Global Forecast System) data, WRF (Weather Research Forecasting model) meso-scale weather model and SWAN (Simulating WAves Nearshore) wave model. The outputs from the GFS-WRF-SWAN model were checked against the observations and used to know the spatial distribution of waves along the Joetsu Coast concerned.

3. Main Results

(1) Figures 1, 2 and 3 show the results of occurrence probabilities of peak significant wave heights, durations and wind speeds, and fitted theoretical distributions. Such theoretical distributions were obtained for all parameters. There are correlations between some parameters, as shown in Fig.5.
(2) Figure 6 show an example of comparison between the observed wave height and predicted one. Figure 7 shows the spatial distribution from which the wave height and direction for the concerned area can be estimated.
(3) The Monte Carlo method was employed to reproduce long-tem time series of wave and wind. Figure 8 shows a result, considering the probability distributions and correlations. The long-tem time-series are served as the input for shoreline change model, from a statistical sense.

[1]Disaster Prev. Res. Inst., Kyoto Univ., Uji, Kyoto 611-0011, Japan, hmase@engan.mbox.media.kyoto-u.ac.jp
[2]Disaster Prev. Res. Inst., Kyoto Univ., Uji, Kyoto 611-0011, Japan, yasuda@engan.mbox.media.kyoto-u.ac.jp
[3]Disaster Prev. Res. Inst., Kyoto Univ., Uji, Kyoto, 611-0011, Japan, nobuhito.mori@hy5.ecs.kyoto-u.ac.jp

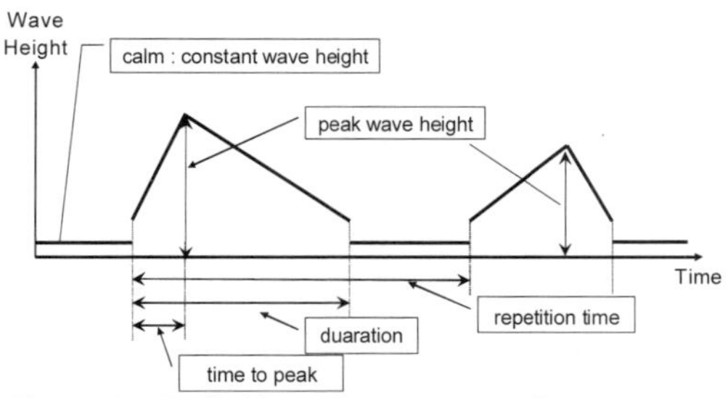

Figure 1 Definition of parameters of storm waves

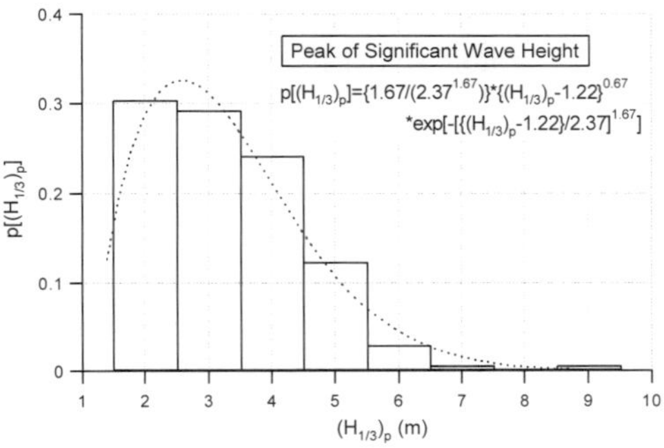

Figure 2 Distribution of peak wave heights

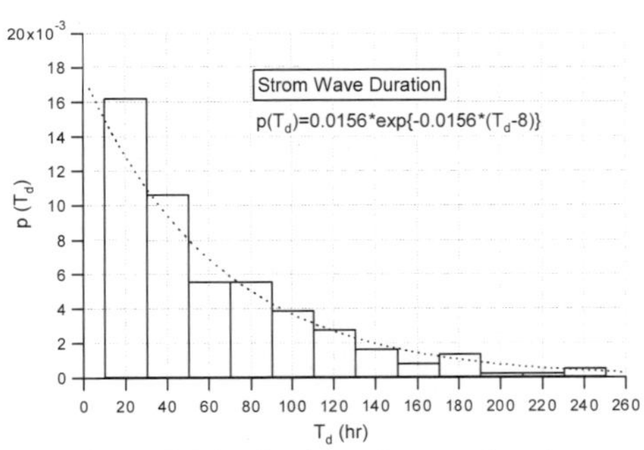

Figure 3 Distribution of storm duration

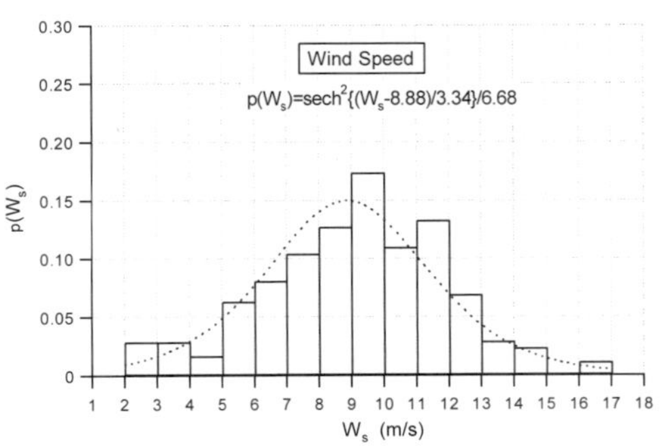

Figure 4 Distribution of wind speed

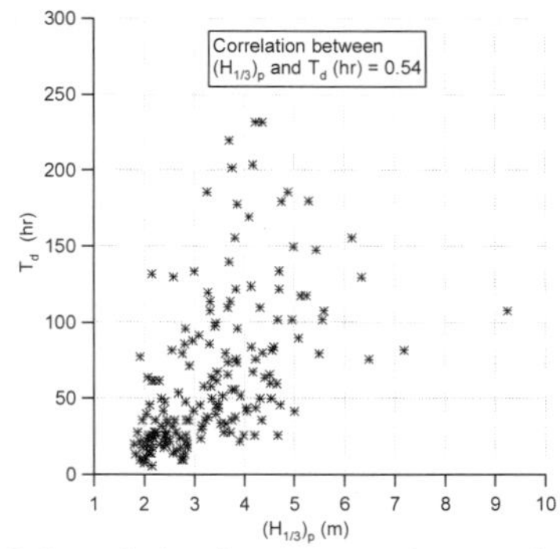

Figure 5 Correlation between peak wave height and wave period

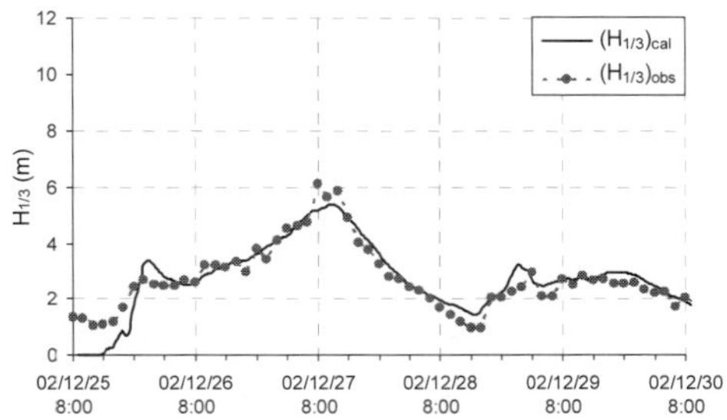

Figure 6 Comparison of predicted wave heights with observed ones

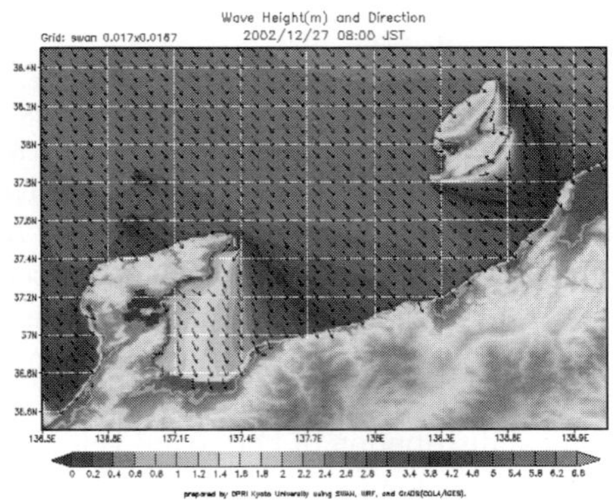

Figure 7 Spatial distribution of waves

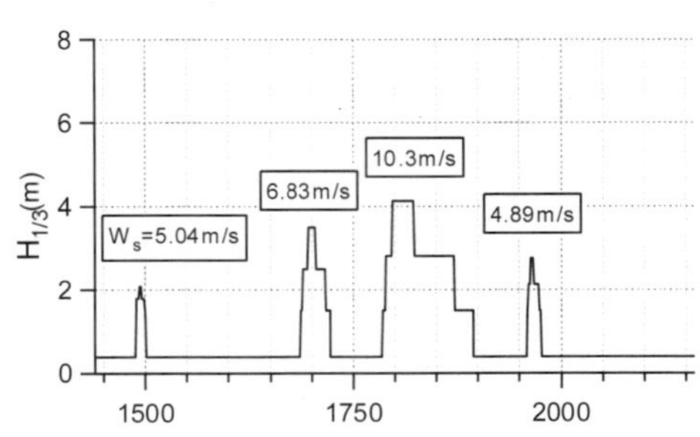

Figure 8 Time series of wave and wind

Coastal Dynamics 2009
Paper No. 32

EXPERIMENTAL STUDY OF SEDIMENT TRANSPORT CAUSED BY TSUNAMI

Yoshii Takumi[1], Masaaki Ikeno[2], Masafumi Matsuyama[3]

1. Introduction

Tsunami has created considerable interest in recent years for its potential to transport a large volume of sea sand. Several investigations have been conducted on the applicability of several major conventional sediment transport rate formulae to tsunami experimentally (i.e. Takahashi et.al, 2000). However, the experiments conducted using water tunnel with steady flow to get high Shield's number as same level as tsunami might differ from actual phenomenon because the effect of water surface elevation was neglected. Furthermore, the process of sediment transport is not well known yet because time variation of that has not been measured. The purpose of this study is to examine the characteristics of sediment transport caused by tsunami from laboratory experiment with wave basin.

2. Experiment Method

The horizontal velocities and suspended sediment concentrations caused by one solitary wave assumed as tsunami passing on the plane sand bed without breaking and splitting were measured. The experiment was carried out in wave basin (length = 78.0 m, width = 0.9m, and depth = 1.2 m) shown in Fig. 1. Water depth on sand bed was 9 cm. To get the higher Shield's number, the narrowed area was set in the flume to increase wave height of incident wave. One solitary waves of 30 seconds period with the wave height up to 6 cm and two sand grain sizes of 0.08 mm and 0.2 mm were used in this study. Time variation of vertical distribution of horizontal velocities and suspended sediment concentrations were measured at x = 7.0 m from 1 cm to 8 cm over the bottom. At the end of sand bed, the micromesh sack of 1 mm height was set to catch the bed load.

3. Results and Discussion

The sediment transports under Shield's number up to 1 have been measured in the experiment. The process of bed materials entraining into the water, diffusing upward, and settling under unsteady flow were observed from the measurement of vertical distribution as Fig. 2 shows. The sediment concentration did not increase quickly with a rise in horizontal velocity. In the case of grain size of 0.08 mm, the sediment concentration decreased near the bottom, but increased in upper layer when the horizontal velocity decreased. The time variation of horizontal velocity was synchronized with that of water level. The regression curve with exponential function shows a good agreement with the vertical distribution of suspended sediment concentration. The regression curve with log-linear model agreed with vertical distribution of horizontal velocity. Time variation of sediment flux was calculated by using both regression curves. The time lag was also appeared between Shield's number and sediment flux calculated. Therefore, the sediment flux formulae expressed by Shield's number such as Brown's formula disagreed with sediment flux observed as Fig. 3 shows.

4. Conclusions

1. The sediment flux formulae expressed by Shield's number can not explain the unsteady processes of sediment transport.

[1]Environmental Science Research Lab., Central Research Institute of Electric Power Industry, 1646 Abiko, Abiko-shi, Chiba 270-1194, Japan. takumi@criepi.denken.or.jp

[2]Environmental Science Research Lab., Central Research Institute of Electric Power Industry, 1646 Abiko, Abiko-shi, Chiba 270-1194, Japan. ikeno@criepi.denken.or.jp

[3]Civil Engineering Lab., Central Research Institute of Electric Power Industry, 1646 Abiko, Abiko-shi, Chiba 270-1194, Japan. matsu@criepi.denken.or.jp

2. The vertical distribution of suspended sediment concentration can be expressed by exponential function.
3. The Ashida-Michiue's bed-load formula showed good agreement with results of the experiment.

References

Takahashi, T., Shuto, N., Imamura, F., Asai, D., 2000. Modeling sediment transport due to tsunamis with exchange rate between bed load layer and suspended load layer. *Proceedings of the 27th International Conference on Coastal Engineering,* 276: 1508-1519.

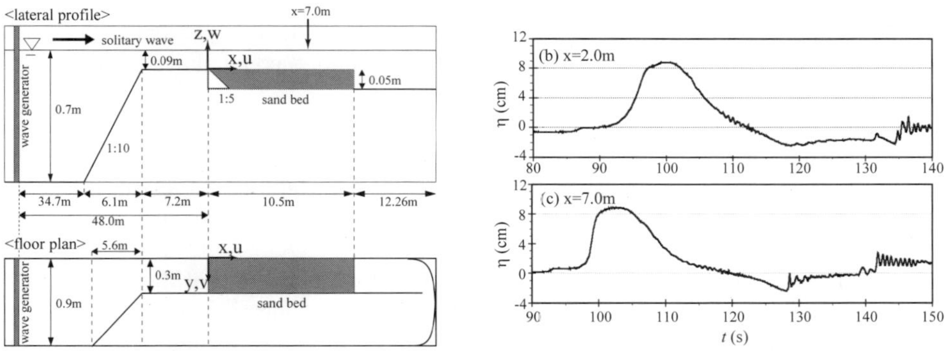

Figure 1. Experimental flume (left) and time series of water level at x = 2.0 m and x = 7.0 m (right)

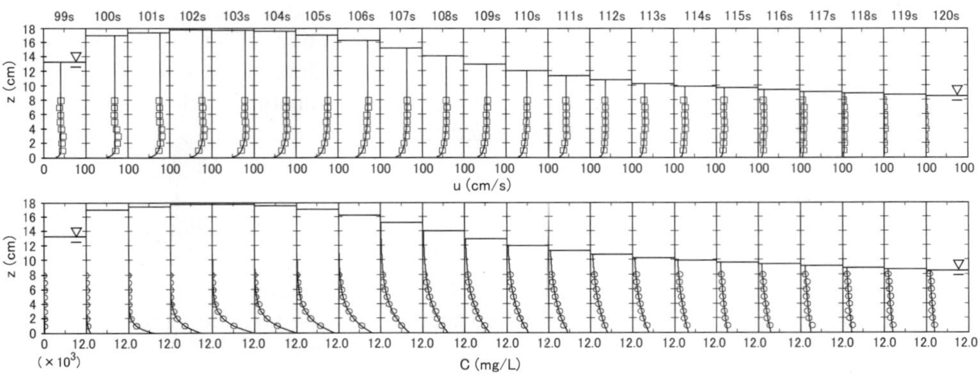

Figure 2. Time variation of vertical distributions of horizontal velocity (top) and sediment concentration (bottom) at x = 7.0 m (input wave height was 6 cm and grain size was 0.08 mm). The solid line shows regression curve.

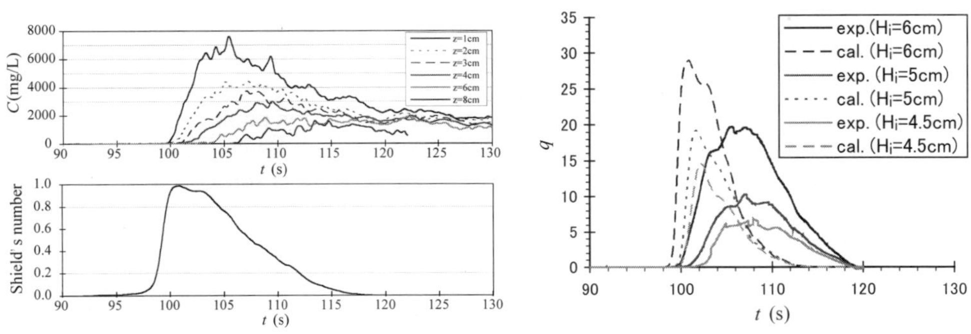

Figure 3. Time variation of sediment flux and Shield's number (left) and sediment fluxes measured and calculated using Brown's formula (right).

Coastal Dynamics 2009
Paper No. 33

REAL-TIME TSUNAMI PREDICTION EMPLOYING OFFSHORE OBSERVED DATA

Tomohiro Yasuda[1], Tomotsuka Takayama[2] and Hajime Mase[3]

1. Introduction

In the case of offshore earthquake, tsunami is usually generated and causes additional damages after devastating damages caused by ground shaking. The Tonankai and Nankai earthquake which generates giant tsunami occurs at the interval of 100 or 150 years, and its occurrence is predicted in the next quarter of century with high probability (50%). It can be said that a giant tsunami is an extremely rare phenomenon for human society. It is extremely difficult to prevent such a rare disaster by the hardware countermeasures, and it should be taken into account the disaster mitigation by software measures.

It is important to estimate the propagating tsunami in real time or to forecast just after the earthquake to help a decision making for evacuation advisory or order. A real-time tsunami prediction method by inverse analysis was established and evaluated its applicability in this study.

2. Inverse Method

The inverse analysis of offshore observed tsunami data estimates the initial displacements of water surface in the ocean. The estimated initial displacements predict tsunami profiles in inshore points of interests. Optimum conditions for the accuracy improvement were also examined aiming at practical use.

Tsunami profiles are supposed to be superposed of unit tsunamis which are 1 m uplift displacement of the small elements according to the linear long wave theory. The initial sea surface displacement of tsunami source is estimated immediately by fitting the summation of the wave profiles of unit tsunamis to the measured tsunami at offshore observation sites with the least squares method. As the initial sea surface displacement of tsunami source is estimated, tsunami profiles along coasts can be predicted by the superposition of unit tsunamis immediately.

3. Conditions of Numerical Experiments

Numerical experiments were conducted for Japanese coastal area along the Pacific Ocean where the Tonankai and Nankai Earthquake will occur. The assumed tsunami source area is divided into 156 small units are shown by gray shaded areas in Figure 1. The propagation of unit tsunami generated by 1 m uplift of each small segment is calculated in advance. Grid size is 1.35km and source unit size is 27km, and time step is 2 seconds. Virtual offshore observation sites with GPS buoys are setup at 9 locations and 4 coastal locations are employed to evaluate the model's ability.

4. Results and Discussion

The upper figure in Figure 2 shows assumed initial tsunami displacements of the test case, and lower figure shows inversely analyzed results using 30 minutes observed data, respectively. Initial tsunami sources are well estimated as compared in these figures. Accuracy of the inversion method by fitting the summation of the wave profiles of unit tsunamis is confirmed in Figure 3 which compares the observed and inverse analyzed tsunami profile at offshore observation points.

Figure 4 shows initial tsunami displacement of the Tonankai and Nankai Earthquake model by the Cabinet Office, Government of Japan. As shown in Figure 5, tsunami sources were estimated well by using the 15 minutes observed data at three GPS buoys, simultaneously. It was also confirmed that the real-time prediction can apply to predict the tsunami at Shimotsu Port in Wakayama Prefecture as well as in Osaka Bay as shown in Figure 6.

---

[1]Disaster Prev. Res. Inst., Kyoto Univ., Uji, Kyoto 611-0011, Japan, yasuda@engan.mbox.media.kyoto-u.ac.jp
[2]Coastal Dev. Inst. of Tech., 3-16-6F, Hayabusa-cho, Chiyoda-ku, Tokyo, 102-0092, Japan, takayama@cdit.or.jp
[3]Disaster Prev. Res. Inst., Kyoto Univ., Uji, Kyoto, 611-0011, Japan, hmase@engan.mbox.media.kyoto-u.ac.jp

5. Concluding Remarks

The analyzed results showed overall good performance from the viewpoint of accuracy. Tsunami sources were well estimated compared with the initial displacement of tsunami generated by the Tonankai and Nankai Earthquake model. The tsunami profiles can be predicted with high accuracy in inshore points of interests and information is updated by using longer observation data.

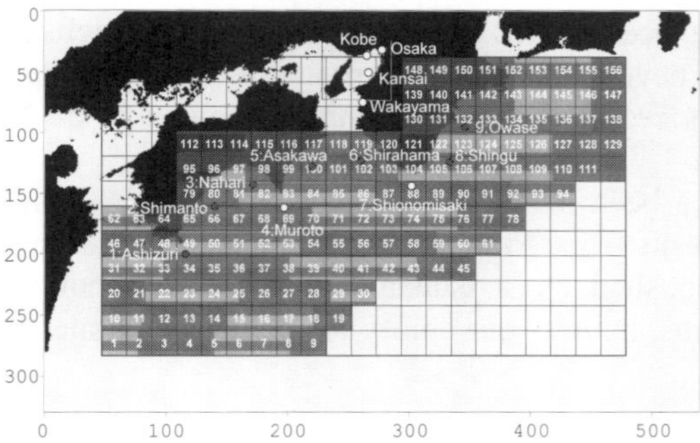

Figure 1 Segmented units of assumed tsunami source

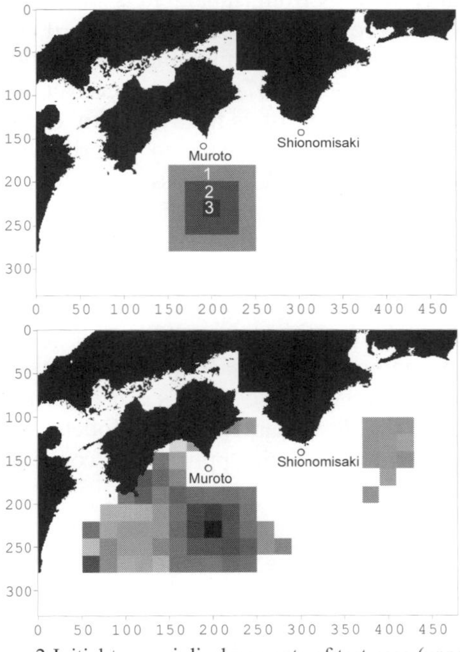

Figure 2 Initial tsunami displacements of test case (upper), and inversely analyzed results by 30min. obs. data (lower)

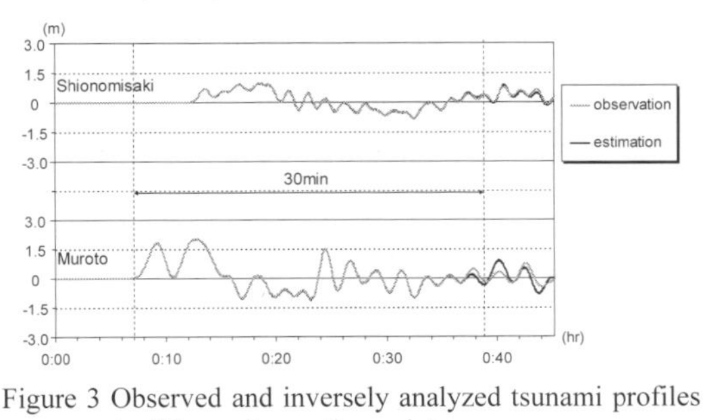

Figure 3 Observed and inversely analyzed tsunami profiles at offshore observation points

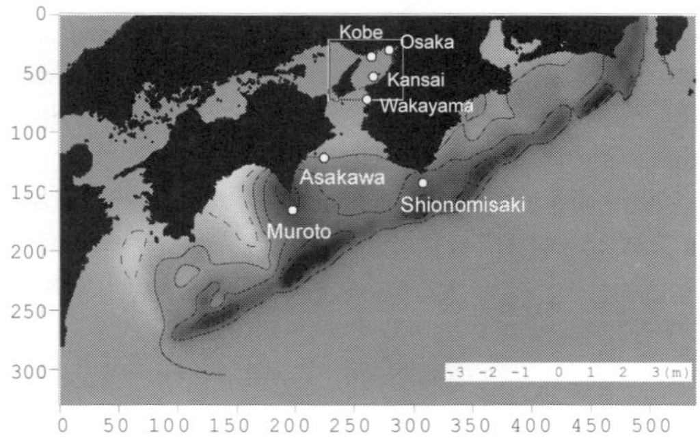

Figure 4 Initial tsunami displacement of the Tonankai and Nankai Earthquake model by the Cabinet Office, Government of Japan

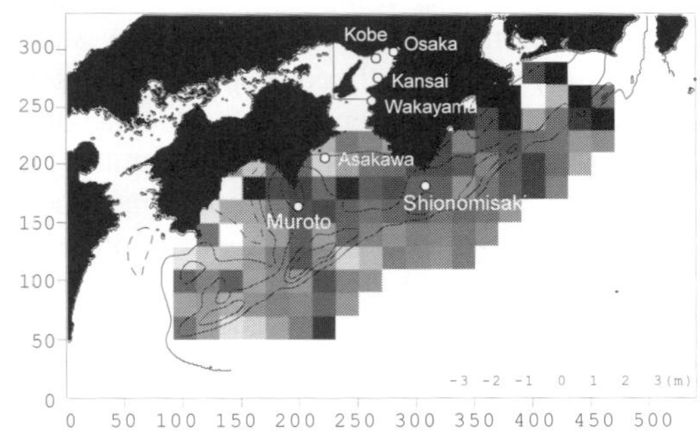

Figure 5 Inversely analyzed initial tsunami displacement of the Tonankai and Nankai Earthquake using 15min. obs. data at 3 observation points

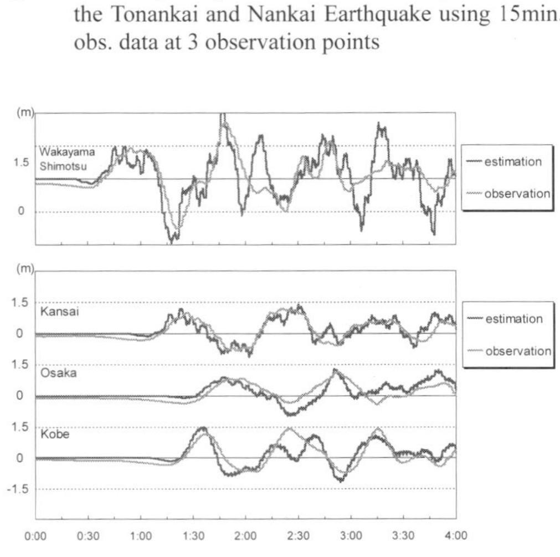

Figure 6 Observed and inversely analyzed tsunami profiles at inshore locations by 15min. obs. data from 3 buoys (upper) at Wakayama, (lower) inside the Osaka Bay

Coastal Dynamics 2009
Paper No. 34

MODELING TIME-VARYING TSUNAMI SEDIMENT DEPOSITION

Alex Apotsos[1], Bruce Jaffe[2], Guy Gelfenbaum[3], Edwin Elias[4]

1. Introduction

Tsunamis are infrequent and unpredictable events that can cause enormous human and economic losses. The 26 December 2004 tsunami in the Indian Ocean killed over 237,000 people and left more than a million homeless. Unfortunately, our ability to develop management programs that can properly plan for and mitigate losses owing to tsunamis is restricted by our limited knowledge of their historical reoccurrence and magnitude.

Over the past two decades significant efforts have been made to lengthen the historical tsunami record using information preserved in sediment deposits (e.g., Nanayama et al., 2007). However, our incomplete understanding of how sediment is deposited by a tsunami, and how the dominant depositional processes are affected by local conditions, inhibits our ability to estimate accurately the flow characteristics of a tsunami from deposits.

2. Tsunami Deposits and Models for Deposition

The stratigraphy of a tsunami deposit can vary significantly not only from tsunami to tsunami, but also spatially within the sediment deposited by a single tsunami. Previous studies have observed deposits ranging from a single, normally graded, well sorted layer to deposits with multiple layers containing inverse grading and complex internal structures (Peters et al., 2007). This large variance in stratigraphy significantly complicates efforts to estimate flow characteristics from sediment deposits.

Recently, both mathematical and conceptual models have been formulated to describe tsunami sediment deposition. To simplify the numerics, the mathematical models often assume that either all sediment settles out of suspension when the tsunami velocity goes to zero (Jaffe and Gelfenbaum, 2007) or that sediment begins to settle out as soon as the wave inundates over land (e.g., Smith et al., 2007). While both models have been shown to agree well with other estimates of flow characteristics, their simplifying assumptions preclude their applicability under a wide range of conditions. The conceptual models (e.g., Morton et al., 2008) reconstruct the hydraulic conditions of a specific tsunami by associating the different observed layers and internal structure with different phases of inundation, including inflow, stagnation, and backwash. However, these models are often tsunami specific, or do not estimate quantitatively the magnitude of the flow characteristics. Models that both quantitatively estimate flow characteristics and are not limited in their applicability could greatly improve the accuracy of historical tsunami flow predictions.

3. Model Simulations

Delft3D, a three dimensional numerical model, has been previously shown to model tsunami propagation and inundation well (Vatvani et al., 2005a, 2005b) and qualitatively match the sediment deposition (Gelfenbaum et al., 2007) from the 2004 Indian Ocean tsunami. Recently, Delft3D has also been used to assess the effects on tsunami inundation and deposition owing to mangrove forests, breakwaters, and reefs (Gelfenbaum et al., 2007; Vatvani et al., 2008). Here tsunami flow and sediment deposition is simulated using the flood solver of the Delft3D-Flow module. The model is driven by realistic offshore water levels obtained from propagating tsunami waves into shallow water using large scale ocean models. Sediment deposition and erosion is recorded in 1 second intervals as

[1] United States Geological Survey, Menlo Park, CA, aapotsos@usgs.gov
[2] United States Geological Survey, Santa Cruz, CA, bjaffe@usgs.gov
[3] United States Geological Survey, Menlo Park, CA, ggelfenbaum@usgs.gov
[4] United States Geological Survey, Santa Cruz, CA, eelias@usgs.gov

the tsunami inundates over three alongshore-uniform model domains (Fig 1A) and one alongshore-varying domain (Fig 1B).

The time and spatial variations of sediment deposition and erosion are examined to determine which phases of inundation dominate deposit formation. Specific attention is paid to the erosional effects of receding and subsequent waves, the relative contributions of suspended and bed load transport, the depositional processes related to each phase of inundation, and the effects of spatial gradients in the flow velocity owing to along- and cross-shore varying topography. To determine how the depositional processes are affected by local conditions, the sensitivity of the model predictions to nearshore bathymetry, topographic variations, sediment source characteristics (including grain-size, composition, amount, and location), roughness, and tsunami wave form is examined.

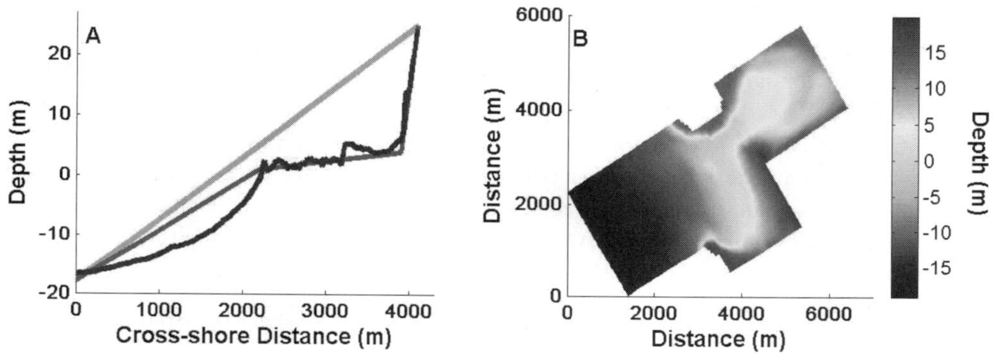

Figure 1: The cross-shore depth profile for (A) idealized linear beach (green), idealized linear beach with flatter topography (red), and bathymetry and topography measured at Kuala Muerisi, Sumatra (blue), and (B) surface view of bathymetry and topography measured at Jantang, Sumatra.

References

Gelfenbaum, G., Vatvani, D., Jaffe, B., and Dekker, F., 2007. Tsunami inundation and sediment transport in the vicinity of coastal mangrove forest, Conference Proceedings Coastal Sediment '07, doi 10.1061/40926(239)86.

Jaffe, B. and Gelfenbaum, G., 2008. A simple model for calculating tsunami flow speed from tsunami deposits, *Sedimentary Geology*, 200: 347-361.

Morton, R.A., Goff, J.A., and Nichol, S.L., 2008. Hydrodynamical implications of textural trends in sand deposits of the 2004 tsunami in Sri Lanka, *Sedimentary Geology*, 207: 56-64.

Nanayama, F., Furukawa, R., Shigeno, K., Makino, A., Soeda, Y. and Igarashi, Y., 2007. Nine unusually large tsunami deposits from the past 4000 years at Kiritappu marsh along the southern Kuril Trench, *Sedimentary Geology*, 200: 275-294.

Peters, R., Gelfenbaum, G., and Jaffe, B., 2007. Distribution and sedimentary characteristics of tsunami deposits along the Cascadia margin of western North America, *Sedimentary Geology*, 200: 372-386.

Smith, D.E., Foster, I.D.L., Long, D., and Shi, S., 2007. Reconstructing the pattern and depth of flow onshore in a palaeotsunami from associated deposits, *Sedimentary Geology*, 200: 362-371.

Vatvani, D., Nieuwenhuis, O.E., de Graaf, R., Waryono, D., Kuijper, M., and van der Plas, T., 2008. Tsunami Flood Defense strategy for Banda Aceh; Numerical Model Investigation, International Conference on Coastal Engineering, Hamburg, Germany.

Vatvani, D., Schrama, E., and van Kester, J., 2005a. Hindcast of tsunami flooding in Aceh-Sumatra, Proc. of the 5th International Symposium on Ocean Wave Measurement and Analysis, Madrid, Spain.

Vatvani, D., Boon, J., and Ramanamurty, P.V., 2005b. Flood risk due to tsunami and tropical cyclones and the effect of tsunami excitations on tsunami propagations, Proc. of the IAEA Workshop on External Flooding Hazards at NNPS, Kalpakkam – Tamil Nadu, India.

Coastal Dynamics 2009
Paper No. 35

A STUDY FOR THE SEDIMENT TRANSPORT BY THE 2004 INDIAN OCEAN TSUNAMI ALONG THE NATURAL COAST OF HAMBANTOTA, SRI-LANKA

Takeshi Nishihata[1], Kazuhisa Goto[2], Yoshimitsu Tajima[3], Tomoyuki Takahashi[4] and Fumihiko Imamura[5]

1. Introduction

Devastating sediment transport accompanied by a tsunami is often observed in a coastal zone and some field observations with bathymetry survey data have been reported. For instance, Chilean tsunami in Kesennuma bay and Indian ocean tsunami in Kirinda fishing port caused huge topographic change and local intensive erosion along the narrow straight portion or breakwaters. On the other hand, less survey data of topographic change by tsunamis are submitted for natural coasts with parallel depth contour.

Remarking the existence of the depth sounding data along the natural coast in Sri-lanka just before the 26th/Dec/2004 Indian ocean tsunami, we conducted the bathymetry survey and compared the depth before and after the tsunami to grasp the geological change induced by the tsunami. In this study, we also applied numerical simulation to the event and objected to consider sediment transport mechanism along the natural coast.

2. Outline of Survey

Echo soundings on the sea and level survey behind the shoreline were executed in the natural coast of the Hambantota city which located in the south of Sri-lanka before and after the 2004 Indian ocean tsunami. Grain size analysis and bore-hole sampling were also conducted with the bathymetry survey. The outline of survey is summarized in the table 1. Erosion and accretion distribution during the investigation with survey lines are shown in the figure 2. Huge erosion by the tsunami is observed in the coast.

3. Reproduction of Topography Change with Numerical Model

Reproduction of topographic change with numerical analysis consists of 3 computation steps as, (1)tsunami propagation analysis, (2)topographic change analysis due to tsunami and (3)evaluation of bathymetry change with usual waves.

Tsunami propagation analysis based on the non-linear shallow water theory is applied to the region from the initial source of the tsunami to the considering site. Using the results of propagation computing, we executed topographic change analysis according to Nishihata et.al.(2006) whose model describes sediment transport due to tsunamis including not only bedload transport but also suspended sediment one. To grasp the inclination of the beach process in the interval of each survey, evaluation of bathymetry change with usual waves is also conducted by the local sediment flux.

[1] Institute of Tech., Penta-Ocean Co., Ltd., 1534-1 Yonku-Cho, Nasu-Shiobara, Tochigi 329-2746, Japan. Takeshi.Nishihata@mail.penta-ocean.co.jp

[2]Disaster Control Research Center, Tohoku University, 6-6-1 Aoba, Sendai, Miyagi 980-8579, Japan. kgoto@tsunami2.civil.tohoku.ac.jp

[3] Dept. of Civil Eng., University of Tokyo, 7-3-1 Hongo, Bunkyo-ku, Tokyo 113-8656, Japan yoshitaji@coastal.t.u-tokyo.ac.jp

[4]Dept. of Civil and Environmental Eng., Akita University, 7-3-1 Hongo, Bunkyo-ku, Tokyo 113-8656, Japan. tomo@ce.akita-u.ac.jp

[5] Disaster Control Research Center, University of Tohoku, 6-6-1 Aoba, Sendai, Miyagi 980-8579, Japan. imamura@tsunami2.civil.tohoku.ac.jp

4. Computed Results and Discussions

Tsunami propagation analysis well agrees with the observed arrival time and maximum inundation height in the Hambantota city, then we presume that the same scale tsunami is reproduced in computation. Figure 2 shows the computed distribution of erosion and accretion by the tsunami, which indicates same tendency that erosion is dominant entirely and especially at the specific landform as the river mouth and capes as well as the observation. The computed results suggest that huge topographic changes are more conspicuous in the tsunami than usual waves in the place where shoreline discontinuations are observed because wave refraction and concentration effect becomes larger and cross-shore sediment transport is more strengthen than usual waves in such landforms. We also find that the eroded coast is recovered during the beach process of the usual waves after the tsunami.

Acknowledgements

Parts of the investigation of this study were supported by the scientific research fund(No.18201033). Here, we express sincere gratitude.

References

Nishihata, T., Tajima, Y., Moriya, Y., and Sekimoto, T., 2006. Topography change due to the Dec 2004 Indian ocean tsunami – field and numerical study at Kirinda port, Sri-Lanka-. *Proceeding of 30th International Conference on Coastal Engineering,* 1456-1468.

Table 1. Outline of survey

survey times	duration	survey lines	objectives
1st	Nov/2004-Dec/2004	46	To gain the topography before the tsunami
2nd	Nov/2005-Jan/2006	49	To gain the topography after the tsunami
3rd	Oct/2007-Nov/2007	15	To grasp the topography change due to usual waves

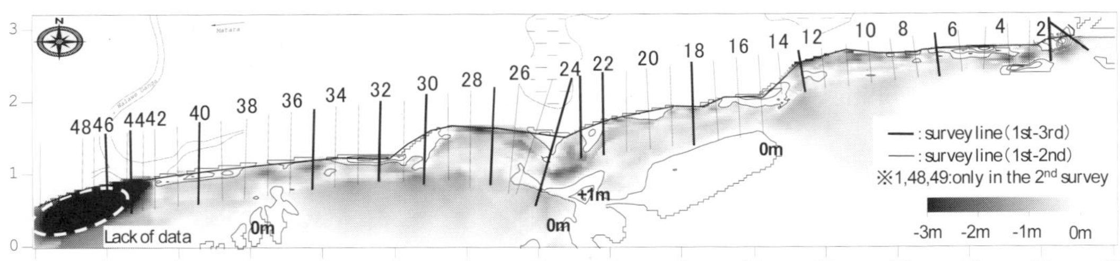

Figure 1. Distribution of erosion and accretion with survey lines(1st survey – 2nd survey)

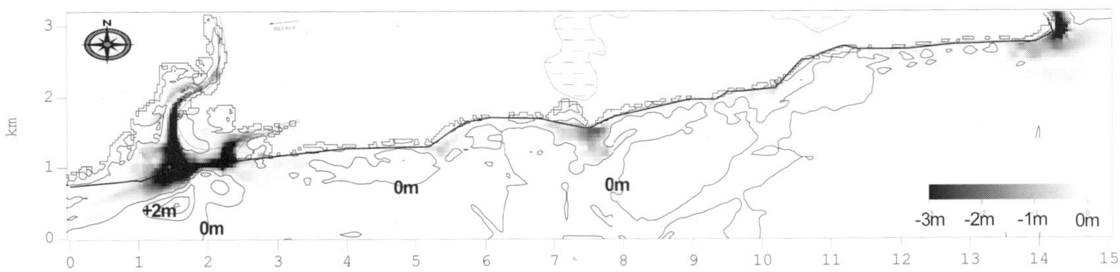

Figure 2. Computed topographic change by the tsunami

Coastal Dynamics 2009
Paper No. 36

A MECHANSIM INHIBITING RIP CHANNEL FORMATION FOR OBLIQUE WAVES

Roland Garnier[1] , Nick Dodd[2] , Albert Falqués[3] and Daniel Calvete[4]

1. Introduction

The reorganization of a straight parallel sand bar into regular series of crescentic bars associated with rip current systems is observed worldwide. These rip channel systems are present along the largest part of the south west French Atlantic coast, however, they seem to disappear where the ambient alongshore current is the strongest (Castelle, 2004). It is increasingly recognized that these systems emerge from self organization processes of the coupling between the hydrodynamics and the evolving topography.

A number of numerical modelling studies simulate the formation or the initial stages of the evolution of rip channels; the nonlinear regime and a dynamical equilibrium have been reached for the first time by Garnier et al. (2008). These studies reveal that the growth rate of these instabilities decreases when the wave angle increases (Castelle, 2004; Calvete et al., 2005; Garnier et al., 2008). However, the reasons for this are unexplored.

Although the mechanisms of the growth saturation are not understood either, a clue has been given by Garnier et al. (2006) who introduced the concept of `global analysis' in the case of transverse bars that appear on a planar beach. This consists of analyzing the beach dynamics from the whole domain, i.e. by using integrated variables over the cross-shore and the alongshore axis. The objective of this work is to understand the stability of rip channels for oblique waves by applying a global analysis.

2. Methodology

The numerical model MORFO55 (Garnier et al., 2006, 2008) solves the wave and depth averaged nonlinear shallow water equations with wave driver, sediment transport and bed updating. A decomposition of the sediment transport formulation gives two variables: the `production' ($P>0$) from the advective part and the `damping' ($\Delta>0$) from the bed-slope transport. If we call $||h||$ the bar norm of the developing instabilities, we show that the growth rate σ of the instabilities can be approximated by the relationship:

$$\sigma = \frac{P-\Delta}{\|h\|^2} \qquad (1)$$

3. Results and further work

The instability of an initially alongshore uniform barred beach perturbed by a Dirac function is studied, for different incidence wave angle θ (Figure 1a and 1b). For oblique wave incidence, the bars migrate down-current and the maximum growth rate of the instabilities decreases with the angle (Figure 1c). For $\theta>6°$, the beach system is stable: rip channels do not appear.

A very useful method to analyze the behavior of the global variables is to represent them as a function of $||h||$. Particularly, Figure 2a shows that, for a fixed $||h||$ corresponding to the linear regime (i.e. let's say $||h||<0.05$ m), the decrease of σ for larger angles is due to $P/||h||^2$, while $\Delta/||h||^2$ does not depend essentially on the wave angle. This is shown in detail in Figure 2b for $||h||=0.01$ m.

[1]Applied Physics Dept, Universitat Politécnica de Catalunya, Barcelona, Spain, 08034, rgarnier@fa.upc.edu
[2]Dept of Civil Eng., University of Nottingham, Nottingham, UK, NG7 2RD, Nick.Dodd@nottingham.ac.uk
[3]Applied Physics Dept, Universitat Politécnica de Catalunya, Barcelona, Spain, 08034, falques@fa.upc.edu
[4]Applied Physics Dept, Universitat Politécnica de Catalunya, Barcelona, Spain, 08034, calvete@fa.upc.edu

Thus, the global analysis allows us to compute a global growth rate of the instabilities. It can be used to identify the saturation processes (decrease of σ with $||h||$) but also to identify mechanisms governing the stability or the instability.
Further work will reveal that the decrease of $P/||h||^2$ with the wave angle is due to the decrease of (1) the maximum cross-shore current u and (2) the phase shift between u and the bed level variations h.
It will also be discussed that crescentic bars are sometimes observed in beaches where the longshore current is strong (e.g. in Noordwijk, The Netherlands or in the Gold Coast, Australia).

References

Calvete, D., Dodd, N., Falqués, A., and van Leeuwen, S. M., 2005. Morphological development of rip channel systems: Normal and near normal wave incidence. *J. Geophys. Res.*, 110(C10006).

Castelle, B., 2004. Modélisation de l'hydrodynamique sédimentaire au-dessus des barres sableuses soumises à l'action de la houle : application à la côte aquitaine. Ph.D thesis, Université Bordeaux 1, France.

Garnier, R., Calvete, D., Falqués, A., and Caballeria, M., 2006. Generation and nonlinear evolution of shore-oblique/transverse sand bars. *J. Fluid Mech.*, 567: 327-360.

Garnier, R., Calvete, D., Falqués, A., and Dodd, N., 2008. Modelling the formation and the long-term behaviour of rip channel systems from the deformation of a longshore bar. *J. Geophys. Res.*, 113(C07053)..

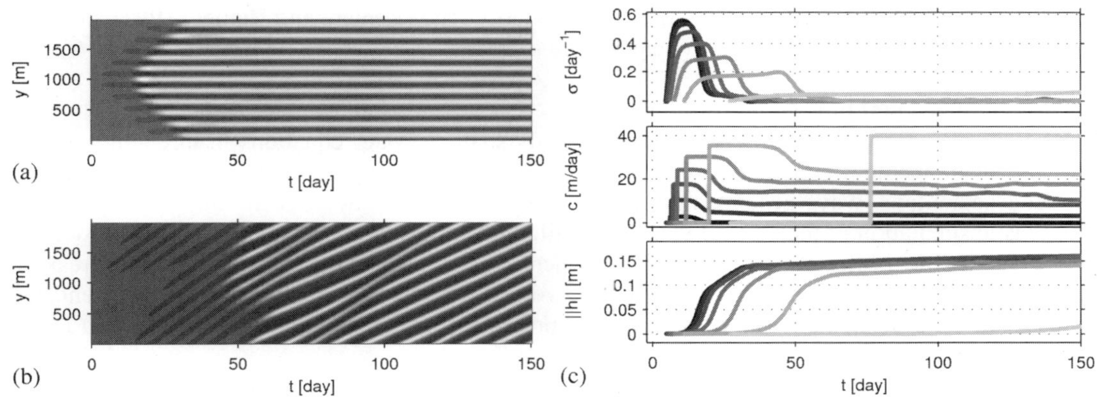

Figure 1. (a, b) Time series of the bed level along the longshore section x = 50 m, for (a) normal (θ=0°), and (b) oblique (θ=5°) wave incidence. (c) Time series of the growth rate σ, the migration celerity c and the bar amplitude $||h||$. The darker is the line, the smaller is the incident wave angle (varying from 0° to 6°).

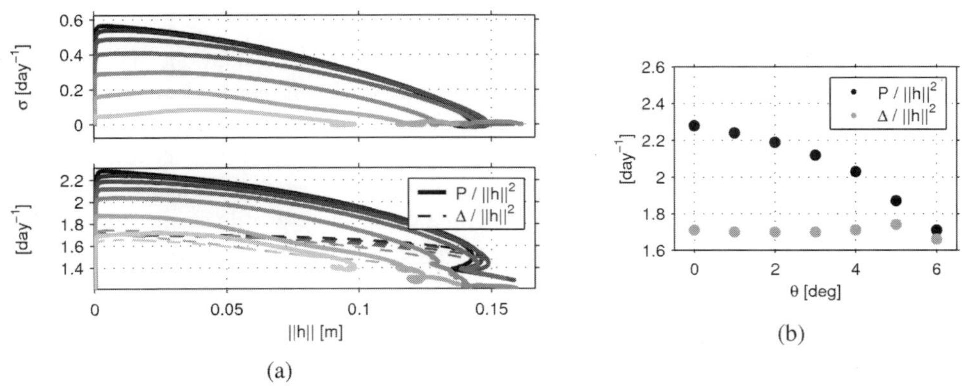

Figure 2. (a) σ, $P/||h||^2$ and $\Delta/||h||^2$ as a function of $||h||$ for different θ (the darker is the line, the smaller is the incident wave angle, varying from 0° to 6°) (b) $P/||h||^2$ and $\Delta/||h||^2$ as a function θ, for $||h||$=0.01 m.

Coastal Dynamics 2009
Paper No. 37

3D CALCULATION OF WAVE-DRIVEN LONGSHORE CURRENTS IN THE SURFZONE

Roald B. Treffers[1], Jebbe J. van der Werf[2], Maarten van Ormondt[3], Dirk-Jan Walstra[1,4]

1. Introduction

Obliquely incident waves induce a mean current parallel to the coastline. These longshore currents are responsible for longshore transport of sediments and therefore determine largely the morphology of coastal areas. However to the authors' knowledge, most of the studies concerning wave-driven *longshore* flows are based on depth-averaged modelling approaches (1DH or 2DH) whereas little research has been devoted to three-dimensional (3D) modelling. Depth-averaged approaches neglect important processes (e.g. wave-breaking induced production of turbulent kinetic energy) which are responsible for the vertical distribution of the longshore flow field to deviate from the standard logarithmic profile. A good representation of these vertical current profiles can only be obtained with a 3D model.

The purpose of the present paper is to compare 3D and 2DH simulated wave-driven longshore currents in a validation study using both field and laboratory measurements. Furthermore, model improvements are intended to more accurately calculate the 3D wave-driven longshore currents in the surfzone. The calculations will be made using the process-based numerical modelling program Delft3D and the results will be validated using laboratory data (Reniers and Battjes, 1997) and field data obtained at Duck, North Carolina (Reniers et al., 2004).

2. Three-dimensional flow computations

In 3D flow calculations, Delft3D solves the unsteady shallow water equations in three-dimensions. The vertical momentum equation is reduced to the hydrostatic pressure equation by neglecting the vertical accelerations. Wave-generated turbulence, wave forcing and Stokes drift are included, aiming at a realistic simulation of the vertical velocity profile. In contrast to a 2DH approach, these processes can be applied at the location in the vertical where they actually take place (e.g. wave breaking induced turbulence near the water surface). The vertical distribution of the longshore current that results from the 3D calculation is of particular importance for current-related suspended load in the longshore direction.

In 2DH modelling the vertical distribution of the longshore current is assumed to be logarithmic. However, (Svendsen and Lorenz, 1989) suggested that due to the presence of wave breaking the velocity profile may become more depth-uniform, since wave breaking-induced turbulence strongly enhances vertical mixing. With 3D calculations this can be verified by comparing calculated results with measurements.

3. Laboratory test validation

To validate the 3D calculations and to compare them with 2DH calculations the measurements obtained during the laboratory tests by (Reniers and Battjes, 1997) are used. These experiments were performed in a large, 40 m long and 25 m wide, multidirectional wave-basin. The test condition involved random waves with a wave height and period of 0.07 m and 1.2 seconds, respectively, propagating over a barred, immobile beach. The angle of incidence was 30 degrees normal to the shoreline. The longshore currents were measured along a cross-shore profile at one-third of the water depth, roughly in agreement with the depth-averaged velocity under the assumption of a logarithmic

[1] Delft University of Technology, faculty of Civil Eng. and Geosciences, section Hydraulic Eng., and Deltares The Netherlands, rtreffers@hotmail.nl
[2] Deltares (formerly WL|Delft Hydraulics), Rotterdamseweg 185, 2600 MH Delft, The Netherlands, Jebbe.vanderWerf@deltares.nl
[3] Deltares, Rotterdamseweg 185, 2600 MH Delft, The Netherlands, Maarten.vanOrmondt@deltares.nl
[4] Deltares, Rotterdamseweg 185, 2600 MH Delft, The Netherlands, DirkJan.Walstra@deltares.nl

velocity profile. Apart from the prescription of vertical layers in the 3D model, the settings of the 2DH and 3D computations are identical.

Figure 1 compares the simulated and measured longshore currents. Both the 2DH and 3D computations agree reasonably well with the measurements. In particular, the location of the maximum longshore current is in good correspondence. However, the 2DH (especially) and 3D model underpredict the longshore current in the bar trough and the 3D model overpredicts the longshore current at the seaward end of the bar.

4. Conclusions

3D calculations are more accurate near the bar crest and in the bar trough compared with 2DH calculations. In particular the location and magnitude of the maximum longshore current correspond better to the measurements. However, the longshore current is underestimated inside the bar trough and overestimated seaward. The exact cause of this is not yet known and therefore this will be further studied.

In the final paper both modelling approaches will be applied to field data from Duck, North Carolina (USA) (Reniers and Battjes, 1997) to verify if the same over- and underpredictions occur. The Duck dataset is in particular interesting since also measurements over the vertical are made. This gives us the opportunity to validate the 3D approach in proto type conditions. These two validations will enable us to thoroughly assess and improve the Delft3D performance in simulating wave-driven longshore currents.

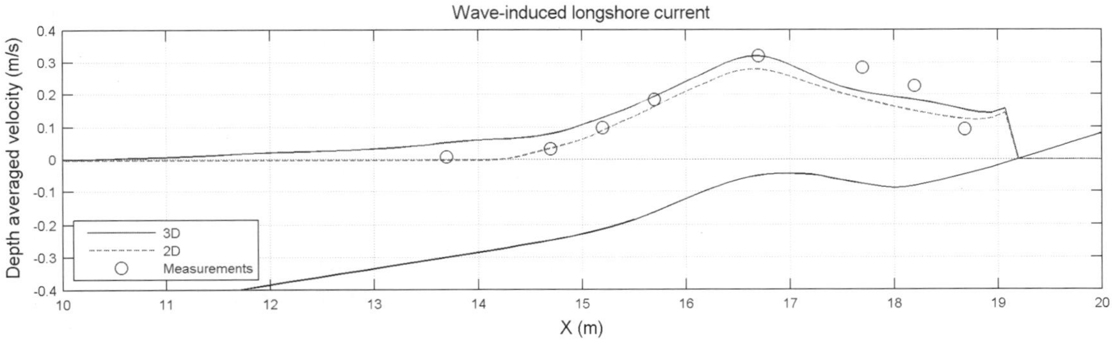

Figure 1: 3D and 2DH calculated longshore current compared with laboratory measurements obtained by Reniers. The black solid line denotes the bed level

Acknowledgement

This research is carried out within the Kustlijnzorg-project of Deltares.

References

Reniers, A. and Battjes, J.A., 1997. *A laboratory study of longshore currents over barred and non-barred beaches*. Coastal Engineering, 30(1-2): 1-21.

Reniers, A.J.H.M., Thornton, E.B., Stanton, T.P. and Roelvink, J.A., 2004. *Vertical flow structure during Sandy Duck: observations and modelling*. Coastal Engineering, 51: 237-260.

Svendsen, I.A. and Lorenz, R.S., 1989. *Velocities in combined undertow and longshore currents*. Coastal Engineering, 13(1): 55-79.

Coastal Dynamics 2009
Paper No. 38

LITTORAL DRIFT BEHAVIOR AROUND A SMALL ISLAND PORT

Takayuki SASAKI[1], Kentaro HAYASHI[2], Hiroyuki MAKINO[3], Akihiro YAMAMOTO[4], Hideo KONDO[5], and Akira KAWAMORI[6]

1. Introduction

Kunnui Fishing Port is located at the inner of the Uchiura Bay in Hokkaido, Japan. This port is the first artificial island fishing port in the littoral drift environment in Japan and connects to the land with a bridge. The construction was started in 1984 and was finished in 1993. The basic objects were to prevent the entrance of port from the shoaling and to minimize the effect on the coast, in the other word, to keep the continuity of longshore sediment transport.

In this study, the filed observation data are analyzed to investigate the wave and current condition around the artificial island fishing port. Also, the morphological characteristic after the construction is examined based on the historical bathymetric surveys. The longshore sediment transport rate is estimated by a simple method with the bathymetric change at the neighbor fishing port and analyzed how much sand volume pass through the back of the artificial island fishing port. In addition, a general numerical simulation is applied to estimate the current field behind the artificial island port and the validation is examined.

2. Field Observation

For one month in the winter of 2000, a wave and current measurement was conducted at 4 points around Kunnui Fishing Port. The maximum of significant wave height was obtained up to 2.9 meters. The dominant wave direction (at St.1) is ESE (88%) and most of the observed wave came toward the shore line straightly. According to the observed data during the typical storm, the time- averaged current at the channel is about 0.14m/s (Kishi et al., 2000). As the result of the surface water elevation (Figure 1), during comparatively high wave, the difference in phase and surface elevation was measured. Since 1984, total 26 bathymetric surveys have been conducted. An accumulated trend is existed behind the artificial island fishing port and an eroded trend is seen at the entrance of port. In addition, the volumetric analysis is applied around the fishing port to the whole bathymetric survey data. Figure 4 shows the accumulated sand volume behind the port (Yamamoto et al. 2005). As shown in Figure 2, for 5 years after the construction started, the volume has greatly increased and then has gradually increased after the port construction finished. The average accumulated rate is 5,000m^3/y. In this analysis, the region behind the artificial island fishing port has been an equilibrium condition.

3. Estimate of Littoral Drift

In order to estimate the longshore sediment transport rate at Kunnui Fishing Port, a simple method, which is proposed by Kondo and Kawamori (2005), is applied in this study. The method is that the longshore sediment transport is estimated by the bathymetric volume change around a jetty. The cross-shore sediment transport distribution is determined by the theory, which was proposed by Dean and Dalrymple (2002). In this study, the longshore sediment transport at Kuroiwa Fishing Port (westward 7km) and Onaka Fishing Port (eastward 7km) are used to estimate that around Kunnui Fishing Port. The result of this estimate is summarized. According to Table 1, the net longshore sediment transport at Kunnui Port is about 20,000m^3/y. In comparison with the accumulated sand volume in Figure 2, about 75% of the longshore sediment transport is estimated to go through behind the artificial island port.

[1~3,5,6] Alpha Hydraulic Engineering Consultants Co., Ltd., 9-14 Hassamu Nishi-ku, Sapporo, 063-0829, Japan.
[4] Hokkaido Prefecture, Japan.

4. Numerical Simulation

In order to simulate the current field behind the artificial island port, a general numerical model is applied. This model is a coupling model with the hydrodynamic module for simulating unsteady two-dimensional flows in one layer fluids and advection-diffusion module for the sediment transport process. The boundary condition at both sides is given by the observed surf beat (surface water elevation at St.2 and St.4), which is discussed in Figure 1. The sediment condition is set up as the settling velocity based on the observed grain size and distribution data, which was reported by Kishi et al. (2000). Figure 3 shows the results of this simulation. The upper figure means the distribution of flow velocity. The lower figure describes the distribution of the suspended sediment transport rate at the same time as the upper figure. As shown in the upper figure, a certain strong current exists behind the artificial island port. The maximum flow velocity reaches about 1.4m/s. We guess that this phenomena works for keeping the continuity of longshore sediment transport behind the port.

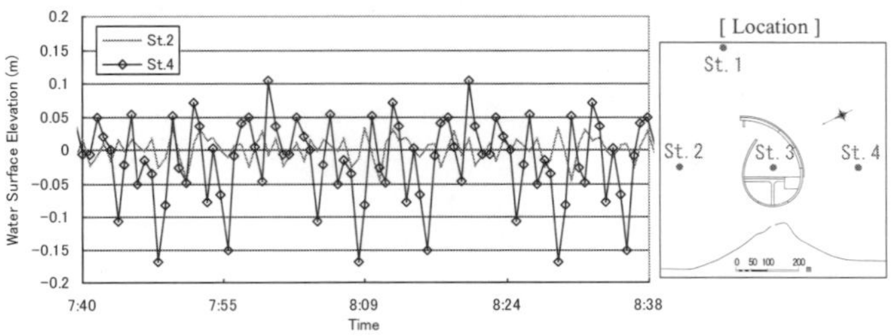

Figure.1 Moving-averaged surface elevation during storm

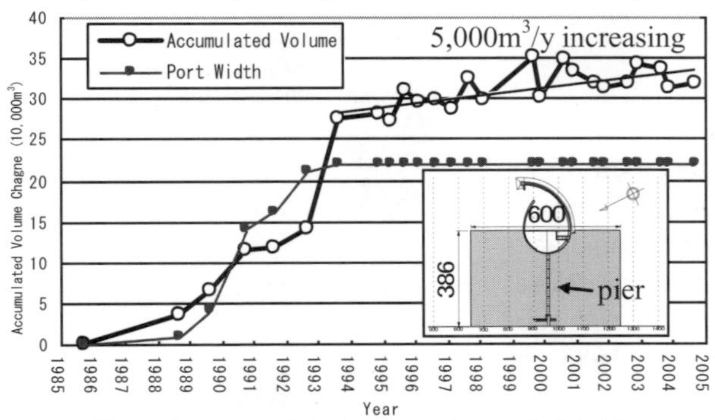

Figure.2 Accumulated volume change behind the port

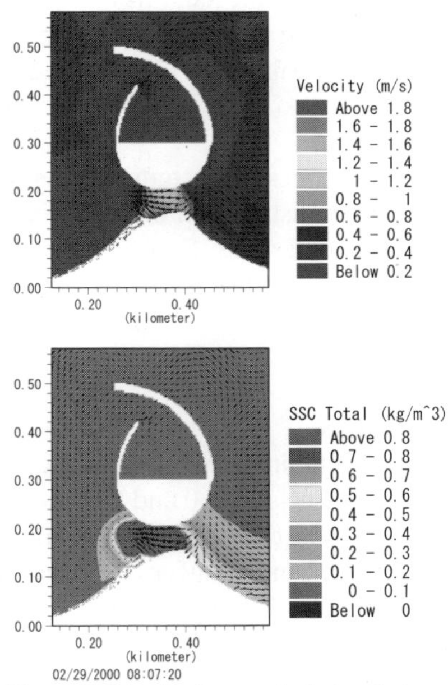

Figure.3 Result of numerical simulation

5. Conclusion

A comprehensive study on the littoral drift behind the artificial island fishing port is carried out. According to the field observation, the difference in phase and surface elevation was measured at the both sides of port, during comparatively high wave. In addition, it is investigated that the longshore sediment transport passes through the back and a certain strong current exists by estimating the longshore sediment transport rate and applying a general numerical simulation.

References

Dean,R.G. and Dalrymple, R.A. 2002. *Coastal Process with Engineering Applications*, Cambridge Univ. Press.

Kishi, H., Kanbara K., Nagano, A., Ujii, K. and Sasaki, T. 2000. *Artificial Island Fishing Port in Littoral Drift System and Evaluation*, Coastal Engineering, JSCE, Vol. 47, pp. 571-575. (in Japanese)

Yamamoto, A., Nakanishi A., Numata, T., Kashiba, N., and Kondo. T. 2005. *Nearshore Deformation by Artificial Island Fishing Port and Validation of Movable Bed Model Study*. Proceedings of Coastal Engineering, JSCE, Vol. 52, pp. 476-480. (in Japanese)

Kondo, H. and Kawamori, A. 2005. *Estimation of Littoral Drift Factors in Hokkaido Coasts*. Proc. of Hokkaido Br. of JSCE, 11-35, Vol. 61. (in Japanese)

Coastal Dynamics 2009
Paper No. 39

PREDICTION OF DEVELOPMENT OF SAND BODY AND MOVEMENT OF FINE SAND OFFSHORE OF DETACHED BREAKWATERS

Masaya Fukuhama[1], Takaaki Uda[2], Koji Yamada[3], Masumi Serizawa[4], Toshiro San-nami[4] and Toshinori Ishikawa[2]

1. Introduction

On the Shizuoka coast facing Suruga Bay, many detached breakwaters have been constructed since the 1970s as a measure against severe beach erosion caused by the decrease in fluvial sand supply from the Abe River. Since the construction of these detached breakwaters, the northward movement of a sand body with a velocity of 250 m/yr has been observed (Uda *et al.*, 2008, ICCE). In contrast, the movement of fine sand has also been observed seaward of the detached breakwaters. Here, a model for predicting both the development of a sand body and the movement of fine sand offshore of the detached breakwaters is developed.

2. Method

Beach changes were studied using bathymetric survey data. The depth distribution of the grain size of seabed materials was also measured. Based on these field data, the contour-line-change model considering grain size changes (Kumada *et al.*, 2006, ICCE) was applied to predict the beach changes.

3. Results

The longitudinal profiles and the depth distribution of d_{50} along transects X=3.0 and 0.5 km on the coast are shown in **Fig. 1**. It is clear that gravel and fine sand accumulated shoreward and seaward of the detached breakwaters, respectively. **Figure 2** depicts the changes in the -6m contour and the shoreline relative to those in 1983. In addition to the movement of the sand body as shown in **Fig. 2(b)**, the sand accretion zone expanded at the depth of 6 m before the movement of the sand body. The mean velocity of this offshore sand accretion zone since 1983 was 375 m/yr, which is 1.50 times the average velocity of the sand body of 250 m/yr.

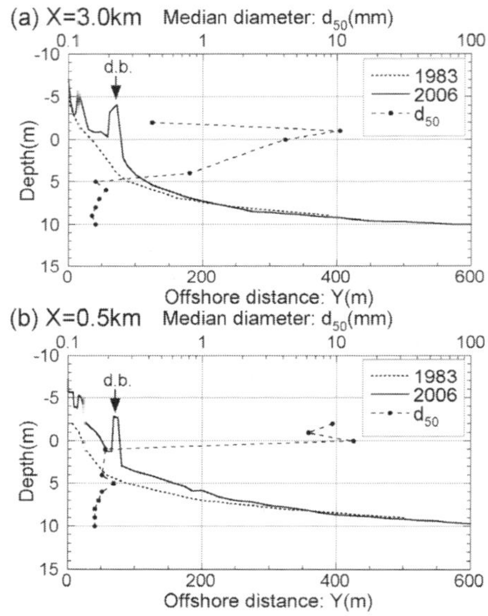

Figure 1. Longitudinal profiles and depth distribution of d_{50}

The bathymetry in 1983 before the development of the sand body was transformed into parallel contours using an expanded coordinate system. The initial longitudinal profile was determined from the measured profiles: 1/10 in the zone shallower than 4 m, 1/30 between 4 and 6 m, and 1/70 in the zone deeper than 6 m. Three characteristic grain sizes were assumed: fine sand with a grain size of 0.2 mm, gravel with a grain size of 2 mm, and pebbles with a large grain size of 100 mm, which is larger than the threshold grain size under the wave action. The initial contents of each grain size were

[1] Head, Kurobe River Office, Hokuriku Regional Development Bureau, Ministry of Land, Infrastructure, Transport and Tourism, 173 Tenjinshin, Kurobe, Toyama 938-0042, Japan. fukuhama-m92fc@hrr.mlit.go.jp

[2] Public Works Research Center, 1-6-4 Taito, Taito, Tokyo 110-0016, Japan. uda@pwrc.or.jp

[3] Coast Division, National Institute for Land and Infrastructure Management, MLITT.

[4] Coastal Engineering Laboratory Co., Ltd., 1-22-301 Wakaba, Shinjuku, Tokyo 160-0011, Japan. coastseri@nifty.com

assumed to be μ_1=0.0, μ_2=0.0, and μ_3=1.0, and their corresponding equilibrium slopes were 1/35, 1/10, and 1/3, respectively. Furthermore, we assumed that waves of H_0'= 3 m with a wave period of 9 s and probability of 5% are obliquely incident to the coastline with an angle of θ_w=10°, and that fine sand and gravel flow into the calculation domain at a rate of 7.0*10^4 m^3/yr through the south boundary. A free boundary condition was set at the north boundary. The diffraction effect of the detached breakwaters was evaluated by the angular spreading method. **Figure 3** shows the predicted movement of a sand body. Sand carried by longshore sand transport through the south boundary is blocked by the detached breakwaters, resulting in the formation of a salient on the lee side of each detached breakwater. Because the seabed slope between the shoreline and 4 m depth is 1/10, gravel with a large equilibrium slope is deposited and accumulates on this steep slope, whereas fine sand with a smaller equilibrium slope is selectively deposited and moves downcoast with a greater velocity than that of the sand body offshore of the detached breakwaters. The predicted velocity of the sand body is 252 m/yr, which is in perfect agreement with the measured velocity of 250 m/yr. The rate of expansion of the sand accretion zone in the offshore zone is 372 m/yr, which is 1.48 times the velocity of the sand body. This is also in good agreement with the measured ratio of 1.50. **Figure 4** shows the changes in the longitudinal profile at X=4.88 km and the cross-shore distribution of the content of each grain size. The formation of a steep slope crossing the detached breakwater due to the deposition of gravel and a gentle slope due to the deposition of fine sand are accurately predicted.

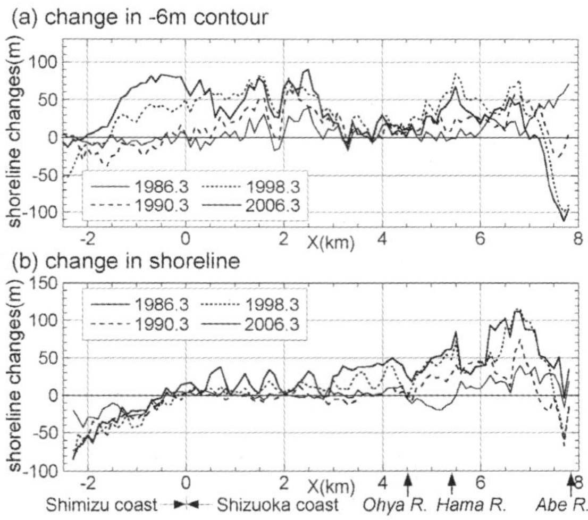

Figure 2. Changes in -6m contour and shoreline

4. Conclusion

The applicability of the present model to predict development and the movement of a sand body as well as the prediction of the movement of fine sand offshore of the detached breakwaters was confirmed.

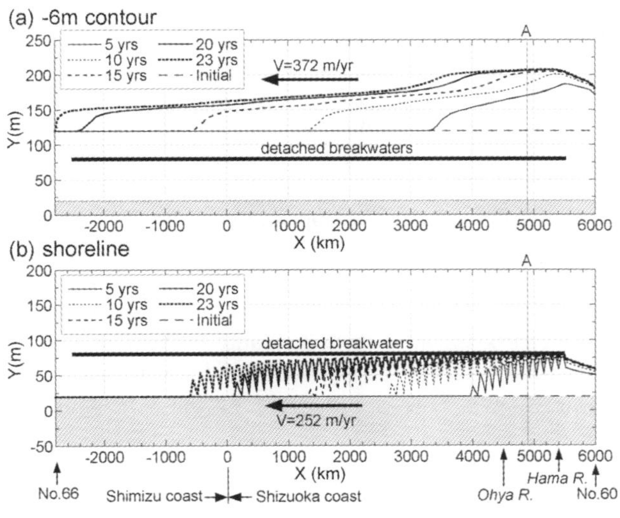

Figure 3. Predicted movement of a sand body

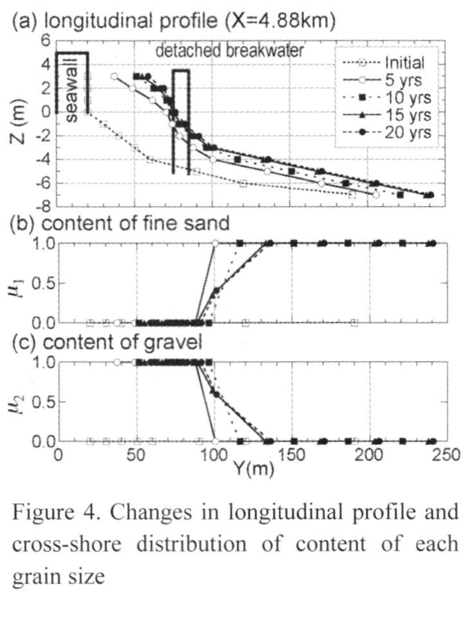

Figure 4. Changes in longitudinal profile and cross-shore distribution of content of each grain size

Coastal Dynamics 2009
Paper No. 40

Modeling of flow through a vertical perforated pipe in the beach and the morphodynamic interpretation– the PEM system

Jørgen Fredsøe[1], Peter Engesgaard[2] and Per Sørensen[3]

1. Introduction

It has been suggested that vertical perforated tubes placed below the beach surface will increase the drainage of the beach, and hence increase the deposition of sand on the beach. The system is called the PEM-system, Pressure Equalization System, and the Danish company SIC is doing the marketing. Although it for a coastal engineer seems obvious that such a device can't drain the beach, SIC has succeeded in installing the system in more than 75 locations around the world (according to the company) In Denmark a full scale experiment at the exposed west coast has been performed through 2005-08, and a similar Dutch test is going on right now at Egmond, Holland. In this paper, we model the flow in the beach taking into account the presence of (high-permeable) tubes and demonstrate that the drainage effect is negligible. Further, the morphodynamic behaviour of the coast in relation to the Danish field test is described.

2. Description of the system.

The principle in the PEM-system is as follows: An array of vertical perforated tubes is drilled down in the beach sand.

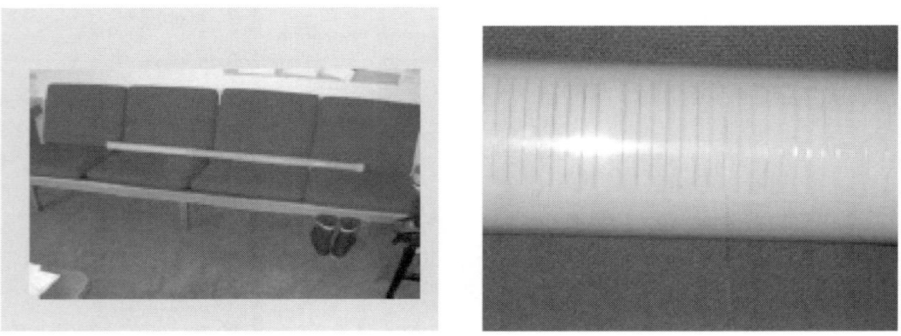

Figure1: Photo of tube in full length, approximately 1.60 m and inner diameter equal 6 cm. The tube is without slots in the upper half, but it is ventilated at the top, so air can go through. To the right: Blow up of the tube: Slots of 0.2 mm width are cut in the lower half of the tubes, so water can flow in or out.

3. Flow through the tubes.

The fluid must be either water from the Sea (usually salty) or from inland (usually fresh). The flow must be either up- or downwards directed. If you have a flow in the tube, and it is upward, then the water must enter the tube at the bottom (or lower half) and escape at the top (upper half). If on the other hand side, the flow is downwards inside the tube, the opposite must be the case. For this reason it is surprising that the tube only is permeable at the lower part. Figure 2 shows an example on tidal induced flow. At I, the flow is directed down, and it is easier to flow through the tube than outside in the surrounding soil: in the tube there are nearly no flow resistance, and with small flow velocities, the pressure within the tube can be taken to be hydrostatic. In the soil you need an excess pressure gradient (in this case negative) to force the flow through the soil, where there is a considerable flow resistance (the Darcy law). This is illustrated by the schematic

[1] Dept. Mechanical Eng., Technical Univ. Denmark jf@mek.dtu.dk.
[2] Dep of Geography and Geology, University Copenhagen pe@geol.ku.dk.
[3] Danish Coastal Authority per.soerensen@kyst.dk.

pressure distribution in figure 2. The continuity equation for the tube requires (in a quasi-steady flow) that the flow into the tube equals the flow out. This requirement determines the water level within the tube relative to the water level outside the tube.

Refined modeling:
Figure 3 show the result of a numerical analysis of the flow (high tide). Supply of fresh water from hinterland is included. The flow pattern, strength and direction are shown in figure 3, which also includes salinity variation.

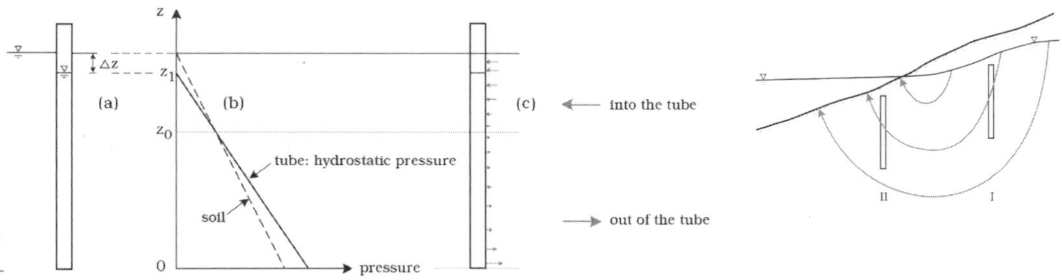

Figure 2: Schematic picture of tidal induced flow and pressure distribution out-and inside the tubes.

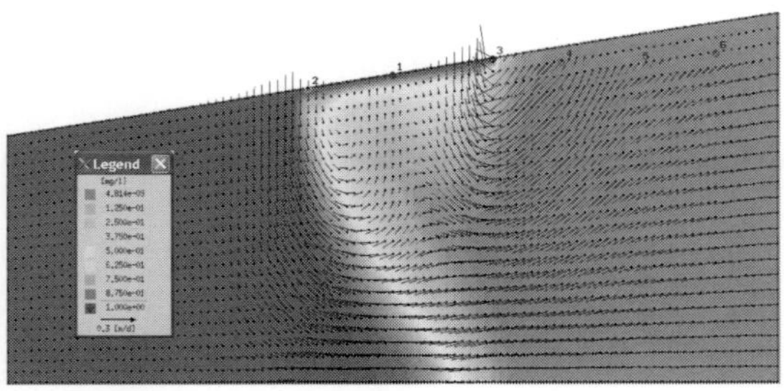

Figure3: refined flow modeling including frewsh/salt water

4. Morphodynamic changes.

An 11 kilometer long stretch was allocated to the test. The stretch was divided in 2 sections with tubes: a 4500 meter long section and another much shorter (900 meter). Three Reference sections, each 1800 meters long, are sections without tubes, for comparison. The changes in beach and shoreface volume was measured during all three years. The impact of the tubes are discussed and compared to natural fluctuations.

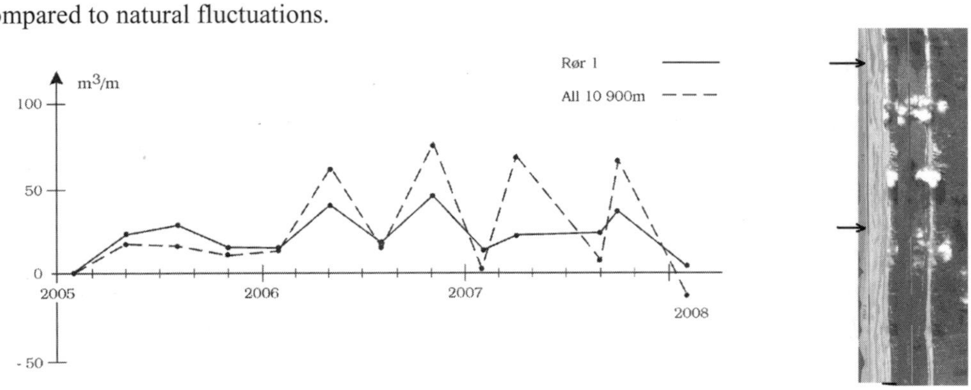

Figure 4: Temporal variation in beach volume through three years. To the right: interaction between the bar system and the beach changes is very clear in the test.

Coastal Dynamics 2009
Paper No. 41

EXAMINATION OF EFFICACY OF SUBMERGED GROIN SYSTEM AGAINST BEACH EROSION THROUGH NUMERICAL SIMULATION

Yoshimi Goda[1], Nobuyuki Ono[2], and Yoshiyuki Uno[3]

1. Introduction

There have been a few reports that a group of submerged groins is effective for enhancing sand accretion and preventing beach erosion. Holmberg (2001) demonstrates several successful cases in the United States but without any explanation of sand accretion mechanism. Goudas et al. (2001) also presents a number of field applications and try to explain the mechanism based on the 2-D current analysis.

The authors propose a mechanism such that a group of submerged groins introduce the artificial roughness against the longshore currents and reduce the current speeds. Within the zone of weakened longshore currents, suspended sediment settles down quickly compared to the case without submerged groins. Increase of sediment settlement would enhance sand accretion around the area of submerged groins.

The validity of the proposed mechanism is examined with numerical simulation of longshore currents induced by directional random waves with and without submerged groins. Transport and settlement of suspended sediment are also simulated for examination of the mechanism.

2. Models of Longshore Current Generation and Suspended Sediment Motion

Longshore currents by obliquely incident random waves are computed by taking account of radiation stresses and surface rollers. The 1-D model employed by Goda (2008) for planar beaches is extended to the 2-D space. Sediment pickup from the seabed by random breaking waves is evaluated by the model by Katayama and Goda (2000).

3. Beach Profile and Submerged Groin System

Computation is made for Dean's equilibrium beach profile for the sand with the median diameter of d_{50} = 0.3 mm. Incident deepwater waves are given the height of H_0 = 2.0 m, the period of $T_{1/3}$ = 8.0 s, and the incident angle of θ_0 = 20°. Seven submerged groins extended to the depth of –2 m are set at the spacing of 48 m. The groins have the height of 0.5 m above the seabed, but their crests are kept at –0.1 m near the shore. Figure 1 shows the beach profile and a submerged groin. The grid size of computation is $\Delta x = \Delta y = 4$ m.

The friction coefficient of the seabed is set at C_f = 0.01. The artificial roughness of submerged groins is represented with a local enhancement of the friction coefficient to C_f = 0.05 for the three meshes at the down-drift side of each groin.

The boundary condition at the shoreline was so set to allow a finite speed there.

4. Current Profiles and Sediment Transport

Figure 2 shows comparison of current profiles at the middle location between groins with and without submerged groins. Attenuation of current speeds by groins is clearly noticed. Figure 2 also shows the comparison of sediment concentration and transport rate profiles for the cases with and without groins. Suspended sediment by the wave breaking is almost the same at the center of groins and sediment

[1]ECOH CORPORATION, 2-6-4 Kita-Ueno, Taito-Ku, Tokyo 110-0014, Japan. goda@ecoh.co.jp
[2]ECOH CORPORATION, ditto. n-ono@ecoh.co.jp
[3]ECOH CORPORATION, ditto. uno@ecoh.co.jp

transport rate is attenuated in proportion to current profiles. Figure 3 shows the spatial distribution of morphological changes after 24 hours of wave actions. Accretion on the up-drift side of groins and erosion on the down-drift side take place with the total accretion volume exceeding the erosional one.

5. Conclusions

The validity of the proposed mechanism for sediment accretion by submerged groins has been confirmed through numerical simulations.

References

Aminti, P., Cammelli, C., Cipriani, L. E., and Pranzini, E., 2001. Evaluating the effectiveness of a submerged groin as soft shore protection. *Proc. 1st Conf. Soft Shore Protection*, Patras, Greece, 151-158.

Goda, Y. 2008. Wave setup and longshore currents induced by directional spectral waves: Prediction formulas based on numerical computation results. *Coastal Engineering Journal*, **50** (4),397-440

Goudas, C. L., Katsiaris, G. A., Labeas, G., Karahalios, G., and Pnevmatikos, G., 2001. Soft protection using submerged groin arrangements – dynamic analysis of system stability and review of application impacts, *Proc. 1st Conf. Soft Shore Protection*, Patras, Greece, 167-186.

Holmberg, D., 2001. Alternative to traditional ways of treating shoreline erosion, *Proc. 1st Conf. Soft Shore Protection*, Patras, Greece, 139-150.

Katayama, H. and Goda, Y., 2000. A sediment pickup rate formula based on energy dissipation rate by random breaking, *Proc. 27th Int. Conf. Coastal Eng.*, Sydney, ASCE, 2859-2872.

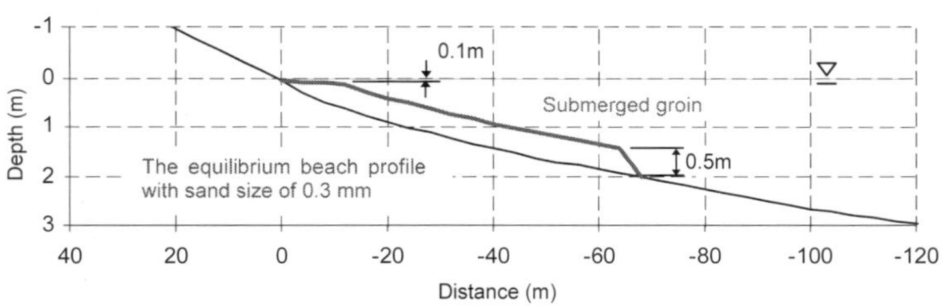

Figure 1. Beach profile and a submerged groin

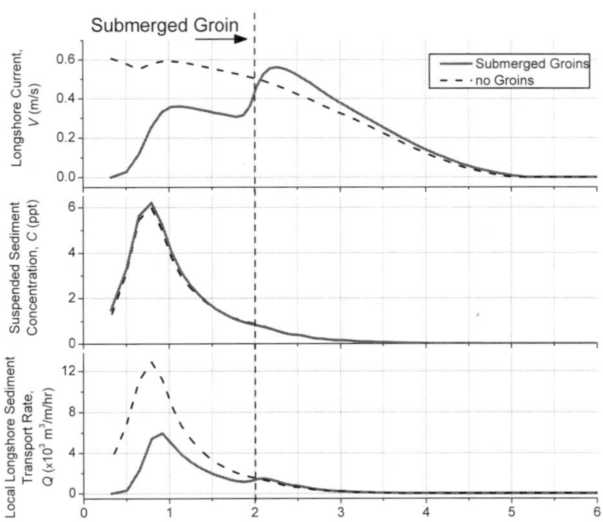

Figure 2. Comparison of sediment transportation, longshore current and sediment transport rate for the cases with and without submerged groins.

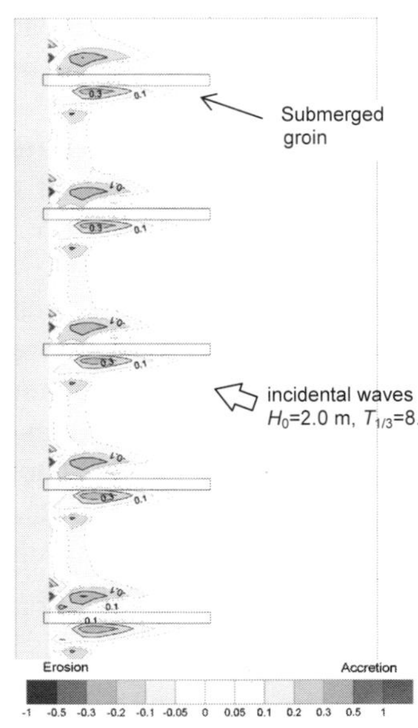

Figure 3. Spatial Distribution of morphological changes around submerged groins

Coastal Dynamics 2009
Paper No. 42

PHYSICAL ENVIRONMENT CHANGE OF LONG SANDY BEACH BY HEADLAND SYSTEM FOR EROSION CONTROL

Akio Okayasu[1] and Takashi Nemoto[2]

1. Headland system and Kashimanada Coast

The headland system, a group of headland groins, is a variation of coastal erosion control works creating equilibrium shoreline forms against seasonal changes of wave directions. However, it separates a long shoreline into small beaches which are individual in terms of sediment transport. Kashimanada Coast is 70km long sandy beach at the Pacific side of Japan. After the completion of Kashima Port at the middle of it, coastal erosion had become a serious problem. It had been therefore planned to install 40 headland groins for every 1km, and 33 of them have been completed since 1985.

It is found in the previous studies that the headland groins are effective against the erosion, but influences of them to the nearshore physical environment are not enough studied. In the present study, time variations of shoreline positions, foreshore slopes, sand grain size and nearshore bathymetry including longshore bar profiles were investigated for the north Kashimanada Coast, a 40km long stretch of coast located between Oarai Port and Kashima Port.

2. Shoreline Change and Grain Size Distribution

Figure 1 shows measuring lines for beach profiles together with positions of the headland groins and sediment sampling. From the change of coastline position obtained from aerial photographs in 1975 and 1984 and bathymetrical surveys from 1984 to 2004, it was found that the beaches located just next to Oarai Port and Kashima Port had been depositional and the areas next to those depositional beaches had been heavily eroded. From the data of 1984 (before the installation of headlands) to 2004, it was found that the headlands had recovered almost half of the eroded shoreline change, which means the headland system was effective for coastal protection.

Longshore variations of grain size at shoreline in 1982 and 2003 are given in Fig. 2 which shows a tendency that grain size became coarser in the eroded areas and that the areas were expanding for about 10km in 20 years. It was found that foreshore slopes tends to be steep at the coarser sand area.

3. Beach Profiles and Nearshore Bathymetry Changes

From the beach profiles, changes of cross-shore positions for the depth of 2m were investigated. Figure 3 gives those for x (longshore axis given in Fig. 1) =1, 4, 24, 39km. The erosive areas which have coarser sand shows retreat of the -2m contour, especially from the year around 2000. Cross-shore beach profiles at x =16km (in the depositional area) and x =30km (in the erosive area) are shown in Figs. 4 and 5, respectively. It can be seen that offshore sandbars were diminished in both areas from 1984 to 2004. At x =16km, the amount of accretion near shoreline is 73.8m^3 per width, whereas the decrement of sand around offshore sand bars is 226 m^3. From analyses of other profiles, the offshore sandbars have been diminished in almost 2/3 of the total stretch in the period of 1984 to 2004.

This gives an idea that under the unchanged condition of sediment supply, the recovery of shoreline position brought by the headland system could be made by cross-shore sediment movement. The headland groins have 200m in the cross-shore direction with each of 1 km coastal stretch, which

[1]Dept. of Ocean Sciences, Tokyo University of Marine Science and Technology, 4-5-7, Konan, Minato, Tokyo 108-8477, Japan. okayasu@kaiyodai.ac.jp

[2]Ibaraki Prefecture Freshwater Fisheries Research Center, Tamatukuri-ko 1560, Namegata, Ibaraki, 311-3512, Japan. nemoto@ignaisuishi.ecnet.jp

gave arch geometry instead of the original long straight shoreline. The sediment movement caused by the headland system as well as the change of beach geometry due to separation of long straight beach may result in the extinction of the offshore sandbars. At the north part of Kashimanada Coast, the decrement of bottom elevation in the offshore area is widely observed even in the reegions showing accretion at shoreline except in the vicinity of the ports.

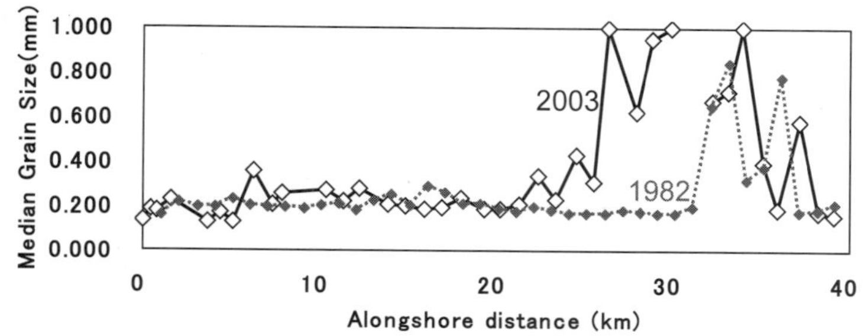

Figure 2. Longshore variations of grain size at shoreline in 1982 and 2003

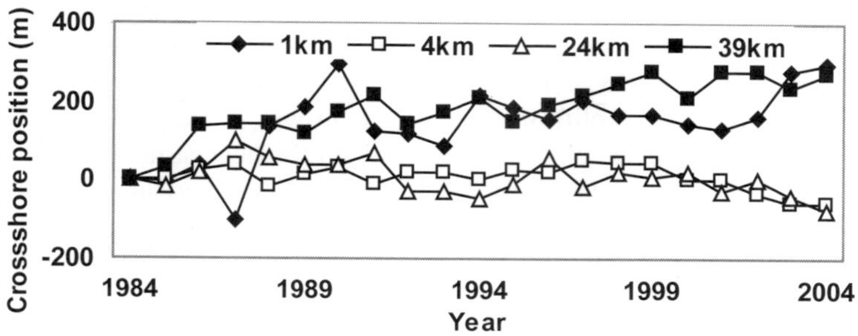

Figure 3. Time variations of cross-shore positions for depth of -2m at x=1, 4, 24, 39km.

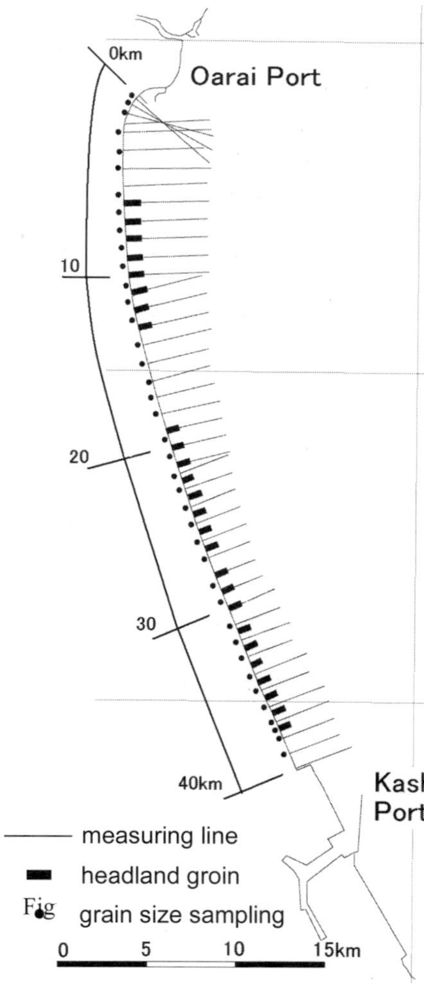

Fig

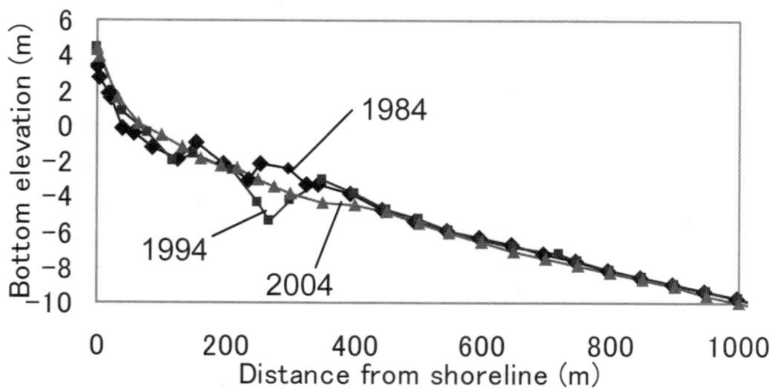

Figure 4. Bathymetrical profiles of depositional beach at x=16km for years of 1984, 1994 and 2004.

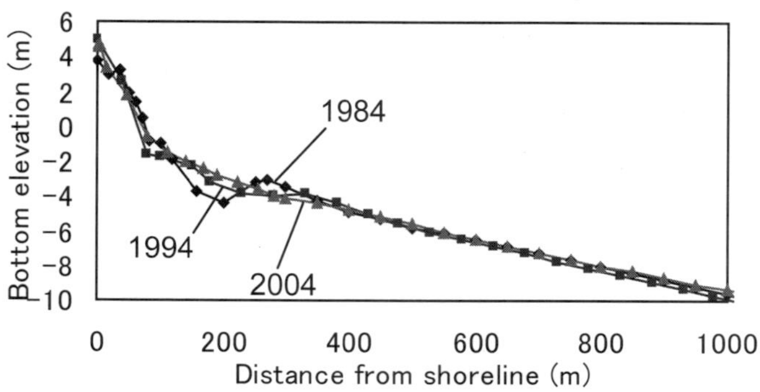

Figure 5. Bathymetrical profiles of erosive beach at x=30km for years of 1984, 1994 and 2004.

Coastal Dynamics 2009
Paper No. 43

Numerical modeling of hydrodynamics on a salt marsh

Tomohiro Suzuki[1], Taro Arikawa[2] and Marcel J.F. Stive[1]

1. Introduction

It has been widely recognized that ongoing climate change may accelerate sea level rise and increase storm frequency, which would be a threat especially for low-lying countries like the Netherlands. To deal with these threats, it is important to explore sustainable and cost-effective methods to protect the shoreline. Salt marshes could form such a protection, and have large ecological values. Specifically, plants on salt marsh play a role in wave attenuation, current attenuation and sediment deposition.

In spite of its significance, hydrodynamics on a vegetated salt marsh has not been clarified yet. Suzuki et al. (2008) conducted a physical model test and numerical simulation about wave attenuation by vegetation and cliff of salt marsh. However, to clarify the process of its erosion and deposition, it is necessary to reveal bottom velocity and turbulence structure around vegetation. To this end, numerical simulation is supposed to be useful. As a numerical model, the three-dimensional VOF model CADMAS-SURF/3D (Arikawa et al., 2005; Arikawa et al., 2007) developed by P.A.R.I. was used and its applicability to this study was tested by comparing previous numerical and experimental results.

2. Approach and Results

To verify the applicability of the numerical model to salt marsh study, we have conducted four cases of calculations as below:

1) Wake after single cylinder: Wake after single square cylinder was tested to confirm the basic hydraulic characteristics by this model. Figure 1 shows velocity distribution after 100 s and velocity time series in Y-direction. Wake started to oscillate around 40 s after the flow started and the Strohal number, which is non-dimensional indication of vortex, corresponds to Tamura (2003).

2) Wake after multiple cylinders: Vegetation on a salt marsh consists of many cylindrical structures. Hence, it is necessary to elaborate hydraulic characteristics in multiple cylinders. As shown in the time series in Figure 1, formation of vortices in multiple cylinders is weaker than that in the single cylinder case. This is because of mutual interference among small vortices produced by single cylinders and this prevents the formation of large vortices.

3) Hydraulic phenomena around a circular cylinder under wave motion: Hydraulic phenomena around a circular cylinder under wave motion were confirmed as shown in Figure 2. When the wave passes through the cylinder, a horse-shoe vortex can be found. The velocity distribution including the vortex around the circular cylinder is supposed to be related to sedimentation in vegetation.

4) Comparison between numerical and experimental wave height in a salt marsh: Numerical results were compared to physical model test results done by Suzuki et al. (2008). In the numerical simulation, the computation area was limited as shown in Figure 3 since three-dimensional computation costs huge computation time. Figure 4 shows wave height in the case of with vegetation and without vegetation. As can be seen in Figure 4, numerical simulation results show good agreement with the experimental results. One of the advantages to use a three-dimensional model is to be able to obtain bottom velocity directly. Bottom velocity under a wave condition was investigated as shown in Figure 5.

---

[1]Dept. of Hydraulic Eng., Delft University of Technology, Stevinweg 1, 2628 CN Delft, The Netherlands. t.suzuki@tudelft.nl
[2]Maritime structure div., Port and Airport Research Institute, 3-1-1 Nagase, Yokosuka 239-0826, Japan. arikawa@pari.go.jp

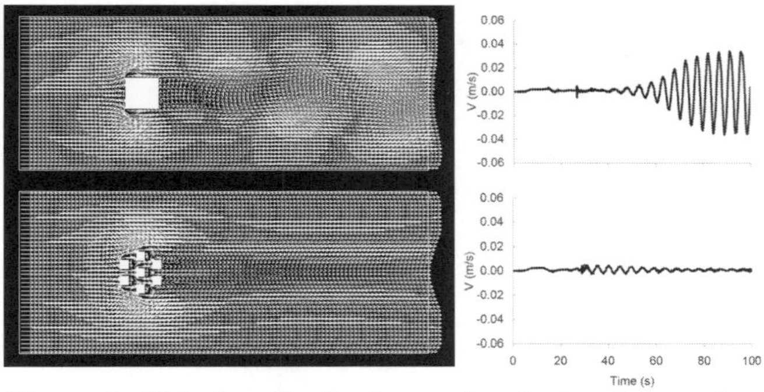

Figure 1. Velocity distribution and velocity time series. Upper figures: under a single cylinder, lower figures: under multiple cylinders.

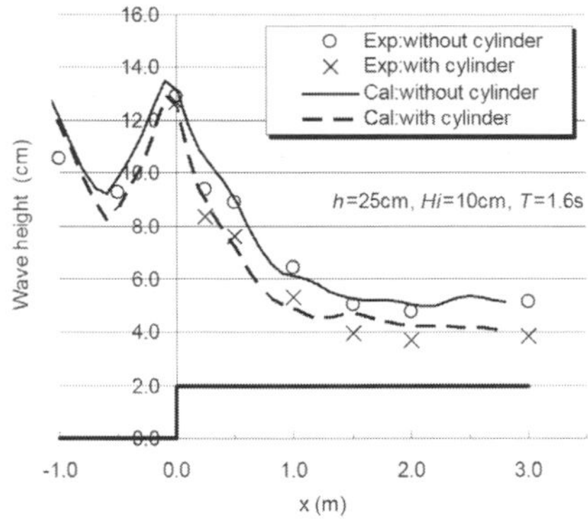

Figure 4. Distribution of wave heights in the case of with cylinders and without cylinders

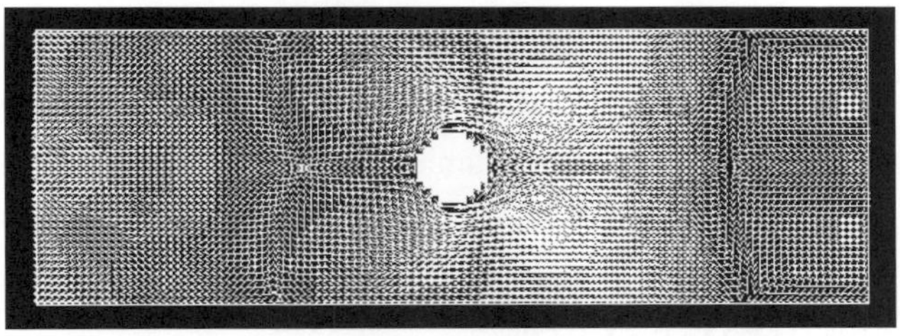

Figure 2. Velocity distribution under a wave condition

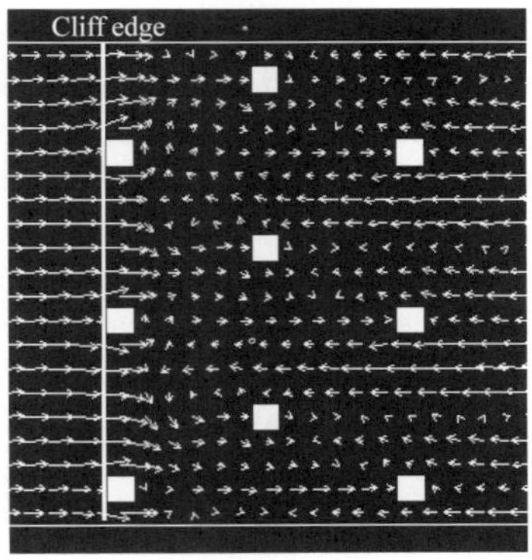

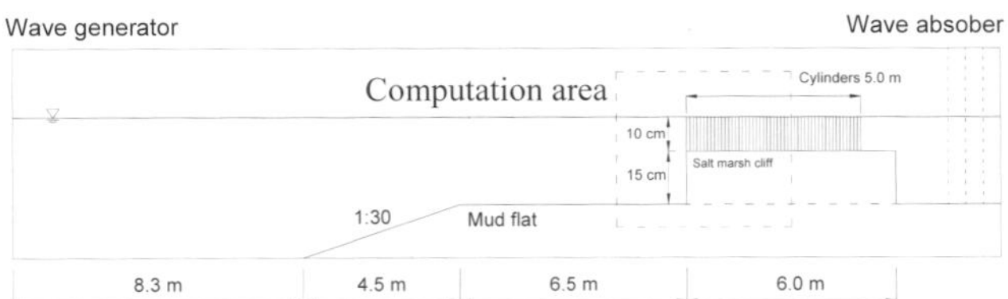

Figure 3. Experiment set-up and computation area.

Figure 5. Calculated bottom velocity distribution in the case of with vegetation.

3. Conclusions

Applicability of a numerical wave channel to hydrodynamics on a salt marsh was tested in this paper. The numerical model results show good agreement with the previous and experimental results; hence, this model can be a useful tool to investigate hydrodynamic phenomena around the cliff in detail.

Acknowledgements

The first author (T. Suzuki) was supported by a scholarship awarded by the Rotary Foundation and currently is supported by Het Lammingafonds. These financial supports are gratefully acknowledged.

References

Arikawa, T., F. Yamada, M. Akiyama. 2005. Study of the Applicability of Tsunami Wave Force in a Three-Dimensional Numerical Wave Flume. *Annual Journal of Coastal Engineering*, JSCE, Vol.52, 46-50. (in Japanese)

Arikawa, T., T. Yamano and M. Akiyama. 2007. Advanced Deformation Method for Breaking Waves by using CADMAS-SURF/3D, *Annual Journal of Coastal Engineering*, JSCE, Vol54, 71-75. (in Japanese)

Tetsuro Tamura (2003) Flow past a Square and a rectangular cylinder - Turbulence effects on its aerodynamics, *Nagare* 22, 7-13. (in Japanese)

Suzuki, T., J. Dijkstra and Marcel J.F. Stive. 2008. Wave dissipation on a vegetated salt marsh. *Proceedings of 31th International Conference on Coastal Engineering*, ASCE. (in press)

Coastal Dynamics 2009
Paper No.44

NUMERICAL EXAMINATION ON COUNTERMEASURE AGAINST WAVE OVERTOPPING DISASTER AROUND ROAD REVETMENT USING TWO-DIMENSIONAL NUMERICAL WAVE FLUME BASED ON VOF METHOD

Koji Kawasaki[1] and Masami Kiku[2]

1. Introduction

Wave overtopping around coastal revetments has caused human and property damages in Okinawa Main Island which is one of the typhoon-prone area in Japan because a lot of major roads and residential areas have been developed behind the revetments. Therefore, immediate countermeasures against wave overtopping around revetments should be required for overtopping disaster prevention. On the other hand, comprehensive viewpoints including not only disaster prevention but also the water environment, economy and coastal utilization are needed to design coastal facilities, and various new types of seawalls have recently been adopted in place of vertical or block mound seawalls. However, a precise estimation procedure for wave overtopping rate of complex-shaped seawalls has not been established yet. The purpose of this study is to numerically examine countermeasures against wave overtopping around road revetment at Okinawa Main Island by conducting numerical simulation based on a VOF (Volume Of Fluid) method changing the shape and crest height of seawalls.

2. Two-Dimensional Numerical Wave Flume "CADMAS-SURF"

A two-dimensional numerical wave flume "CADMAS-SURF" (SUper Roller Flume for Computer Aided Design of MAritime Structure) (Coastal Development Institute of Technology, 2001) based on the VOF method was employed in this study in order to compute wave overtopping phenomena.

The wave overtopping phenomena around the road revetment at Okinawa Main Island under the typhoon 200423 on October 23rd, 2004 were simulated to validate the CADMAS-SURF qualitatively. Figure 1 illustrates the cross-section of the bottom topography and the revetment, which were modeled on the base of the detailed field survey. As shown in Fig. 2, the numerical results were in agreement with the video images, and the CADMAS-SURF can, therefore, represent violent wave overtopping phenomena at actual field with accurate accuracy.

3. Numerical Examination of Countermeasures against Wave Overtopping Disaster

In this study, three different types of revetments were taken up as shown in Fig. 3, in which (a), (b)

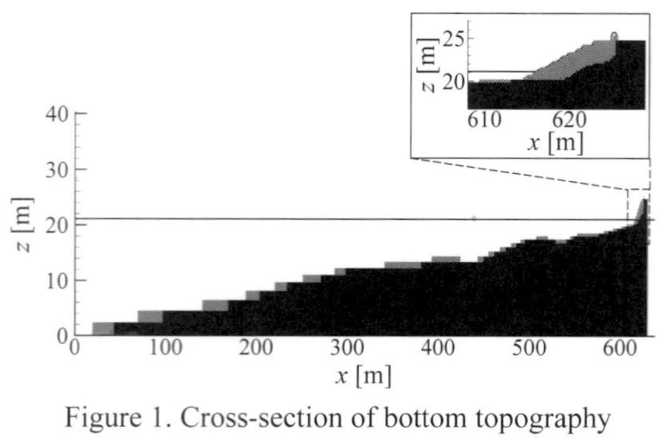

Figure 1. Cross-section of bottom topography

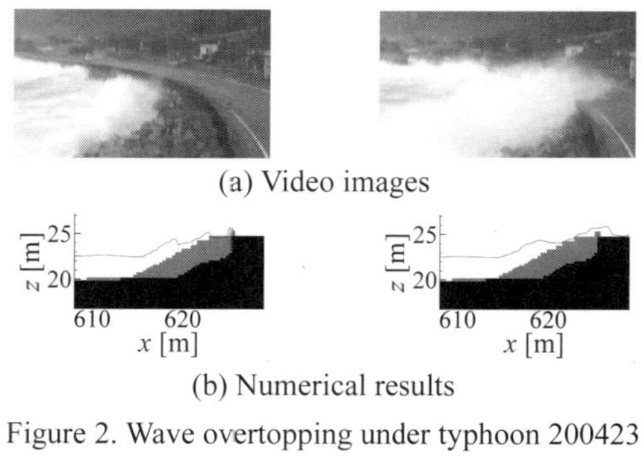

Figure 2. Wave overtopping under typhoon 200423

---

[1]Dept. of Civil Eng., Nagoya University, Furo-cho, Chikusa-ku, Nagoya 464-8603, Japan. kawasaki@civil.nagoya-u.ac.jp

[2]Dept. of Civil Eng., Nagoya University, Furo-cho, Chikusa-ku, Nagoya 464-8603, Japan. kiku@nagoya-u.jp

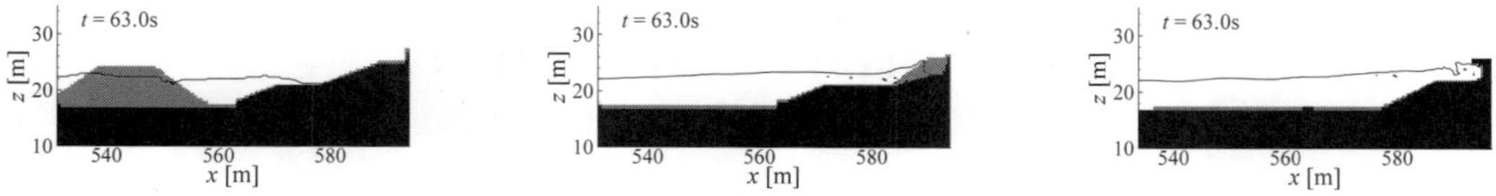

(a) Structure 1 (b) Structure 2 (c) Structure 3

Figure 3. Numerical results

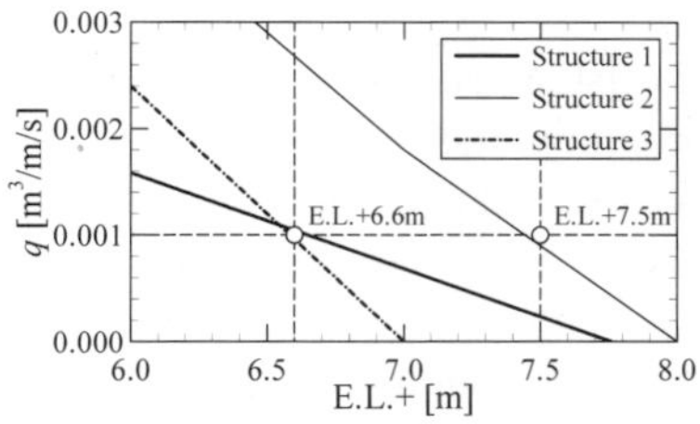

Figure 4. Estimation of optimum crest height of revetments

Table 1. Comprehensive evaluation of revetments

	Structure 1	Structure 2	Structure 3
Crest height	A E.L.+7.5m	B E.L.+8.0m	A E.L.+7.5m
Economy	B	A	A+
Environment	C	B	A+
Workability	B	A	B+
Total	B	A	A+

* A+ : Superior, A : Above Average, B : Average, C : Below Average

and (c) indicate an upright seawall with an offshore breakwater (Structure 1), a seawall covered with wave-dissipating blocks (Structure 2) and a flaring-shaped seawall (Structure 3), respectively. The height and period of the incident regular wave were 6.2m and 15.39s respectively, based on the 50-year design wave condition.

From Fig. 3 showing the numerical results at 63.0s after the start of the computation, the CADMAS-SURF can be said to adequately compute wave deformation and velocity fields induced by the different configurations of the revetments, such as wave attenuation due to the offshore breakwater or the blocks (Figs. 3(a) and (b)) and wave reflection by the seawall (Fig. 3(c)). The optimum crest heights of each revetment were calculated by numerical computation changing the crest heights of each revetment by 1.0m so that the wave overtopping rate can be below the allowable rate (1×10^{-3} $m^3/m/s$). The design crest heights of each revetment were, furthermore, decided by adding the extra crest heights (0.5m ~ 1.0m) to the optimum crest heights shown in Fig. 4.

The comprehensive evaluation of each revetment was implemented in terms of the function, cost economy and workability of revetment construction and the coastal environment, as summarized in Table 1. Table 1 shows that Structure 3 would be the most suitable revetment for the devastated area, which was taken up in this study.

4. Conclusions

This study has numerically examined countermeasure against wave overtopping disaster around the road revetment in Okinawa using the numerical wave flume "CADMAS-SURF". As a result, the utility of the CADMAS-SURF was qualitatively confirmed in comparison with the on-site situation of wave overtopping. A numerical wave flume such as the CADMAS-SURF was, furthermore, found to be useful in comprehensively discussing the prevention of wave overtopping disaster.

Acknowledgements

The authors would like to thank Mr. Hiroshi Shimada (Pacific Consultants Co., LTD.) for the support of the present work.

References

Coastal Development Institute of Technology, 2001. *Research and Development of Numerical Wave Flume, (Super Roller Flume for Computer Aided Design of MAritime Structure)* (in Japanese).

Coastal Dynamics 2009
Paper No. 46

WAVE AMPLIFICATION AND AIR-GAP RESPONSE UNDER A MULTI-COLUMN PLATFORM

Jing Li[1] Soon Keat Tan[2] Zhenhua Huang[3] and Adi Kurniawan[4]

1. Introduction

Waves are the major environmental loadings affecting coastal and offshore structures. In the case of coastal structures such as a jetty or wharf with overhang skirting or continuous decking systems, the heights of incident waves can be significantly amplified under the deck resulting in the loss of air-gap, leading to negative loading or under deck slamming. In some cases, deck overtopping could take place. It is the consensus that both wave amplification and related air gap are necessary and critical considerations in the design of large span deck structures. Wave amplification (i.e. relative wave amplification) is normally depicted in terms of the maximum and minimum water surface elevations or defined as the ratio of the standard deviation of resulting waves to the standard deviation of incident waves, while air-gap is the corresponding clearance under the deck (Kriebel, 2001). This paper aims to study the effects of trapped waves under a multi-column floating platform, and numerically investigate the corresponding air-gap response. Finally the study would establish the extreme distribution of air-gap under the platform, determine the critical locations and evaluate the air-gap behavior, in terms of the potential loadings underdeck.

2. Methodology and Approach

A suitable numerical method was developed and adopted to analyze the wave amplification and associate air-gap response. Initially, the incident random wave was treated as Gaussian. Considering the effects of structural response, instantaneous relative wave elevation along the structure was expressed as local wave elevation and vertical displacement of the floating structure (rigid body). The corresponding air-gap, defined as the difference between the static air-gap and maximum wave record at the same point, at selected strategic locations was evaluated. In particular, the extreme air gap observed in each wave train was extracted by means of the wave-by-wave analysis. The three-parameter distribution - Pearson type III distribution was introduced here to perform the extreme distribution of air gap (IMSL, 1987).

3. Results and Discussions

A case study was presented herein to illustrate the proposed method. WAMIT diffraction analysis will be performed for three wave headings (head seas, quartering seas, and beam seas) at five critical locations of 6-column-model (WAMIT, 2006).

After the target wave spectrum has been established, the computation will be performed of two sea states (storm A&B) (Adi et al., 2009). Figure 1 shows the wave amplification of the model in head sea, while Figure 2 shows the extreme surface elevations (maximum and minimum) at each location compared with incident wave. It can be seen that the local wave elevation at position 4 will be significantly amplified while the wave field at other locations do not experience significant changes.

[1]DHI-NTU Centre, Nanyang Technological University,50 Nanyang Avenue,639798,Singapore, Jingli@ntu.edu.sg
[2]School of Civil and Environmental Engineering , Nanyang Technological University, 50 Nanyang Avenue, 639798,Singapore, ctansk@ntu.edu.sg
[3]School of Civil and Environmental Engineering, Nanyang Technological University, 50 Nanyang Avenue, 639798,Singapore, zhhuang@ntu.edu.sg
[4]Maritime Research Centre, Nanyang Technological University, 50 Nanyang Avenue, 639798, Singapore, Kuniawan@ntu.edu.sg

The observed largest amplification, exceeding 1.4 (in storm A) and 1.1 (in storm B) occur at point 4. The corresponding values at all other locations are less than 1 even under severe ocean environment (storm B).

Extreme air-gap distributions at certain critical locations are shown for the case of head seas in two sea states as well. Generally, the probability distribution of the simulated air–gap was found to show good agreement with the proposed extreme distribution. It was also found that extreme air-gap response appear to decrease with the exasperating environment. The extreme distribution of air-gap response at different locations revealed and pin-pointed the most critical position of the floating platform.

4. Conclusion

This paper presented the findings of a numerical investigation of the wave field along a 6-colomn floating platform. The effect of wave on the structure in terms of the wave amplification was presented. Several potential locations were examined to point out the critical position under each wave heading. Furthermore, a novel extreme distribution of air-gap response was also presented to secure the prediction of extreme air gap demand precisely. The results showed that the wave field might be significantly amplified at certain critical loading point. Examining the appearance of local wave amplification together with the air-gap response will be a necessity to yield insight to the design of floating systems.

References

IMSL(International Mathematical and Statistical Libraries), 1987. STAT/LIBRARY User's Manual, Version1.

Kriebel., D, and Wallendorf, L., 2001, *Air Gap Model Tests on a MOB Module*, Proc. 11th ISOPE Conf., pp. 286-293.

Sweetman, B., and Winterstein, S. R., 2003, *Non-Gaussian Air Gap Response Models for Floating Structures*, ASCE Journal of Engineering Mechanics, 129(3):302-309.

WAMIT User Manual (2006) WAMIT, Inc. (available from the website www.wamit.com).

Kurniawan A., Huang. Z., Li J., Liu C., Wang X., Hao Z., Tan s. and Nah E., 2009, *A Numerical Analysis of the Response and Air Gap Demand for Semi-submersibles,* Proceedings of the 29th International Conference on Offshore Mechanics and Arctic Engineering (OMAE2009)

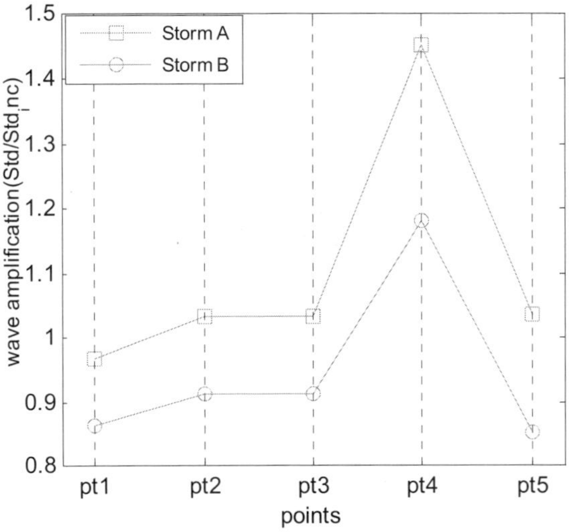

Fig.1. Wave amplification in terms of standard deviation for head seas

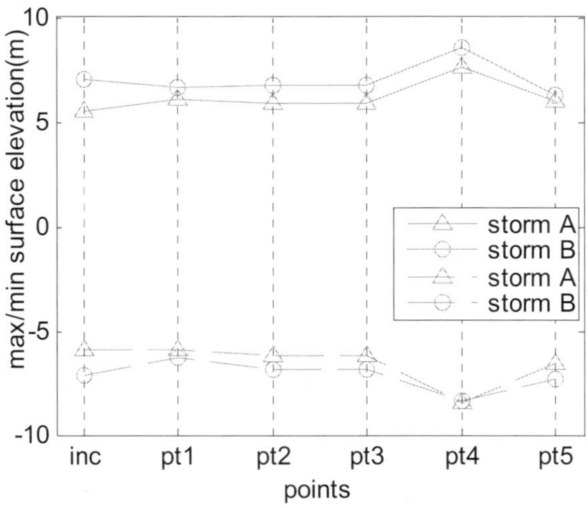

Fig.2. Wave amplification in terms of maximum and minimum surface elevations for head seas

Coastal Dynamics 2009
Paper No.47

RESEARCH AND DEVELOPMENT OF FLAP GATE-TYPE WAVE BREAKER FOR PROTECTION AGAINST TSUNAMI AND STORM SURGE

Hideyuki Niizato[1], Yuichiro Kimura[2], Kyouichi Nakayasu[3] Takashi Fujita[4] and Osamu Kiyomiya[5]

1. Introduction

Recently, the flap-gate-type wave breaker which can blockade a harbor entrance at short time attracts attention as the protective structure which prevent it from raising of sea level in the coastal region against the tsunami by the earthquake and storm surges in typhoon approaching.

The flap-gate-type wave breaker is placed on the bottom of the sea. It have a mechanism that opens and closes the gate based on pin structure.

Generally, as shown in figure1(a), the flap gates are stored in the seabed. Therefore, it doesn't hinder steamer under way. Also, it is the coastal structure which is good for the environment because it doesn't hinder the tidal exchange and seawater circulation like the stationary-type wave breaker and moreover it doesn't compromise a view of the harbor and water channel.

On the other hand in emergency situations like tsunami or storm surge, as shown in figure1(b), the flap-gates instantly rise from the bottom of the sea to seawater surface using natural power and buoyancy. Moreover at the case of tsunami, as shown in figure1(c), the gates raise itself up perpendicularly because of raising of water level by natural tsunami power. They blockade a harbor entrance and the water channel, and prevent them from approaching the instant wave and the flow of seawater. At the case of undertow as shown in figure1(d), the flap gates try to sink into the sea with the undertow. Though, they are supported with the safety stopper and don't sink below than the constant gradient. Therefore, the sea level in the harbor can be kept constant and it is applicable with the 2nd and the 3rd wave of the tsunami. In the past paper, Shirai (2006) showed that a modified flap type gate was effective as a tsunami countermeasure. Kiyomiya (2006) experimented two-dimensional hydraulic model tests on the scale of 1:50. This paper describes about the hydraulic and hydrodynamic characteristics of 'the flap-gate-type wave breaker' that we have researched and developed before now.

2. Architecture of flap gate-type wave breaker

As shown in figure2, the flap-gate-type wave breaker is mainly composed of gates, tension rods, bottom structure, stabilizer plates, air supply and exhaust ventilation systems, safety stoppers. Then, it is the wave breaker's structure that the gate is placed on one line along crosswise direction.

The gate consists of the hollow main pillars and the skin plate. The tension rod consists of short and long three steel bars and transmits to bottom structure the hydraulic loading which acts on the gate. The bottom structure on the seabed takes charge of the hydraulic loading which is transmitted from the gates. The stabilizer plate reduces an impact load against tidal wave. The air supply and exhaust ventilation systems control the gate's buoyancy when the gates open and close. The safety stopper supports the gate which tries to sink into the seabed in case of the undertow.

3. Hydraulic model test

We installed a flap-gate-type wave breaker model with reduced scale ratio 1/35 in the two-dimensional water tank with wave generator and experimented about the gate motions in waves, in or

[1]Hitachi Zosen Corporation, 2-2-11 Funamachi, Taisho-ku, Osaka 551-0022, Japan. niizato@hitachizosen.co.jp
[2]Hitachi Zosen Corporation, 2-2-11 Funamachi, Taisho-ku, Osaka 551-0022, Japan. kimura@hitachizosen.co.jp
[3]Hitachi Zosen Steel Structures Corporation,1-5-1 Chikko shinmachi, Nishi-ku, Sakai 592-8331, Japan. nakayasu@hitachizosen.co.jp
[4]Hitachi Zosen Corporation, 2-2-11 Funamachi, Taisho-ku, Osaka 551-0022, Japan. Fujita_ta@hitachizosen.co.jp
[5]Waseda University, 3-4-1 Okubo, Shinjyuku-ku, Tokyo 169-8555, Japan. k9036@waseda.jp

der to investigate about the hydraulic and hydrodynamic characteristics of the gate behavior.

The axial force acting on tension rod in waves is shown in figure3. From the figure3, the axial force which acts on the rod in the irregular wave in comparison to the axial force in the regular wave becomes small.

4. Conclusion

As the result except figure 3, when the tsunami acts on the gates, the wave pressure and rotation moment which acts on the gates can be reduced by means of the stabilizer plate to have installed in the tension rod.

The flap-type gate for protection against storm surge and tsunami doesn't compromise a view of the harbor, and is a good wave breaker.

References

Shirai, S.,etc, 2006. Experimental study of the flap type gate as a tsunami countermeasure. *Jornal of Civil Engineering in the Ocean*,22: 577-582.

Kiyomiya, O., etc, 2006. Characteristics of the edge of tsunami acting on the flap type gate. *Jornal of Civil Engineering in the Ocean*,22: 571-576.

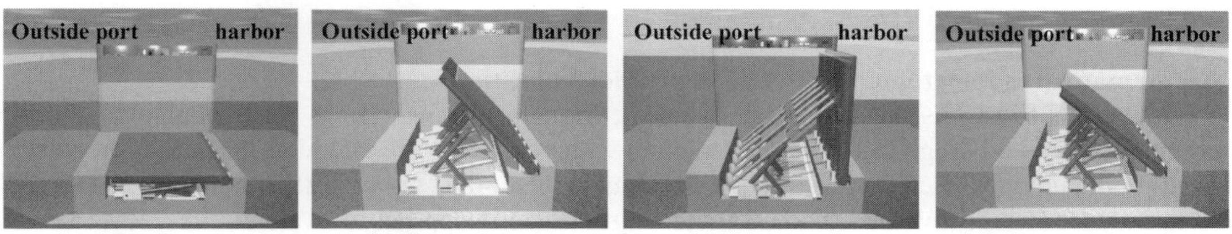

(a) Stored mode (b) Floated mode (c) Erect mode (d) Supported mode

Figure 1. States of flap gate

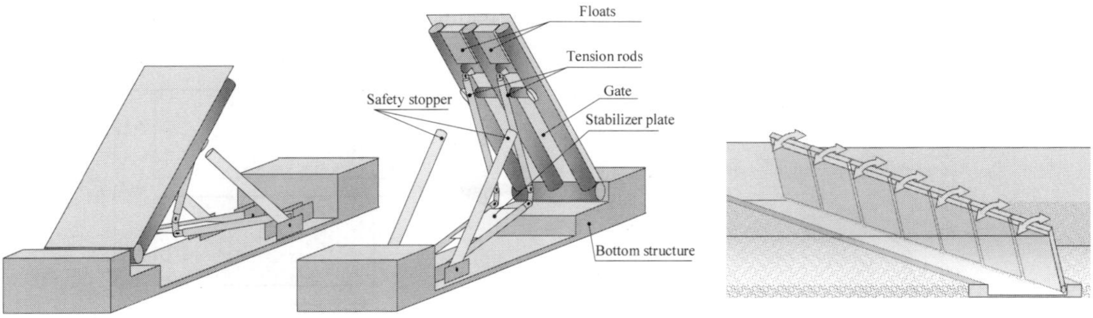

Figure 2. Architecture of flap gate-type wave breaker

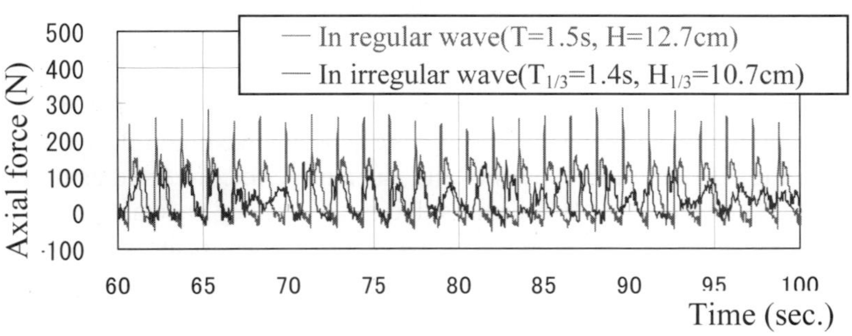

Figure 3. Axial force acting on the tension rod in waves

Coastal Dynamics 2009
Paper No.48

TSUNAMI FLOW VELOCITY BEHIND THE COASTAL FOREST WITH AN OPEN GAP– EFFECTS OF TSUNAMI CONDITION AND TREE DENSITY

Nguyen Ba Thuy[1], Norio Tanaka[2] and Katsutoshi Tanimoto[3]

1. Introduction

The effect of an open gap in a coastal forest on tsunami run-up has been studied by both numerical simulations and laboratory experiments (Tanimoto et al. 2008, Thuy et al. 2009). Experimental results by Thuy et al. (2009) showed that the exit flow velocity from the gap was significantly higher compared to the flow velocity without vegetation. Tanimoto et al. (2008) and Thuy et al. (2009) systematically investigated the effect of an open gap in a forest of *Pandanus odoratissimus* on tsunami run-up by numerical simulations and found that a specific gap width (15 m) causes the [1]highest velocity under their calculated conditions. This study investigates the effect of tsunami height, tsunami period and tree density for flow velocity behind the vegetation patch and at the gap exit by numerical simulations. A coastal forest of *Pandanus odoratissimus*, is selected for the simulation, because it is a representative coastal vegetation in South Asia and reported especially effective in providing protection from tsunami damage due to its high density and complex aerial root structure (Tanaka et al. 2007).

2. Mathematical model, topography and vegetation condition

Governing equations are based on two-dimensional non-linear long wave equations. Bottom friction, diffusion and drag force due to interaction with vegetation are included. The drag force due to vegetation including vertical stand structure is evaluated with the method by Tanaka et al. (2007). The *SDS* turbulence model given by Nadaoka and Yagi (1998) is applied to evaluate the eddy viscosity with modifications related to bottom friction and vegetation drag.

A uniform coastal topography with a straight shoreline with the cross section perpendicular to the shoreline as shown in Fig. 1(a) was selected as a model case. The coastal forest width B_F and length L_F show in Fig. 1(b) are fixed as 200m. The gap width b_G is 15m at the middle of the forest length.

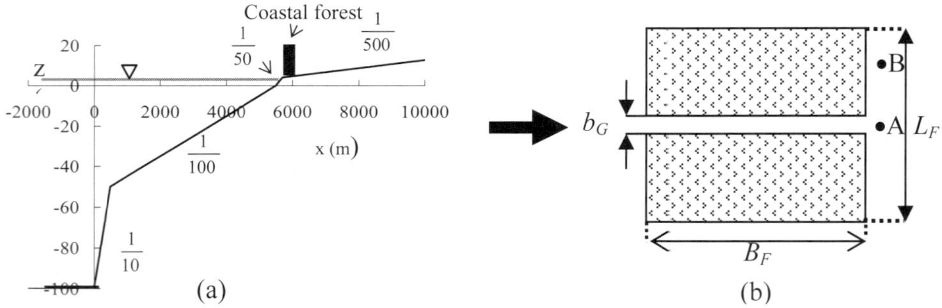

Fig. 1. Schematic of the topography for numerical simulation, (a) cross section of topography, (b) sketch of gap arrangement.

3. Results and discussion

Figs. 2(a)-(c) show the reduction rate of maximum flow velocity (V_{max}/V_1) at point A and B by tsunami height, tsunami period and forest density, where V_1 is the maximum flow velocity for the case of no vegetation. In Fig. 2(a), the forest density and tsunami period are fixed as 0.2 trees/m^2 and 20 minutes, respectively. The velocity at B decreases due to the drag of vegetation, while at point A, it increases due to the enhancement of inflow from vegetation sides to the gap, except for the case

[1] Graduate School of Science and Engineering, Saitama University, 255 Shimo-okubo, Sakura-ku, Saitama, 338-8570, Japan. [1]s07de059@mail.saitama-u.ac.jp, [2]tanaka01@mail.saitama-u.ac.jp, [3]tanimotok@coda.ocn.ne.jp

tsunami height is less than 2 m. However, the change rate is not large as the tsunami height exceeds 3 m. In Fig. 2(b), the conditions of tsunami height and forest density are fixed as 6 m and 0.2 trees/m^2. When tsunami period becomes long, the variation of reduction rate of velocity at point B is small. On other hand, at point A the rate increases as tsunami period increases. In Fig.2(c), the conditions of tsunami height and tsunami period are fixed as 6 m and 20 minutes respectively. As the forest density increases, the velocity at point B decreases and increases at point A. Fig. 3 shows the normalized increase of velocity at point A relative to point B ($[V_{max(A)}-V_{max(B)}]/[V_{maxB}]$) against the tsunami height for the forest densities (γ) of 0.1 and 0.4 trees/m^2. The amplification of flow through the gap is quite clear, and is strongly dependent on the forest density and weakly on wave height.

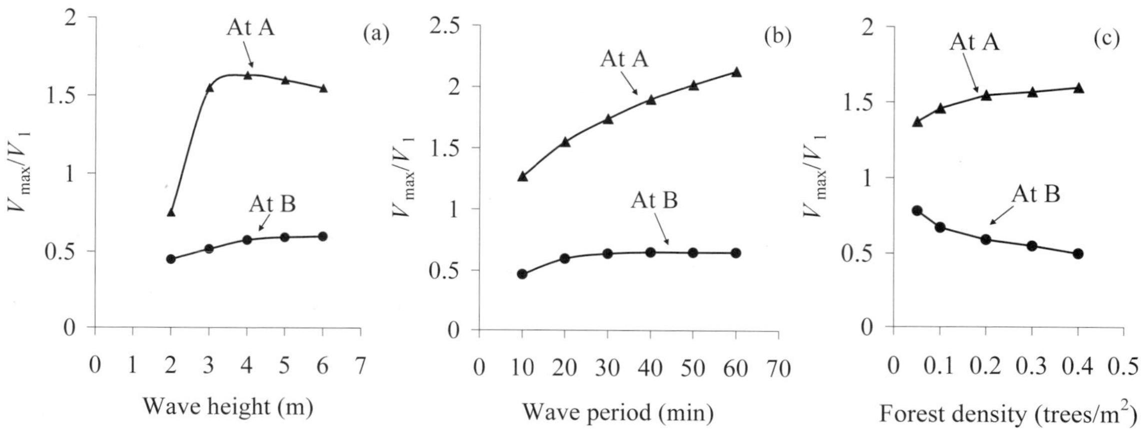

Fig. 2. Reduction rate of maximum velocity. (a) Effect of tsunami height, (b) effect of tsunami period, and (c) effect of tree density.

4. Conclusions

The effects of tsunami height, tsunami period and forest density on the tsunami flow velocity behind the coastal forest with the presence of an open gap were discussed in this study. It is found that the change rates of velocity behind the gap and vegetation patch are greatly different. When tsunami period becomes large, the increase of velocity at the gap exit becomes large but behind the vegetation patch the velocity does not so vary. The enhancement of velocity at the gap exit is strongly dependent on the forest density but it is weakly influenced by the tsunami height.

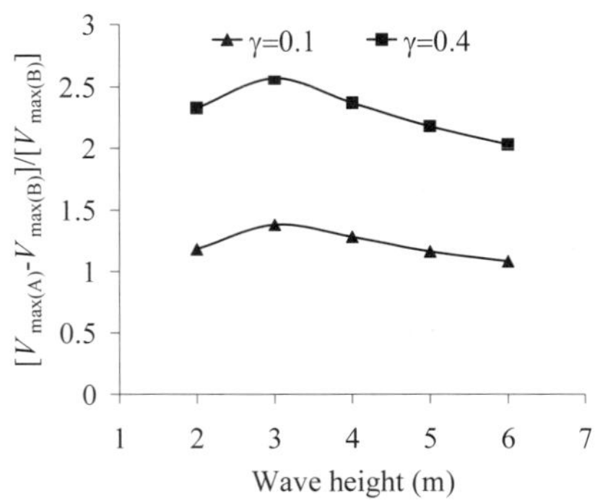

Fig. 3. Variation of normalized exit velocity increase.

References

Nadaoka, K., Yagi, H., 1998. Shallow - water turbulence modeling and horizontal larger-eddy computation of river flow. *Journal of Hydraulic Engineering*, 124(5), 493-500.

Tanaka, N., Sasaki, Y., Mowjood, M.I.M., Jinadasa, K.B.S.N., 2007. Coastal vegetation structures and their functions in tsunami protection: Experience of the recent Indian Ocean tsunami, *Landscape and Ecological Engineering*, Vol. 3, pp. 33-45.

Tanimoto, K., Tanaka, N., Thuy, N. B., Nandasena N. A. K., Iimura, K., 2008. Effect of open gap in coastal forest on tsunami run-up – investigation by 2-dimensional numerical simulation, *Annual Journal of Civil Engineering in the Ocean*, JSCE, Vol. 24, pp. 87-92. (in Japanese).

Thuy, N. B.,Tanimoto, K., Tanaka, N., Harada. K., Iimura, K.. Effect of open gap with on tsunami run-up-Investigation by experiment and numerical simulation, 2009. *Journal of Ocean Engineering*, (Submitted).

Coastal Dynamics 2009
Paper No.49

DIRECT BED SHEAR STRESS MEASUREMENTS UNDER SOLITARY TSUNAMI-TYPE WAVES AND BREAKING TSUNAMI WAVE FRONTS

Jaya Kumar Seelam[1,2] and Tom E. Baldock[3]

1. Introduction

Tsunamis have been known for destruction and devastation along the coastlines due to release of the enormous energy they possess. However, their effect on the seabed and submarine structures is not completely understood. Tsunamis undergo physical transformation when they encounter different water depths and feel the bottom more rapidly than wind waves and hence may have greater effect on the seabed at greater depths compared to wind waves. It is often assumed that tsunamis take the form of solitary waves (e.g., Goring, 1979; Synolakis, 1986). A distinct feature of all the submarine regions around the world are the continental shelf and continental slope, where there is a sudden change in the seabed slopes.

Most of the offshore oil exploration and exploitation activities around the globe are located on relatively shallow (up to 200 m depth viz., continental shelf), as well as deeper regions (beyond 200 m water depth viz., continental slopes, abyssal plains, ocean floor). Transportation of this oil & gas is carried out through submarine pipelines across the continental slopes and shelf regions. Tsunami induces shear stresses and pressures on the seabed are normally ignored in the design of submarine pipelines, as there is no specific guideline to account for these. These shear stresses and pressures generated by tsunamis might trigger submarine land slides which are detrimental to the submarine pipelines. Therefore investigations on the tsunami induced shear stresses and pressures are required. This paper presents preliminary results of such investigations being carried out at the University of Queensland.

2. Methodology

Experiments are being conducted at The University of Queensland's Gordon McKay Hydraulics Laboratory to investigate shear stresses and bottom pressures generated due to wave forms that represent Tsunamis. A shear cell to measure the shear stresses, pressure transducers to measure the bottom pressures and ultrasonic wave gauges to measure the wave heights are employed. The setup of the laboratory experiment is presented in Fig. 1. Solitary waves, bores, and leading elevation/depression 'N' waves are generated utilizing a piston type wave generator. Waves of various amplitude-to-water depth ratios were generated and the resulting seabed pressures and shear stresses were measured. Various scenarios representing the continental shelf and slope regions will be studied in the laboratory. Non-breaking and breaking conditions are studied.

Simulations of propagation and hydrodynamics of the waves in wave flume are being carried out using depth-averaged hydrodynamic models ANUGA (Nielsen et al, 2005) and MIKE21-HD (Anonymous, 2007), to ascertain the friction factors corresponding to the laboratory conditions. Preliminary modeling of tsunami propagation for the north-west Australian coastal region has been carried out.

3. Results

Tsunamis represented by solitary waves, bores and leading elevation or depression waves in the

[1] The University of Queensland, School of Engineering, Division of Civil Eng., St Lucia, QLD-4072, Australia. j.seelam@uq.edu.au
[2] National Institute of Oceanography (CSIR), Goa – 403 004, India. jay@nio.org
[3] The University of Queensland, School of Engineering, Division of Civil Eng., St Lucia, QLD-4072, Australia. t.baldock@uq.edu.au

laboratory are generated and the total shear stresses, pressures and velocities near the bed under these wave are measured. Typical result of the wave characteristics generated and corresponding shear stress, pressures and velocities is shown in Fig. 2. Further analysis of the data is under progress to validate theoretical aspects on tsunami kinematics and shear stresses. Numerical modeling of the tsunami wave propagation using 2D depth averaged hydrodynamic numerical models, ANUGA and MIKE21 is under progress both to validate the numerical model as well as to ascertain its usefulness in further studies related to tsunami kinematics and shear stresses. Preliminary results showed that both the models provide satisfactory results in terms of the wave form except close to the wavemaker. The importance of the total stress acting on the bed, which includes pressure gradient forces, is also considered.

Acknowledgements
Authors acknowledge the support and facilities provided by the University of Queensland, Australia. JKS is an Endeavour IPR scholar and he also acknowledges the support of his parent organization, the National Institute of Oceanography (CSIR), Goa, India. This work is supported by the CSIRO wealth from ocean flagship collaboration cluster on subsea pipelines.

References

Anonymous, 2007. *MIKE21 Flow model – Hydrodynamicmodule Scientific Documentation*, DHI Water and Environment, Denmark.

Goring, D. G., 1979. *Tsunamis-The propagation of Long waves onto a shelf*, PhD Thesis, California Institute of Technology.

Nielsen, O., Sexton, J., Schneider, J., 2005. *The ANUGA tsunami inundation model*. In: Proceedings of ModSIM 2005. Australian National University, Australia, pp. 518–523.

Synolakis, C. E., 1986. *The runup of Long*, PhD Thesis, California Institute of Technology.

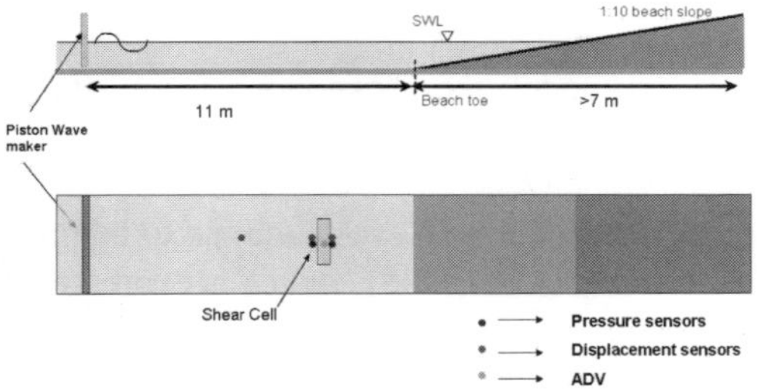

Figure 1. Experimental setup for seabed pressures and shear stresses

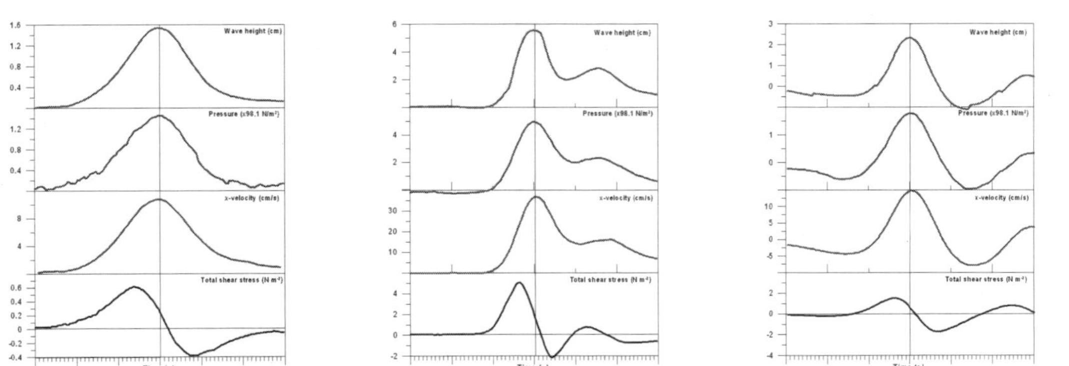

Figure 2. Wave height, on-bottom pressure, cross-shore velocity component and total shear stress for (a) solitary wave (b) bore (c) leading depression N wave

Coastal Dynamics 2009
Paper No.50

NUMERICAL MODEL FOR BAR MIGRATION AT HASAKI, JAPAN

Yoshiaki Kuriyama[1]

1. Introduction

Longshore bars are frequently observed on sandy beaches and significantly influence currents and morphological variations in the nearshore zone. Bars migrate seaward and shoreward in short terms. However, medium-term bar migrations are frequently in the seaward direction.

Although a lot of numerical simulations on bar migration have been developed, most of them are for bar migrations in short terms, two or three days to several days. Ruessink et al. (2007) have succeeded to develop a numerical model to reproduce bar migrations for several weeks. However, the validity of the model for reproducing cyclic seaward bar migrations has not been fully examined. The objective of this study is to develop a one-dimensional numerical model for bar migration and to investigate the validity of the model with beach profiles measured in the field almost every day for 2 years.

2. Model

The one-dimensional numerical model for bar migration is composed of four sub-models for wave transformation, undertow, velocity and acceleration skewnesses, and beach profile change. The wave transformation and undertow sub-models are based on Kuriyama and Nakatsukasa's model (2000), in which the vertical distribution of velocity in a surface roller was assumed to be triangular with celerity at the top and zero at the bottom. The velocity skewness and acceleration skewness, which is also called velocity atiltness, are estimated using the wave nonlinearity parameter proposed by Goda (1983) and the formulae for velocity skewness and asymmetry proposed by Doering and Bowen (1995).

In the beach profile change sub-model, the cross-shore sediment transport rate was assumed to consist of four contributions due to sediment suspension and undertow Q_s, velocity skewness $Q_{b,skew}$, velocity atiltness $Q_{b,atilt}$ and beach slope $Q_{b,slope}$.

As for Q_s, the amount of suspended sediments was assumed to be proportional to the surface roller energy dissipation rate as in Kobayashi et al. (2008), and Q_s is expressed by

$$Q_s = \alpha_1 \frac{D_r}{\rho g(s-1)w_f} U \tag{1}$$

where α_1 is a coefficient, D_r is the roller energy dissipation rate, s is the sediment specific gravity, w_f is the sediment fall velocity and U is the time-averaged undertow velocity.

The value of $Q_{b,skew}$ was assumed on the basis of Bailard (1981).

$$Q_{b,skew} = -\alpha_2 (\sqrt{\beta_1})_u u_{b,rms}^3 \cos\theta \tag{2}$$

where α_2 is a coefficient, $(\sqrt{\beta_1})_u$ is the velocity skewness, $u_{b,rms}$ is the amplitude of near bottom velocity and θ is the wave angle.

The value of $Q_{b,atilt}$ is expressed by Eq. (3) as in Hoefel and Elgar (2003).

$$\begin{aligned} &Q_{b,atilt} = -\alpha_3\left((\beta_3)_u a_{b,rms} - a_{cr}\left|(\beta_3)_u\right|/(\beta_3)_u\right)\cos\theta, && \left|(\beta_3)_u a_{b,rms}\right| > a_{cr} \\ &Q_{b,atilt} = 0, && \left|(\beta_3)_u a_{b,rms}\right| \le a_{cr} \end{aligned} \tag{3}$$

where α_3 is a coefficient, $(\beta_3)_u$ is the velocity atiltness, $a_{r,rms}$ is the amplitude of near bottom

[1]Port and Airport Res. Inst., Nagase 3-1-1, Yokosuka, Kanagawa 239-0826, Japan. kuriyama@pari.go.jp

acceleration and a_{cr} is a threshold value (= 0.2 m/s^2).

The values of $Q_{b,slope}$ is also based on Bailard (1981), but the coefficient α_4 for $Q_{b,slope}$, which is equal to α_2 for $Q_{b,skew}$ in Bailard (1981), was assumed to be different from α_2.

$$Q_{b,slope} = \alpha_4 (\tan\beta / \tan\phi)\left(u_{b,rms}^3 + u_{bl,rms}^3\right) \tag{4}$$

where $\tan\beta$ is the beach slope, ϕ is the internal friction angle of the sediment (= 30 degree) and $u_{bl,rms}$ is the amplitude of long period near bottom velocity.

3. Results and Conclusion

The model was calibrated and compared with beach profiles measured on the Hasaki coast of Japan every weekday during a 2-year period from 1989 to 1990, when the period of the cyclic seaward bar migration is about 1 year. The comparison shows that although the predicted bars are developed more than the measured ones (Figure 1), the model almost predicted the cyclic seaward bar migration with a period of about 1 year (Figure 2).

References

Bailard, J.A., 1981. An energetics total load sediment transport model for a planar sloping beach. *Journal of Geophysical Research*, 86 (C11), 10938-10954.

Doering, J.C. and Bowen, A.J., 1995. Parameterization of orbital velocity asymmetries of shoaling and breaking waves using bispectral analysis. *Coastal Engineering*, 26, 15-33.

Goda, Y., 1983. A unified nonlinearity parameter of water waves. *Report of Port and Harbour Research Institute*, 22(3), pp.3-30.

Hoefel, F. and Elgar, S., 2003. Wave-induced sediment transport and sandbar migration. *Science*, 299, 1885-1887.

Kobayashi, N., Payo, A. and Schmied, L., 2008. Cross-shore suspended sand and bed load transport on beaches,Sand suspension, *Journal of Geophysical Research*, 113(C07001), doi:10.1029/2007JC004203.

Kuriyama, Y. and Nakatsukasa, T., 2000. A one-dimensional model for undertow and longshore current on a barred beach. *Coastal Engineering*, 40, 39-58.

Ruessink, B. G., Kuriyama, Y., Reniers, A. J. H. M., Roelvink, J. A. and Walstra, D. J. R., 2007. Modeling cross-shore sandbar behavior on the timescale of weeks. *Journal of Geophysical Research*, 112, F03010, doi:10.1029/2006JF000730.

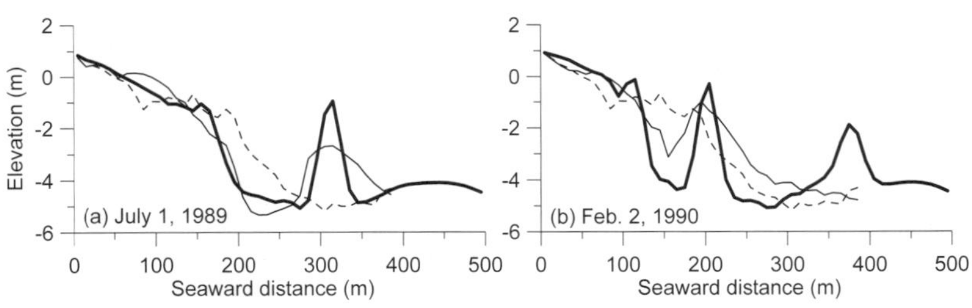

Figure 1. Beach profiles predicted (thick solid lines) and measured (thin solid lines). The dashed lines show the initial beach profile, measured on January 4, 1989.

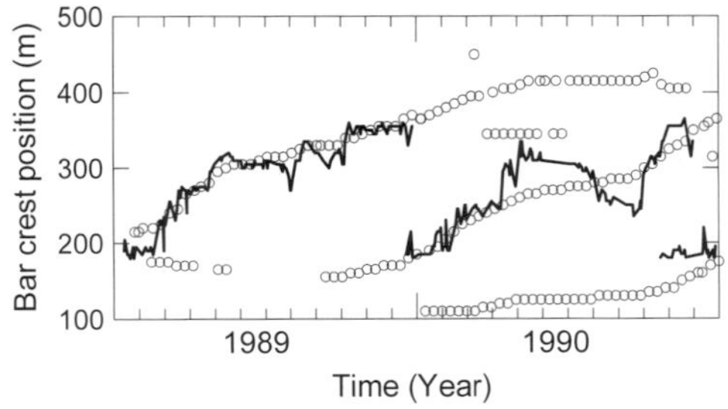

Figure 2. Temporal variations of bar crest positions predicted (open circles) and measured (solid lines).

Coastal Dynamics 2009
Paper No.51

CONCEPTUAL MODELLING OF A DOUBLE CRESCENTIC BARRED COAST (LEUCATE-PLAGE, FRANCE)

Pierre Ferrer[1], Raphael Certain[1] and Jean-Paul Barusseau[1]

The coastal environments have strong social-economic stakes. In a scientific way, it is necessary to understand the natural processes which take place there to predict the future of this environment. They are submitted to erosion and/or accretion. A beach classification has been made by Short & Aagaard (1993). It allows to distinguish beaches with one or more sandbars and beaches with different types of sandbars: longshore bars trough (LBT), rhythmic bars and beach (RBB), transverse bars and rips (TBR) and ridge-runnel or low tide terrace (LTT). In France, a lot of studies took place in meso-macrotidal environment, on the Aquitanian coast (Lafon *et al.*, 2003; Castelle *et al.*, 2007). In the world, these studies also took often place in tidal environments (Wright and Short, 1984 ; Lippman and Holman, 1990…), but some studies were made in microtidal environments (Rivière *et al.*, 1961 ; Goldsmith *et al.*, 1982).

Leucate-Plage coast is located in the south part of the Gulf of Lion. It has few human activities and presents an interest for the understanding of microtidal natural sandbars systems. This place is interesting for the presence of double crescentic sandbars, rhythmic features which are parallel to the shoreline.

The objective of this study is to propose a conceptual model of morphodynamic evolution for Leucate-Place shoreface (emerged beach and nearshore).

To realise this work, a monthly survey has started in 2005. It allows us to look after the monthly evolution of the beach thanks to topographic surveys, of the marine beach thanks to bathymetric surveys, and of coastline variations with GPS surveys. These surveys are completed with hydrodynamic data acquired during intensive campaigns (FEST) or with the Datawell buoy located offshore Leucate-Plage (DRE Languedoc Roussillon).

The first field observations of the coast, realised in 2000, showed a system with two crescentic sandbars, an inner bar near the coast and an outer bar offshore. The outer bar is located between -5 and -7m deep, at a distance of 400m of the coastline. It presents a wavelength of 500m (fig. 1). The inner bar is located between -2 and -4m deep, at a distance of 125m of the coastline. Its wavelength is about 250m (fig. 1). The topographic map of the emerged beach showed two depressions where the coastline is in retreat. The coastline is characterized by bays and horns oriented offshore. The wavelength of these megacusps is closed of the one of the outer bar.

During our survey, we don't have observed storm events with significant height upper than 5m. We can say that the outer bar is a long-lasting system in time. The rhythm and the depth are closed during all the survey. Locally, the horns undergo erosion or accretion. The inner bar is more active. Indeed, more of the time it appears in an intermediate step between RBB and TBR. After a storm event, the crescentic features are broken and the system is more closed of TBR, whereas after small swell events, the system is closer of RBB. The interaction between the inner and the outer system exists, we could observe that the outer bar is more or less linked to the inner bar. When the crest of the inner bar crescents break, the horns of the outer bar are in accretion. The relationship between the outer bar and the shoreline is less evident, they look like independent in spite of their similar wave length. However, the inner bar has a role of plug between these systems. These bar system seem more stable in time than others system described in literature (Wright & Short, 1984; Lippman & Holman, 1990).

[1]IMAGES, University of Perpignan Via Domitia, 52 Av. Paul Alduy, 66860 Perpignan Cedex, France. Mail: pierre.ferrer@univ-perp.fr

Acknowledgements

Access to Datawell buoy data were kindly provided by the DRE Languedoc Roussillon. Thank you very much to Fanny Adloff, Laurent Cance and Mathieu Gervais for their collaboration in this work.

References

Castelle, B., Bonneton, P., Dupuis, H., and Sénéchal, N., 2007. Double bar beach dynamics on the high-energy meso-macrotidal French Aquitanian Coast: A review. Marine Geology, 245, 141-159.

Goldsmith, V., Bowman, D. and Kiley, K., 1982. Sequential stage development of crescentic bars: Hohaterim beach, southeastern Mediterranean, Journal of Sedimentary Petrology, 52,233-249.

Lafon, V., De Melo Apoluceno, D., Dupuis, H., Michel, D., Howa, H., and Froidefond, J.-M., 2004. Morphodynamics of nearshore rhythmic sandbars in a mixed-energy environment (Sw France):i. mapping beach changes using visible satellite imagery. Estuarine Coastal and Shelf Sciences, 61 (2), 289–299.

Lippmann, T.C., & Holman, R.A., 1990. The spatial and temporal variability of sand bar morphology. J Journal of Coastal Research, 95, 575–590.

Rivière, A., Arbey, F., and Vernhet, S., 1961. Remarque sur l'évolution et l'origine des structures de plage à caractère périodique, Comptes Rendus Hebdomadaires des Séances de l'Académie des Sciences, 252, 767–769.

Short, A. D. & Aagaard, T., 1993. Single and multi-bar beach change models. Journal of Coastal Research, Special Issue No. 15, 141-157.

Wright, L. D., & Short A. D. (1984), Morphodynamic variability of surf zones and beaches: A synthesis, Marine Geology, 56, 93– 118.

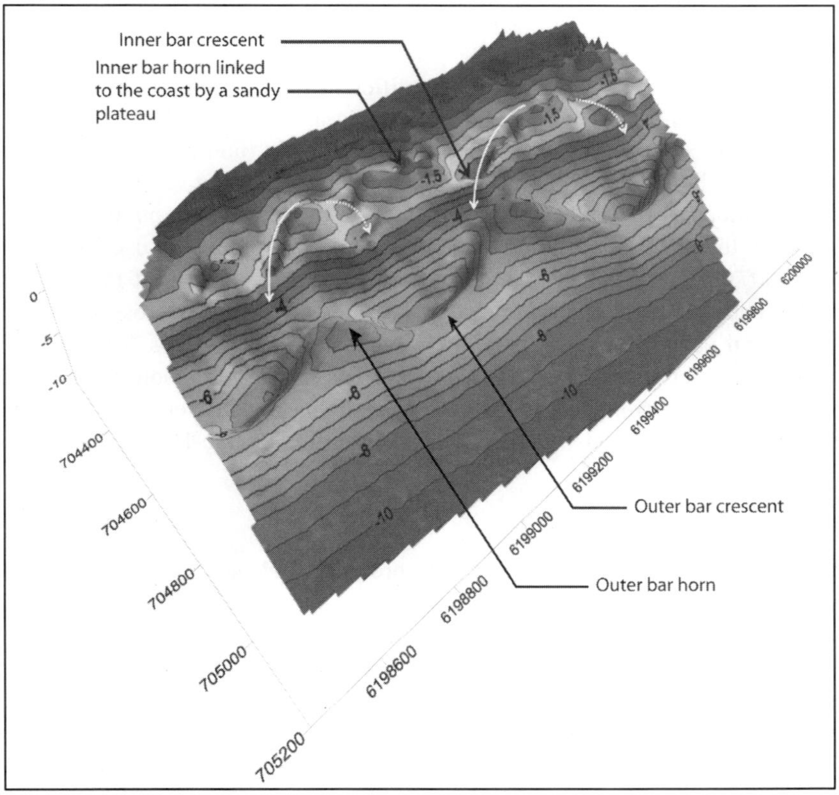

Figure 1 : topo-bathymetric map of Leucate-Plage coast from the 20/01/2007. On this map, the beach is violet whereas the marine system have more clolors corresponding to the depth variations. The shoreline is red and the coastal channels are white.

Coastal Dynamics 2009
Paper No.52

MODELS AND SCALES FOR NEARSHORE SANDBAR BEHAVIOR

Leo Pape[1], Gerben Ruessink[1] and Yoshiaki Kuriyama[2]

1. Introduction

Alongshore sandbars are often present in the nearshore zones of storm-dominated micro-to mesotidal coasts. The interactions between water motion due to waves and currents and submerged sandbars in the nearshore zone are generally understood and modeled in terms of wave and sediment transport processes represented on temporal scales of seconds to hours. Predicting sandbar behavior on timescales of months to years has proven to be difficult with models based on small- and short-scale processes because of exponential accumulation of errors made in the process description or state. However, mechanisms have been proposed that describe how the combination of small- and short-scale processes drives the nearshore zone toward a stable configuration and corresponding sandbar location (e.g. Plant *et al.* (1999)). If such a stable state indeed evolves from the combination of small- and short-scale processes, long-term sandbar behavior might better be described on the spatial and temporal scales of the sandbar itself then on the scale of the underlying processes. In the present work we perform an empirical investigation into the performance of different representations and corresponding models for cross-shore sandbar behavior on timescales of months to years.

2. Data and Models

The data we use for testing the performance of the different models consists of cross-shore profiles measured daily in the period between 1987 and 2002 along a 400 m long pier situated at the Hasaki Oceanographic Research Station (HORS) in Japan. Additionally, hourly-observed water level and daily-averaged wave height and wave period data are included as forcing variables. The observations, extracted sandbar positions and wave heights are depicted in the top and bottom panels of Figure 1.
As representative of the small- and short-scale modeling paradigm we use the Unibest model (Ruessink *et al.*, 2007). The other end of the modeling spectrum consists of models that represent cross-shore sandbar behavior as migration toward a wave-height dependent cross-shore location. Here, we use a NARX neural network (Pape *et al.*, 2007) as instance of this class. The ability to describe sandbar behavior on smaller timescales and other locations has been demonstrated before for both models, but here we perform (for the first time) a comparison of long-term predictive capacity for representative models from two modeling paradigms, based on a unique 15-year long high-resolution dataset.

3. Results

Predictions are made in blocks of continuous sandbar presence. Relevant morphology variables (entire profiles (Unibest) or sandbar crest locations (NARX)) are supplied to the model at the start of each block. The models are then iterated forward with hydrodynamic variables and previously predicted profiles or sandbar locations. The neural networks were trained on all blocks not used for prediction, while the parameters found by Ruessink et al. (2007) were used for the Unibest runs. Predictions for both models together with original sandbar crest observations are depicted in the two center panels in Figure 1. For both models we found a rather poor performance of 87 (Unibest) and 52 (NARX) meter

[1]Faculty of Geosciences, Dept. of Physical Geography, Universiteit Utrecht, Heidelberglaan 2, 3584 CS Utrecht, The Netherlands. {l.pape, g.ruessink}@geo.uu.nl
[2]Marine Environment and Engineering Dept., Port and Airport Research Institute, Nagase 3-1-1, Yokosuka, Kanagawa 239-0826, Japan. kuriyama@pari.go.jp

root-mean-squared difference between observed and predicted sandbar crest location.

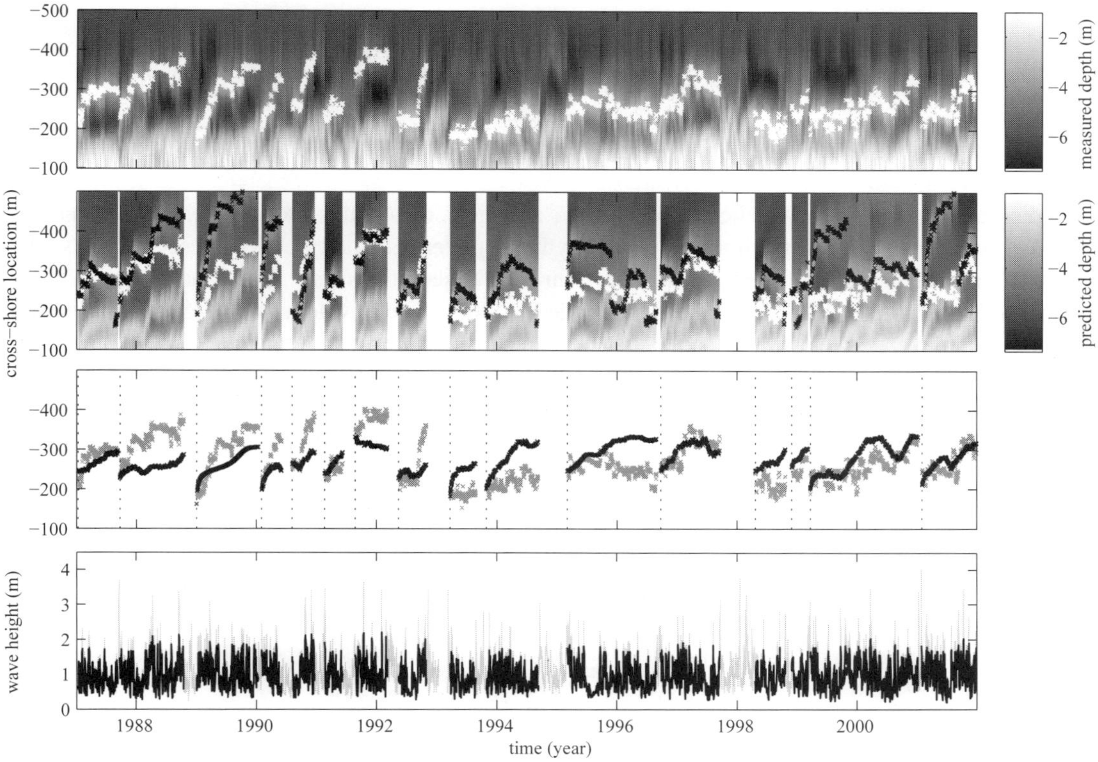

Figure 1. From top to bottom: observed depth with white crosses indicating sandbar crest positions; Unibest predicted depth, with white (black) crosses indicating observed (predicted) sandbar crest positions; NARX predicted (black) and observed (gray) sandbar crest locations; observed offshore (gray) and local (black) wave heights.

4. Discussion and Conclusion

The Unibest model predicts offshore migration and subsequent new sandbar formation at the shore, a feat that is not easily achieved with the NARX model. The neural network model also correctly predicts annual and interannual offshore migration. However, in both models the slope of this offshore trend deviates significantly from the observations, which is the main cause for the rather large error values. We find that the NARX model yields more accurate predictions than Unibest, which provides initial empirical support for our hypothesis that long-term sandbar migration can better be described on the level of sandbar crest locations and wave height. In the full paper we will describe the details of our approach, a more elaborate discussion of the differences between the model types, a more detailed inspection of the performance on different timescales, the results of a hybrid model based on additional profile information, and the results found for an improved Unibest calibration procedure.

References

Pape, L., Ruessink, B. G., Wiering, M. A., and Turner, I. L., 2007. Recurrent neural network modeling of nearshore sandbar behavior. *Neural Networks*, 20:509-518.

Plant, N. G., Holman, R. A., Freilich, M. H., and Birkemeier, W. A., 1999. A simple model for interannual sandbar behavior. *Journal of Geophysical Research*, 104:755-776.

Ruessink, B. G., Kuriyama, Y., Reniers, A. J. H. M., Roelvink, J. A. and Walstra, D. J. R., 2007. Modeling cross-shore sandbar behavior on the timescale of weeks. *Journal of Geophysical Research*, 112:F03010.

Coastal Dynamics 2009
Paper No. 53

THE USE OF LINEAR STABILITY ANALYSIS TO CHARACTERIZE THE VARIABILITY OF MULTIPLE SANDBAR SYSTEMS

Giovanni Coco[1] and Daniel Calvete[2]

1. Introduction

Multiple sandbar systems have been observed in a variety of settings worldwide. The changes in multiple sandbar systems sometimes result in rhythmic patterns, usually called crescentic bars. The nature of their occurrence has been studied both through observations (e.g. Short and Aagaard, 1993) and numerical studies (e.g. Klein and Schuttelaars, 2006). The development of crescentic bars in multiple sandbar systems is part of sequential beach changes (Short and Aagaard, 1993) displaying a range of configurations that can involve coupling between the sandbars and, sometimes, also the shoreline (Ruessink et al., 2007). Here we use a numerical model based on linear stability analysis (Calvete et al., 2005) to study the role of hydrodynamic conditions and initial cross-shore sandbar profile on the evolution of the double sandbar system.

2. Numerical Model

The numerical model includes depth- and time-averaged momentum and continuity (1) equations coupled with wave-energy and phase (2) equations:

$$\frac{\partial u_i}{\partial t}+u_j\frac{\partial u_i}{\partial x_j}=-g\frac{\partial z_s}{\partial x_i}-\frac{1}{\rho D}\frac{\partial}{\partial x_j}\left(S'_{ij}-S''_{ij}\right)-\frac{\tau_{bi}}{\rho D}\qquad \frac{\partial D}{\partial t}+\frac{\partial}{\partial x_j}\left(Du_j\right)=0 \tag{1}$$

$$\frac{\partial E}{\partial t}+\frac{\partial}{\partial x_j}\left[\left(u_j+c_{gi}\right)E\right]+S^i_{ij}\frac{\partial u_j}{\partial x_i}=-Diss;\qquad \frac{\partial \Phi}{\partial t}+\omega+u_j\frac{\partial \Phi}{\partial x_j}=0 \tag{2}$$

where $x_{i,j}$ refers to the cross- and along-shore coordinates, t is time, g is gravity, ρ is water density, z_s is the mean sea level, D is the total mean depth, u is the depth averaged current vector, τ_b refers to the bed shear stress, E and Φ indicate wave energy density and wave phase respectively, ω is the intrinsic frequency, c_g is the group velocity vector, S' and S'' are the radiation and Reynolds stresses respectively, *Diss* refers to the wave energy dissipation model. The hydrodynamic system is coupled to a sediment conservation equation:

$$\frac{\partial z_b}{\partial t}+\frac{1}{1-p}\frac{\partial q_j}{\partial x_j}=0\qquad \vec{q}=\alpha\left(\vec{u}-\gamma\vec{\nabla}h\right) \tag{3}$$

where z_b is the mean bed level, q is the sediment flux vector, p is sediment porosity (equal to 0.4 in all simulations), α is a stirring function, γ is a diffusion coefficient and h is the seabed perturbation (see Calvete et al., 2005 for model details). The system of equations is assumed to be in equilibrium with respect to the hydrodynamic forcing conditions when alongshore uniformity is assumed. The solution of the system's equations is perturbed and equations linearized as in any standard linear stability analysis. For a given set of forcing conditions (wave height, period) and cross-shore profile, outputs of the analysis are the characteristics of the fastest-growing instability of the system: the growth time

[1]National Institute of Water and Atmosphere (NIWA), Gate 10 Silverdale Road, P.O. Box 11-115, Hamilton, New Zealand. g.coco@niwa.co.nz

[2]Departament de Física Aplicada, Universitat Politécnica de Catalunya, Barcelona, Spain. calvete@fa.upc.edu

(the e-folding time) and the alongshore pattern periodicity.

3. Preliminary Results

Figure 1 shows typical output-morphologies of the linear stability analysis. The two initial cross-shore profiles primarily differ for the depth of the offshore bar with wave height larger for the case of the shallower offshore bar. The fastest growing mode resulting from the linear stability analysis is radically different in the two cases: the morphology with a shallower offshore sandbar crest (and H_{rms}=2 m) develops into a coupled system of inner-outer crescents; the morphology with a deeper offshore sandbar crest (H_{rms}=1 m) results in a coupled system between the inner bar and the area close to the shoreline (the offshore sandbar is inactive). Also, the spatial scale of the inner bar periodicity is twice that of the shoreline features. Numerical simulations show various other configurations that have been observed in the field and that can be related to a combination of offshore hydrodynamic conditions and initial cross-shore beach profiles.

4. Acknowledgements

Giovanni Coco funded by the (New Zealand) Foundation for Research, Science and Technology (WRHC092). The authors wish to thank MMC for providing DS-time necessary to complete this work.

References

Calvete, D., Dodd, N., Falques, A. and Van Leeuwen, S.M., 2005. Morphological development of rip channel systems: normal and near-normal wave incidence, *Journal of Geophysical Research*, 110, C10006, doi:10.1029/2004JC002803.

Klein, M.D. and Schuttelaars, H.M., 2006. Morphodynamic evolution of double-barred beaches, *Journal of Geophysical Research*, 111, C06017, doi: 10.1029/2005JC003155.

Ruessink, B.G., Coco, G., Ranasinghe, R. and Turner, I.L., 2007. Coupled and noncoupled behavior of three-dimensional morphological patterns in a double sandbar system, *Journal of Geophysical Research,* 112(C7), 10.1029/2006JC003799.

Short, A.D. and Aagaard, T., 1993. Single and multi-bar beach change models, *Journal of Coastal Research*, 15 (2), 141-157.

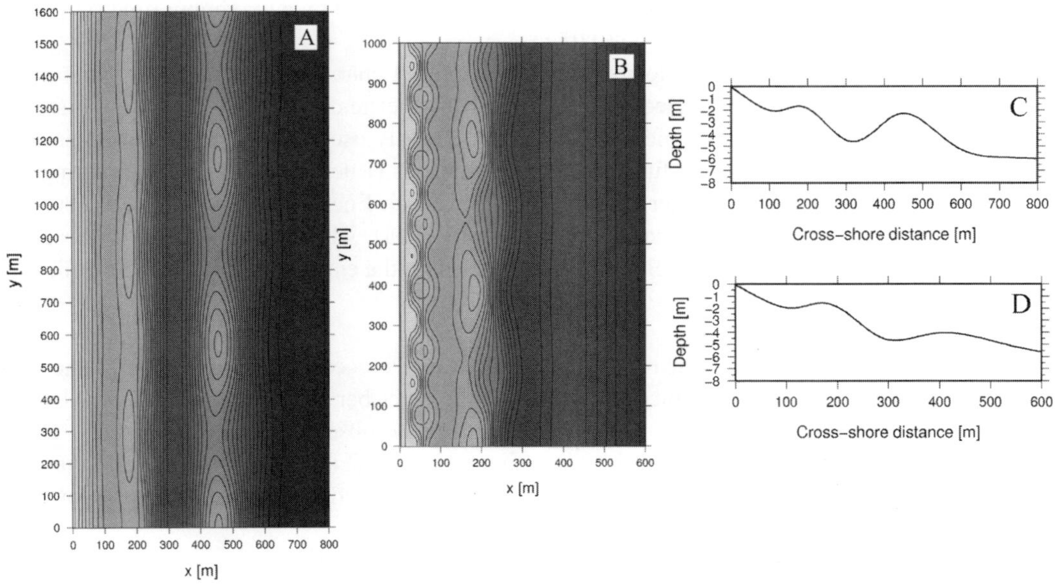

Figure 1. Topographic perturbations (A and B) and the associated initial cross-shore profiles (C and D). The cross-shore and along-shore positions are x and y, respectively (shoreline is at x=0). Wave height is equal to 2.0 and 1.0 m for A and B, respectively. Wave period is 10 s in both cases. C and D have a different cross-shore scale.

Coastal Dynamics 2009
Paper No. 54

PROCESS-BASED MODELING OF CYCLIC BAR BEHAVIOR ON YEARLY SCALES

Dirk-Jan Walstra[1,2] and Gerben Ruessink[3]

1. Introduction

In general, nearshore bars have a multi-annual lifetime, during which they behave in a cyclic, offshore directed manner with often a strong longshore coherence (Wijnberg and Terwindt, 1995). This implies cross-shore processes are dominant over longshore processes. Cycle periods vary worldwide between approximately 1 and 15 years. The net offshore bar migration is the result of gradual onshore movement during calm periods and episodic strong offshore movement during storms (Figure 1). As the bar moves offshore it increases in height and width before finally decaying in the outer surf zone (Figure 2).

In this paper, the process-based morphological profile model UNIBEST-TC (Ruessink et al., 2007) is first used to investigate the bar migration trends and the cyclic behavior at Noordwijk, the Netherlands by aggregating 40 years of annually measured and 4 years of daily modeled data. Next, the simulated profiles are used to investigate the details of the bar migration during its multi-annual life time. The multiple bar system at Noordwijk in the Netherlands was selected because it has a relative short bar cycle of about 3-4 years. This two-step approach provides details which are imperative for further understanding cyclic behavior and the observed worldwide variation of the cycle period.

2. Approach

The model is used to simulate a 1207 day period from 1981 to 1984 which approximately encompasses a complete bar cycle at Noordwijk. The model is forced with measured wave and tide timeseries covering the entire period without applying any input reduction. The model settings were left at the default values as determined by Ruessink et al. (2007). The initial bottom profile was obtained from the Jarkus database without applying any longshore averaging. For this study the model is extended with an automatic time stepping method which is limited by a maximum time step (1 day) combined with a maximum profile change (0.05 m).

As the main aim of this study is to investigate the overall characteristics of cyclic bar behavior rather then directly comparing predicted and measured bathymetric data we have applied a number of analysis techniques and aggregation methods that are commonly used to analyze measured data such as the complex empirical orthogonal function (CEOF) analysis (Ruessink et al., 2003) and migration statistics on weekly, seasonal and inter-annual time scales (van Enckevort and Ruessink, 2003). We analyzed the annual profile measurements (from 1964 to 2003), which covers about 10 bar cycles. This will allow us to evaluate the model performance based on the entire set of measured profiles.

3. Results

Results from the analysis reveal that the model is able to give acceptable predictions of the actual cycle period, bar height and bar distance. The decay of the outer bar and formation of a new bar after about 900 days (bottom plot of Figure 1) is in the range of the observed natural variability. The bar height and bar distance as a function of the cross-shore position (Figure 2) show that the maximum bar height is under-predicted, but that the growth (near shore) and decay (offshore) is represented

[1] Deltares, Department of Hydraulic Engineering (HYE), PO Box 177, 2600 MH Delft, The Netherlands. Phone: + 31 (0)152858887, fax: +31 (0)152858714, email: DirkJan.Walstra@Deltares.nl.

[2] Delft University of Technology, Faculty of Civil Engineering and Geosciences, The Netherlands.

[3] University of Utrecht, Institute for Marine and Atmospheric Research, Department of Physical Geography, PO box 80115, 3508 TC Utrecht, The Netherlands, email: G.Ruessink@geo.uu.nl

accurately. Distance between the bars is under-predicted somewhat but the increase in distance when bars are migrating offshore is also found by the model.

In the paper a more detailed discussion on the applied model and analysis methods combined with a thorough discussion on the relevance of various physical processes and environmental conditions that influence cyclic bar behavior will be presented.

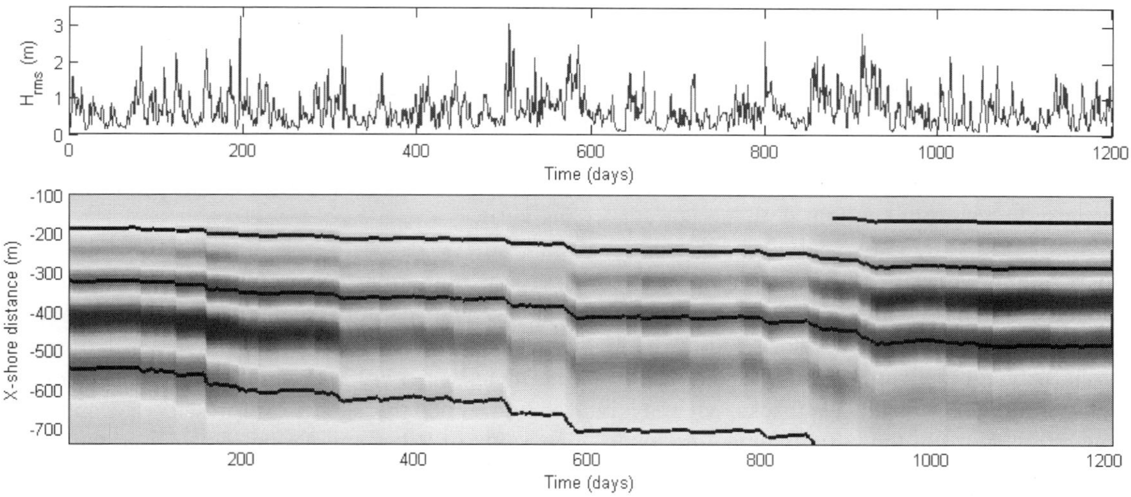

Figure 1. Time series of offshore wave height (top) and time stack of the first CEOF mode of the predicted profiles (bottom), black solid lines indicate the derived bar crest position. The red colors correspond to bar crests, blue colors correspond to bar troughs.

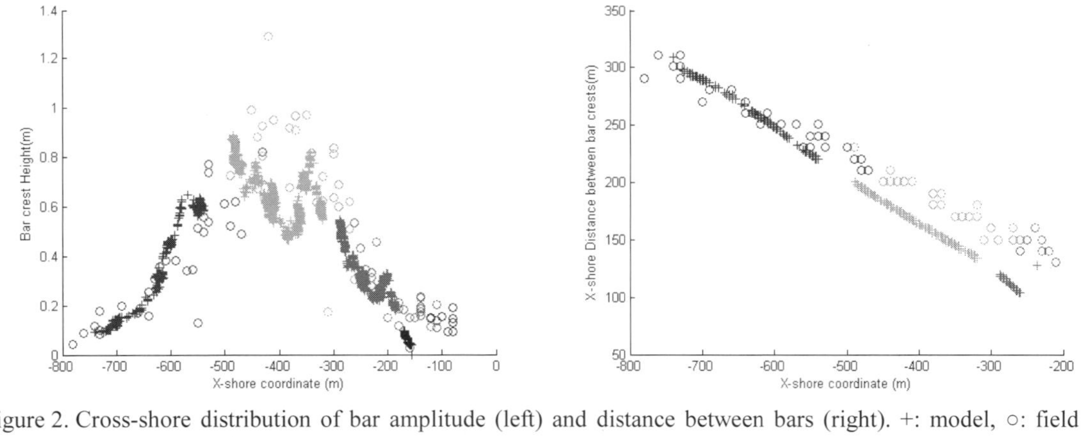

Figure 2. Cross-shore distribution of bar amplitude (left) and distance between bars (right). +: model, ○: field data. Colors; blue: initial outer bar, green: middle bar, red: initial inner bar, black: new inner bar.

References

Ruessink, B.G., Y. Kuriyama, A.J.H.M. Reniers, J.A. Roelvink, and D.J.R. Walstra (2007), Modeling cross-shore sandbar behavior on the timescale of weeks, J. Geophys. Res., 112, F03010, doi:10.1029/2006JF000730.

Ruessink, B.G., K.M. Wijnberg, R.A. Holman, Y. Kuriyama, and I.M.J. van Enckevort (2003), Intersite comparison of interannual nearshore bar behavior, J. Geophys. Res., 108(C8).

Van Enckevort, I.M.J. and B.G. Ruessink (2003), Video observations of nearshore bar behaviour. Part 1: alongshore uniform variability. Continental Shelf Research 23.

Wijnberg, K.M. and J.H.J. Terwindt (1995), Extracting decadal morphological behaviour from high-resolution, long-term bathymetric surveys along the Holland coast using eigenfunction analysis, Mar. Geol., 126.

Coastal Dynamics 2009
Paper No.55

THE ROLE OF SURFACE ROLLERS ON THE FORMATION OF SURFZONE TRANSVERSE SAND BARS

Francesca Ribas[1], H.E. de Swart[2], Daniel Calvete[3] and Albert Falqués[4]

1. Introduction

Patches of transverse sand bars have been observed in the surf zone of several beaches. These bars are spaced with a remarkable alongshore periodicity (from 20 to 200 m). A patch of transverse bars with a wavelength of *45* m, observed during September 2002 at Noordwijk beach (the Netherlands), is shown in Figure 1a (Ribas and Kroon, 2007). The bar crests deviated some 30° from the shore-normal against the longshore current flow (up-current orientation).

A possible explanation for the formation of this type of up-current oriented bars is based on the concept of morphodynamic self-organization. Topographic perturbations superimposed on an alongshore uniform beach induce hydrodynamic perturbations, which can lead to convergence of sand transport over the bars, hence producing a positive feedback. Stability analysis is a convenient tool to investigate the possible feedbacks, yielding information about the shape, the growth rate and the migration speed of the initially emerging modes. The study of Ribas and Kroon (2007) used the field observations at Noordwijk to test the predictions of the existing idealized self-organization models for transverse bar formation. Considering that Noordwijk bars were up-current oriented and that they emerged during periods of clearly oblique wave incidence, only the 'bed-flow mechanism' described by Ribas et al. (2003) remained as a viable explanation for their formation. Their crucial assumption is related to the cross-shore distribution of the depth-averaged volumetric sediment concentration, C_{da}. Only when C_{da} decreases seaward in the inner surf zone, up-current oriented bars can emerge.

The aim of the present contribution is to model and understand the role of surface rollers on the formation of up-current oriented surfzone bars using a more realistic model. Rollers can play a crucial role because they create turbulent bores that can lead to significant sediment resuspension in the inner surf zone. This may give a cross-shore distribution of C_{da} that could explain the formation of up-current bars. We use the up-to-date self-organization model described in Calvete et al. (2005), based on stability analysis, after extending it to include rollers and bore resuspension in the inner surf zone.

2. Model

The model describes the feedback between wave and roller dynamics, depth-averaged currents and bed evolution, so that self-organized processes can develop. Wave heights are supposed to be randomly distributed. Wave evolution is described using linear theory and a wave energy balance equation (with wave-current interactions). The roller energy balance equation is modelled following Reniers et al. (2004). The large-scale fluid motions are governed by the momentum and mass conservation equations, where the radiation stresses due to wave and roller propagation are included. All equations and quantities are wave- and depth-averaged. Finally, the depth-averaged sediment transport in the bed evolution equation is modelled following the widely accepted Soulsby and van Rijn formula. We have extended it to include the extra contribution to C_{da} due to the sediment stirring

---

[1] Dept. Appl. Physics, Univ. Politècnica de Cat., Jordi Girona 1-3, 08034 Barcelona, Spain. cesca@fa.upc.edu
[2] IMAU, Utrecht Univ., Princetonplein 5, 3584 CC Utrecht, the Netherlands, h.e.deswart@uu.nl
[3] Dept. Appl. Physics, Univ. Politècnica de Cat., 08034 Barcelona, Spain. calvete@fa.upc.edu
[4] Dept. Appl. Physics, Univ. Politècnica de Cat., 08034 Barcelona, Spain. falques@fa.upc.edu

created by the bore induced turbulence. We follow Roelvink and Stive (1989), who assumed that this extra C_{da} was proportional to the dissipation of roller energy.

Firstly, the equilibrium solution is computed using a reference beach profile measured at Noordwijk. Then, stability analysis is applied in a standard way. Starting from arbitrary initial conditions, the dynamics after some time will be dominated by the fastest growing mode.

3. Results and conclusions

The model is applied to the specific wave and bathymetric conditions measured at Noordwijk. Realistic positive feedback leading to formation of bars like those observed only occurs if the stirring of sediment due to bore turbulence is included in the model. In that case, the depth-averaged sediment concentration C_{da} decreases seaward across the inner surf zone. This, in combination with an offshore-directed flow over the up-current oriented bars, leads to accretion of sediment in the crest areas. Thereby, our results confirm that the bed-flow interaction, described in Ribas et al. (2003), can cause the development of transverse bars in Noordwijk.

The up-current oriented shape (with an orientation of some 60°), the wavelength (around 60 m) and the growth rate (of the order of half a day) of the modelled bars are in good agreement with observations at Noordwijk. However, modelled migration speeds (from tens to a few hundreds of meters per day) are significantly higher than those measured in the field. Figure 1c shows the fastest growing mode obtained for the conditions measured during the event shown in Figure 1a.

Acknowledgements

Funding from the Spanish research project `Morfodinámica de playas: predicciones en las grandes escalas espacio-temporales' (contract CTM2006-08875/MAR) is acknowledged.

References

Calvete, D., Dodd, N., Falqués, A., and van Leeuwen, S.M., 2005. Morphological development of rip channel systems: Normal and near normal wave incidence. *J. Geophys. Res.*, 110:C10006, doi:10.1029/2004JC002803.

Reniers, A.J.M.H., Roelvink, J.A., and Thornton, E.B., 2004. Morphodynamic modeling of an embayed beach under wave group forcing. *J. Geophys. Res.*, 109: C01030, doi:10.1029/2002JC001586.

Ribas, F., and Kroon, A., 2007. Characteristics and dynamics of surfzone transverse finger bars, *J. Geophys. Res.*, 112: F03028, doi:10.1029/2006JF000685.

Ribas, F., Falqués, A., and Montoto, A., 2003. Nearshore oblique sand bars, *J. Geophys. Res.*, 108: C43119, doi:10.1029/2001JC000985.

Roelvink, J.A., and Stive, M.J.F., 1989. Bar-generating cross-shor flow mechanisms on a beach. *J. Geophys. Res.*, C94: 4785-4800.

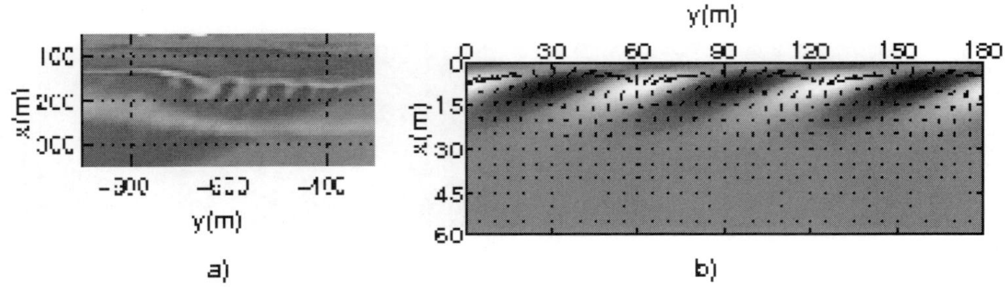

Figure 1. a) Observed transverse sand bars, in a 10-minute average of video images that shows a clear white signal over the bars due to predominant wave breaking; b) Modelled transverse sand bars, in an image with the topographic perturbations (white areas stand for shallow regions) and the current perturbations in small arrows.

Coastal Dynamics 2009
Paper No.56

MODELLING TRANSITIONS BETWEEN BARRED BEACH STATES ON A STRAIGHT COAST

Darrell Strauss[1] and Rodger Tomlinson[1]

1. Introduction

The identification of hazardous beach conditions is aided by an understanding of the complex processes capable of altering the morphology of barred beaches. Transitions between individually classified beach states are readily observed in sequential time averaged video images. Recognising the existence of distinct antecedent beach morphologies and analysing wave data co-incident with observed morphological change highlights conditions which can trigger a transition between beach states. Several intermediate beach state transitions and the wave conditions surrounding the change of state are analysed in this manner to produce conceptual models of beach state transitions. A regional wave model is constructed to simulate the propagation of waves derived from a deepwater global model (Wavewatch III) to a near shore hydrodynamic and sediment transport model (Delft 3D). The wave model was calibrated with directional wave buoy data and beach state transitions were simulated using the deepwater wave model hindcast parameters.

2. Monitoring Beach States

The widely used beach state model of Wright and Short (1984) provides a means of classifying distinct surf zone morphologies. Beaches which exhibit primarily intermediate beach states are the most likely to present a hazard to beach users in the form of rips or strong longshore currents. The long, straight coastline of the northern Gold Coast, Australia is subject to significant net northward littoral transport and exposed to a highly variable wave climate. The beach regularly exhibits the higher energy intermediate beach states of longshore bar and trough, rhythmic bar and beach and transverse bar and rip as described by the Wright and Short classification. Gold Coast City Council (GCCC) implemented video based monitoring with an Argus coastal imaging system in July 1999 (Turner and Leyden, 2000). Daily timex images provide a quick visual identification of the beach morphology and occurrences of transition between the intermediate beach states.

2005-01-11

Figure 1. Argus image of Linear Bar and Trough morphology (GCCC, WRL UNSW).

2005-01-16

Figure 2. Argus image of Transverse Bar and Rip morphology (GCCC, WRL UNSW).

[1]GCCM, Griffith University, Gold Coast, 4222, Australia. D.Strauss@griffith.edu.au

3. Modelling Results

The regional wave model (SWAN) was forced at the offshore boundary with significant wave height, peak period and direction from Wavewatch III global wave model hind casts (NOAA/NCEP). The nearest grid point to the study site was approximately 31 km offshore (-28°S, 153.75°E). Figure 3 compares significant wave height of the boundary input, wave rider buoy measurements and model output at the wave-rider location. The wave model provides a good representation of the near shore wave heights. In particular there are several episodes of high offshore wave heights from a southerly direction which are strongly attenuated due to sheltering by significant reefs and the general topography of the coastline. The 2DH hydrodynamic and sediment transport model with initial longshore bar and trough morphology was forced with the conditions of an observed scenario corresponding to a change in wave direction from oblique to shore-normal incidence.

4. Conclusions

SWAN modelling of wave propagation using Wavewatch III hindcast boundary conditions provides a good proxy for measured offshore wave height data. Modelled wave, flow and sediment transport responses associated with episodes of LBT to TBR beach state transitions agree qualitatively with morphological signatures of visually observed transitions. Offshore directed flows emerged and result in alongshore non uniform wave dissipation. Offshore transport of sediment occurs within rip channels and shoreward transport on the bars between rip features (Figure 4).

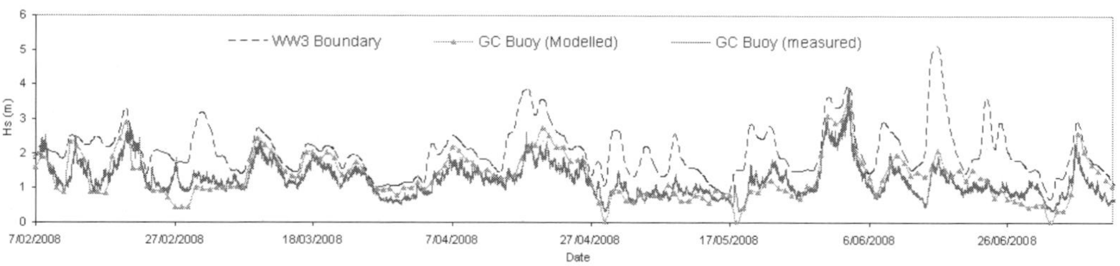

Figure 3. Comparison of model input, output and measured significant wave height.

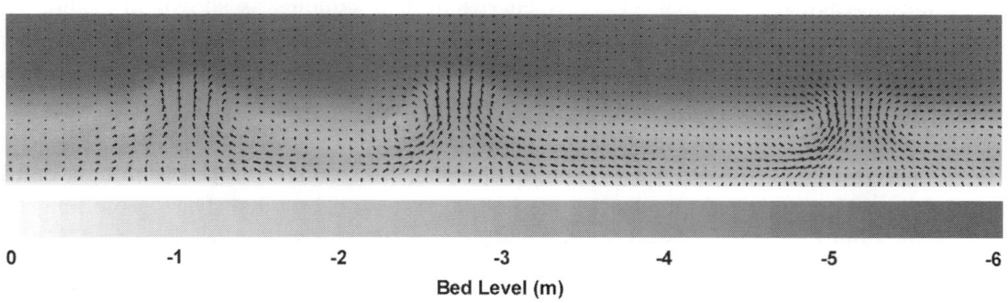

Figure 4. Emergence of rip currents in response to shore-normal wave incidence.

References

Turner, I.L. and Leyden, V., 2000. System Description and Analysis of Shoreline Change: August 1999 - February 2000. *Report 1: Northern Gold Coast Coastal Imaging System. WRL Technical Report 00/12*, Water Research Laboratory, University of New South Wales.

Wright, L.D. and Short, A.D., 1984. Morphodynamic variability of Surf Zones and Beaches: A synthesis. *Marine Geology*, 56, 93-118.

Coastal Dynamics 2009
Paper No.57

LONGSHORE SEDIMENT MOVEMENT ALONG THE ENSHUNADA COAST INFERRED FROM FELDSPAR THERMOLUMINESCENCE

Haijiang Liu[1], Yuuki Ogawa[1], Tomohiro Takagawa[1] and Shinji Sato[1]

1. Introduction

Traditionally, using fluorescent or dye-marked tracers to study the nearshore sediment transport is applied in field works (Hiramatsu *et al.*, 2008). Nevertheless, the artificially injected tracers are different from natural coastal sands in physical and chemical properties, which may cause unreliable field measurements. Recently, Rink (2003), Liu *et al.* (2009) suggested the use of natural residual thermoluminescence (TL) properties of quartz or feldspar grains as a possible transport indicator in the study of various coastal depositional processes. Applying this approach, sediment particles constituting the investigated object (natural sand), themselves, are used as self-tracers to monitor the nearshore sand movement. This allows more reliability related to the field investigation on the nearshore processes. In this study, such technique was applied to estimate the longshore sediment movement pattern in a relatively wide coastal area.

2. Research Area and Methodology

In total, about 50 TL samples were collected along the shoreline of Enshunada Coast, which is located in the west of Shizuoka Prefecture, Japan and has a twin-arc shoreline profile with Tenryu River mouth extruding in the middle (Figure 1). Tenryu River, considering its large sediment discharge, is assumed to be the original and dominant sediment source for the nearshore sands at the Enshunada Coast. Sample collection also went eastward to the nearshore region along the Suruga Bay.

After field sampling, conventional pretreatment was conducted to extract the feldspar fraction in the range 180-300 μm. Afterward, a monolayer aliquot was applied in measurement to avoid the possible inhomogeneous features in luminescence signals from individual grains. Measurements were performed based on the TL test protocol proposed in Liu *et al.* (2009).

3. Primary Results and Discussions

Figure 2 illustrates normalized feldspar TL glow curves from 5 samples as shown in Figure 1, in which sample T was collected at Tenryu River mouth and samples E1, E2, E3 and E4 were from the east side of Enshunada Coast with an increasing distance from the river mouth. Average over four aliquots were considered for every collected sample. It is seen that the normalized natural residual TL signals are gradually decreasing with increasing distance from their primary source at the Tenryu River mouth. This is in consistent with the measurement result from Rink (2003). It is thought that the more significant the solar radiation, the smaller the natural residual TL signal. During the longshore sediment transport, sand particles undergo light exposure with respect to moving time and distance, thus depopulate their stored TL signal in the nearshore process. Hence, a decreasing trend of TL signal is expected in the downdrift direction as demonstrated from Figure 2.

Spatial distribution of the TL intensity by considering the longshore distance away from the Tenryu River mouth is illustrated in Figure 3. TL intensity is evaluated by considering average of the normalized TL value over temperature range 260-380^{0}C, which covers dominant TL signals as shown in Figure 2. It is found that around the river mouth (sediment source), similar to the previous observation, a decreasing trend of TL intensity can be observed on both sides. However, in the river

[1]Dept of Civil Eng., University of Tokyo, 7-3-1 Hongo, Bunkyo-ku, Tokyo 113-8656, Japan. hjliu@coastal.t.u-tokyo.ac.jp; ogawa@coastal.t.u-tokyo.ac.jp; takagawa@coastal.t.u-tokyo.ac.jp; sato@coastal.t.u-tokyo.ac.jp

mouth-far area, *e.g.*, 30 km away, such decreasing trend terminates for both sides. TL intensity keeps a rather constant value, *i.e.*, about 1.6 at west side and 0.7 along east side. The constant value reveals the hard-to-bleach TL component remained in feldspar grains, which can not be thoroughly depleted under natural environment. Difference between west and east sides, *e.g.*, longshore TL intensity decreasing rate and remaining TL level, reveals different nearshore movement mechanisms with particle light exposure (TL signal depopulation) being more significant for the east side. In case of the same longshore distance, sand particles take a longer travelling time (more sunlight bleaching) along the east side, which further demonstrates that considering the Enshunada Coast, the general longshore sediment transport rate should be smaller at the east side of the Tenryu River mouth.

References

Hiramatsu, H., Tomita, S., Sato, S., Tajima, Y., Aoki, S. and Okabe, T., 2008. Sediment movement mechanism around the Magome River mouth based on color sand tracer experiments. *In: Proceedings of Coastal Engineering*, JSCE, 55(1), 696-700. (in Japanese).

Liu, H., Kishimoto, S., Takagawa, T., Shirai, M. and Sato, S., 2009. Investigation of the sediment movement along the Tenryu-Enshunada Fluvial system based on feldspar thermoluminescence properties. *Journal of Coastal Research.* In press.

Rink, W.J., 2003. Thermoluminescence of quartz and feldspar sand grains as a tracer of nearshore environmental processes in the southeastern Mediterranean Sea. *Journal of Coastal Research.* 19(3), 723-730.

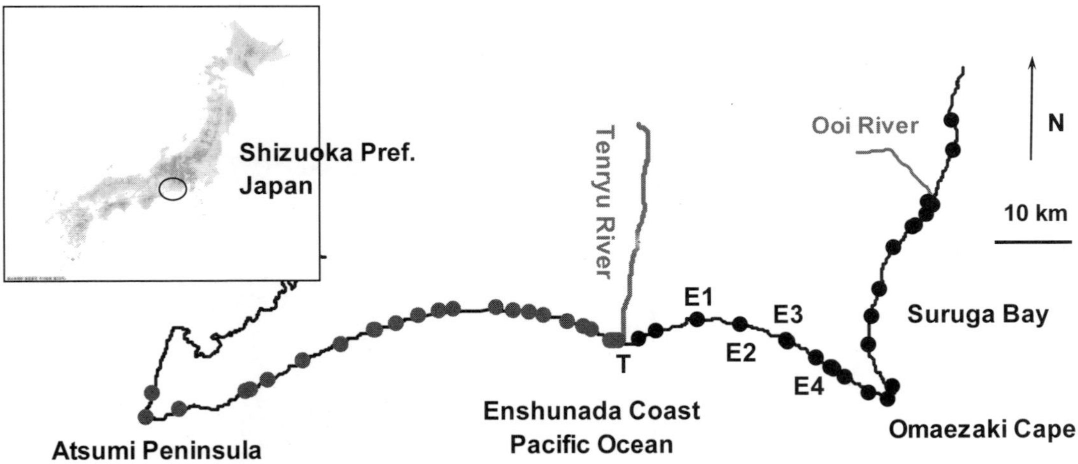

Figure 1 Research area and TL sample locations

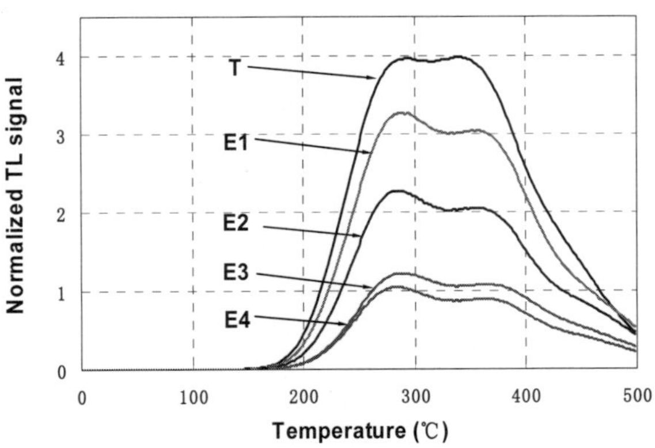

Figure 2 Normalized feldspar TL signals

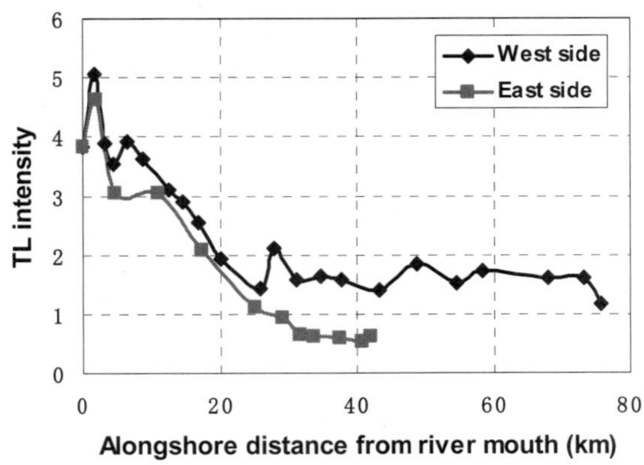

Figure 3 Spatial distribution of TL intensity

Coastal Dynamics 2009
Paper No.58

OPTICALLY STIMULATED LUMINESCENCE DATING OF A SANDY TIDAL FLAT IN THE DANISH WADDEN SEA – TOWARDS ESTABLISHING A SEDIMENT BUDGET

Anni T. Madsen[1, 2], Thorbjørn J. Andersen[1], Andrew S. Murray[2] and Morten Pejrup[1]

1. Introduction

Sediment budgets for inter-tidal areas have been estimated for several compartments in the Danish part of the Wadden Sea (e.g. Pejrup et al., 1994; Petersen and Bartholdy 2006). However, these budgets are based exclusively on ^{210}Pb dating of fine-grained sediments, and so only describe the accumulation of fine-grained sediments. However, in coastal-plain estuaries such as the Wadden Sea, most of the deposited sediment is sand. Thus information on sand budgets for tidal lagoons is very important, especially when trying to predict the future morphological development of shallow tidal lagoons in a period with a rapidly rising sea-level. This study uses Optically Stimulated Luminescence (OSL) dating to determine sedimentation rates and aims to look into the spatial and temporal variability of sedimentation on a sandy tidal flat.

2. Study area

The study site consists of the tidal flats enclosed by the barrier island Rømø and the mainland coast in the northern part of Lister Dyb tidal area, Denmark. Nine sediment cores have been retrieved on a sandy tidal flat along two transects perpendicular to the mainland coast (Figure 1).

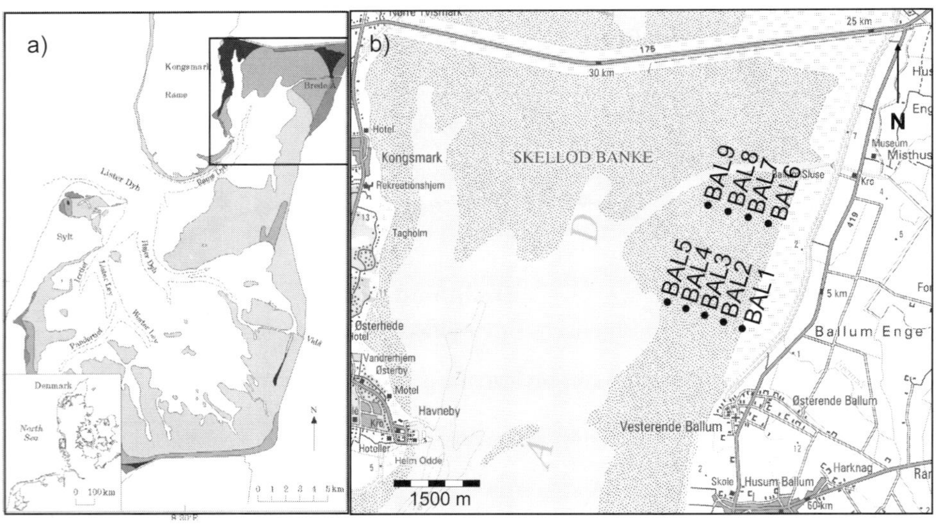

Figure 1. a) Sedimentological map displaying the study site; light grey areas are sand flats (>95% sand); the dark grey areas are mixed mudflat (10–50% sediment < 63 μm) and black areas are mudflat (>50% sediment < 63 μm). b) Sample locations in the northern part of Lister Dyb tidal area.

[1]Dept. of Geography and Geology, University of Copenhagen, Øster Voldgade 10, 1350 Copenhagen K, Denmark. atm@geo.ku.dk, tja@geo.ku.dk, mp@geo.ku.dk
[2]Nordic Laboratory for Luminescence Dating, Dept of Earth Sciences, University of Aarhus, Risø DTU, 4000 Roskilde, Denmark. anmu@risoe.dtu.dk

3. Methodology

OSL dating was chosen as the primary chronological tool to establish a time framework in this study. The technique assesses the period of time elapsed since the sediment grains were last exposed to sunlight as a result of sediment transport in the coastal environment. The method is based on the ability of certain minerals (e.g. quartz and feldspar) to trap and accumulate charge in crystal defects for long periods. The trapped charge is derived from energy absorbed from ionization radiation from naturally occurring radio-nuclides and cosmic radiation, i.e. the dose rate. More details on the OSL dating methodology as applied to young sediments are given in Madsen and Murray (2008, *in press*).

4. Results

The sandy tidal flats adjacent to the mainland coast have experienced significant bioturbation by lugworms. Moreover, reworking by waves and currents on the fairly exposed tidal flat will have resulted in re-suspension and erosion events, which together with bioturbation must have smoothed the age-depth trends in the cores to some degree. It is, however, not known how complete the mixing has been and the importance of vertical mixing on the OSL ages is therefore difficult to interpret. The effect of bioturbation is generally to make the OSL ages younger, as younger material is transported to a greater depth and older material is delivered at the surface, where it is solar reset before reburial.

Despite the bioturbation, an absolute OSL chronology, with ages ranging from a few years up to ~14,000 years, has been obtained for the sediment cores. A considerable temporal variability of inter-tidal sedimentation rates is observed across the tidal lagoon. The lengths of inter-tidal sequences are furthermore very variable at the nine sites. One may speculate why the length/extent of the inter-tidal sequence differ from site to site and whether this variation is related to the spatial variation in surface elevation at the various sites, and thus different times for the sea-level to rise to a level where the different sites get inundated during flood tides.

Inter-tidal sedimentation rates are found to be similar to the rate of relative sea-level rise, which is estimated at ~0.7 $mm.a^{-1}$ in this area (Szkornik et al., 2008). Despite some spatial variability across the tidal lagoon, with some eroding and some accumulating areas, in general the sedimentation rates on the sandy tidal flat seem to have kept up with the sea-level rise.

5. Conclusions

This application of OSL dating shows that it is possible to date the coarse-grained fraction of the sediment in a tidal lagoon, so that establishment of sediment budgets for sandy tidal flats now is possible. In contrast to previous fine-grained sediment budgets based on ^{210}Pb, budgets based on OSL dating are able to estimate the total sedimentation in coastal lagoons, and so data interpretation is significantly more robust than in earlier studies. We conclude that sea-level oscillations are very important controls on inter-tidal deposition in the Lister Dyb tidal area, and it is likely that this result applies to many other shallow inter-tidal areas.

Acknowledgements

This study is financed by a PhD-grant from the Faculty of Science, University of Copenhagen, Denmark and an 'EliteForsk'-grant from the Danish Agency for Science, Technology and Innovation.

References

Madsen, A. T., Murray, A. S., 2008 (*in press*). Optically stimulated luminescence dating of young sediments: a review. *Geomorphology*.

Szkornik, K., Gehrels, W. R., Murray, A. S., 2008. Aeolian sand movement and relative sea-level rise in Ho Bugt, western Denmark, during the 'Little Ice Age'. *The Holocene* 18, 951−965.

Pedersen, J. B. T, Bartholdy, J., 2006. Budgets for fine-grained sediment in the Danish Wadden Sea. *Marine Geology* 235, 101−117.

Pejrup, M., Larsen, M., Edelvang, K., 1994. A fine-grained sediment budget for Sylt-Rømø tidal basin. *Helgoländer Meeresuntersuchungen* 51, 263−268.

Coastal Dynamics 2009
Paper No.59

FIELD INVESTIGATION OF STRATEGIC BORROW SITES OFF THE MEDITERRANEAN COAST OF SPAIN

Luis Moreno[1] and Ana García Fletcher[2]

1. Introduction

The Spanish National Coastal Authority has started the systematic geophysical investigation of the coast of Spain between depths of 40 m and 80 to 100 m.

The European Union funded project "Eurosion" suggested that the Member States should investigate the so-called "*strategic borrow sites*" to identify sources of sand suitable for beach renourishment projects on the long term and at regional scales.

Within this framework, field investigations of possible strategic borrow sites off the Mediterranean coast of Spain have been carried out. This abstract deals with the brief description of the methodologies and equipment used in relation to a very successful case study.

2. Study Area

The study area is located off the East coast of Spain, in the Mediterranean Sea at depths comprised between 60 and 80 m. The geophysical study previously performed showed the presence of sand dune type of features at around those depths. This led to the prospective investigation of those features by means sixty 6 m depth vibrocoring tests scattered around a large seabed area. The promising results obtained in many of the tests as far as grain size and thickness of good quality sand layers is concerned led to the definition of more precise areas to perform further in-depth field investigation.

3. Methodology

A specific field campaign was designed comprising 280 vibrocoring stations, being the vibrocoring equipment itself able to penetrate the seabed up to 6 m in the absence of rock. The location of the stations was carefully selected taking into account the previously existing bathymetric and geophysical information of the area. Figure 1 shows the 43.5 m long research vessel and Figure 2 shows the 3200 kg vibrocoring equipment.

Immediately before the execution of the vibrocoring test, seabed-profiling information was collected by means of an Ore 3.5 kHz piece of equipment to compliment the information obtained by the vibrocoring test. Also, photographic information of the seabed at the vibrocoring stations was taken as well as video recordings in order to help identify benthic communities. Water quality and marine currents were also monitored during the length of the vibrocoring field campaign.

The vibrocoring columns were split into 0.5 m long segments, and sediment samples were taken in the laboratory and grain size analyses of all 2744 samples was performed. Other type of laboratory analyses such as mineralogy assessments, investigation of possible pollution of sand, organic matter content and chemical characterization and heavy metals determination of the upper sediments, etc., were performed over selected samples.

4. Results

A strategic borrow site with approximately $100 \cdot 10^6$ m^3 of high quality suitable sand for

[1]Coastal Engineering Division, Intecsa-Inarsa, S.A., Santa Leonor, 32, 28037-Madrid, Spain. luis.moreno@snclavalin.com
[2]National Coastal Authority, Ministry of Environmental Protection, Plaza San Juan de la Cruz, s/n, 28071-Madrid, Spain. agfletcher@mma.es

renourishment of beaches has been identified and characterized down to the vibrocoring penetration depth. The seabed profiling results showed that the available volume of sand below the penetration depth may be even larger.

Some sectors of the borrow site are covered with silt and / or mud, but no pollution has been detected thus making the range of potential beneficial uses of this material wide.

Figure 1. Research vessel

Figure 2. Vibrocoring equipment on board the research vessel

Acknowledgements
The authors wish to thank the Spanish National Coastal Authority for granting permission to publish this information.

Coastal Dynamics 2009
Paper No.60

COMBINING GEOLOGIC AND NUMERICAL MODELS TO ASSESS SAND RESOURCES ON THE INNER CONTINENTAL SHELF OF NORTHEAST FLORIDA

Gary A. Zarillo[1]

Over the past few decades beach quality sand has been recovered from nearshore resources to help restore eroding beaches along the Atlantic and Gulf Coasts of Florida and along other areas of the U.S. East Coast. Sand ridges as far as 10 nautical miles offshore in some areas of the Florida coastal ocean may have crest elevations as shallow as -11m. In order to conduct an efficient and cost effective search for beach quality sand on the inner continental shelf it is necessary to first apply the Geologic Model for sand ridges to assess which shoal systems are most likely to yield beach quality sand. Conceptual models of sand ridge genesis place their origin in the littoral to nearshore zone where migrating tidal inlets coupled with rising sea-level result in large ebb shoal features that expand over time and eventually become detached from the shoreface. Once part of the inner continental shelf, these sand ridges are maintained by energetic wave and current events that can rework the upper one or two meters of the ridges. Bedforms on modern sand ridges and similar features found in the geologic record shown that the crest areas of sand ridges are characterized by relatively coarse cross-bedded sands and sandstones that indicate a high energy environment (Figure 1). These sands, originating in the littoral environment and maintained in the shallow marine environment can provide ideal material for beach fill.

Since sand ridges and multi-ridged sand banks are common on the inner continental shelf of northeast Florida and other areas off the U.S. East Coast a method for limiting exploration surveys to a manageable level is required. To accomplish this goal the Coastal Modeling System (CMS) was applied in a series of six model grids to simulate sand transport and topographic change along Florida coast. Developed by U.S. Army Engineer Research and Development Center (ERDC), Coastal and Hydraulics Laboratory (CHL), the CMS is a group of numerical models for calculating waves, circulation, sediment transport, and morphology change.

For characterization of the Florida sand ridges with respect to potential for holding clean beach quality sand in the crest area, the CMS was forced with measured time series of water elevation, wind stress, and directional wave data. A wave-produced radiation stress field and orbital velocities predicted from a the CMS wave model were included in the model forcing. A steering module in the model interface allows the wave field to be updated in order to apply a time varying wave field to a coupled circulation and sand transport model. Model calibration was performed by comparing predicted data to measured water surface elevations and measured directional wave data.

Model results showed that strong wave and current action over the crest of sand ridges and sand banks is restricted to extreme events when wave height and direction is altered as long period waves propagate across the crest zone of sand ridges. Simulations of sand transport and topographic change were conducted for periods as long as 2-years. Analysis of model results indicate that predicted reworking (erosion and deposition) was linked to the passage of local storms or due to waves arriving in the area from storms in distal fetch areas of the North Atlantic Ocean. The model simulations predicted that up to 1 meter of erosion and deposition may occur over a sand ridge crest as a result of a single event (Figure 2). Sub-surface seismic reflection data and core borings confirm that sand ridges predicted to be most susceptible to reworking by storms contain clean, silt free sand in their upper sedimentary units at higher crest elevations (Figure 3).

[1] Dept of Marine and Environmental Systems, Florida Institute of Technology, Melbourne, USA zarillo@fit.edu

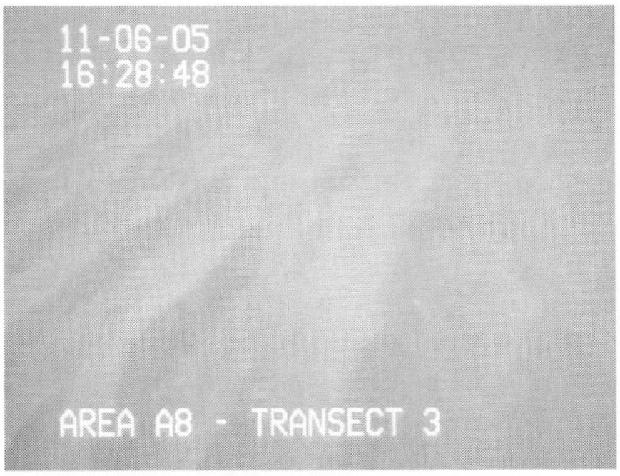

Figure 1. Sand waves viewed under daylight conditions on the crest of shoal A8 at 12m depth off the coast of NE Florida

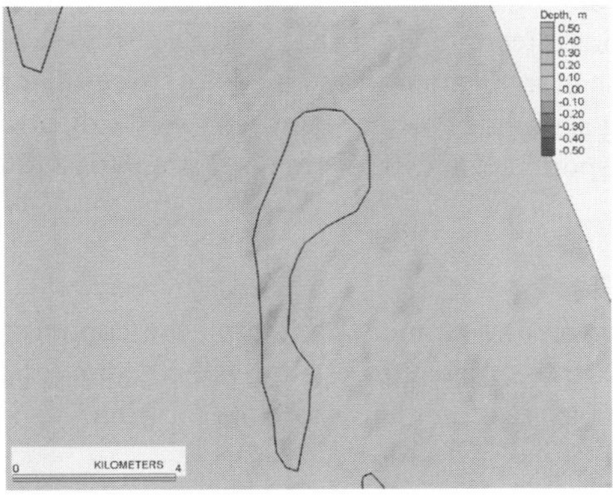

Figure 2. Sand waves viewed under daylight conditions on the crest of shoal A8 at 12m depth off the coast of NE Florida

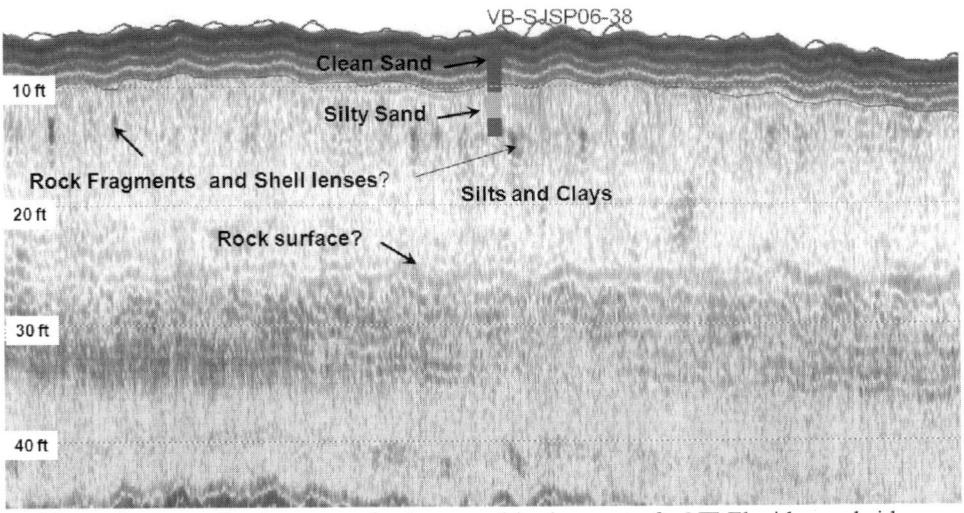

Figure 3. Acoustic signature of clean sand in the crest of a NE Florida sand ridge.

Coastal Dynamics 2009
Paper No.61

ONE-LINE MODEL USING THE COMBINATION OF POLAR AND CARTESIAN COORDINATES FOR CRENULATE SHAPED BAY

Thamnoon Rasmeemasmuang[1] and Sutat Weesakul[2]

1. Introduction

When predominant waves obliquely propagate to the shoreline in-between two headlands, the shoreline will gradually transform into a concave so-called "crenulate shaped bay". The characteristic of this shape is the hooked form in the vicinity of upcoast headland. The typical one-line model using Cartesian coordinates cannot simulate the fitted shape around the hooked region because of the limitation of coordinates system. The polar-and-Cartesian combined coordinates model proposed by Weesakul and Rasmeemasmuang (2002) was applied to reproduce the crenulate shaped bay in many cases of different wave direction and the results were verified with laboratory data and the parabolic bay-shaped equations (Hsu, Silvester and Xia, 1987).

This paper presents a further application of the model of Weesakul and Rasmeemasmuang (2002) to reproduce the time-varying shoreline changes between two headlands with three different equations of alongshore sediment transport, i.e. CERC formula, Kamphuis (1991)'s formula and the newly-modified formula.

2. Numerical Model

One-line model is a shoreline predictive model based on the continuity equation of sand inside the volume of which the certain depth of closure, D_s and the equation of alongshore sediment transport rate Q. The continuity equation governing the shoreline position, y_s in Cartesian coordinates $x-y$ and time t is expressed by

$$\frac{\partial y_s}{\partial t}+\frac{1}{D_s}\left(\frac{\partial Q}{\partial x}\right)=0 \tag{1}$$

The present study modified the typical one-line model that executes on Cartesian coordinates with the combination of polar and Cartesian coordinates. The shoreline between two headlands will be divided into two zone; hooked and unhooked zones, as shown in Fig. 1. All governing equations in Cartesian coordinates were modified to be compatible with polar coordinates. To determine the rate of alongshore sediment transport, the so-called CERC formula was applied as an elemental algorithm.

3. Simulations and Results

The present model was run to reproduce the shoreline evolution between two headlands with the conditions same as the conditions of previous laboratory experiments in which the incident wave angles of 10 to 60 degrees were studied. Figure 2 shows the example of the results of simulation compared with the laboratory data in case of the incident wave angle of 45 degrees.

Furthermore, Kamphuis (1991)'s alongshore sediment transport rate based on three-dimensional hydraulic model experiments was used to substitute CERC formula. It is found that Kamphuis's formula that does not consider the effect of current velocity gave the larger eroded shoreline, particularly inside the hooked zone. In this study, Kamphuis's formula was modified by accounting for the immersed weight sediment transport related both wave energy flux and current velocity. The

[1]Department of Civil Engineering, Faculty of Engineering, Burapha University, Saensuk, Chonburi 20131, Thailand. thamnoon@buu.ac.th
[2]Water Engineering and Management Program, Asian Institute of Technology, P.O. box 4, Klong Luang, Pathumthani 12120, Thailand. sutat@ait.ac.th

results of shoreline changes reproduced by the present model with the newly-modified formula of sediment transport rate were analyzed and discussed.

4. Conclusion

The combination of Cartesian and polar coordinates was used in one-line model to overcome the limitation of Cartesian coordinates in which a concave or a hooked shoreline, so-called crenulate shaped bay, is not applicable. The present numerical model is capable of reproducing the time-varying shoreline evolution which is impossible in case of the application of the parabolic equation of the equilibrium shoreline proposed by, for example, Hsu, Silvester and Xia (1987). Three different formulas of the alongshore sediment transport rate were applied into the present model. The computational results reproduced by the model with the newly-modified formula based on Kamphuis's formula are more consistent with those of the laboratory data than the results reproduced by the model with the original Kamphuis' formula.

References

Hsu, J. R. C., Silvester, R., and Xia, Y. M., 1987. New Charateristics of Equilibrium Shaped Bay. *Proc. of the 8th Aust. Conf. Coastal and Ocean Eng.*: 140-144.

Kamphuis, J. W., 1991. Alongshore Sediment Transport Rate. *Journal of Waterway, Port, Coastal, and Ocean Eng.*, ASCE, Vol. 117: 624-640.

Weesakul, S., and Rasmeemasmuang, T. 2002. Numerical Computation of Crenulate Bay Shape, *Proc. of the 28th International Conference on Coastal Engineering*, Cardiff, UK: 3259-3271.

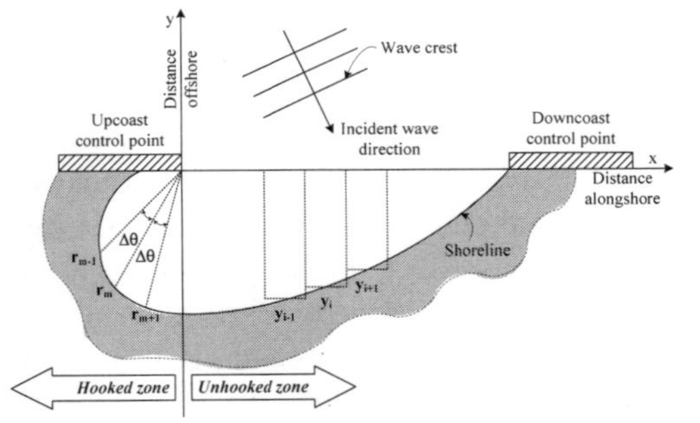

Figure 1. Schematic diagram of one-line model using the combination of polar and Cartesian coordinates for shoreline changes between headlands

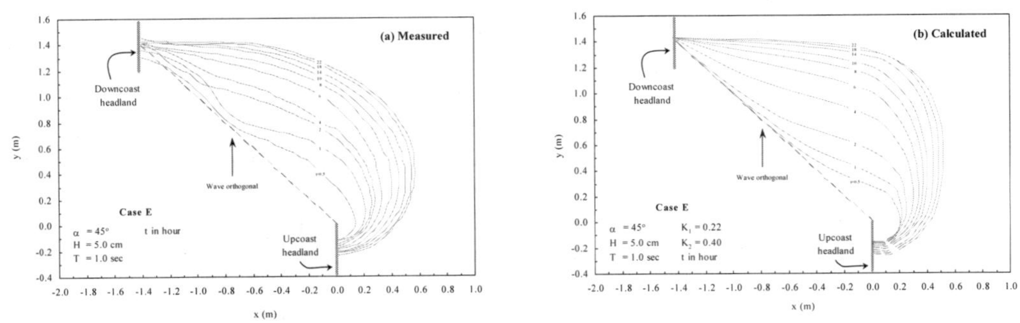

Figure 2. Time-varying shoreline changes in the case of the incident wave angle of 45° (a) measured and (b) calculated results.

Coastal Dynamics 2009
Paper No.62

2D-BEACH CHANGE MODEL INCLUDING THREE TRANSPORT PROCESSES

Shingo Sasaoka[1], Kenji Noguchi[1], Yoshio Suwa[1], Hidetaka Kobayashi[2*],
Akira Watanabe[2] and Toshimitsu Takagi[2]

1. Introduction

Coastal sand transport is often classified into the longshore transport and cross-shore transport for convenience. For the prediction of long-term beach evolutions, the longshore transport is generally significant and the cross-shore transport is ignored mainly because of its reversibility. The cross-shore transport is predominant, however, in the processes of sever beach erosion due to stormy waves during a short period and of gradual accretion after the storm. Hence beach deformation due to the cross-shore transport is extremely important for disaster prevention, but its prediction method is not established yet owing to its complex mechanism.

In the nearshore zone wave-induced near-bed oscillatory flow causes sediment transport. In addition in the surf zone the sediment is suspended by high turbulence of breakers and is transported by mean flow such as the undertow. It is also reported that noticeable shoreward transport occurs in the swash zone where bore-like waves vigorously carry sediment with eddy and turbulence.

The present study sets up a beach evolution model including these three sediment transport processes and examines its validity through comparisons between computations and measurements on the beach erosion and accretion.

2. Outline of 2D-Beach Change Model

Nearshore wave deformation is computed using extended Boussinesq equations. A model by Rattanapitikon and Shibayama (2000) is adopted to calculate the undertow velocity that is important for the transport of suspended sand in the surf zone. We assume that the beach change is governed by the three transport processes under the effect of the gravity as follows: (a) net transport due to asymmetric near-bed oscillatory flow, (b) carriage of suspended sediment in the surf zone, (c) transport in the swash zone caused by bore-like waves and (d) the gravity effect.

For (a), we apply a formula of sheet-flow transport rate under the asymmetric oscillatory flow based on the work by Dibajnia and Watanabe (1996). For (b), concentration of the suspended sand in the surf zone is estimated by a formula of Shibayama and Jayaratne (2004) and is combined with the undertow velocity. The process (c) is incorporated by modifying Dibajnia-Watanabe's formula on the basis of Honda et al. (2003). For (d), the effect of gravity is treated by the method of Watanabe et al. (1986). The relative magnitudes of these factors determine whether a beach is eroded or accreted.

3. Comparison with Measurements in a Large-scale Wave Tank

Computations with the present 2D-beach change model are compared with beach profile change measurements in a large-scale wave tank conducted by Shimizu et al. (1985). They have concluded that the beach profile change will be classified into five types: 1) erosion: type-A, 2) erosion: type-B, 3) transition, 4) accretion: type-A and 5) accretion: type-B. Type-B is characterized by the bimodal profiles, whereas Type-A is unimodal. Figure 1 shows the comparisons for the types 1), 2), 4) and 5) after several hours wave action, and it demonstrates the high capability of computing rather complicated beach profile changes of the present model that includes the three transport processes.

---

[1]Coast Division, River Department, National Institute for Land and Infrastructure Management, 1 Asahi, Tsukuba City, Ibaraki Prefecture 305-0804, Japan. kaigan@nilim.go.jp

[2]Coast Division, INA, 8-4-1 Toshima, Kita-ku, Tokyo 114-0003, Japan. *Contact: hi-koba@ina-eng.co.jp

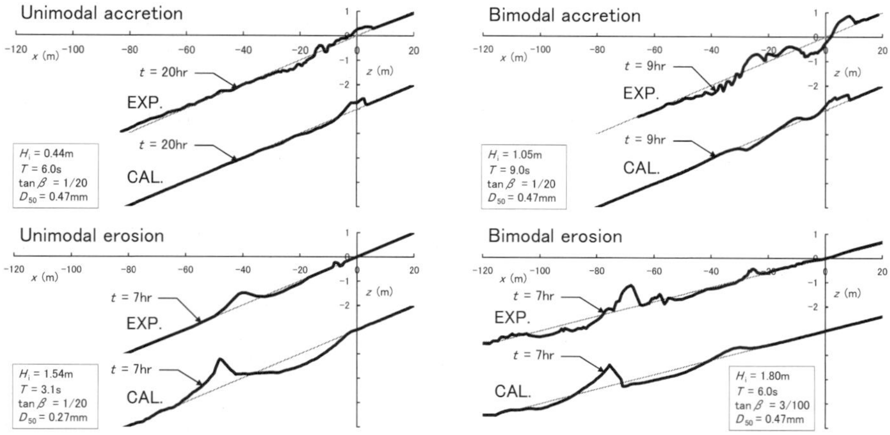

Figure 1. Comparison between measured and computed beach profile changes

4. Characteristics of Beach Profile Change

Examples of computed beach profile changes in Figure 2 may also confirm the reliability of the present model, because effects of the wave steepness, sediment grain size and initial bottom on the beach profiles agree well with the past studies.

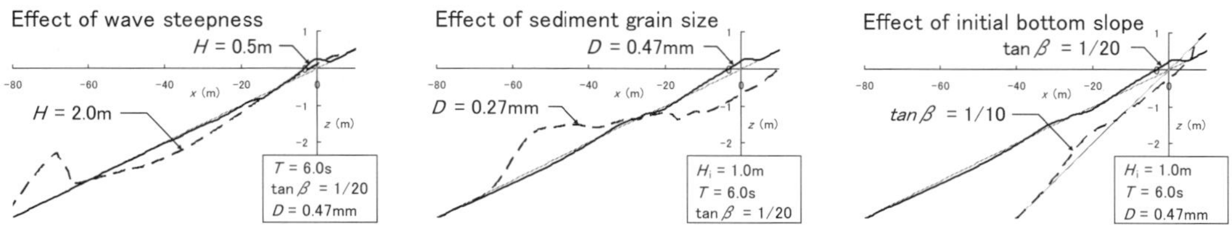

Figure 2. Example of computed beach profile changes (after 24 hours)

5. Conclusions

a) With consideration of the three significant sediment transport processes, a 2D-beach change model has been developed.

b) The capability of the present model for reproducing complicated beach profile changes has been confirmed through the comparison with the large-scale wave tank measurements.

c) Whether a beach is eroded or accreted depends on the balance of the three transport processes and the gravity effect.

References

Dibajnia, M. and A. Watanabe, 1996. A transport rate formula for mixed-sized sands. *Proc. 25th Int. Conf. on Coastal Eng.*, pp. 3791-3804.

Honda, T., S. Sato, A. Watanabe, M. Isobe, 2003. A numerical model of three-dimensional beach deformation due to graded sediment transport in nearshore and swash zone. *Proc. 50th Conf. on Coastal Eng.*, pp. 511-515 (in Japanese).

Rattanapitikon, W. and T. Shibayama, 2000. Simple model for undertow profile. *Coastal Eng. Jour.*, 42(1): 1-30.

Shibayama, T. and M. P. R. Jayaratne, 2004. An evaluation method of suspended sediment concentration in and outside the surf zone. *Proc. 29th Int. Conf. on Coastal Eng.*, pp. 1715-1727.

Shimizu, T., S. Saito, K. Maruyama, H. Hasegawa, R. Kajima, 1985. Modeling of offshore-onshore sand transport rate distribution based on the large wave flume experiments. *CRIEPI Rep.*, No. 384028(in Japanese).

Watanabe, A., K. Maruyama, K. Shimizu, T. Sakakiyama, 1986. Numerical prediction model of three-dimensional beach deformation around a structure. *Coastal Eng. in Japan,* 29: 179-194.

Coastal Dynamics 2009
Paper No.63

BEACH PROFILE MODELING AT DIFFERENT SCALES

Pieter Koen Tonnon[1], Claartje Hoyng[2] and Leo C. van Rijn[1,3]

1. Introduction

Within the EU-sponsored SANDS project, a series of laboratory scale experiments using identical wave conditions based on Froude scaling were carried out a on three scales: large-scale Hannover wave flume experiments (beach slope of 1 to 15), medium scale Barcelona wave flume experiment (beach slope of 1:15) and small-scale Delft wave flume experiments (beach slopes of 1 to 10, 15 and 20). The experimental set-up consisted of a horizontal bed followed by a plane sloping beach. In the experiments, both erosive and accretive wave conditions were applied. The main sediment and hydrodynamic parameters for the erosive wave conditions are presented in Table 1.

The SANDS project aims to advance the knowledge of cross-shore processes and mobile bed hydraulics under waves to a higher quantative level and with a special focus on the reduction of errors due to scale effects in physical scale models. A basic element of the SANDS research is the numerical modeling of the beach volume changes under erosive and accretive wave conditions to study the different scales at which the dominant cross-shore processes are operating in the three laboratory experiments. This will help to better identify and to model the scale effects related to cross-shore processes.

Table 1. SANDS model scale experiments in wave flumes of Hannover, Barcelona and Delft

Parameter	Hannover	Barcelona	Delft
beach slope	1 to 15	1 to 15	1 to 10, 15 and 20
waves (H_{m0} [m], T_p [s]	0.97,5.6	0.53, 4.14	0.167, 2.33
sediment (d_{10}, d_{50}, d_{90} [mm])	0.14,0.27,0.51	0.16,0.25,0.37	0.1, 0.13, 0.17
depth, sediment and distortion scale	1, 1, 1	1.8, 1.08, 1	6.4, 2.08, 1.5 to 0.5

2. Analysis

Beach erosion was dominant under relatively high waves (storm waves) and beach accretion was dominant under relatively low waves (post-storm waves). Figure 1 shows the dimensionless beach erosion volumes (ratio of beach erosion volume and offshore wave height squared) as a function of time, which was used to study the morphological time scale of the experiments involved. The deviation of the small-scale Delft experiments from both Hannover and Barcelona experiments are attributed to scale effects. A similar plot will be presented for beach accretion. Numerical modeling of beach erosion and accretion at three different scales can help to identify the most important wave and sediment parameters influencing the scale errors by means of a detailed sensitivity study.

[1]Deltares, P.O.Box 177, 2600 MH Delft, The Netherlands. pieterkoen.tonnon@deltares.nl
[2]Royal Haskoning, Rotterdam, The Netherlands
[3]Dept. of Physical Geography, University of Utrecht, Utrecht, The Netherlands

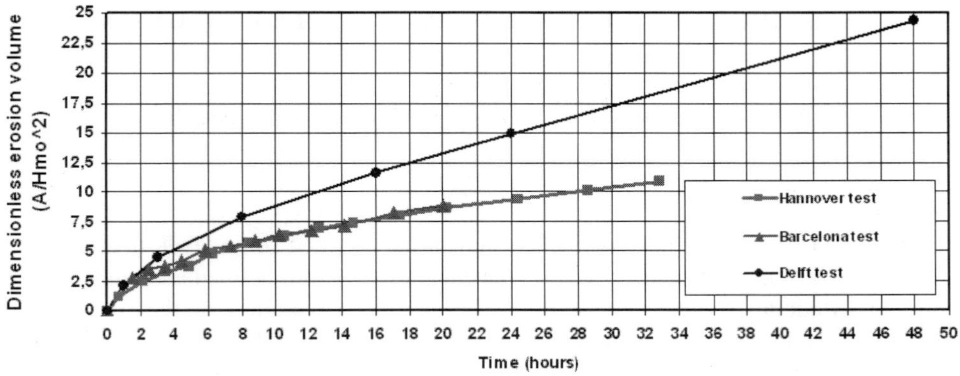

Figure 1. Measured dimensionless beach erosion volume as function of time; Hannover, Barcelona and Delft

3. Modeling

This paper focuses on the process-based modeling of the SANDS laboratory experiments using three cross-shore process-based models: CROSMOR (see Van Rijn et al. 2003), UNIBEST-TC (see Roelvink and Stive 1989 and Bosboom et al. 1997) and DELFT3D (see Lesser et al. 2004).

A basic model calibration is carried out focusing on the reproduction of measured wave heights, velocities and beach erosion volumes resulting in a single set of calibration parameters per model for all experiments. The model results are evaluated in terms of (dimensionless) erosion/accretion volumes and beach profile characteristics (beach slopes and bar migration) as a function of time to both improve the model formulations and the scaling laws for physical model experiments.

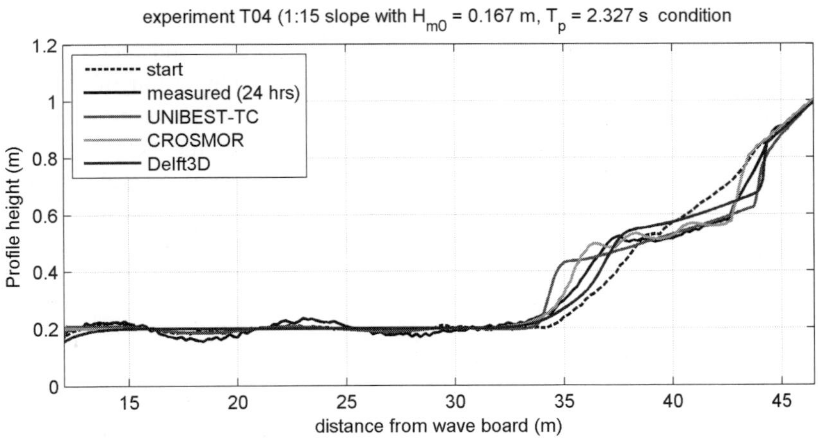

Figure 2. Modeled profile development for Delft 1:15 slope

References

Bosboom, J., Aarninkhof, S.G.J., Reniers, A., Roelvink, J.A., Walstra, D.J.R., 1997. UNIBEST-TC 2.0, Overview of model formulations. Report H2305, WL | Delft Hydraulics.

Lesser, G., Roelvink, J.A., van Kester, J.A.T.M., and Stelling, G.S. 2004. Development and validation of a three-dimensional morphological model. *Coastal Eng., 51, 883-915.*

Roelvink, J.A. and Stive, M.J.F. 1989. Bar generating cross-shore flow mechanisms on a beach. *Journal of Geophysical Research 94, 4785-4800.*

Van Rijn, L.C., Walstra, D.J.R., Grasmeijer, B.T., Sutherland, J., Pan, S. and Sierra, J.P., 2003. The predictability of cross-shore bed evolution of sandy beaches at the time-scale of storms and seasons using proces-based profile models. *Coastal Eng., 47, 295-327.*

Coastal Dynamics 2009
Paper No.64

A PREDICTIVE MODEL FOR BEACH PROFILE EVOLUTION

Bradley Johnson[1], Mark Gravens[1], and Nobuhisa Kobayashi[2]

1. Introduction

Hydrodynamic modeling of the nearshore environment has reached some level of maturity over the past several decades. This is primarily because the equations are well defined and numerical solutions are established. Modeling sediment transport and beach profile evolution, on the other hand, has not yet approached a similar level of accuracy. Many presently used models for shoreline and beach-face evolution change rely at least partially on empirical relationships to predict changes. Recent approaches have included a physical basis but are much more complex and demonstrate no greater skill. The predictive model CSHORE was originally developed as a cross-shore model but was recently extended to include longshore current (see Kobayashi et al. 2009). It is limited to longshore uniform cases, however, and thus remains a one-dimensional model. This simple formulation is well suited to life-cycle modeling (Gravens et al. 2007) where efficiency is required. CSHORE has been extensively tested using laboratory oscillating tunnels, flumes, and directional wave basins. Comparisons to large-scale field data have been lacking, however. This paper focuses on the comparison of the model predictions to a broad set of field data collected in the US.

2. Model for Beach Profile Change, CSHORE

The numerical model CSHORE has been under development for the past several years, approaching a practical and accurate code that predicts beach evolution over the nearshore region, and a full description of the model development is available in Kobayashi et al. 2009. The combined wave and current model operates under the assumption of longshore uniformity and includes the effects of a wave roller and quadratic bottom shear stress. The numerical integration of the depth-averaged energy, momentum, and continuity equations results in predictions of wave height, water level, and wave-induced steady currents. The majority of effort has been in the development of new and physically defensible sediment transport algorithms for a nearshore breaking wave environment. The model accounts for wave and current interaction, bedload, suspended load, and wave-related sediment transport. In a departure from conventional models that directly or indirectly relate transport to bottom shear, the new model represents total suspended volume V_s with the expression

$$V_s = \frac{e_B D_B + e_f D_f}{\rho g (s-1) w_f} P_s \qquad (1)$$

where e_B = empirical efficiency of breaking, D_B = Dissipation due to breaking, e_B = empirical efficiency of bottom friction, D_f = Dissipation due to bottom dissipation, ρg = weight of water, s = specific gravity, w_f = fall velocity, and Ps = probability of sand suspension related to local turbulence levels. Eqn 1 is intuitively satisfying in that the transport is dependent on both bottom shear through D_f and the important wave breaking process through D_B. Although there continues to be reliance on empirical factors, this effort will demonstrate that the values can be assumed constant over markedly different profile shapes and wave climates and still generate accurate predictions.

---

[1]Coast. and Hyd. Lab., Eng. Res. and Dev. Ctr. Vicksburg, MS, 39180 USA.
Bradley.D.Johnson@erdc.usace.army.mil, Mark.B.Gravens@erdc.usace.army.mil
[2]Dept of Civil Eng., University of Delaware, Newark, DE 19716 nk@coastal.udel.edu

3. Predicted Storm Response

The model CSHORE has been developed and tested against many small-scale laboratory data sets. Comparisons with large-scale laboratory data and field data have a vastly different spatial and time scale and can reveal deficiencies in formulation. This effort will include model and data comparisons for storm conditions over a range of scales including laboratory and approximately fifty sets of field data including profiles from the east and west US coasts. All of the data sets used in the validation of the US Army Corps model SBEACH (Wise et al. 1996) will be included. An example of the predictions are provided in Figure 1, which shows model results for the Dewey Beach, Delaware storm of 1992 and the Del Mar, California storm of 1988. In each case the model is run with default transport parameters and the reasonable agreement with measured data is indicative of the predictive capability for the entire data set

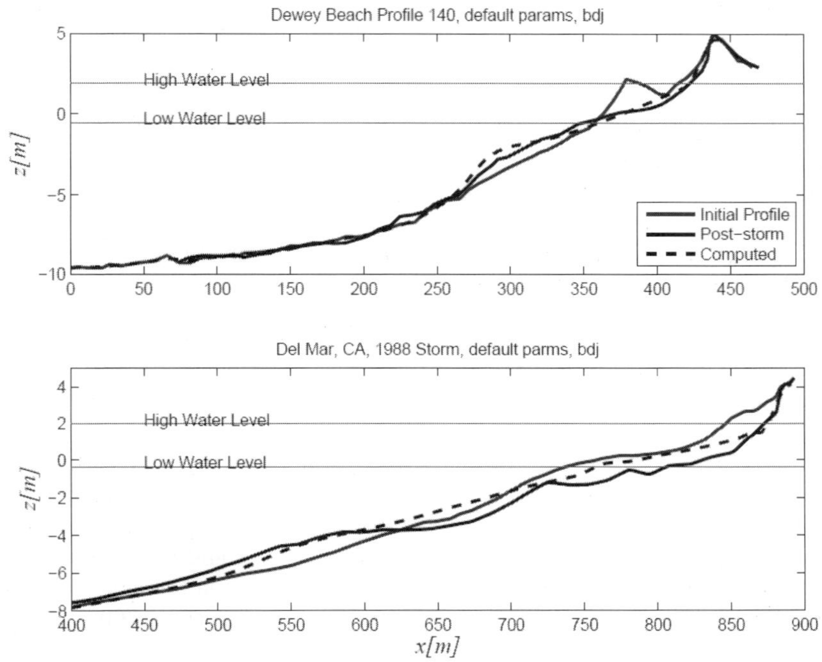

Figure 1. Example profile predictions for storm conditions.

Acknowledgements

We acknowledge the Chief of Engineers, U.S. Army Corps, for permission to publish this effort.

References

Kobayashi, N., Buck, M., Payo, A. and Johnson, B.D. (2009). 'Berm and Dune Erosion during a Storm.' *Journal of Waterway, Port, Coastal and Ocean Engineering*, ASCE, 135(1), 1-10.

Gravens, M.B., Males, R.M., Moser, D.A. (2007). 'Beach-fx: Monte Carlo Life-Cycle Simulation Model for Estimating Shore Protection Project Evolution and Cost Benefit Analyses.' *Shore & Beach*, 75(1), 12-19

Wise, R. A., Smith, S. J., and Larson, M. (1996). 'SBEACH: Numerical model for simulating storm-induced beach change; Report 4: Cross-shore transport under random waves and model validation with SUPERTANK and field data.' *Technical Report CERC-89-9*, U.S. Army Engineer Waterways Experiment Station, Coastal Engineering Research Center, Vicksburg, MS

Coastal Dynamics 2009
Paper No.65

A MODEL FOR PREDICTING MORPHODYNAMIC RESPONSE OF ESTUARIES

Harshinie Karunarathna[1], Dominic Reeve[2] and Mark Spivack[3]

1. Introduction

Estuarine systems are highly complex morphological features. They are extremely dynamic land forms and continuously react to changes in marine and fluvial environment. Likely causes of changing natural marine and fluvial forcing are increasing sea level, changes in storminess and changes in precipitation as result of global climate change. In addition, many estuaries are subjected to morphological perturbation followed by human interference such as dredging, construction works, land reclamation and flood protection. The aim of this research is to develop a morphodynamic model for prediction of morphological changes in estuaries in medium to long term time scales. The paper describes the development and application of a hybrid between top-down and bottom-up estuary morphology models.

2. The Model

A brief description of the formulation of the model is given here. The governing equation of the model takes the form of two dimensional diffusion formulation with a source function. If the bottom bathymetry of the estuary relative to a reference water level is given by $h(x,y,t)$ and the sediment diffusion coefficients in the x and y coordinate directions are K_x and K_y respectively where $K_x = K'_x + k_x$ and $K_y = K'_y + k_y$, then the rate of change of h can be given by (Pelnard-Considère, 1956; Karunarathna et al., 2008):

$$\frac{\partial h}{\partial t} = K'_x \frac{\partial^2 h}{\partial x^2} + K'_y \frac{\partial^2 h}{\partial y^2} + G(x,y,t) \tag{1}$$

K'_x and K'_y are reference values of diffusion coefficient in the x and y coordinate directions respectively. $G(x,y,t)$ is a two-dimensional source function. Spatially varying components of the diffusion coefficient k_x and k_y are embedded in G. The source function represent an aggregate of the effects of all processes other than those described by diffusion. If x and y are rescaled in that the coefficients of the spatial derivatives are equal, then the inverse solution for source function can be obtained from Equation (1) as (Spivack and Reeve, 2000; Karunarathna et al., 2008)

$$G(x,y) = \frac{1}{\tau}\left[\exp\left(-\frac{\tau D}{2}\right)h(t_m+\tau) - \exp\left(\frac{\tau D}{2}\right)h(t_m)\right] \tag{2}$$

Where D is the operator $(\partial/\partial x^2 + \partial/\partial y^2)$. The source function in Equation (2) can be established using consecutive pairs of historic estuary bathymetry survey maps. Once the time varying source function is constructed using historic bathymetry data then, equation (2) can be solved forward to determine future bathymetries as:

$$h(t_m+\tau) = \tau G(x,y)\exp\left(\frac{\tau D}{2}\right) + \exp(\tau D)h(t_m) \tag{3}$$

Equation (3) serves as the governing equation for the estuary morphology prediction model.

[1]Dept of Civil Engineering, University of Glasgow, Glasgow G12 8LT, UK karunarathna@civil.gla.ac.uk
[2]School of Engineering, University of Plymouth, Plymouth PL4 8AA, UK reeve@plymouth.ac.uk
[3]Dept of Applied Mathematics and Theoretical Physics, University of Cambridge, Cambridge CB3 0WA, UK *ms100@hermes.cam.ac.uk*

3. Model Application

The model described above is applied to predict future bathymetries of the Humber Estuary, UK. Figures 1 and 2 show a map of the Humber Estuary and a map of bathymetry (co-ordinate transferred) of the outer and middle estuary in 2000 respectively. Traditionally, the estuary had been depositional during the first ¾ of the 20th century. With the reversal of sedimentation trend afterwards, the estuary had become erosional during the last quarter of the 20th century.

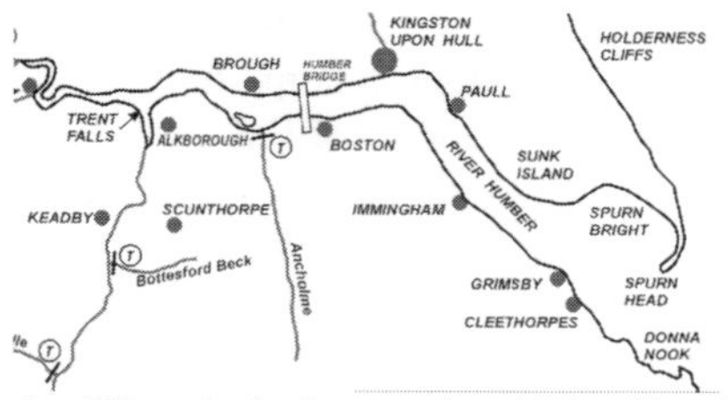

Figure 1 – Historic bathymetries of the Humber Estuary

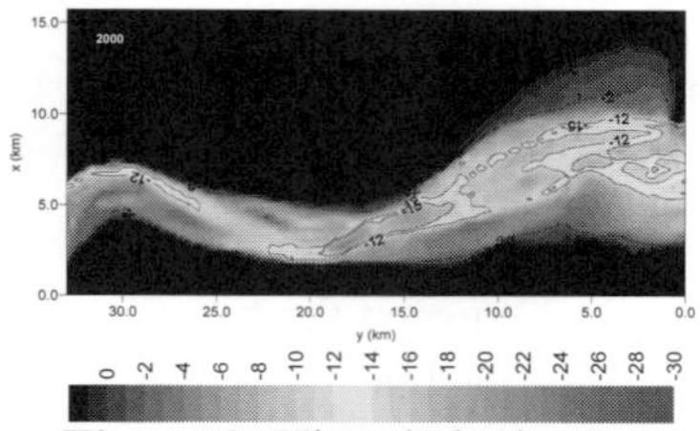

Figure 2- Historic bathymetry map

Figure 3 shows a source function constructed using the bathymetry pair 1998-2000. Nineteen pairs of historic bathymetries from 1851 to 2000 were used to produce 19 source functions similar to that in Figure 2. The time varying source function was then used in equation (3) to predict future bathymetries of the Humber Estuary.

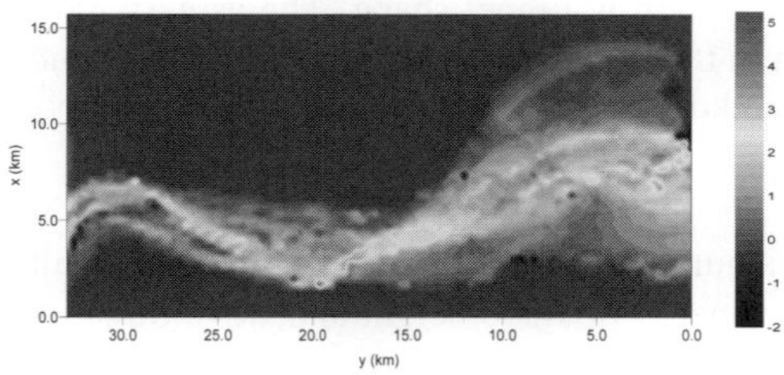

Figure 3 – Source function

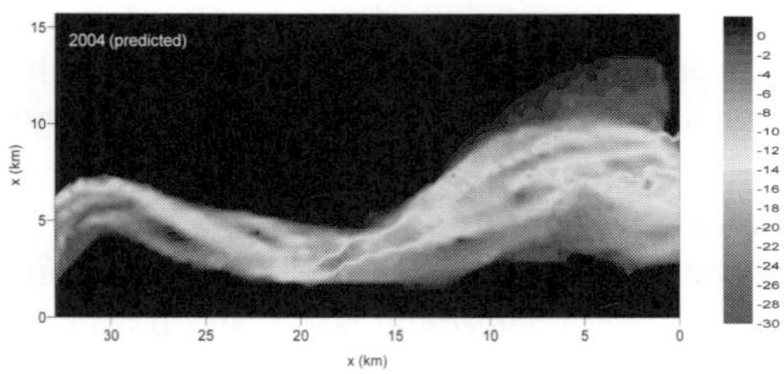

Figure 4 – Predicted Humber Estuary bathymetry for 2004

Predicted bathymetry of the Humber Estuary in 2004 is shown in Figure 4. The predicted bathymetry was in close agreement with the measured bathymetry in 2004 (not shown). Despite its simplicity, sedimentation pattern within the estuary is well captured by the numerical model.

4. Conclusions

A simplified estuary morphology model is used to predict future bathymetries of the Humber Estuary. The predicted bathymetries in years 2002 and 2004 are in close agreement with the measured bathymetries with less than 14% root mean square error. When considered the simplicity of the model, time and spatial scale involved and complex nature of the problem, the results give reasonable confidence on the ability of the model in predicting future morphologies of the estuary.

References

Karunarathna, H. and Reeve, D.E., 2008, A Boolean approach to prediction of long term evolution of estuary morphology, *J. Coastal Research*, 24(2B), pp.51-61.

Pelnard-Considere, R., 1956, Essi de theorie de l'evolution des formes de ravage en plages de sables er de galets, *Sciete Hydrotechnique de France, IV'eme Journee de L'Hydraulique Question III, Rapport* 1, 74-1-10.

Spivack, M. and Reeve, D.E., 2000, Source reconstruction in a coastal evolution equation, *J. Comp utational Physics*, 161, pp.169-181.

Coastal Dynamics 2009
Paper No.66

MODELLING OF SEDIMENT TRANSPORT AT EXMOUTH ESTUARY, DEVON, UK

Shunqi Pan[1], Yanliang Du[2], Yongping Chen[3] and Sam Reed[4]

1. Introduction

The Exe Estuary is located on the south-west coast of England, adjacent to the Teign Estuary, as part of Lyme Bay within northwest part of English Channel, which can be characterised as a shallow (~30 m) and gentle sloping. The River Exe created a low-laying valley along a "dip" in geological beds, between Dawlish in the west and Straight Point in the east, which formed the Exe Estuary due to the inundation and infilling, see Fig 1.

The spring tidal range at the Exe Estuary is approximately 3.8 m, with the maximum tidal current up to 3.0 m/s, while the tidal current in the Lyme bay is relatively weak, of 0.5 m/s. The waves in the area are predominately the swell waves from Atlantic in the south-west direction, with yearly mean significant wave height of 2.0 m (max 4.0 m) and for 1 in 50 year return period, the mean significant wave height is 2.7 m (max 5.3 m). The estuary is also influenced by the large river flow, Halcrow (2007).

The sediment transport outside the estuary is predominately alongshore, sourced from cliff and shoreline erosion. The estuary is characterized by the meandering channel and large mud flat and sandbanks, which are partly exposed at low tides, with a narrow funnel shape. The estuary mouth is mainly confined by the sand spit from Dawlish Warren in the west. The estuary is highly dynamic and is bounded by offshore sandbanks and large ebb and flood deltas.

2. Model Description

The process-based model used in the study consists of a number of fully interactive modules, mainly a wave module to determine wave-period averaged wave energy or wave height and wave direction for the wave transformation from offshore to nearshore; and a current module to compute the depth-integrated current velocity and water surface elevation under both tide and wave actions; and a morphological module to compute the sediment transport rate using equilibrium formulae, as well as the resulting bed level changes, Pan et al (2005). The model includes modules for bedform predictions to allow for the bed shear-stresses dynamically computed; the full wave-current interaction; and full hydrodynamic-morphological interaction, Pan et al (2007). The model setup included the part of Exmouth estuary and the adjacent coasts, as shown in Figure 2. The computational domain covered an area of 5.5 km by 4.0 km and the grid size was 50 m by 50 m.

3. Results

Model results of the computed tidal and wave-induced currents are shown in Figure 2(a), and the total bed level changes after an 80-hour period, including storm events are shown in Figure 2(b). The estuary responded the storm conditions with bed level changes up to ±2.0 m, particularly in the areas near the tip of the spit sand and main channel. Figure 3 shows the comparisons of the computed bed level changes with the measurements along Profiles 7 & 21, indicating a fairly good agreement in a range of ±0.4 m. The final paper will also include the results on the beach response to the extreme conditions for investigations of risk of possible shoreline breach, as well as the effect of groynes.

[1]School of Engineering, University of Plymouth, Drake Circus, Plymouth, UK. Shunqi.Pan@plym.ac.uk
[2]School of Engineering, University of Plymouth, Drake Circus, Plymouth, UK. Yanliang.Du@plym.ac.uk
[3]School of Engineering, University of Plymouth, Drake Circus, Plymouth, UK. Yongping.Chen@plym.ac.uk
[4]School of Engineering, University of Plymouth, Drake Circus, Plymouth, UK.

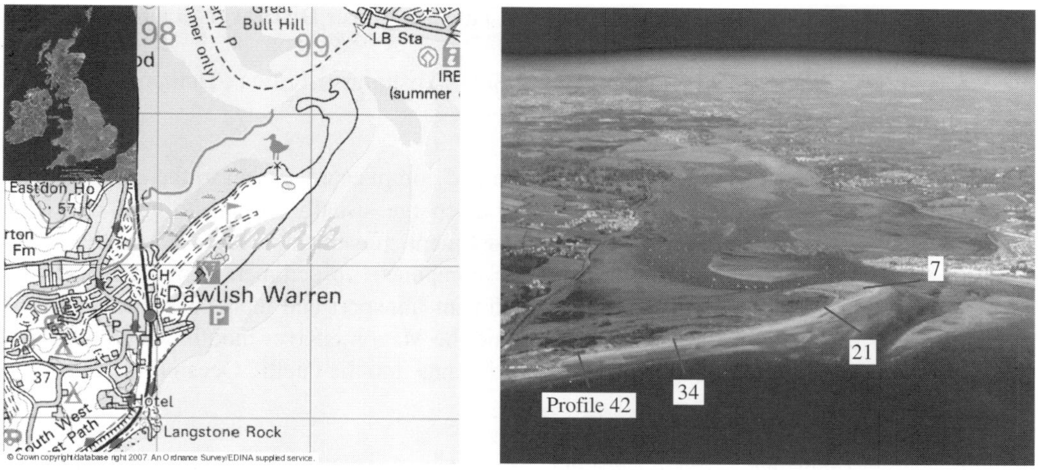

Figure 1. Exmouth Estuary

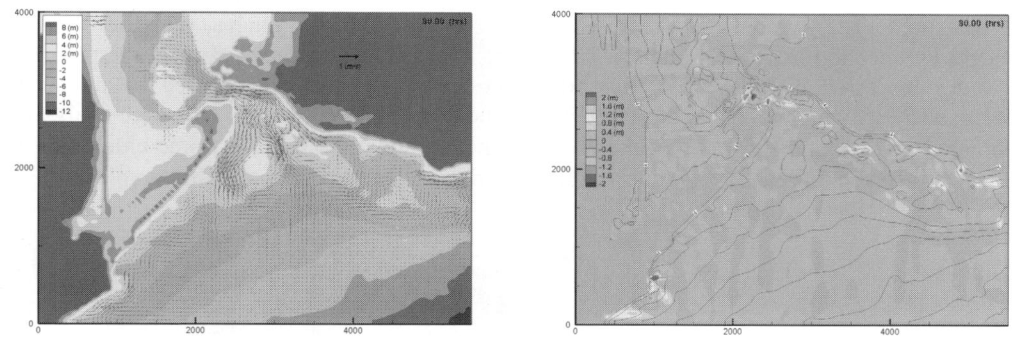

(a) (b)

Figure 2. Computational domain and (a) velocity vectors; (b) bed level changes

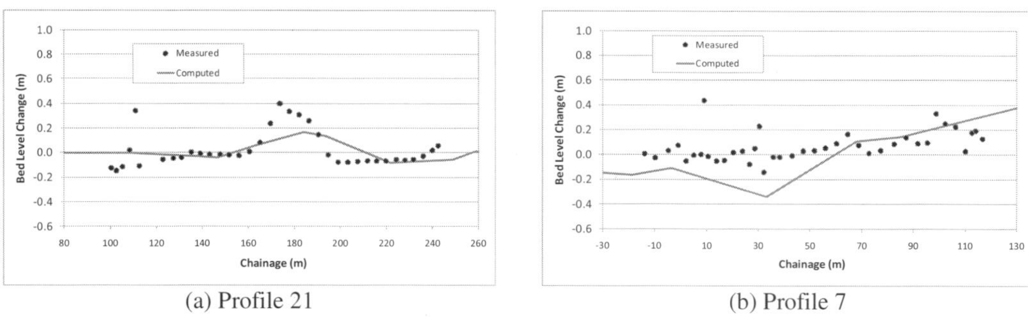

(a) Profile 21 (b) Profile 7

Figure 3. Comparisons of computed and measured bed level changes: (a) Profile 21; (b) Profile 7

References

Pan, S., Fernando, P., Li, M., Zhu, Y., O'Connor, B., Vincent, C., Taylor, J., Dolphin, T. Bacon, J., 2005, Effect of shore parallel breakwaters on coastal morphology under storm conditions. In: N. W. H. Allsop ed(s) Coastlines, Structures and Breakwaters. Thomas Telford, the Institution of Civil Engineers, London, 64-73

Pan, S., MacDonald, N., Williams, J. J., O'Connor, B. A., Nicholson, J., Davies, A. M.,2007, Modelling the hydrodynamics of an offshore sandbank, *Continental Shelf Research*, 27: 1264-1286

Halcrow, (2007), Exe Estuary Coastal Management Study: Preliminary Understanding of Historic to modern day evolution, East Devon District Council

Coastal Dynamics 2009
Paper No.67

Sediment Movement under Combined Waves, Tide and River Discharge in a River Mouth

Xiaojing Niu[1], Satoshi Ueyama[1], Shinji Sato[1], Yoshimitsu Tajima[1], Haijiang Liu[1]

Introduction

Sediment movement around river mouth is intensive and complicated, owing to the combined effects of short waves, long waves, tide currents and river discharge, usually leading to great and frequent topography change. This study aims to trace the sediment movement in a river mouth, further to distinguish the contribution of each hydrodynamic component to sediment movement, and to get better understanding on the nonlinear effects to sediment transport and their interactions. Both field survey and numerical modeling are carried out around the Magome River mouth, which is located on the Enshunada Coast, Shizuoka Prefecture, Japan and drains into the Pacific Ocean (Fig. 1).

Field survey

Two kinds of color sands were used to trace the sediment movement. They were placed at the blue and yellow points illustrated in Figure 2 during low tide on Sep 13th 2008. In the following two days, we took sand samples inside the river mouth hourly. At the same time, flow velocities and wave heights were measured. The symbol "H" represents the location of wave gauges, and the symbol "V" represents the location of velocity meters. After that, we continued surface sediment sampling once in two weeks for three months. Colored sands in collected samples are counted by an automated counting machine, which takes sand sample image and recognizes color sand by image analysis. Then we obtained the spatial distribution of the colored sand as shown in Figure 3. Some blue sands moved upstream and stayed there, and only a few yellow sand particles were found in the samples.

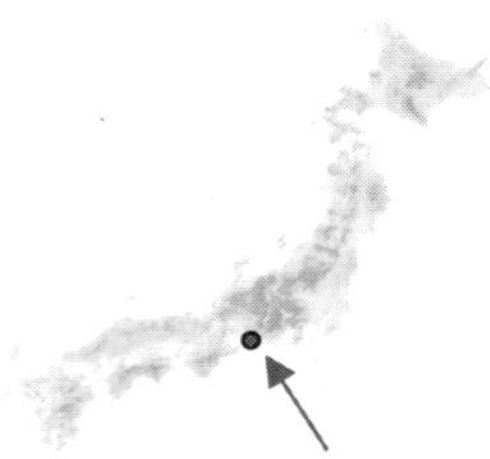

Fig. 1 Location of Magome river

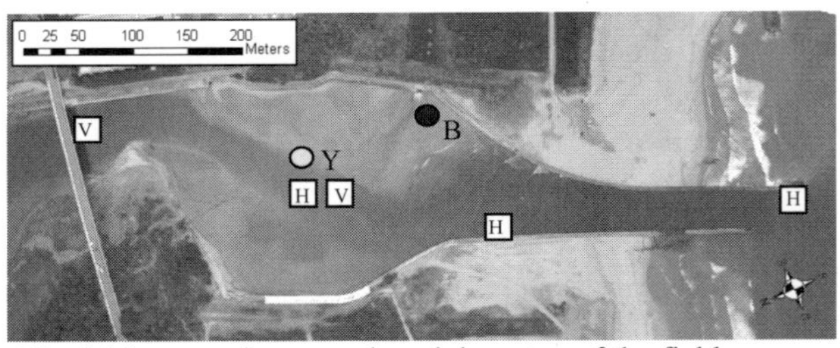

Fig. 2 Magome river mouth and the setup of the field survey

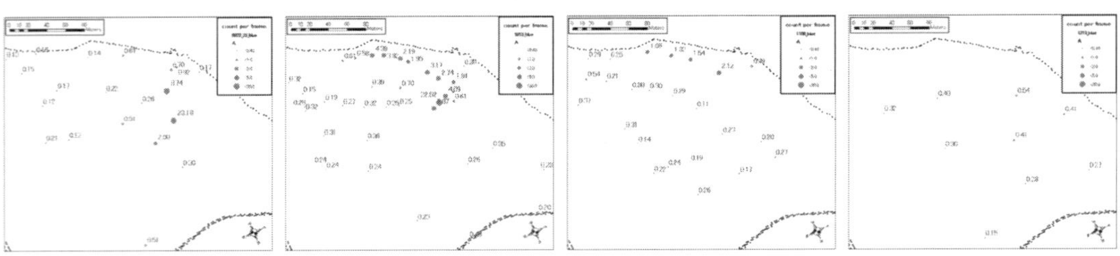

(a) After one week (b) After one month (c) After two months (d) After three months

Fig.3 Colored sand distribution (blue sand)

Numerical modeling and discussion

The short waves are estimated based on the spectral action balance equation with the effect of current. Radiation stress is included to consider the effect from short waves. Shallow water equation is then applied to calculate the flow due to tide, river discharge, and long waves. The movement of sediment

[1] Dept of Civil Eng., The University of Tokyo, 7-3-1 Hongo, Bunkyo-ku, Tokyo 113-8656, Japan. nxj@coastal.t.u-tokyo.ac.jp

particles is associated with the bottom shear stress under combined waves-current field, and is numerically traced by Lagrangian method. Sediment particles are assumed moving with the same velocity as currents if the local Shields number is larger than the criterion for general movement. The sediment movement is terminated if the local Shields number is smaller than the critical value.

Inside the river mouth basin, waves are refracted due to the topography, as shown in Figure 5. This induces a circulation in the river basin, which can be clearly observed when tide current is small, *e.g.*, during high tide period as illustrated in Figure 6 (a). Figure 6 presents the current field at a specified time. Seaward flow during ebb tide is stronger than landward flow during flood tide, because of the contribution of river discharge. Figure 7 shows the calculated velocity compared with the measured flow velocity at the jetty channel. Simulated results are acceptably agreed with measurement.

Two sediment particles, initially located at two locations where color sands were placed, were traced numerically for one tide period from low tide. Figure 8 shows simulated particle trajectories. The yellow particle almost moves out of the river mouth directly; which is consistent with observation that yellow sand could barely be found. Blue particle moves upstream and then downstream, which accords with the fact that some blue sands were found around that area. Distribution of color sands by tracing large number of particles will be presented in full paper.

Conclusion

Field survey inside Magome river mouth was carried out. Color sands were used to trace the sediment transport and distribution of color sands inside the river basin was obtained. A numerical model was developed to simulate the sediment movement under combined wave, tide and river discharge. Good agreement was found between field survey and numerical results.

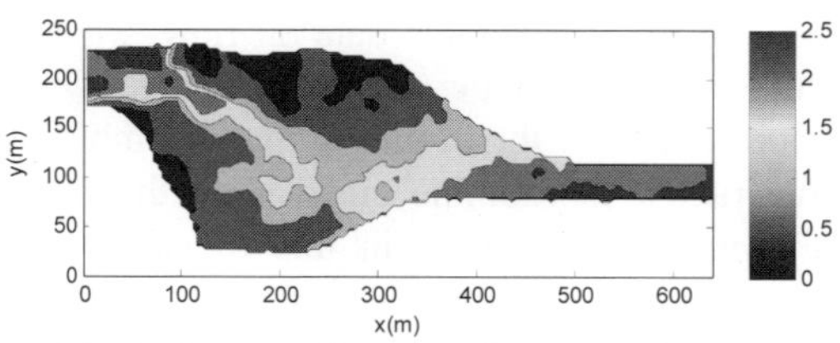

Fig.4 Magome River mouth topography

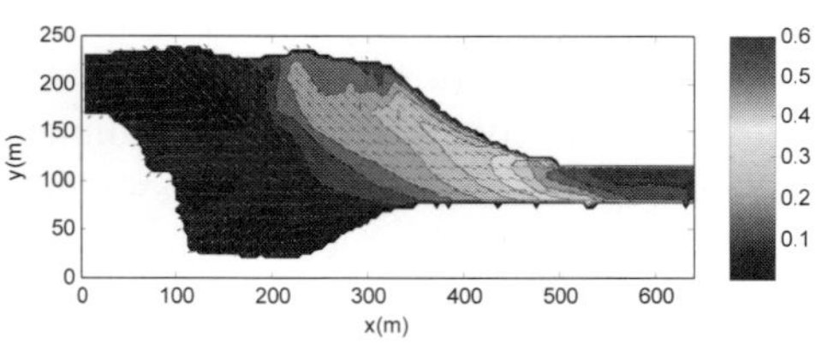

Fig.5 Wave height distribution and wave direction

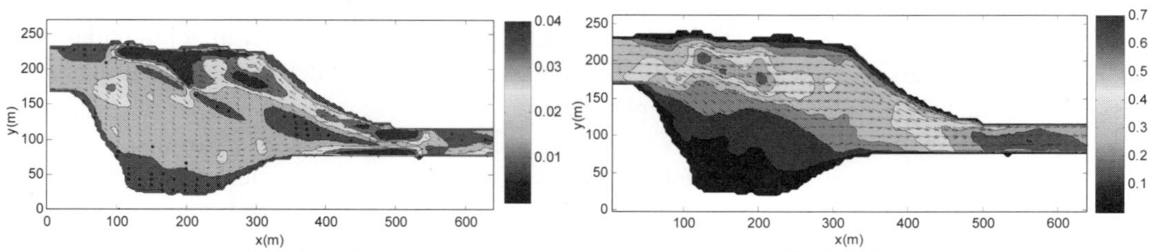

(a) High tide (b) Ebb tide

Fig.6 Tide current field

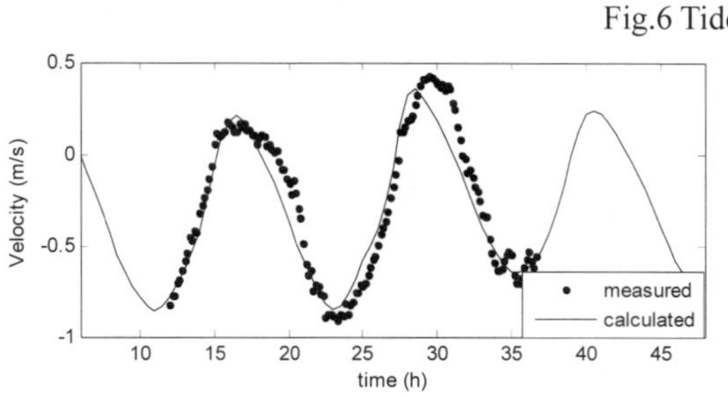

Fig.7 Velocity at the outlet (x=600m)

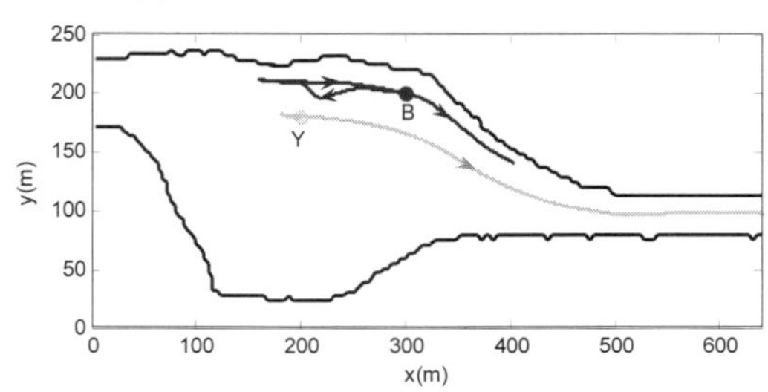

Fig.8 Movement of sediment

Coastal Dynamics 2009
Paper No.68

MODELING PROCESSES CONTROLLING SEDIMENT TRANSPORT AT THE MOUTH OF THE COLUMBIA RIVER

Edwin Elias[1] and Guy Gelfenbaum[2]

1. Introduction

The Columbia River is the largest river on the Pacific coast of North America and is a major supplier of sediment to the adjacent coast. However, human interventions such as the construction of dams along the river for hydro-electric power, dredging, and construction of entrance jetties have altered the system significantly. The reversal from an accretional adjacent coast in the past to severe localized erosion at present is not fully understood, but likely is associated with anthropogenic influences in and around the river. The ultimate goal of our work is to improve understanding of the long-term consequences of management decisions on the sediment budget of the Mouth of the Columbia River Mouth (MCR) and adjacent beaches. As a first step we aim to identify the relationships between tidal and local topographic flow, waves, and river discharge on sediment transport patterns.

2. Method and Model

Neither field measurements nor modeling alone can provide a comprehensive understanding of sediment transport at the MCR due to the large size of the study area and the complexity of the forcing processes: a severe wind and wave climate, a large tidal range and estuarine circulation associated with major freshwater inputs. Therefore we used a combination of field measurements and numerical model experiments to examine the processes controlling the sediment transport in the region. Five tripods were deployed in a transect across the MCR to collect data on flow, waves and sediment transport were collected from 8 August 2005 to 9 September 2005 (the Mega Transect Experiment), and a hydrodynamic and sediment transport model for the Mouth of the Columbia River (MCR) using the Delft3D modeling system was constructed to examine and isolate the physical processes responsible for sediment transport and morphological change in this dynamic estuary entrance. Modeling sediment transport at the MCR is a challenging task as the numerical model must at least be able to account for the interaction of highly variable stratification, include waves and wave-current interaction, reproduce internal tidal asymmetry and related mean flow in the system, and must allow for wetting and drying as large tidal flats and wetlands in the lower estuary affect the tidal prism, tidal currents and density stratification. In principle the Delft3D modeling system contains the necessary physical processes to capture these phenomena.

Model simulations of sediment transport and flow forced by measured time-series of wind, waves and tides provide high spatial and temporal resolution model data in the MCR. The model is forced by tidal constituents and wave measurements at the seaward side, while discharge measurements provide upstream river boundary conditions. Mega Transect data are used to validate and calibrate the numerical model. These short term model results are not suitable for resolving processes on morphological time scales of years, but they do provide insight into the processes governing flow and sediment transport patterns in the instrumented and un-instrumented areas, and sensitivity analyses can be used to identify dominant processes and mechanisms under a range of forcing conditions.

---

[1] Deltares, Delft, 2629 DH, the Netherlands. Edwin.elias@deltares.nl
[2] U. S. Geological Survey, 345 Middlefield Road, , Menlo Park, CA 94025, USA. ggelfenbaum@usgs.gov

3. Model results and Conclusions

Time-series of measured forcing factors (tide, river discharge, ocean swells, and meteorological conditions) for representative summer and winter conditions were used as model forcing. Hydrodynamic calibration and validation of the model results with detailed field observations show that the model is able to reproduce the dominant features in the tidal flow, salinity and wave fields. Our results show that a spatially and temporally varying balance of physical processes governs the transport of sand through the MCR. During the Mega Transect campaign tidal and non-tidal (density-driven) contributions roughly offset each other: in the absence of salinity gradients, tidal flows induce a large sediment export as both the tidal residual and tidal asymmetry components are ebb-dominant. Including salinity affects the near-bed residual flow considerably, and the resulting changes in the vertical flow distribution reduce net sediment export at the entrance to the CRM to near zero. Loss of momentum and dissipation of wave energy due to wave breaking generates wave-driven currents and associated sediment transports outside the CRM. Wind only played a minor role (Figure 1).

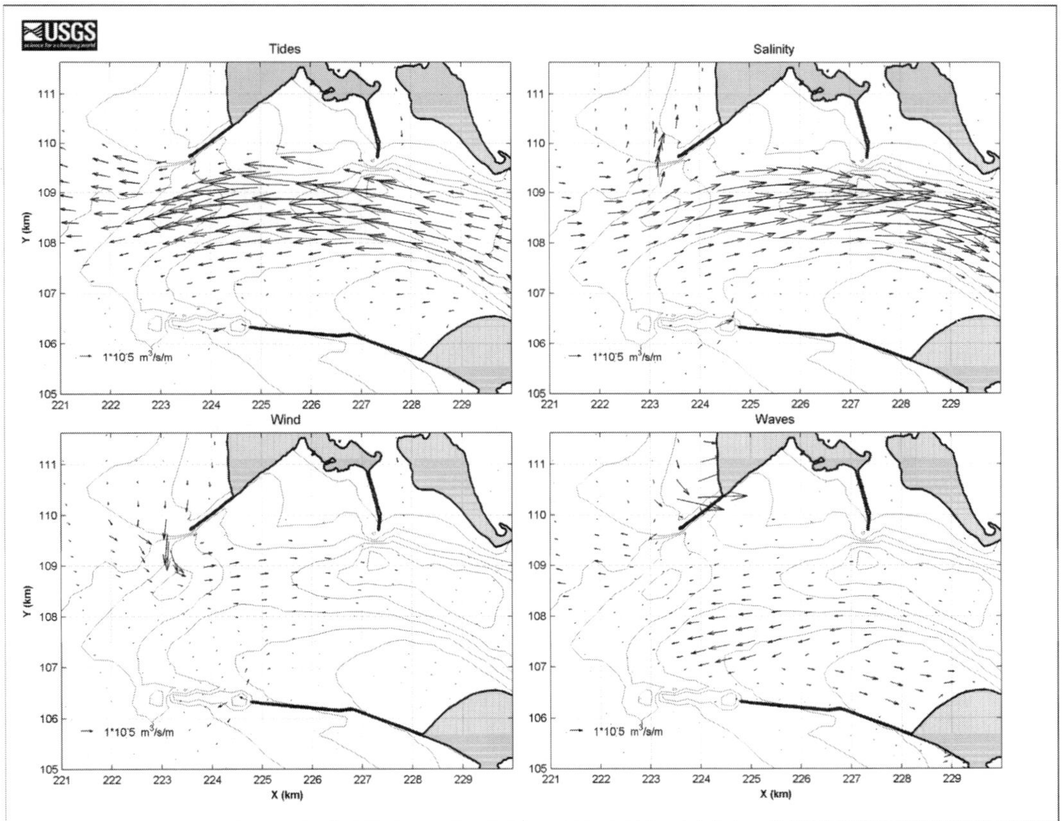

Figure 1. Overview of the contribution of various forcing processes (tides, salinity, wind and waves) to the mean sediment transport rates (in $m^3/s/m$) at the MCR.

Acknowledgements
The work herein was carried out in cooperation between the US Geological Survey and Deltares. Funding for this study was provided by NOAA, USGS and USACE. We thank Giles Lesser for his early contributions to the MCR model and Hans Moritz for his leadership in obtaining the Mega Transect data.

Coastal Dynamics 2009
Paper No.69

ANTHROPOGENIC REDUCTION OF THE NATURAL SUPPLY OF SEDIMENTS TO THE COASTS OF WASHINGTON, OREGON, AND CALIFORNIA

Orville T. Magoon[1] and Donald D. Treadwell[2]

Introduction

Particularly within the last 100 to 150 years, modifications of coastal areas due to the general increase of human activities and due to the exploitation of the hydraulic and sediment resources of rivers have caused extensive coastal erosion. The true costs and impacts of this erosion are difficult to assess and have not been reflected in the price of providing commodities such as sand and gravel. These impacts and their estimated costs are illustrated using information from the states of Washington, Oregon, and California on the Pacific Coast of the United States of America.

Rivers and streams are the main sources of sediment for the Pacific Coast. Along the northern portion of the coast, in Washington, Oregon and northern California, the rivers and streams tend to be large watershed systems, like the Columbia, the Klamath, the Eel, and the Sacramento. These systems provide many millions of cubic meters of sediments to the coast. In Southern California, the watersheds are smaller but are still the key contributors of new beach material.

The long-term sustainability of Pacific Coast beaches depends on periodic deliveries of sand and gravel from coastal rivers and streams (Magoon et al, 2004). The main activities that have altered fluvial sediment regimes and contributed to coastal erosion are: (1) sand and gravel mining operations that remove sediment from the fluvial system, (2) dams that intercept sediment within the fluvial system, and (3) debris basins that intercept sediments before they can reach the fluvial system.

Sand and Gravel Mining

Many beaches are impacted by reduction of sediment delivery to the coastal zone caused by gravel and sand mining within coastal watersheds. Streams and rivers are the transportation systems that deliver sediments to the coast. The rivers act as "conveyor belts" that move sediment from areas of weathering and erosion in the headwaters regions through middle reaches where little erosion or deposition occurs to regions of deposition in the lower reaches of rivers and then ultimately to coastal sites.

The time scale of sediment movement down these river systems is measured in terms of decades to centuries. Movement of sediments is not constant, but rather is controlled by episodic peak flows during extreme rainfall events that often trigger floods. Additionally, the reduction of peak flows by dams reduces the ability of the systems to move sediment.

The concept of "safe yield" of aggregate mines encompasses the argument that as long as the volume of sand and gravel that is mined annually from river channels is less than the annual replenishment of sediment from natural erosion then the effect on river channels is negligible. This argument may hold true for local reaches of rivers close to mining sites. However, the volume of sediment in the fluvial system is reduced and thus less volume is ultimately delivered to the coast. For the present, the total sediment loss to the coastlines attributable to sand and gravel mining in California, Oregon, and Washington is estimated to be about 1.2 billion cubic meters since 1950.

Dams

The reduction of supply of sediments to the California coast due to the construction and operation of dams as treated by Willis and Griggs (2003) notes that "The long-term sustainability of California's

[1] Consulting Engineer, San Francisco, California, USA; omagoon@sbcglobal.net
[2] Consulting Engineer, Sausalito, California, USA; ddtreadwell@comcast.net

beaches depends on periodic deliveries of sand and gravel from coastal rivers and streams. To assess the long-term health of California's beaches, this study characterized the current state of fluvial sediments delivery and quantified on a littoral cell basis, the cumulative impacts of dams on decreasing annual discharge.

As a conservative estimate, this study assumes that the reduction has been about 2 million cubic m eters annually, due to dams and regulated dredging for navigation. Komar (1998) notes that "other tha n the effects of the dams on the Columbia River, this is not a particular issue on the coasts of Oregon a nd Washington." The total sediment loss to the coastlines attributable to dams in the coastal watershe ds of California, Oregon, and Washington is estimated at about 195 million cubic meters since 1950.

Debris Basins

Debris basins are typically used in southern California to protect urban development from the effects of intense rainfall. As stated in Ellis et al (2001), "Debris basins are designed to trap sediments being transported by debris flows.... As of 2000, 162 debris basins trapped a total of more than 13,761,900 cubic meters of debris... Assuming a 50 percent sand content for these deposits, the basins have trapped about 6,881,000 cubic meters of sand. It is assumed that little of this sand is returned to the drainage system, and therefore this impoundment represents a loss of sand from the coastal budget." The estimated sediment loss since 1950 attributable to debris basins is about 7.6 million cubic meters.

Economic Impacts

The information to be presented herein indicates that 1,500,000,000 cubic meters or more of sediment has been lost to the coastlines of Washington, Oregon, and California since 1950. The replacement value of these lost sediments is conservatively estimated to be at least US $22,500,000,000 (22.5 billion dollars). About 80 percent of the loss is attributable to sand and gravel mining in coastal watersheds.

Acknowledgements

The authors acknowledge the contributions and encouragement of many colleagues in this effort, including Jeff Williams, Linda Lent, Jim Richmond, Lesley Ewing, Tony Pratt, Scott Douglass, Billy Edge, Katherine Stone, Ron Noble, Patrick Barnard, Gary Griggs, Tom Kendall, Cope Willis, Paul Komar, Monte Hampton, and Bob Wiegel. The support and enthusiasm of friends and family is also very much appreciated.

References

Ellis, J. T., Bass, A. C.W., Barron, K. M; Hansen, D. J., and Sherman, D. J. (2001), "Sediment Impoundment by Coastal Debris Basins in California.", California Shore and Beach Preservation Association, 2001 Annual Conference.

Komar, P. (1998), "The Pacific Northwest Coast: Living with the Shores of Oregon and Washington", Duke University Press, Durham, North Carolina.

Magoon, O. T., Williams, S. J., Lent, L. K., Richmond, J. A., Treadwell, D. D., Douglass, S. L., Edge, B. L., Ewing, L. C., and Pratt, A. P. (2004), "Economic Impacts of Anthropogenic Activities on Coastlines of the United States", Proceedings, ICCE 2004, 29th International Conference on Coastal Engineering, ASCE, Lisbon, Portugal, pages 3022-3035.

Willis, C. M. and Griggs, G. B. (2003), "Reductions in Fluvial Sediment Discharge by Coastal Dams in California and Implications for Beach Sustainability", Journal of Geology, Volume 111, pages 167-182.

Coastal Dynamics 2009
Paper No.70

SHORELINE ROTATION ON POCKET BEACHES CAUSED BY ANTHROPOGENIC FACTORS

Takaaki Uda[1], Mitsuo Takamura[2], Kazuo Fujimoto[2], Fumihito Nakanishi[2], Shinji Iki[3] and Toshinori Ishikawa[4]

1. Introduction

The west coast of Miura Peninsula facing Sagami Bay has a complicated shape with a number of rocky headlands and pocket beaches. These pocket beaches have long been stable, and are well known places for surfing and swimming in summer. However, erosion has occurred in recent years on these pocket beaches, adversely affecting the use of the coast and the scenery of the natural sandy beaches. For each pocket beach, the cause of the changes has been studied individually, but the exact cause has not yet been clarified. In this area, waves incident from the southwest prevail because of the shape of Sagami Bay, which is open to the Pacific Ocean in the south. In the planning of ports in this area, therefore, it was decided to construct the port breakwater at the south end of the pocket beach, which had an advantage of creating a calm wave zone behind the breakwater, and the breakwater was extended in accordance with this concept. However, this had a significant adverse effect on the nearby coast, resulting in a trade-off problem. We investigate this type of change of the pocket beaches, while particularly focusing on anthropogenic factors, taking four coasts located on the western side of Miura Peninsula as examples.

2. Analysis of Aerial Photographs and Bathymetric Changes

Four coasts, the Zushi, Hori-uchi and Akiya located on the western side of Miura Peninsula and the Hatsuse coast located at the tip of the peninsula, as shown in **Fig. 1**, were selected as examples, and beach changes on these coasts were investigated using aerial photographs and bathymetric survey data, as well as by the grain size analysis of seabed materials. For example, **Fig. 2** shows the bathymetry of the Hori-uchi coast in 2006 and the bathymetric changes between 1999 and 2006, which is located between Hayama Port and the mouth of the Morito River at the north and south ends, respectively. This figure clearly indicates that the beach in the zone shallower than -5 m in the north part of the coast was eroded, and that sand accumulated in the zone shallower than -3 m next to the mouth of the Morito River. In **Fig. 3**, which shows the shoreline changes since 1954, the shoreline retreated in the north part and advanced in the south part, although a cuspate foreland was formed behind the detached breakwater built in the north part of the coast. This distribution of accretion and erosion zones is in good agreement with the results in **Fig. 2**, indicating the clockwise rotation of the shoreline. In this case, the construction of a river mouth jetty at the south end of the pocket beach triggered the shoreline rotation. **Figure 4** shows the bathymetry of the Hatsuse coast in 2007 and the bathymetric changes between 2003 and 2007. The beach was eroded in the north part far from the breakwater, and simultaneously, sand was deposited behind the breakwater in the south part. These beach changes are in good agreement with the shoreline changes since 1954, as shown in **Fig. 5**. Consequently, it is clear that the clockwise rotation of the shoreline also occurred on this coast after the construction of the fishing port breakwater at the south end of the pocket beach. In the case of the Zushi and Akiya coasts, where beach changes were triggered by the construction of the port breakwater at the south end of the

[1] Dr. Eng., Executive Director, Public Works Research Center, 1-6-4 Taito, Taito, Tokyo 110-0016, Japan.
[2] Yokosuka Public Works Office, Kanagawa Prefectural Government, 1-56- Kugocho, Yokosuka, Kanagawa 238-0022, Japan.
[3] Asia Air Survey Co., Ltd., 1-2-2 Manpukuji, Asao, Kanagawa 215-0004, Japan.
[4] Public Works Research Center, 1-6-4 Taito, Taito, Tokyo 110-0016, Japan.

pocket beach, clockwise rotation also occurred. Thus, on all four coasts, similar shoreline changes were triggered by the extension of a port breakwater or a river mouth jetty at the south end of the pocket beach. Southward longshore sand transport was induced from outside to inside the wave shelter zone, resulting in the erosion and accretion in the north and south parts, respectively.

3. Conclusions

On the west coast of Miura Peninsula, the predominant wave direction is from the southwest because of the geographical shape of Sagami Bay, which faces the Pacific Ocean. It was, therefore, reasonable to build a port at the south end of one of the pocket beaches to create a wide wave shelter zone behind the breakwater. This was effective for enhancing the safety of navigation inside the harbor, but the wave diffraction effect extended to the whole of the pocket beach, resulting in the rotation of the shoreline, i.e., deposition inside the port and shoreline recession far from the wave shelter zone. Another factor must be pointed out; a small river flows into each pocket beach at the south end because of the natural wave sheltering effect of the headland against southwest swells, and the river has been stabilized against wave action. The fluvial sediment supplied from the river was trapped by the extended port breakwater, resulting in the further deposition of sand around the port. This required further maintenance in the form of dredging the navigation channel and also caused insufficient sand to be supplied to the pocket beach. Thus, the pocket beaches located on the west coast of Miura Peninsula have complicated trade-off problems, because it is impossible to demolish the ports. To soundly maintain these pocket beaches in the long run, sand back pass from inside to outside the fishing port is necessary along with the construction of some auxiliary facilities to prevent longshore sand transport from returning to the fishing port.

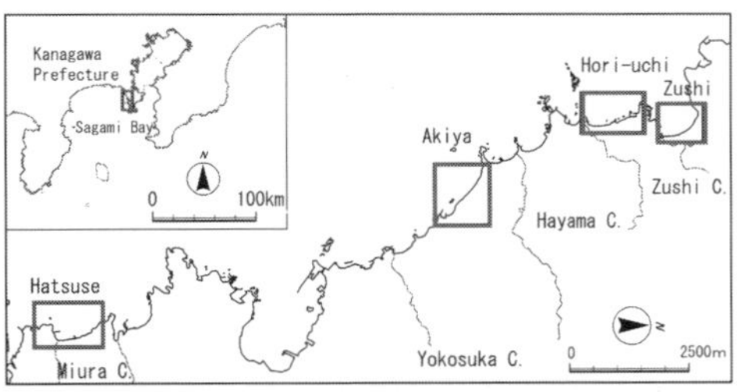

Figure 1.
Location of study areas on west coast of Miura Peninsula (Four coasts, the Zushi, Hori-uchi, Akiya and Hatsuse located on the western side of Miura Peninsula)

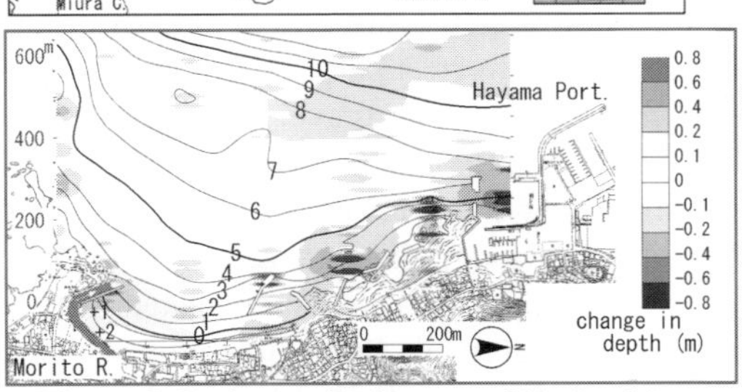

Figure 2. Bathymetric changes of Hori-uchi coast (1999 vs. 2006)

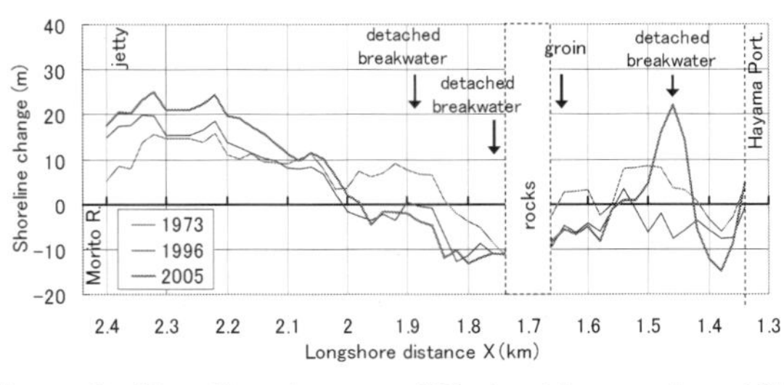

Figure 3. Shoreline changes of Hori-uchi coast since 1954 (the clockwise rotation of the shoreline)

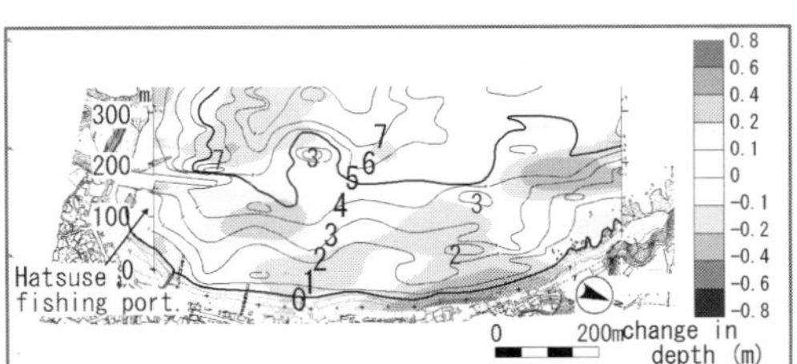

Figure 4. Bathymetric changes on Hatsuse coast (2003 vs. 2007)

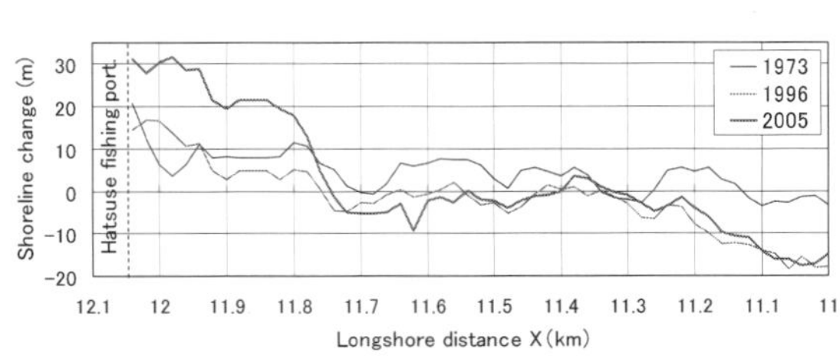

Figure 5. Shoreline changes of Hatsuse coast since 1954 (the clockwise rotation of the shoreline)

Coastal Dynamics 2009
Paper No.71

COMPREHENSIVE MANAGEMENT OF SAND CONSIDERING GRAIN SIZE ON SHONAN COAST

Toshinori Ishikawa[1], Takaaki Uda[1], Genji Aoshima[2], Atsushi Yoshioka[2] and Toshiro San-nami[3]

1. Introduction

On the Shonan coast with a 11 km length extending between the Sagami River and Enoshima Island and facing Sagami Bay, severe beach erosion has occurred from the vicinity of the river mouth since the 1970s, because the fluvial sediment supply greatly decreased up to a level of 3% of the natural sediment supply of this river after the construction of dams in the upstream basin. As a measure against beach erosion, beach nourishment has been considered. In this case, the effective use of the available sand, which is a limited resource in this area, is strongly required for economically maintaining the coast in the long-term basis under the condition that the fluvial sand supply from the Sagami River has been exhausted. The grain size of sand necessary for shore protection differs from place to place depending on the coastal condition whether a coast has been subjected to erosion or not, because the grain size of sand increases after erosion and therefore, in coastal management, not only the sand volume but also the grain size of sand must be sufficiently considered. In this study, the sand budget considering grain size is studied, taking the Shonan coast as an example, and a comprehensive map showing the sand budget considering the grain size is proposed for coastal management.

2. Analysis of Bathymetric Data of Shonan Coast

To investigate the sand movement of the Shonan coast in recent years, the shoreline changes between 1996 after the construction of an artificial headland and 2007 were first analyzed. Using bathymetric survey data, we calculated the characteristic height of beach changes, by which a shoreline change is transformed into the cross-sectional area of the beach. When we multiply shoreline changes with this characteristic height of beach changes, determined from the bathymetric changes in each area, the volume changes of the sandy beach per unit width and year can be calculated. Furthermore, assuming that the longshore sand transport at the east end is 0, the distribution of longshore sand transport is calculated by integrating the volume change in the updrift direction from the east end of the coast. The seabed materials were sampled at 1 m intervals in the depth zone between +3 and -9 m (depth of the closure) along eight transects, and the contents of fine sand, medium sand and gravel were obtained in each transect. Because the grain size of the seabed material significantly changes with depth and the longshore change in grain size is gradual compared with the cross-shore change in grain size, the longshore distribution of the content of each grain size was linearly interpolated, as shown in **Fig. 1**. In the accretion zone, we can assume that the amount of the supplied sand is the same as that of the deposited sand, because previous data on sand deposition are well preserved. Taking this into account, the distribution of longshore sand transport of each grain size was obtained, as shown in **Fig. 2** by multiplying the volume changes with the content of each grain size shown in **Fig. 1** and by integrating the longshore sand transport of each grain size from the downdrift end, given the boundary condition Q=0 at the east end. In the drawing shown in **Fig. 2**, the contributions of changes in sand volume due to dredging, sand back pass, sand bypassing, sand nourishment and windblown sand between 1996 and 2007 were considered. Finally, a comprehensive map of the sand budget considering the grain

[1] Public Works Research Center, 1-6-4 Taito, Taito, Tokyo 110-0016, Japan. ishikawa@pwrc.or.jp
[2] Fujisawa Public Works Office, Kanagawa Prefectural Government, 1-7 Shiomidai, Chigasaki, Kanagawa 253-0 033, Japan.
[3] Coastal Engineering Laboratory Co., Ltd., 301, 1-22 Wakaba, Shinjuku, Tokyo 160-0011, Japan.

size was obtained as shown in **Fig. 3**, where each block was determined based on the structural conditions, such as the presence of an artificial headland and fishing port breakwater and the location of the nodal point of the shoreline change determined by shoreline change analysis. Longshore sand transport passing through the boundary of each block was determined from **Fig. 2**. The ratio of the longshore sand transport of each grain size to the total transport is expressed by the percentage in a circle at the boundary of the blocks shown in **Fig. 3**. The volume change in each block was determined by the shoreline change multiplied by the characteristic height of beach changes, and the content of each grain size was evaluated as the mean value in the calculation domain.

3. Conclusions

Figure 3 shows a comprehensive map indicating not only the volume changes in several blocks and longshore sand transport crossing the boundary between the blocks, but also the initial contents of all grain sizes, as well as the contents of all grain sizes in the sand volume exchange process between the sea and the land, such as windblown sand, beach nourishment and dredging from a navigation channel. This map is useful in determining necessary information, such as the volume and quality of sand, and the grain size at a particular location. We can use this information for the comprehensive sand management of the coast.

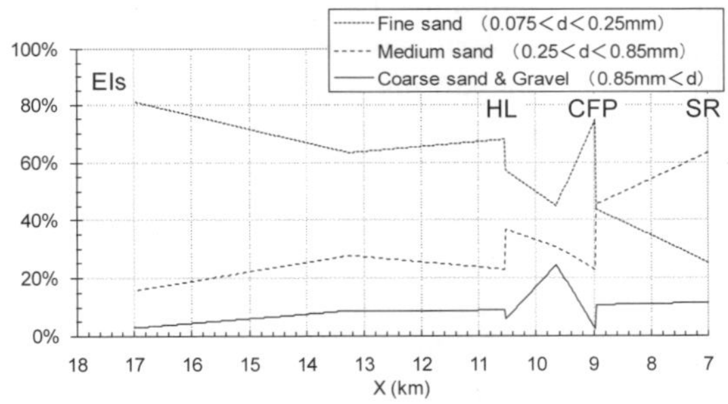

Figure 1. Longshore distribution of content of each grain size

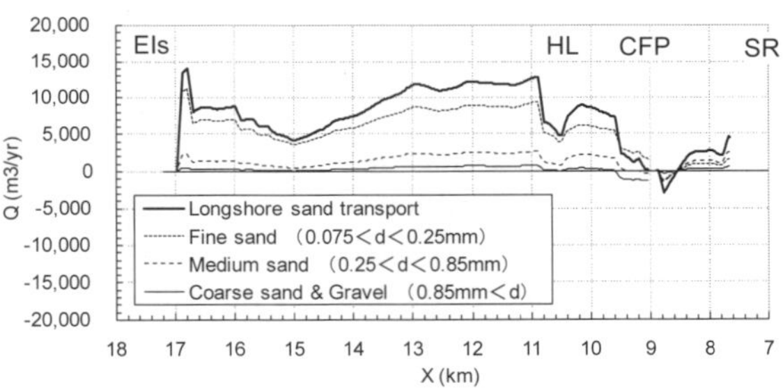

Figure 2. Distribution of longshore sand transport of each grain size

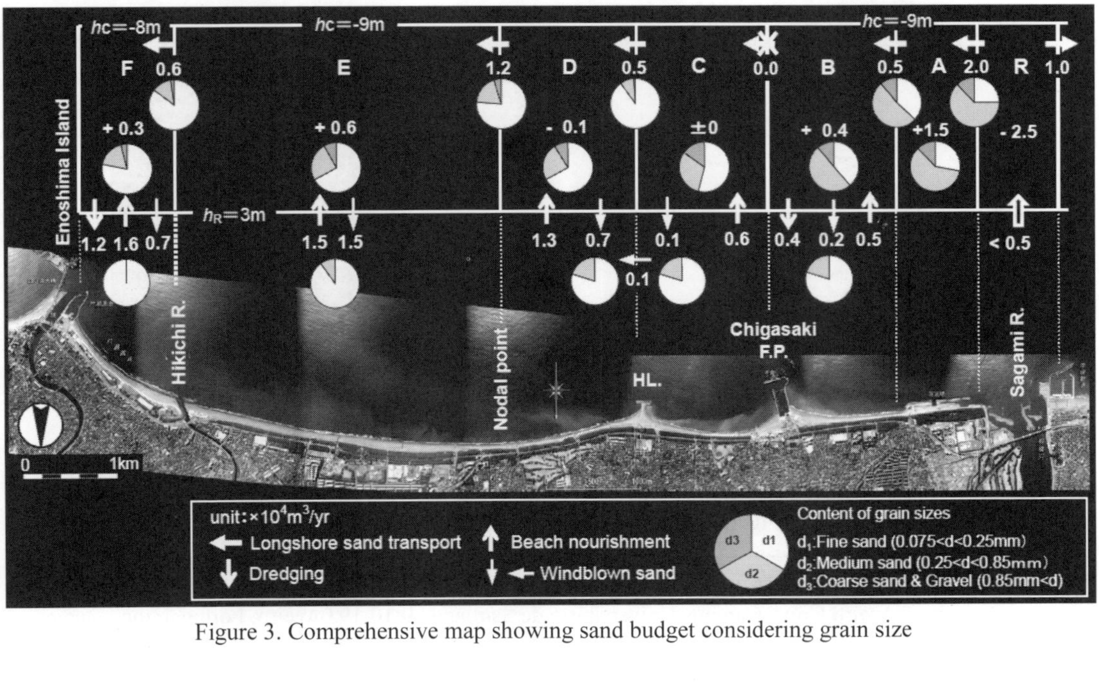

Figure 3. Comprehensive map showing sand budget considering grain size

Coastal Dynamics 2009
Paper No.72

COASTAL EROSION DUE TO ANTHROPOGENIC IMPACTS ON SEDIMENT TRANSPORT IN DOURO RIVER - PORTUGAL

Carlos Coelho[1], Tony Conceição[2] and Bruno Ribeiro[3]

1. Introduction

The coastal erosion causes and their relative importance are site specific, changing from place to place. In the Portuguese Northwest coast case, namely between Douro River and Nazaré, coastal erosion is mainly due to the generalized sea-level rise, the misruled littoral occupation and use, sometimes with natural defenses destruction, the external works in harbors which result in perturbations to the littoral system and the reduction of sediments supply from natural sources. Moreover, the sediment supply reduction is considered as the main cause of the erosion problem felt in this coastal stretch (Silva *et al.*, 2007).

2. Study Area

The Northwest Portuguese coast is essentially a sandy coast approximately N21°E oriented. It is a highly energetic coastal stretch with a wave regime typically from Northwest characterized by a mean significant wave height of 2 m and a mean period of 12 s. During storms, significant wave heights may reach 8 m persisting for up to 5 days. The tide regime is semi-diurnal with a tidal range between 2 m and 4 m in spring tides. The potential alongshore transport mainly due to the wave action is approximately 1-2 million m^3/year. The most contributive sediment sources for this coastal stretch are the Douro River and coastal erosion (Silva *et al.*, 2007).

According to INAG (2001), the Douro River basin (Figure 1a) area is 97603 Km^2, but only 18643 Km^2 (19.1%) are located in Portuguese territory. About 95% of the basin corresponds to rocks with some permeability and the infiltration rates are pointed as 10% of the rain volumes. The average annual rainfall ranges from a maximum of 2500 mm and a minimum of 400 mm. The rainfall average in the Portuguese side is 1030 mm. 30% of the area is for agricultural proposes and 22% is forest without maintenance, supposing important rates of superficial erosion of the soils.

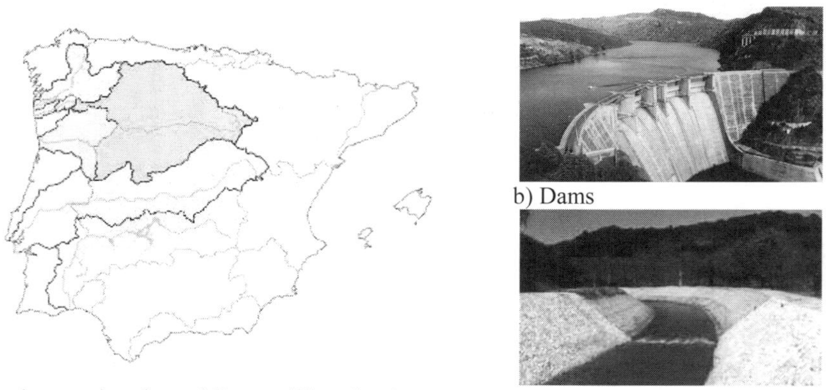

a) Iberian peninsula and Douro River basin

b) Dams

c) Revetments

Figure 1. Study area and it anthropogenic actions

[1]Dept of Civil Eng., Aveiro Univ., Campus Universitário de Santiago, 3810-193 Aveiro, Portugal. ccoelho@ua.pt
[2]Dept of Civil Eng., Aveiro Univ., Campus Universitário de Santiago, 3810-193 Aveiro, Portugal. tony@ua.pt
[3]Dept of Civil Eng., Aveiro Univ., Campus Universitário de Santiago, 3810-193 Aveiro, Portugal. a33811@ua.pt

The flow on the main course of the Douro River depends a lot of the behavior on the Spanish stretch. Changes on the flow are function of the natural water availability and of the water use. Nowadays, medium annual flow in the entire basin is 17100 hm^3, equivalent to 540 m^3/s on Douro's mouth. Without human intervention, those values would increase, corresponding to about 20600 hm^3, or 650 m^3/s. A dry year can represent volumes of 10600 hm^3 and 12700 hm^3, in actual and in natural situations respectively. Historic data on Régua section of the river has peak flow levels above 10000 m^3/s and a maximum of 18000 m^3/s, in 1739 (INAG, 2001).

3. Anthropogenic Actions

The research developed by Conceição (2008) tried to identify the type and chronology of the anthropogenic actions in the Douro River basin that interact with the river sediment transport rates. Those actions can be of two different types: works on the river itself or in its basin, and represent available water volumes and sediment volumes reduction and decrease of flow velocities.

The works on the river have impact on the sediment volumes in transport and interventions on the basin have impact on the production of sediments by soil erosion. Dams for hydroelectric purposes, for water supply, for irrigation and for leisure activities (Figure 1b), dredging to allow navigability, sand extraction for construction and fluvial protection works (Figure 1c) can be referred as the first type. Changes in the land use, on the agricultural techniques and on the soil protection are included in the second type of anthropogenic actions.

4. Results

The Portuguese Northwest Coast suffers from a continued high sedimentary deficit that is primarily due to Douro River sediment supply reduction. In its natural regime the Douro River would supply about 1.8 millions m^3/year (Oliveira, 1997). Different authors refer Douro River sediment supply reduction of about 2000000 m^3/year in the past to about 250000 m^3/year with the present river conditions. Severe measures are needed to face the problem (Silva *et al.*, 2007).

One typified cross-section was considered by Conceição (2008) to quantify the negative sediment balance of each action, to define relative importance on total sediment volume reduction in the Douro River mouth. The distribution of the different causes importance to this was estimated applying some sediment transport methodologies in a single cross section. The main cause of sediment transport reduction is related with the lower flow velocities caused by the reservoirs of the dams on the updrift side and by the dredging. Less important, but still significant is the flow volumes reduction, by flow regularization and for populations supply (mainly in Spain), irrigation and industry.

The studies conclusions show a major importance of the dams constructed in the Douro River basin. To understand the real effects of the dams some laboratory works on a hydraulic flume of the Civil Engineering Department laboratory are being developed. The results of the studies will contribute to the knowledge of major dam's impacts on the sediment transport rates, allowing reduction of coastal erosion problems on the Northwest Portuguese Coast.

References

Conceição, T., 2008. Impacto das Acções Antropogénicas no Comportamento Sedimentar do Rio Douro, *Master Thesis*, Aveiro University (in Portuguese).

INAG. 2001. Plano de Bacia Hidrográfica do Rio Douro, *Relatório Final*, Instituto da Água (National Water Institute) (in Portuguese).

Oliveira, I.B.M., 1997. Proteger ou Não Proteger ou Sobre a Viabilidade de Diferentes Opções Face à Erosão da Costa Oeste Portuguesa, *Colectânea de Ideias Sobre a Zona Costeira de Portugal* (in Portuguese), Associação Eurocoast-Portugal, 205-227.

Silva, R., Coelho, C., Taveira-Pinto, F. and Veloso-Gomes, F., 2007. Dynamical Numerical Simulation of Medium Term Coastal Evolution of the West Coast of Portugal, *Journal of Coastal Research*, SI 50: 263-267.

Coastal Dynamics 2009
Paper No.73

PROFILE EQUILIBRATION AND LONGSHORE SPREADING ASSOCIATED WITH BEACH NOURISHMENT: A CASE STUDY ALONG WEST-CENTRAL FLORIDA COAST

Ping Wang[1] and Nicole A. Elko[2]

1. Introduction

A beach nourishment introduces an artificial perturbation to the nearshore system to which the natural forces modify in both the cross-shore and longshore directions (Dean, 2002). After the beach fill is constructed, the steep profile evolves toward equilibrium shape, and the protruding shoreline adjusts via longshore spreading. This study examines temporal-spatial scales and controlling factors of profile equilibration and longshore spreading based on field data from beach nourishment projects along the west-central Florida coast. Two nourishment projects, using sand of different sizes, were conducted in 2004 and 2006 (Fig. 1). The 2004 Upham Beach nourishment was a short project along a chronically eroding beach downdrift of a structured inlet. The 2006 nourishment was a large project spanning 3 barrier islands, and divided into several sections (Fig. 1). Intensive monitoring during and after construction was conducted, including weekly to monthly profile surveys extending roughly to closure depth in this area (Wang and Davis, 1999).

The study area has a tidal range of less than 1 m with waves typically lower than 0.4 m. Higher waves during the winter are driven by passages of cold fronts, typically every 1 to 2 weeks. These winter storms are responsible for the regional net southward longshore transport. High waves in the summer are rare and mostly generated by tropical storms. The hurricane season of 2004 was active, with passages of 3 major hurricanes after the Upham Beach nourishment. Only one tropical storm, Alberto, passed near the project areas in 2006, during the construction of the Indian Shores section.

2. Summary

Profile equilibration occurred rapidly and was dominated by high wave events, especially the first post-nourishment storm. During the equilibration process, fine sand tends to be transported offshore while coarse sand and gravel tend to be transported onshore. For most of the profiles, equilibration occurred with the formation of a nearshore bar (Fig. 2). The overall beach-profile slope after the first storm approached the pre-nourishment slope for both barred and non-barred beaches. The dynamic bar responded to the subsequent storm impacts and substantially reduced the erosion rate along the dry beach by cross-shore transport. Numerical models including SBEACH and XBEACH were used to simulate the response of the construction profile to the first storm impact.

Longshore spreading occurred over a longer period of time than profile equilibration. For the 2006 projects, longshore spreading occurred through transport and deposition of the fine sand (~ 0.20 mm) along the nearshore bar with little to no change along the shorelines and on the dry beach (Fig. 3). The 2004 project was constructed with bimodal sand (~ 0.52 mm) with a significant amount of coarse shell debris. Spreading occurred through longshore growth of a diffusion spit of the shell debris (Fig. 4), which abruptly changed the shoreline orientation at the ends of the project (Elko and Wang, 2007). The runnel landward of the spit was eventually filled in by overwash resulting in a significant shoreline gain. These different longshore spreading patterns were controlled by different grain sizes. The coarse shell debris in the 2004 nourishment material, which was further concentrated near the shoreline due to the initial profile equilibration, tends to be transported as bedload and deposited near the shoreline. The fine sand used in the 2006 nourishment tends to be transported as suspended load along the nearshore bar. Multi-line numerical models incorporating various sediment grain sizes are

[1]Dept of Geology, University of South Florida, 4202 E. Fowler Ave, Tampa, FL 33620, USA. pwang@cas.usf.edu
[2]Dept. of Envir. Mgmt., Pinellas County, 512 S. Ft. Harrison Ave, Clearwater, FL 33756, USA. nelko@co.pinellas.fl.us

expected to be required to predict the above longshore spreading patterns.

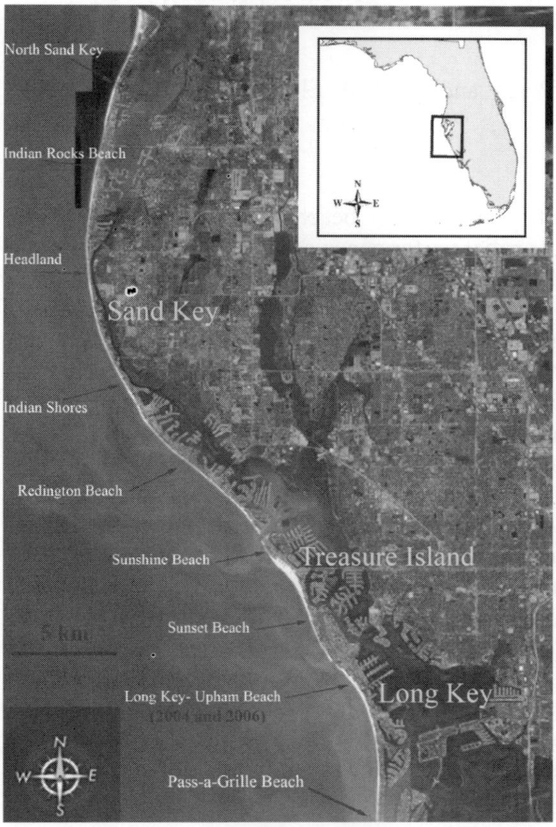

Figure 1. Study area along west-central Florida. Upham Beach was nourished both in 2004 and 2006. The rest were nourished in 2006.

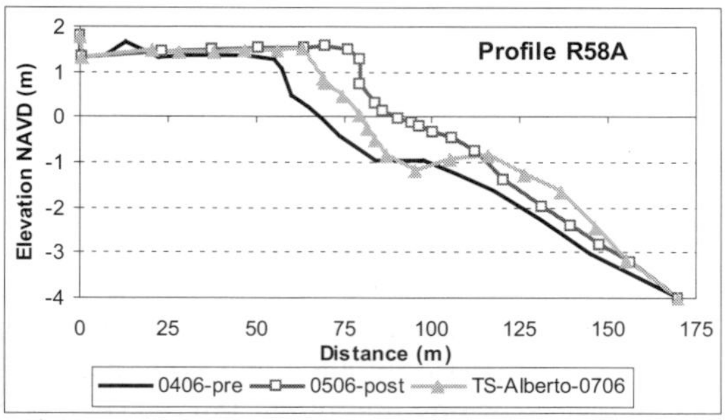

Figure 2. Substantial profile-shape change after the first post-nourishment storm, Tropical Storm Alberto. Note the post-storm profile has similar shape as that of pre-nourishment profile. Example from North Sand Key (Fig. 1).

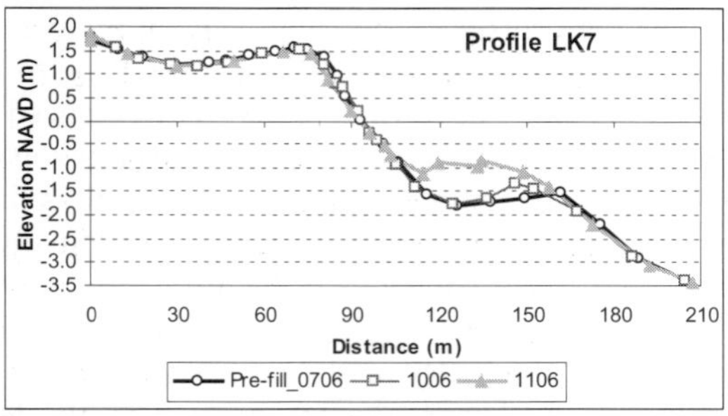

Figure 3. Initial longshore spreading downdrift of the Upham Beach after the 2006 nourishment. Note that most of the sand was deposited in the nearshore zone with little shoreline change.

Figure 4. Initial longshore spreading downdrift of the Upham Beach after the 2004 nourishment. Note the longshore growth of a shelly diffusion spit. The different forms of longshore spreading are controlled by the much coarser shelly sand (0.53 mm) used in the 2004 nourishment as compared to the fine sand (0.20 mm) used in the 2006 nourishment.

Acknowledgements

This study is funded by Pinellas County, Florida and the University of South Florida.

References

Dean, R.G., 2002. *Beach Nourishment, Theory and Practice.* World Scientific, Singapore, 399p.

Elko, N.A. and P.Wang. 2007. Immediate profile and planform evolution of a beach nourishment project with hurricane influences, *Coastal Engineering*, 54: 54-79.

Wang, P. and R.A. Davis. 1999. Depth of closure and the equilibrium beach profile – A case study from Sand Key, West-Central Florida, *Shore and Beach*, 67: 33-42.

Coastal Dynamics 2009
Paper No.74

PHYSICAL PERFORMANCE OF BEACH NOURISHMENT PROJECTS ALONG A MICROTIDAL LOW-ENERGY COAST, WEST-CENTRAL, FLORIDA, USA

Tiffany M. Roberts[1], Ping Wang[1] and Nicole A. Elko[2]

1. Introduction

Detailed physical monitoring is essential to quantify and predict beach nourishment performance (NRC, 1995). This study examines the temporal and spatial performance of several beach nourishment projects constructed in 2006 along three barrier islands in west-central Florida (Fig. 1). Morphologic characteristics of this microtidal low-wave energy coast include 1) a shoreline orientation change of 65 degrees from northwest to southwest; 2) a slightly deeper inner continental shelf in the northern portion of the study area; and 3) a regional southward longshore sediment transport interrupted by two tidal inlets. Waves in the study area are typically sea-type generated by local winds with an average significant wave height of less than 0.4 m. Higher waves are associated with the passages of cold fronts every couple of weeks in the winter and occasional tropical storms in the summer.

The performance of the nourishment projects is evaluated based on monthly surveys of 145 beach-profiles spaced less than 300 m, extending to roughly the closure depth determined by Wang and Davis (1999). The vertical datum is referenced to NAVD88, which is 8 cm above Mean Sea Level (MSL) in the study area. Erosion or deposition on the dry beach, above MSL, above spring low tide, and across the entire profile are calculated as the volume changes above +1 m, 0 m, -1 m NAVD88, and the depth of closure (DOC), respectively. The influence of the project length, width, and orientation, characteristics of the profiles (i.e., with or without a bar), background transport gradient, grain size, bathymetry, and distance to tidal inlets on nourishment performance are analyzed using surveyed bathymetry and wave modeling. This study includes an evaluation of the two-year performance of each project, as well as the temporal and spatial variations of individual profiles within the project.

2. Summary

Significant spatial and temporal variations occurred at individual profiles as well as regionally. Cyclical patterns in nearshore morphology change were measured at most of the profiles. Offshore bar migration and intertidal erosion occurred directly after a high-energy event, such as a strong winter or tropical storm. Subsequent to storms, gradual onshore bar migration with accretion in the vicinity of the shoreline occurred. Profile change dominated by the above cross-shore sediment transport patterns, resulted in little volume change above DOC (Fig. 2). In contrast, profile change dominated by a longshore sediment transport gradient resulted in persistent cumulative volume loss at the four contour levels (Fig. 3). Figs. 2 and 3 are examples of the two end members, whereas most of the profile-volume variations fall in between. The trend is observed for a number of adjacent profiles, indicating the presence of localized longshore transport gradients with a spatial and temporal scale on the order of less than 300 m and one month, respectively.

Background transport rates, localized morphology variations, and project length are key factors that dominated the overall nourishment performance. Significant volume loss was measured at the shorter projects with a persistent longshore transport gradient (Fig. 4). On the other hand, the longer projects with a laterally extensive nearshore bar had less overall volume losses. Examples of localized morphological variations include 1) wave energy focusing at the headland and due to a nearshore dredge pit at Sunset Beach; 2) interruption of longshore sediment transport at Upham Beach by a

[1]Dept. of Geology, Univ. of South Florida, 4202 E. Fowler Ave, Tampa, FL 33620, USA. tmrobert@cas.usf.edu
[2]Dept. of Envir. Mgmt., Pinellas County, 512 S. Ft. Harrison Ave, Clearwater, FL 33756, USA. nelko@co.pinellas.fl.us

structured inlet; and 3) reversal of longshore sediment transport at Sunshine Beach due to wave refraction over the John's Pass ebb delta.

Figure 1. The study area: including individual projects from the 2006 nourishment.

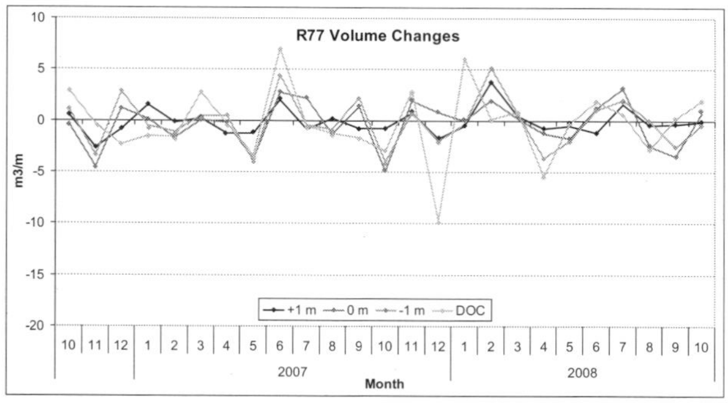

Figure 2. Monthly profile-volume changes at R77, an example from the long project on Sand Key, illustrating variations dominated by cross-shore transport gradients.

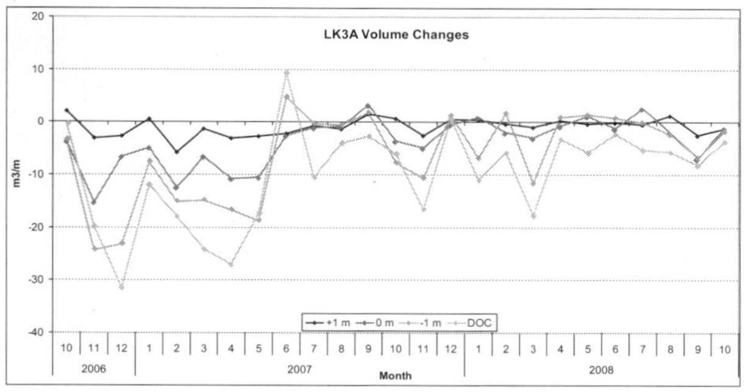

Figure 3. Monthly profile-volume changes at LK3A, an example from Long Key, illustrating variations dominated by longshore transport gradients at all four contour levels.

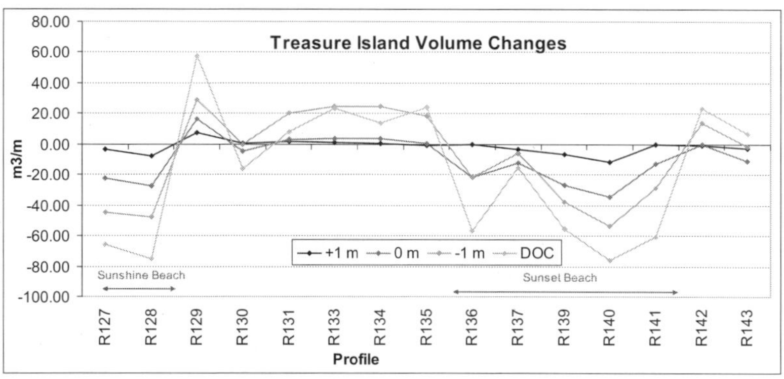

Figure 4. Profile-volume changes along Treasure Island at all four contour levels illustrates the influence of persistent longshore sediment transport gradients resulting in erosion at hotspots, and accretion downdrift.

Acknowledgements

This study is funded by Pinellas County, Florida and the University of South Florida.

References

National Resource Council (NRC). 1995. *Beach Nourishment and Protection*. National Academy Press, Washington, D.C., 334 p.

Wang, P. and R.A. Davis. 1999. Depth of closure and the equilibrium beach profile – A case study from Sand Key, West-Central Florida, *Shore and Beach*, 67: 33-42.

Coastal Dynamics 2009
Paper No. 75

THEORY AND FIELD TEST OF BEACH NOURISHMENT USING COARSE SAND AND GRAVEL

Takayuki Kumada[1], Takaaki Uda[2] and Toshinori Ishikawa[2]

1. Introduction

On the Akiya coast located on the west shore of Miura Peninsula, beaches have been severely eroded, causing wave overtopping and damage to houses. Beach nourishment was planned to recover the beaches while reducing the impact on the ecosystem of the offshore seabed, which can be damaged by the diffusion of fine material. In this study, beach changes on the coast were studied using aerial photographs taken since 1973 and the effect of beach nourishment was investigated using the contour-line-change model considering grain size changes (Kumada *et al.*, 2006, ICCE). Finally, nourishment using 7,600 m^3 of gravel was conducted and the subsequent changes were monitored.

2. Measured Beach Changes

Fig. 1 shows an aerial photograph of the study site in 2004 and the bathymetry in 2003. Kuruwa fishing port is located at the east end, and a pocket beach with 1.3 km length extends. In the offshore zone, an exposed rock bed obliquely extends from the fishing port. Taking the x- and y-axes to be along and normal to the shoreline in 1973, the beach width was measured along the y-axis, and the change in the shoreline position since 1973 was determined (**Fig. 2**). In the wave shelter zone behind the port breakwater, the shoreline advanced, with a maximum shoreline advance of 45 m, and the shoreline far from the port retreated over time. The cause of beach changes is due to the wave sheltering effect by the construction of port breakwater.

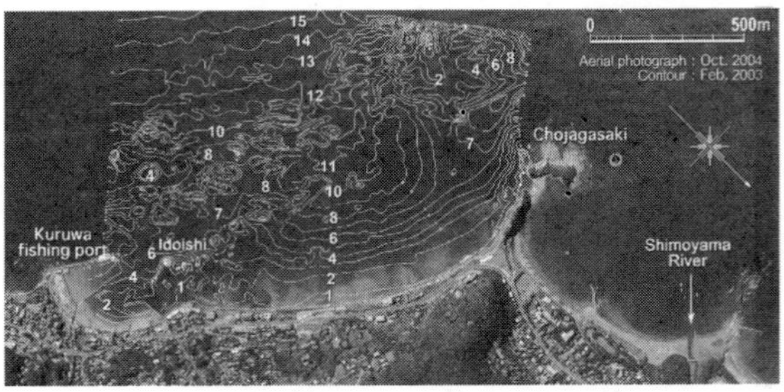

Fig. 1. Aerial photograph and bathymetry of Akiya coast.

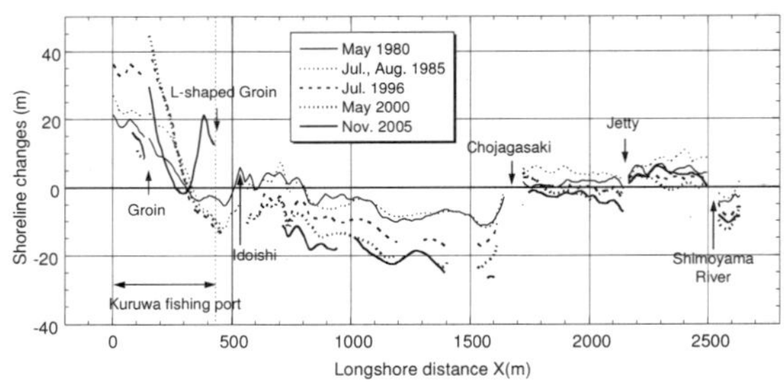

Fig. 2. Shoreline changes on Akiya coast.

3. Prediction of Beach Changes

First, the natural seabed topography in 1973 was reproduced. The changes in wave height and direction due to offshore reefs were taken into account by applying the angular spreading method, given the offshore reef and the parallel initial contours. Then, the stable bathymetry was calculated. The mean slope was assumed to be 1/10 in the zone shallower than -1 m, 1/100 in the zone deeper than 2 m. For wave conditions, the energy-mean wave height was used: H_b =1.0 m, T=5 s and a wave direction of SSW. The depth range of beach changes is between the berm height h_R=1 m and the depth of closure h_c=2 m. The depth of closure is smaller than that observed on other open coasts, because of the existence of exposed rock bed. To express the grain size distribution, three grain sizes of 0.11, 0.25 and 2 mm were selected based on the bottom sampling. The contents of each grain size are assumed to be 40, 15 and 45% in the foreshore zone and 57, 38 and 5% in the offshore zone, respectively. The equilibrium slope is assumed to be 1/10 in the zone shallower than -1 m and 1/100 in the zone between -1 and -2 m. The bathymetry and grain size distribution in 1996, 2000 and 2005 were

---

[1]Laboratory of Aquatic Science Consultant Co., Ltd., 1-14-1 Kami-ikedai, Ota, Tokyo 145-0064, Japan.
[2]Public Works Research Center, 1-6-4 Taito, Taito, Tokyo 110-0016, Japan.

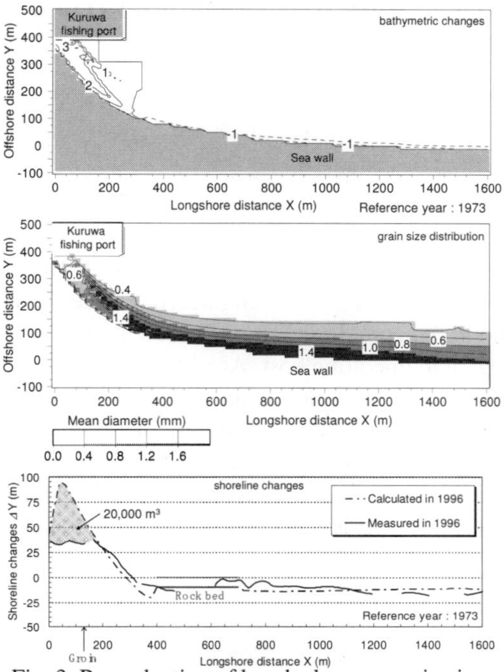

Fig. 3. Reproduction of beach changes, grain size distribution and shoreline changes (1996).

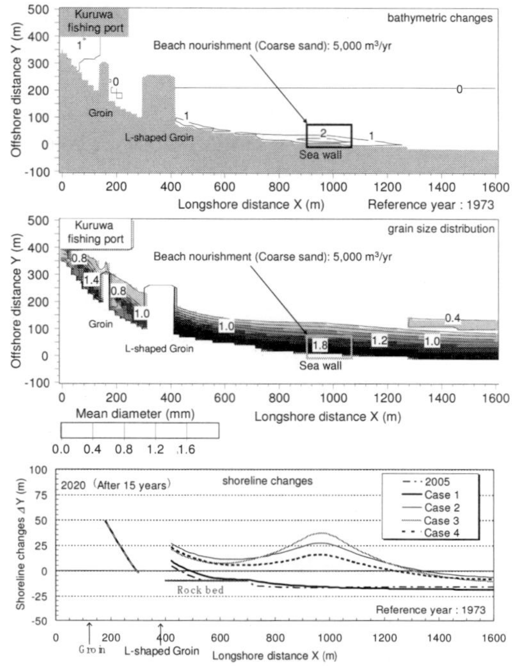

Fig. 4. Prediction of beach changes, grain size distribution and shoreline changes (2020).

reproduced given the bathymetry and grain size distribution in 1973 (**Fig. 3**). In the wave shelter zone, sand accumulated due to longshore sand transport from outside to inside the wave shelter zone, whereas the beach was eroded over an extensive area outside it. Fine sand was selectively transported into the wave shelter zone and deposited there, leaving coarser sand in the erosion zone. The predicted shoreline recession and advance are in good agreement with the measured shoreline. Although the discrepancy between the measured and predicted shorelines is large south of the existing groin, this is due to the dredging of 2.0×10^4 m^3 of sand. For the prediction of beach nourishment, the effect of beach nourishment at a rate of 5,000 m^3/yr using sand of mixed grain size was predicted. Three initial grain sizes were assumed: 0.11, 0.25 and 2 mm with contents of 15, 18 and 67%, respectively. **Fig. 4** shows the calculation results. The nourishment sand was concentrated near the shoreline of the nourishment site, showing the effectiveness of gravel nourishment. Based on this calculation, beach nourishment using 7,600 m^3 of gravel, the grain size of which ranged between 5 and 25 mm, was carried out in 2007. The beach changes were monitored by beach surveys at 20 m intervals. An example of the changes in longitudinal profiles at X=20 m is shown in **Fig. 5**. After nourishment, gravel was deposited in the zone shallower than -1 m, and no movement of the nourishment gravel was observed offshore of -1 m.

4. Conclusions

In a field test, nourishment gravel was deposited along a narrow band near the shoreline. This is in good agreement with the results of a numerical simulation using the contour-line-change model considering grain size changes. Gravel nourishment can prevent fine materials from diffusing and covering the offshore rock bed, thus damage to the sea grass field.

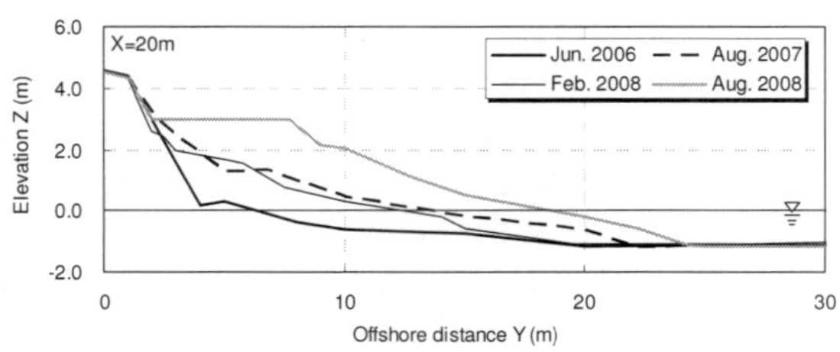

Fig. 5. Changes in longitudinal profiles after gravel nourishment.

Coastal Dynamics 2009
Paper No. 76

EXPERIMENTAL SIMULATIONS OF SHOREFACE NOURISHMENT UNDER STORM CONDITIONS

Florent Grasso[1], Hervé Michallet[1] and Eric Barthélemy[1]

1. Introduction

The large impact of human activities on coastal morphodynamics is now widely admitted. Beach protection and shore reconstruction strategies are at the forefront of coastal engineering issues. One traditional way to maintain beaches consists in beach nourishment. However, shoreface nourishment appears as a good compromise between efficiency, ecology and operating cost. According to Van Duin *et al.* (2004) and Grunnet and Ruessink (2005), the shoreface nourishments may have two different effects, namely the lee and feeder effects. The lee effect refers to the ability to increase wave dissipation offshore for protecting the beach face. The feeder effect comprises the onshore movement of nourished sand. The purpose of the present study involves shoreface nourishments under storm conditions simulated in laboratory experiments. Physical modelling is particularly useful because of high temporal and spatial discretizations.

2. Experimental set-up

Experiments are carried out in the 36 m long LEGI flume equipped with a piston wave generator. The sloping bottom consists of a loose material of low density (1.19 g/cm^3) with a median diameter d_{50}=0.6 mm. The Shields and Rouse numbers for the experiments are of the same magnitude as those of natural environments. Time and length scales ratios are roughly 1/3 and 1/10 (Michallet *et al.*, 2007). Irregular waves are generated according to a JONSWAP spectrum, characterised by a significant wave height (H_s) and a peak period (T_p). A schematic storm is defined as a succession of different wave climates in order to reproduce the three storm phases, respectively characterised by increasing, energetic and decreasing wave conditions. The storm duration is 10.5 hours for an apex (H_s=16 cm, T_p=3 s) lasting 4.5 hours.

3. Results

This study is focused on the effect of three scenarios of shoreface nourishments in the form of localized sediment deposits. These are located: (i) in the bar trough, (ii) on the outer bar and (iii) in the inner zone. The nourishments in the trough (Fig. 1b) and on the bar (Fig. 1c) lead to beach face erosion and bar feeding. In contrast, nourishment in the inner zone (Fig. 1d) leads mostly to shore feeding and beach face reconstruction. Hydrodynamic conditions along beach profiles have been investigated during storm simulations. It can be seen in Figure 1 (circles) that wave breaking on the bar (on the beach face) is stronger (respectively weaker) for case (c) compared to case (a). Hence, bar feeding increases the wave dissipation offshore and reduces wave dissipation on the beach face. Shoreline positions and shore volume variations are estimated throughout the shoreface nourishments. As it is shown in Figure 2, increasing and energetic wave conditions lead to a retreat of the shoreline. As soon as waves decrease in the last phase of the storm the shoreline moves seaward but the initial shoreline position is not totally recovered. Nevertheless, with the successive shoreface nourishments the dynamics of the beach face is weaker and weaker.

In the framework of our experiments, the outer bar acts firstly as a wave filter dissipating the energy of the larger breaking waves, which corresponds to the lee effect. Secondly, the inner zone

[1]Laboratoire des Ecoulements Géophysiques et Industriels (CNRS – UJF – INPG),
BP 53, 38041 Grenoble cedex 9, France. florent.grasso@hmg.inpg.fr

nourishment produces an onshore propagating active bar, which typically characterises the feeder effect.

Acknowledgements

This study was funded by the LITEAU II program of the French Ministry of Ecology and the EC INTERREG-IIIc program under project BEACHMED-e. The first and third authors benefited of the ECOS-CONYCIT C07U01 project. The technical support of J.-M. Barnoud is gratefully acknowledged.

References

Grunnet, N.M. and B.G. Ruessink 2005. Morphodynamic response of nearshore bars to a shoreface nourishment. *Coastal Engineering*, 52(2), 119-137.

Michallet, H., Grasso, F., and Barthélemy, E., 2007. Long waves and beach profiles evolutions. *Journal of Coastal Research*, SI50, 221-225.

Van Duin, M.J.P., Wiersma, N.R., Walstra, D.J.R., Van Rijn, L.C. and Stive, M.J.F, 2004.Nourishing the shoreface: observations and hindcasting of the Egmond case, The Netherlands. *Coastal Engineering*, 51(8-9), 813-837.

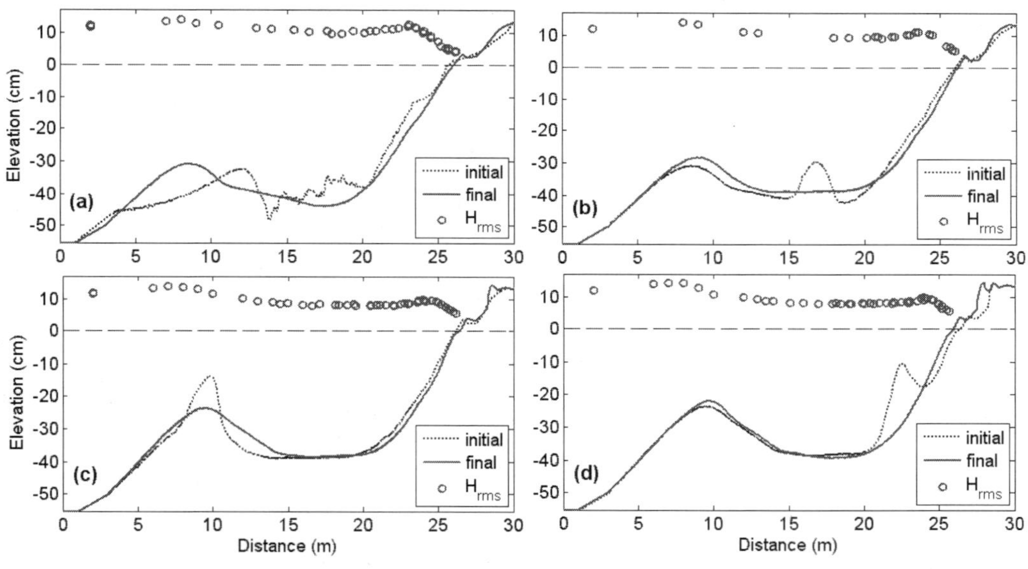

Figure 1. Initial (dotted) and final (solid) beach profiles for storm simulations, circles represent root mean square wave height (H_{rms}) for energetic wave conditions (H_s=16 cm, T_p=3 s). a) No nourishment, b) trough nourishment, c) bar nourishment and d) inner zone nourishment.

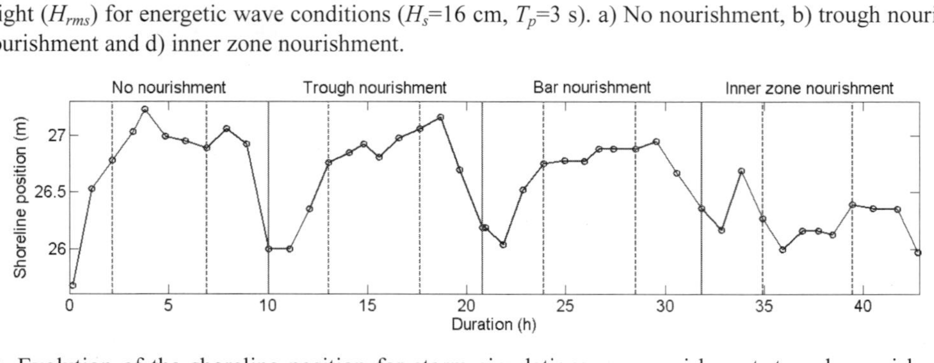

Figure 2. Evolution of the shoreline position for storm simulations: no nourishment, trough nourishment, bar nourishment and inner zone nourishment. Vertical solid lines delineate the four storm simulations and dashed lines delineate the three phases (increasing, energetic, decreasing) in each storm.

Coastal Dynamics 2009
Paper No. 77

ON THE ROLE OF IMPULSIVE PRESSURES INDUCED BY PLUNGING BREAKERS ACTING ON GRAVEL BEACHES

Adrián Pedrozo-Acuña[1a], Alec Torres-Freyermuth[2a], Qingping Zou[1b], Tian-Jian Hsu[2b] and Dominic E. Reeve[1c]

1. Introduction

Gravel beaches are important for shore protection and flood defence, and also provide valuable maritime habitats. They are shoreline features which include: barrier beaches; spits that protect harbours and estuary hinterlands from flooding; and steep terraces that provide protection against erosion of soft cliffs. The existing understanding of the behaviour of gravel beaches is not good. This lack of understanding, together with other constraints, makes the successful and cost-efficient management of gravel beaches one of the most challenging coastal management tasks. Therefore, research efforts looking at this type of beach have been concentrated on their management, principally in the context of coastal defence, i.e. to reduce the risks of flooding and erosion.

On gravel beaches, as a result of the steepness, waves tend to break over a shorter distance and more violently than on a sandy beach. The greater permeability of gravel beaches allows water to be jetted into the pore spaces, accumulating a significant portion of the swash lens and thus creating asymmetry in the swash transport potential. Moreover, during a recent investigation on coarse-grained beaches it was hipothesised that pressures induced by wave-impacts from plunging breakers, should be considered as an additional mechanism for stirring and mobilising the sediment (Pedrozo-Acuña et al., 2007). In a subsequent study, Pedrozo-Acuña et al. (2008) presented measurements of wave-induced impact pressures within a gravel beach profile. It was demonstrated that the biggest impulses are associated with plunging breakers, which are the most common type of breakers during storms.

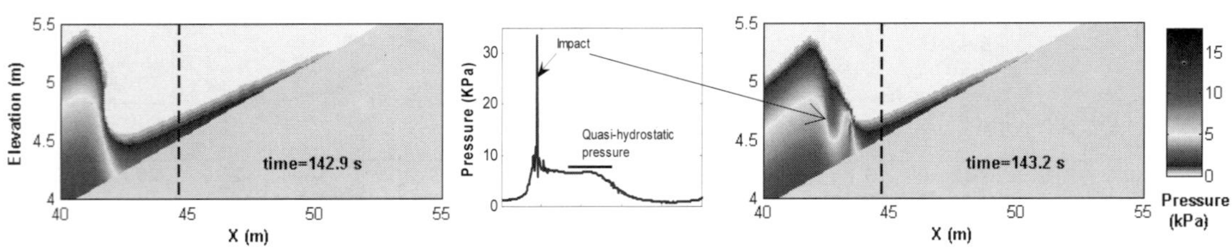

Figure 1. Left- Pressure field from the RANS model before impacting on the gravel beach. Middle – Pressure time curve measured for a given impact from a plunging breaker. Right - Pressure field from the RANS model at the moment of wave impaction on the gravel beach.

Thus, a natural step forward in this research is to investigate the anatomy of the pressure gradient under the action of plunging breakers. Indeed, works related to sandy beaches with mild slopes, have started to investigate pressure gradients relating them to local acceleration evaluated from the near-bed velocity (e.g. Drake and Calantoni, 2001; Hoefel and Elgar, 2003). However, the debate of the

[1]University of Plymouth, Centre for Coastal Dynamics and Engineering, School of Engineering, Drake Circus, Plymouth, PL4 8DF, United Kingdom,(a)apedrozoacuna@plymouth.ac.uk ; (b)qingping.zou@plymouth.ac.uk; (c)dominic.reeve@plymouth.ac.uk
[2] University of Delaware, Center for Applied Coastal Research, Department of Civil and Environmental Engineering, Newark, Delaware 19716, USA

usefulness of this parameterisation remains open. For instance, in a recent study Foster et al. (2006) approximated the pressure gradient using both the local flow acceleration and the advection term $u\ du/dx$. Moreover, a recent numerical investigation concerning the simulation of swash zone fluid accelerations, concluded that given the poor correlation between local acceleration and pressure gradient, it is not likely that local acceleration can be used as a surrogate of pressure gradient (Puleo et al. 2007). Following this debate and given the magnitude of the recorded impulsive pressures, the need for investigation of pressure gradients induced by plunging breakers on gravel beaches is clear.

2. Aim of the research

We present results from a diagnostic investigation (Pedrozo-Acuña et al. in review) involving the comparison of model results against prototype scale laboratory data. The numerical model is based on the Reynolds Averaged Navier Stokes equations (RANS), with a turbulence closure coupled with a free-surface tracking scheme (Lin and Liu 1998a; Lin and Liu 1998b; Lin and Xu 2005). This numerical model is selected for its ability to describe the complex hydrodynamics associated with plunging breaking waves propagating over permeable beaches (see Figure 1 left and right panels). Results from the numerical simulations are compared against measurements obtained in tests at the GWK (Lopez de San Roman-Blanco et al. 2006). Good agreement is found. The numerical model is employed to investigate the impulsive pressures with the aim of identifying the momentum term(s) that contribute to this process. This is done in order to pave the road towards an enhanced parameterisation of sediment transport on gravel beaches.

Acknowledgements
The tests in the GWK of the Coastal Research Centre (FZK), Germany were supported by the European Community (HPRI-CT-1999-00101). Dr. Pedrozo-Acuña, Dr. Zou and Prof. Reeve acknowledge the support of the UK Natural Environment Research Council (NERC) under the FREE Programme (NE/E0002129/1). Prof. Lin is acknowledged for providing the source RANS code.

References

Drake, T.G. and Calantoni, J., 2001. Discrete particle model for sheet-flow sediment transport in the nearshore. *Journal of Geophysical Research*, 106(C9): 19859-19868.

Foster, D. L., Bowen, A.J., Holman, R.A. and Natoo P. 2006. Field evidence of pressure gradient induced incipient motion. *Journal of Geophysical Research* 111(C05): doi:10.1029/2004JC002863.

Hoefel, F. and S. Elgar 2003. Wave-induced sediment transport and sand bar migration." *Science* 299: 1885-1887.

Lin, P. and Liu, P.L.-F., 1998a. A numerical study of breaking waves. *Journal of Fluid Mechanics*, 359: 239-264.

Lin, P. and Liu, P.L.-F., 1998b. Turbulence transport, vorticity dynamics, and solute mixing under plunging waves in surf zones. *Journal of Geophysical Research*, 103(15): 15,677– 15,694.

Lin, P. and Xu, W., 2005. NEWFLUME: a numerical water flume for two-dimensional turbulent free surface flows. *Journal of Hydraulic Research*, 44(1): 79-93.

López de San Román-Blanco, B., Coates, T.T., Holmes, P., Chadwick, A.J., Bradbury, A., Baldock, T.E., Pedrozo-Acuña, A., Lawrence, J. and Grune, J., 2006. Large scale experiments on gravel and mixed beaches: Experimental procedure, data documentation and initial results. *Coastal Engineering*, 53(4): 349-362

Pedrozo-Acuña, A., Simmonds, D.J., Chadwick, A.J. and Silva, R., 2007. A numerical-empirical approach for evaluating morphodynamic processes on mixed and gravel beaches. *Marine Geology*, 241(1-2): 1-18.

Pedrozo-Acuña, A., Simmonds, D.J. and Reeve, D.E., 2008. Wave-impact characteristics of plunging breakers acting on gravel beaches. *Marine Geology*, 253(1-2): 26-35.

Pedrozo-Acuña, A., Torres-Freyermuth, A., Zou, Q., Hsu, T.-J., Reeve, D.E., *in review*. Diagnostic investigation of impulsive pressures induced by plunging breakers impinging on gravel beaches. *Marine Geology*.

Puleo, J. A., Farhadzadeh, A., and Kobayashi N. (2007). Numerical simulation of swash zone fluid accelerations. *Journal of Geophysical Research* 112(C07007): doi:10.1029/2006JC004084.

Coastal Dynamics 2009
Paper No. 78

STORM RESPONSE AND BEACH ROTATION ON A GRAVEL BEACH

Amaia Ruiz de Alegria[1], Gerhard Masselink[1], Kenneth Kingston[1] and Jon Williams[1]

1. Introduction

Gravel barriers are particularly widespread on high-latitude wave-dominated coastlines (Buscombe and Masselink, 2006) and represent sustainable coastal defences that protect low-lying coastal regions from flooding during storms. However, these barriers are relatively narrow and dynamic systems that in conditions of reduced alongshore sediment supply are highly vulnerable to breakdown (Orford et al., 2003).

In the present time coastal scientists and engineers need to deal with the issue of differentiating between short-term shoreline fluctuations and long-term shoreline trends (Morton, 1991). Therefore, there is a need to understand the long-term behaviour of large-scale beach systems in order to assess coastal stability and be able to find out whether the beach is stable, or progressively eroding or accreting (Forbes et al., 2004). Moreover, the ability to model and predict the shoreline evolution associated to wave climate variations is an additional challenge.

2. Study site

Slapton Sands is a 4.5 km long and 100–140 m wide macrotidal gravel barrier located in the southwest of the UK (Fig.1). Inshore hydrodynamics are dominated by southwest and northeast swells characterised by medium-to-high energy wave conditions (mean $H_s \geq 1$ m; storm $H_s \geq 3$ m). The barrier is highly dynamic over a range of time scales and significant damage has occurred during recent storm events due to flooding and overtopping processes. The high shoreline variability in Slapton is induced by irregular waves and significant longshore transport rates within the narrow swash region. Over long- and medium-term time scales (decadal to annual) the highest morphological variability occurs in the alongshore direction, whereas short-term (weeks to months) changes mainly take place in the cross-shore direction (Ruiz de Alegria et al., 2008).

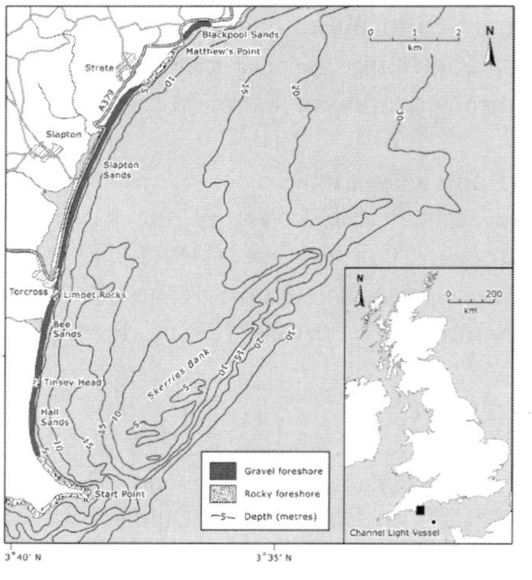

Figure 1. Location of Slapton within the UK.

3. Methodology and preliminary results

A validated Argus video system (Ruiz de Alegria et al., 2008) has been used to derive weekly shorelines from the images over three years (September 2005–2008). The position of the shorelines was related to the inshore wave conditions and long-term climate trends. A comparison of the position of weekly shorelines from September 2005 to the present time shows that the beach has rotated, and that over this 3-year period the northern end of the barrier widened by 35 m, whereas the southern end of the monitored area narrowed by 25 m (Fig. 2).

The short-term storm-induced shoreline retreat and the post-storm recovery were also determined from the Argus images (Fig. 3). As expected, the highest shoreline variability occurred between late autumn and early spring when the main storms took place (Fig. 3 left panel). The post-storm recovery was found to be well correlated to the change in wave direction. For example, the 17th of April 2008

[1]School of Marine Science and Technology, University of Plymouth, Drake Circus, Plymouth, PL4 8AA, UK. amaia.alegria@plymouth.ac.uk

easterly directed storm caused significant erosion but the beach recovered in less than two weeks due to the rapid shift into southerly waves (Fig. 3 right panel). Further work will link morphological response to wave conditions in more quantitative terms.

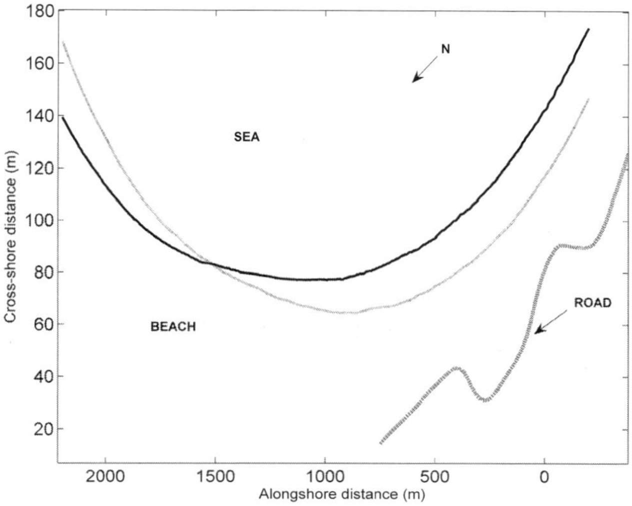

Figure 2. Video derived shorelines for 10Sep05 (straight) and 26Sep08 (dotted) showing beach rotation. The N arrow indicates the direction of the north. A section of the A379 road was realigned after a storm event in 2001; this is shown with an arrow.

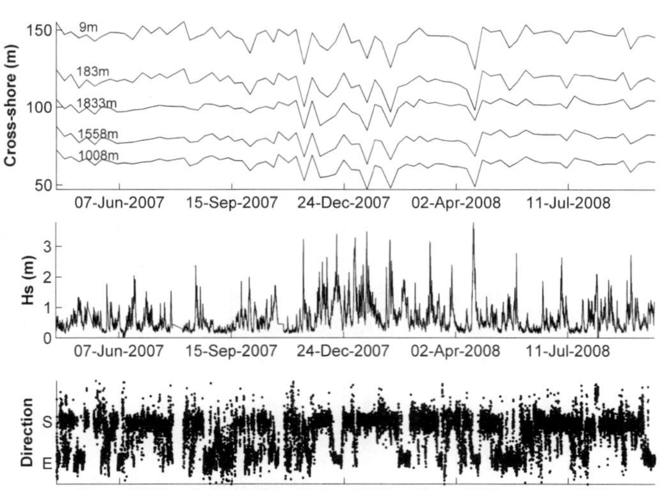

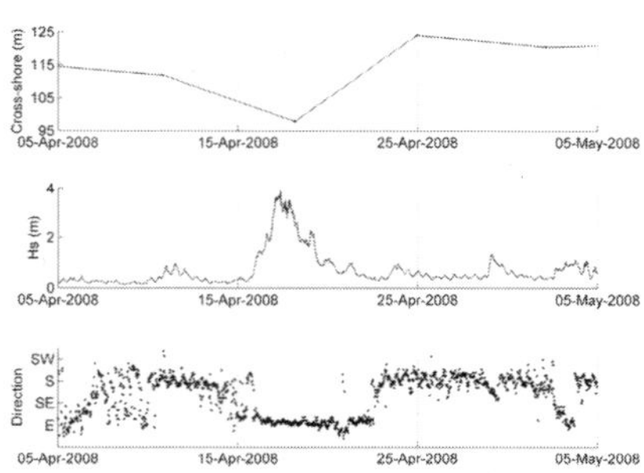

Figure 3. In the left, shoreline evolution in time at several alongshore positions (top panel) compared to the significant wave height (*Hs*) and wave direction (bottom panel). A zoomed section of this figure for the 1558 m alongshore position is shown in the right panel.

References

Buscombe, D. and Masselink, G., 2006. Concepts in gravel beach dynamics. *Earth Science Reviews*, 79(1): 33-52.

Forbes, D.L, Parkes, G. S., Manson, G. K. and Ketch, L.A., 2004. Storms and shoreline retreat in the southern gulf of St. Lawrence. *Marine Geology* 210(1-4): 169-204.

Morton, R.A., 1991. Accurate shoreline mapping: past, present and future. *Coastal Sediments '91*, (1): 997-1010

Orford, J.D., Jennings, S. and Pethick, J., 2003. Extreme storm effect on gravel-dominated barriers, *Coastal Sediments '03*, ISBN-981-234-422-7, 12pp.

Ruiz de Alegria, A., Masselink, G., Kingston, K., Williams, J. and Buscombe, D. 2008. Monitoring storm impacts on a gravel beach using the Argus video system. *International Conference on Coastal Engineering '08*. ASCE.

Coastal Dynamics 2009
Paper No. 80

EFFECTS OF GROUNDWATER TABLE ON WAVE BREAKING IN A GRAVEL BEACH

Kwang-Ho, LEE[1], Norimi MIZUTANI[2] and Toshiaki FUJII[3]

1. Introduction

Wave breaking is one of the most complex phenomena and closely related to various coastal engineering problems. In particular, in designing a coastal structure and planning countermeasure of beach erosion the wave breaking height and depth are key parameters. For this reason, several efforts have been devoted to understand the characteristics of wave breaking. However, few studies for wave breaking in a gravel beach can be founded. Wave fields in the surf zone can be expected to be affected by the exfiltration caused by groundwater discharged from a beach face. Lee et al. (2007) reported that wave field on a gravel beach is especially susceptible to the groundwater table, and the effect of groundwater has to be considered in investigating the wave breaking phenomenon. In this study, an experimental study is performed to elucidate wave breaking in a gravel beach where the groundwater does not follow tidal variations; there are differences between seawater and groundwater levels as well as to provide the quantitative experimental data for the wave breaking height and depth.

2. Experimental Setup

A hydraulic model experiments were conducted in a two-dimensional wave flume at the Coastal and Ocean Engineering Laboratory, School of Engineering, Nagoya University. The wave flume is 30 m long, 0.7 m wide and 0.9 m deep. A gravel beach with a slope of 1/7 was constructed in the wave flume. A water-level control tank made of 1 cm thick acrylic plastic plates was installed behind the gravel beach to control the simulated groundwater. In order to avoid the water level rise by installation of the water-level tank, the wave flume was divided laterally into two parts, one 0.3 m wide and the other 0.4 m wide, using a waterproof painted plywood board that started 2.0 m from the wave generator. These two divided wave flumes are opened to each other at the near the wave generator and behind the water-level control tank. Thus, maintaining sea water level constantly in the working wave flume is possible during the experiments. Water for simulating the groundwater was supplied and drained through pumps between the flume and the water-level control tank, permitting adjustment of groundwater to any desired level while maintaining a constant mean water level in the flume. Figure 1 shows a schematic diagram of the experimental arrangement. We took pictures near the surf zone with a high-speed digital camera for all 469 cases. The wave breaking height and depth were determined by the image analysis (see photo 1).

3. Results and discussion

Figure 2 and 3 show the non-dimensional wave breaking depth and wave breaking height as functions of the deepwater wave steepness, H_o/L_o for the difference between the water depth and groundwater level, Δh. The results computed by the existing empirical formulas (Rattanapitikon et al., 2003; Rattanapitikon and Shibayama, 2006) for predicting wave breaking index for a sandy beach and the regress curves obtained through the present experimental data for a gravel beach are also plotted simultaneously in the figures. It is well known that the breaking water depth increases with H_0/L_0. We

[1]Dept of Civil Eng. , Nagoya University, Furo-Cho, Chikusa-ku, Nagoya, 464-8603, Japan. leekh@civil.nagoya-u.ac.jp
[2]Dept of Civil Eng. , Nagoya University, Furo-Cho, Chikusa-ku, Nagoya, 464-8603, Japan. mizutani@civil.nagoya-u.ac.jp
[3]Dept of Civil Eng. , Nagoya University, Furo-Cho, Chikusa-ku, Nagoya, 464-8603, Japan. akkk160@hotmail.com

can also observe this phenomenon from figure 2. Further, the breaking water depth slightly increases as the difference between water depth and groundwater level Δh increases. The reason might be that the velocity exfiltrated from near the shoreline became strong as Δh increases, and a breaking wave point is moved toward the offshore side. From the comparisons between the experiment results (square filled black) for $\Delta h = 0.0$ and existing empirical formula (solid line) by Rattanapitikon and Shibayama (2006), it is clear that the breaking water depth for a gravel beach is less than that for a sandy beach, and its difference between the gravel beach and the sandy beach increases with H_0/L_0. In figure 3, it is confirmed that wave breaking height increases as deepwater wave height increases and decreases slightly with Δh . As Δh increases, the wave breaking height became relatively small because breaking wave point was moved to the offshore side. Also, the wave breaking height by the empirical formula for a sandy beach overestimates the experimental results.

References

Lee, K.-H., Mizutani, N., Hur, D.-S. and Kamiya, A., 2007. The effect of groundwater on topographic changes in a gravel beach, *Ocean Engineering*, 34: 605-615.

Rattanapitikon, W., Vivattanasirisak, T. and Shibayama, T., 2003. A proposal of new breaking height formula, *Coastal Engineering Journal*, 45: 28-48.

Rattanapitikon, W. and Shibayama, T., 2006. Breaking wave formulas for breaking depth and orbital to phase velocity ratio, *Coastal Engineering Journal*, 48: 395-416.

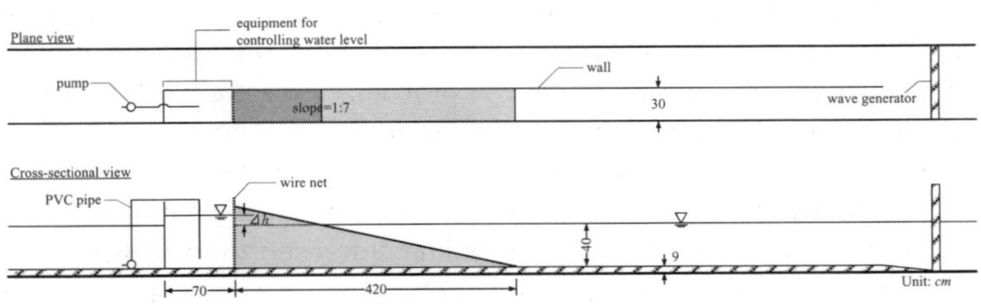

Figure 1. Experimental setup

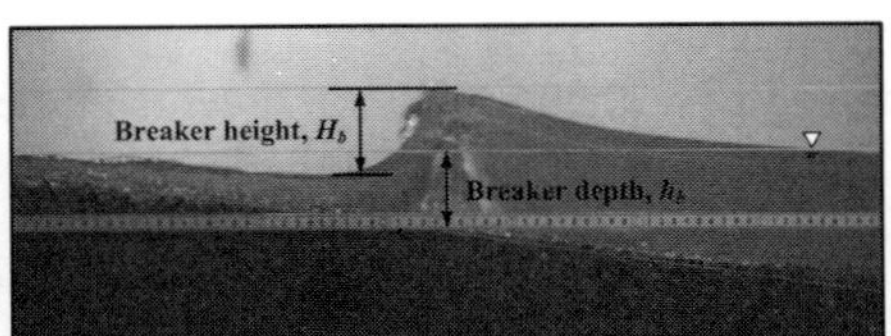

Photo 1. Example of image analysis

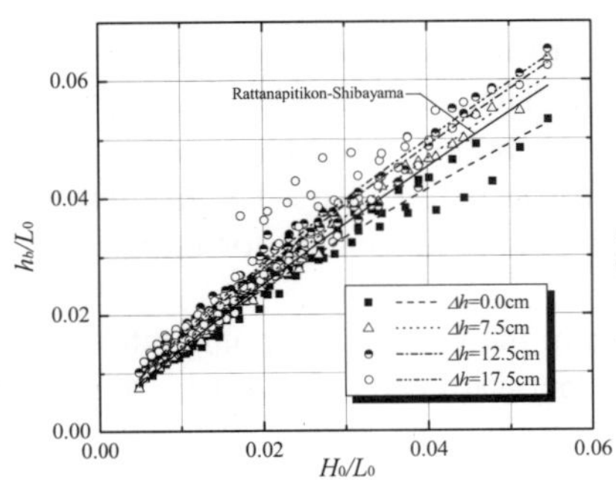

Figure 2. Relationship between H_0/L_0 and H_b/L_b

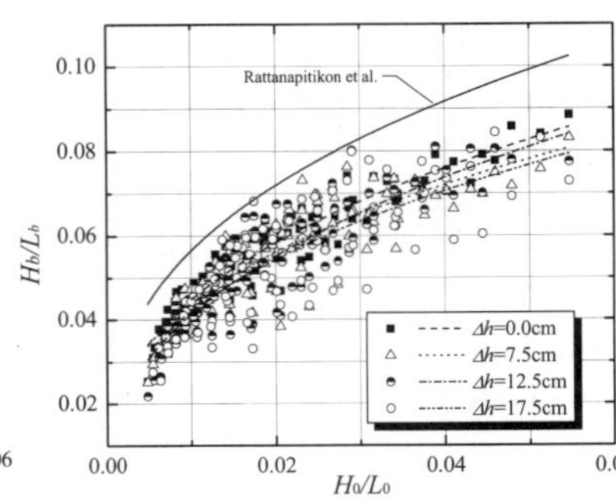

Figure 3. Relationship between H_0/L_0 and h_b/L_0

Coastal Dynamics 2009
Paper No. 81

DUNE EROSION AND OVERWASH

Nobuhisa Kobayashi[1] and Ali Farhadzadeh[2]

1. Introduction

A wide beach with a high dune provides storm protection and damage reduction. Wave overwash of a dune may occur more frequently due to the combined effects of sea level rise and storm intensification due to global warming. No process-based model exists presently for the accurate prediction of wave overwash of a dune and flooding risk. This study aims to develop and evaluate a simple and transparent model for dune erosion and overwash.

2. Numerical Model CSHORE

The cross-shore numerical model CSHORE developed by Kobayashi et al. (2009) consists of the combined wave and current model based on the time-averaged continuity, momentum and wave action equations coupled with the transparent formulas for suspended sand and bedload transport rates which synthesize available data and formulas. CSHORE is extended here to the wet and dry zone above the mean water shoreline to predict wave overwash of a dune. The hydrodynamic model for the wet and dry zone is based on the time-averaged continuity and momentum equations derived from the nonlinear shallow-water wave equations. The probability distribution of the water depth is assumed to be exponential. The concept of the wet probability is introduced to perform the averaging only for the wet duration. The cross-shore velocity and water depth are related locally using a formula for a bore in the presence of a current. This model has been shown to predict the water depths and velocities measured in the wet and dry zone on seven different dikes in 118 tests within a factor of about 2 (Farhadzadeh et al. 2008). The sediment transport formulas for suspended sand and bedload in the wet and dry zone are essentially the same as those used in the wet zone except that the velocity distribution in the wet zone has been assumed to be Gaussian.

3. Comparisons with Experiments

Kobayashi et al. (2009) conducted two small-scale experiments on berm and dune erosion for a high and narrow berm (Experiment H) as well as a low and wide berm (Experiment L). The fine uniform sand volume for the two different berms was approximately the same. The measured berm and dune profiles were compared with CSHORE limited to the wet zone only where scarping was modeled by a linear landward extrapolation of the offshore sediment transport rate on the steep slope exceeding the limiting slope of 0.63. This intuitive scarping algorithm is not used in the extended CSHORE which computes the suspended sand and bedload transport rates in the wet and dry zone. The extended CSHORE is compared with Experiments H and L.

Fig. 1 compares the measured and computed profiles at time t = 2.00, 3.67 and 6.11 h starting from the initial profile at $t = 0$ for Experiment H. The still water level is indicated in each panel where the significant wave height and spectral peak period were approximately 19 cm and 2.6 s, respectively. The degree of agreement is similar for Experiment L. The explicit modeling of wave overwash in the extended CSHORE does not improve the agreement for Experiments H and L with minor overwash.

Fig. 2 shows the computed temporal variations of the wave overtopping rate q_o, bedload transport rate q_b, and suspended sand transport rate q_s at the landward end (x = 14 m) in Fig. 1 for Experiments

[1]CACR, Univ. of Delaware, Newark, DE 19716, USA. nk@udel.edu
[2]CACR, Univ. of Delaware, Newark, DE 19716, USA. alif@udel.edu

H and L. No wave overtopping and overwash occurred for the duration of $t = 0 - 1$ h when the still water level was at zero elevation. The wave overtopping and overwash rates increased as the water level increased and the dune crest was lowered during $t = 1 - 3.67$ h. These rates decreased as the water level was lowered during $t = 3.67 - 6.11$ h. The difference between Experiments H and L is within a factor of 2 because the berm was eroded completely by $t = 2.00$ h. The computed variations in Fig. 2 are consistent with visual observations but no measurement was made because these experiments were intended for minor overwash. The extended CSHORE predicts the dominance of bedload in the thin layer of overtopping flow.

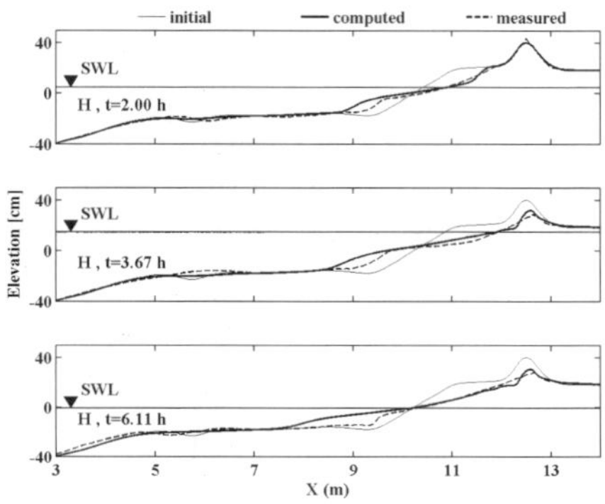

Figure 1. Measured and computed profiles for Experiment H.

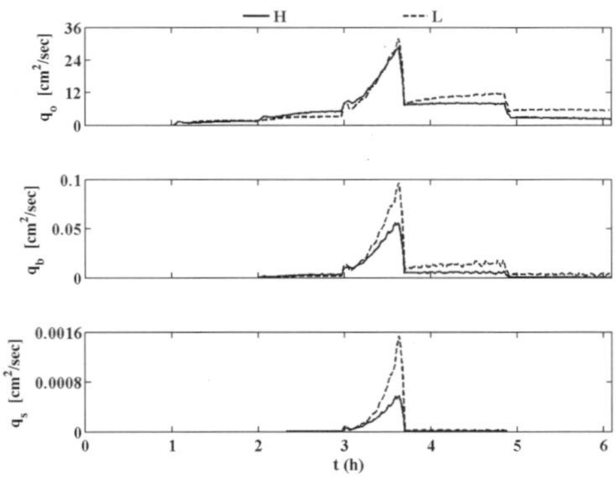

Figure 2. Computed time series at $x = 14$ m for Experiments H and L.

Additional experiments are being conducted for the transition from minor to major overwash. The overtopping rate q_o and total transport rate ($q_b + q_s$) at the landward end will be measured in these experiments. The comparisons of the extended CSHORE with the new data will be presented at the conference.

References

Farhadzadeh, A., Kobayashi, N., and Melby, J.A., 2008. Wave overtopping and overflow on inclined structures. *Proc. 31st Coastal Eng. Conf.* (in press).

Kobayashi, N., Buck, M., Payo, A., and Johnson, B.D., 2009. Berm and dune erosion during a storm. *J. Waterway, Port, Coastal, Ocean Eng.*, 135(1).

Coastal Dynamics 2009
Paper No. 82

MICORE: DUNE EROSION AND OVERWASH MODEL VALIDATION WITH DATA FROM NINE EUROPEAN FIELD SITES (AND BEYOND)

Ap Van Dongeren[1], Javier Benavente, Yann Balouin[3], Paolo Ciavola[4], Rui Taborda[5], Kazimierz Furmańczyk[6], Piet Haerens[7], Dano Roelvink[8], Ekatarina Trifonova[9], Jon Williams[10]

1. Introduction

The European funded project MICORE - Morphological Impacts and COastal Risks induced by Extreme storm events – has as the main objective to develop and demonstrate on-line tools for reliable predictions of the morphological impact of marine storm events in support of civil protection mitigation strategies. Severe storms have historically affected European coastlines and the impact of each storm has been evaluated in different ways in different countries. The project is specifically targeted to contribute to the development of a common probabilistic mapping of the morphological impact of marine storms and to the production of early warning and information systems to support long-term disaster reduction.

2. Approach

The first objective of one of the 7 workpackages is to use, validate and extend a free-ware coastal erosion model called X-Beach (Roelvink, et al. 2008) for various European coastal sites and compare its outputs with off-the-shelf packages such as LITPROF, SMC, SBeach, Transed and Durosta which are now locally used. The coastal sites where field data is taken are shown in Fig 1. They include Lido di Dante-Lido di Classe, Ravenna (Italy), Praia de Faro (Portugal), Urban beaches of Cadiz Bay (Spain), Lido of Sete to Marseille Beach (France), The Dee Estuary (United Kingdom), Egmond Beach (The Netherlands), Mariakerke Beach (Belgium), Dziwnow Spit (Poland) and Kamchia-Shkorpilovtsi Beach (Bulgaria). Each field site has a unique bathymetry/topography and/or wave/tidal climate and will contribute to and test the modelled physics.

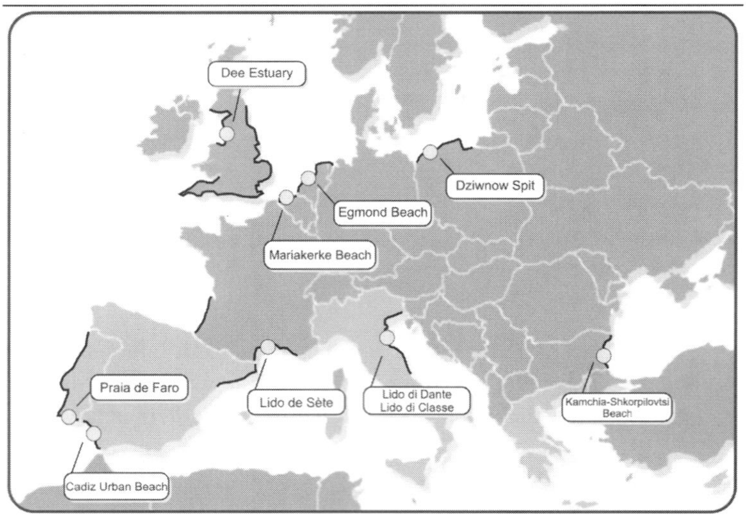

Figure 1. Study sites in the European Union.

[1] Deltares/Delft Hydraulics (The Netherlands) (Ap.vanDongeren@deltares.nl), [2]U. Cadiz (Spain), [3] BRGM (France), [4]U. Ferrara (Italy), [5] U. Lisbon (Portugal), [6]U Szczecin (Poland), [7]IMDC (Belgium), [8]Unesco-IHE (The Netherlands), [9]IO-BAS (Bulgaria), [10]U. Plymouth (UK)

The end goal is to incorporate jointly-developed (between partners) algorithms into one shared operational forecast/predictive model able to predict coastal hazards occurring on a time-scale of storm durations (days). In this way dune breach, dune erosion, wave run-up and overtopping will be predicted for storm and extreme meteorological conditions for a wide variety of coasts. From these physical processes the local socio-economic impact of storms can be established so that a complete set of data will be obtained to produce risk maps. The operational models can thus be used for the design of preventive measures, either coastal defences, mitigation or evacuation schemes.

3. Results

Some preliminary results for the Dutch coast are shown in Fig. 2 wherein the dune erosion calculated from Xbeach is compared to Durosta and Sbeach for a typical profile. The final paper and the presentation will show results for the above European beaches but also US sites which are investigated in a concurrent US Corps of Engineers project.

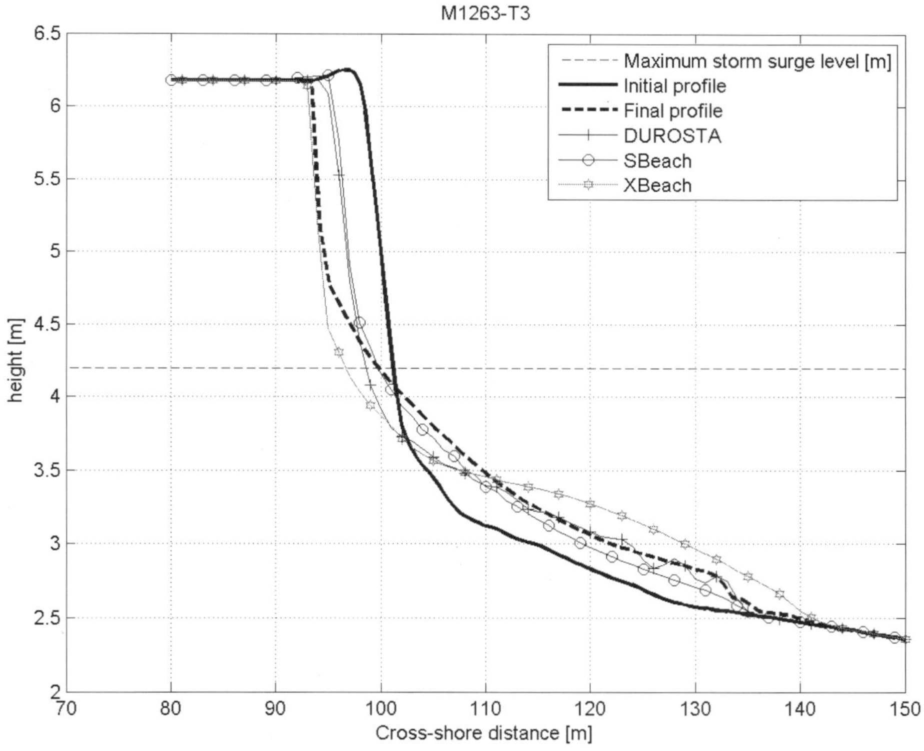

Figure 2. Dune erosion profile for the Dutch coast comparing measured end profile (dashed), Durosta (pluses), SBeach (circles) and XBeach (stars).

Acknowledgements

This research has funded by the European Union as part of Framework program FP7 under project number 202798. Additional funding for the Deltares effort has been granted through Unesco-IHE Delft by the U.S. Government through its European Research Office of the U.S. Army., under Contract no. N62558-06-C-2006.

References

Roelvink, J.A., A.J.H.M. Reniers, A.R. van Dongeren, J.S.M van Thiel de Vries, R. McCall (2008). Modelling storm impacts on beaches, dunes and barrier islands. Proc. ICCE, in press.

Coastal Dynamics 2009
Paper No. 83

WASHOVER DEVELOPMENT ON MIXED-ENERGY, MESOTIDAL BARRIER ISLAND SYSTEMS

Piet Hoekstra[1], Albert Oost[1, 2], Miriam ten Haaf[1] and Pieter Buijs[1]

1. Introduction

Overwash and the development of washover systems are extensively described for micro- and mesotidal, wave-dominated coastal environments, in particular along the US Atlantic and Gulf coasts (Donelly et al., 2006). Many of these studies relate to long elongate and low barrier islands with a small tidal prism and limited number of inlets. Mesotidal, mixed-energy and tide-dominated barrier islands though have a different morphology and are characterized by a drumstick shape, numerous tidal inlets, large tidal basins (and associated tidal prism) and well-developed dune systems. Dune elevation, in combination with the presence of foredunes are major factors that constrain overwash. As a result, overwash is commonly supposed to be a less important process and washovers should be less widespread phenomena. Nevertheless, in recent studies in micro- to mesotidal and fully mesotidal, mixed energy or tide-dominated environments as the Danish and Dutch Wadden Sea and the Algarve overwash is still considered to be an important process for barrier island development and washovers are prominent morphological, sedimentological and ecological features.

The main objective of the present (pilot) study is to reconstruct the morphological development of washovers in a mesotidal environment and to explore the main hydrodynamic factors and conditions that affect overwash in this type of setting. In addition, we will demonstrate the effect of overwash for two extreme storm events in November 2006 and 2007, respectively. The pilot study is part of a larger research programme that focuses on the development of a more dynamic and natural type of shoreline management. This new approach includes the option to reactivate washover systems in order to increase the sediment budget of the barrier islands to cope with sea level rise and to stimulate the ecological development and biodiversity of the beach-dune-salt marsh system.

2. The West Frisian barrier islands in the Netherlands

The West Frisian barrier island system marks the boundary between the North Sea and the Wadden Sea. The system is characterized by relatively short (15-30 km in length) and drumstick-shaped barriers. The barrier width varies between 2 and 5 km and washovers are primarily found on the downdrift, smaller sides of these islands (see Fig. 1) where they alternate with relatively high foredunes (heights up to 10 m). The back-barrier tidal basins have a considerable size and the mixed micro-to mesotidal, semidiurnal regime is characterized by a tidal range between 1.5 and 2.3 m. For the present study we focus on the barrier island Schiermonnikoog; the associated tidal prism is about 270×10^6 m^3. Mean annual significant (offshore) wave height is about 1.1-1.3 m. During storm surges, the maximum recorded set-up in water level is about 3.5-4 m. At present, the north coast of the island has an extremely wide and dissipative beach (> 500 m. of "green beach").

3. Long term morphological development of washovers

The morphological study is based on an extensive analysis of aerial photographs (1927-2007), coastal profile data, satellite images, local maps and a number of dedicated field surveys in 2006 and 2007. Washover development can be summarized by means of a conceptual model. Washover formation starts with run-up overwash and inundation overwash on a bare sand flat. Shortly after, dune-building processes result in an alternation of newly developing dune complexes alternating with both broad and narrow washovers. Simultaneously, inundation overwash creates subaqueous deposition at the basin side, creating washover deltas. As soon as the dune building processes continu to proceed in a seaward direction, sheltered areas become occupied by marsh vegetation. The location of the washovers

[1] Dept. of Physical Geography, Utrecht University, P.O. Box 80.115, 3508TC Utrecht, the Netherlands; p.hoekstra@geo.uu..nl
[2] Deltares, P.O. Box 177, 2600MH Delft, the Netherlands; a.oost@deltares.nl

becomes more or less fixed, tidal creeks at the basin side demonstrate a headward erosion and the washover drainage pattern gradually interacts with the tidal creeks. The former overwash deltas are no longer maintained and gradually disappear.

Comparing past and present developments, we observe that for decades overwash is observed at the same places, located at gaps in the foredunes. As a result, the washover systems have developed a typical funnel-shaped morphology with a narrow throat (< 200 m wide) and a length which largely exceeds the width. In landward direction sandy washover systems gradually transform into muddy marshes with tidal creeks.

4. Inundation overwash during storm surges: morphological response

We determined hydrodynamic processes and conditions during overwash based on monitoring data, field observations and initial modeling with Delft3D (Van Ormondt and Van Dongeren, 2007). Partly as a result of the wide and dissipative beaches overwash has become a very episodic process which may occur once or twice a year. During the pilot study the effect of two major storms (in November 2006 and November 2007, respectively) was recorded and the morphological impact was established afterwards. During both storm surges the basin water levels during rising tide were higher than at open sea due to wind stresses and additional input of water into the tidal basin. Consequently, inundation overwash at the seaward side was simultaneously accompanied by flooding from the basin side and inundation overwash was initially not able to reach the back-barrier basin. Hydrodynamic modeling clearly supports this observation (Van Dongeren and van Ormondt, 2007) and also demonstrates that sea water masses and basin water masses during flood meet each other at the transition zone of washovers and upper marsh. In addition, the modeling also demonstrates that the width of the washover has a pronounced effect on the flow velocities. A wide (small) throat results in low (high) flow velocities since the marsh area rapidly (slowly) fills with water, quickly (gradually) reducing the gradients in water level across the washover system. Flow contraction generally implies that maximum flow is observed on both side of the throat instead of in the middle of the throat.

Video-based field observations clearly demonstrate that dune-erosion is mainly a product of the impact of low-frequency waves. Surprisingly enough though, this erosion is hardly reflected in the amount of washover deposition and most of the sediment is simply deposited in front of the duneface.

5. Conclusions

The lack of deposition due to inundation overwash seems to be the result of the lack of wave-driven longshore and oblique currents that could supply (suspended) sediment during inundation. Apparently, the wide and dissipative (green) beach results in two major effects: 1) the total dissipation of short wave energy; the long waves will only act as a local stirring factor and 2) the lack of mobility of beach sediments due to the presence of small primary dunes (with vegetation) and algal mats.

References

Donelly, C., Kraus, N., and Larson, M., 2006. State of Knowledge on Measurement and Modelling of Coastal Overwash. J. of Coastal Res., vol. 22, 965-991.

Van Dongeren, A. and Van Ormondt, M., 2007. Hydrodynamic aspects of overwash.; framework: H&I Zeereep (report in Dutch). WL Delft Hydraulics, report Z4412, 33 pp.

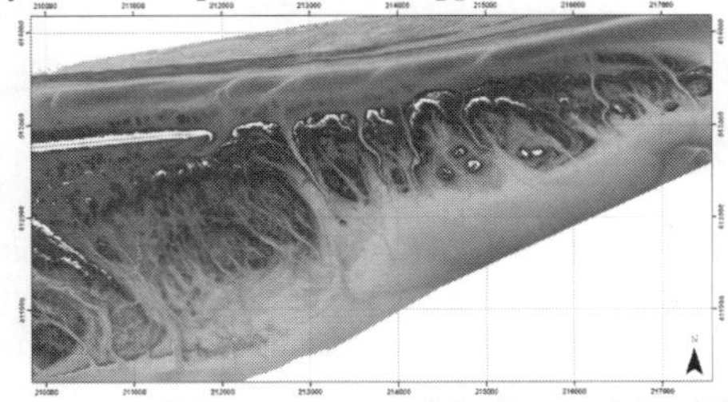

Figure 1: Digital elevation measurements of the east coast of the barrier island Schiermonnikoog with washovers; the washover are connected with tidal marsh creeks in the back-barrier basin.

Coastal Dynamics 2009
Paper No. 84

MEASUREMENTS ON THE INTERACTION BETWEEN DUNES AND DIKES DURING EXTREME STORM EVENTS

Pieter van Geer[1], Marien Boers[1] and Marcel van Gent[1]

1. Introduction

Dunes and dikes are two kinds of primary sea defenses to provide protection against flooding. Dune erosion at an undisturbed dune section has been studied intensively (see also Van Gent et al., 2008). At the location where a dune and a dike connect, we expect that the amount of dune erosion is affected by the presence of the dike (Van Gent, 2008). Until now no datasets are available to analyze these interactions between dunes and dikes in 3D situations. To that end, physical model tests were performed in the Vinjé basin, that has an effective length and width of 21m and 26m respectively. Furthermore, the effects of breaches and dune revetments were investigated during the experiments. For the analysis of three dimensional effects high resolution bathymetry measurements were required. We developed a new technique based on video images to provide these data.

2. Setup of the experiments

Two configurations with dune and dike segments were constructed. The first configuration consisted of a dike with 3 breaches of various widths. This allows for analysis of the dune erosion in these beaches. Figure 1 shows a picture of the resulting bathymetry of one of the experiments with this configuration. The second configuration contained a connection between a dune and a dike. This allows for the analysis of the dune erosion near such a connection. An example of a bathymetry measurement with this configuration is presented in Figure 1 as well. The experiments were carried out at a depth scale of $n_d = 60$ with a prototype wave height of H_{m0}=9m. For each configuration two experiments were carried out with a prototype wave period of T_p = 12s or T_p = 16s, leading to four different experiments. The total duration of each experiment was 90 minutes, corresponding with a 15 hour storm on prototype scale.

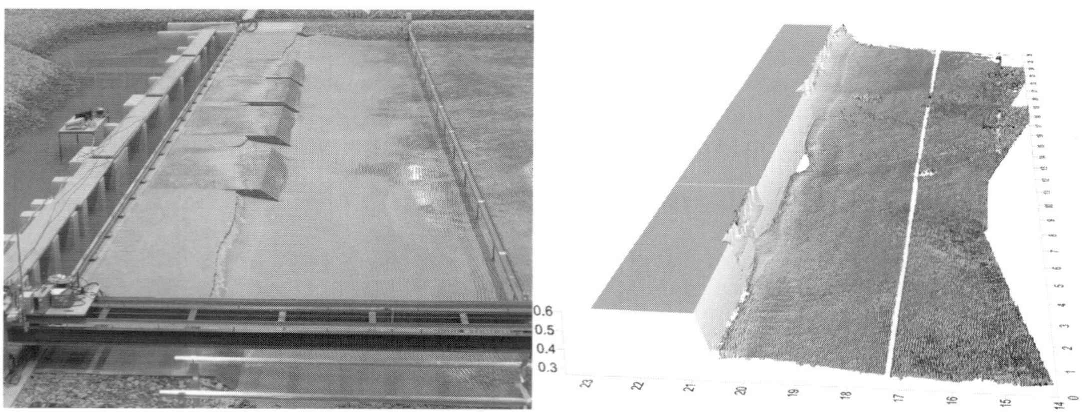

Figure 1. Experiment set-up to analyze the interaction between dunes and dikes. Left: a photograph of the configuration with breaches. Right: Bathymetry measurements of the configuration with a connection between a dike and a dune. Bathymetry measurements were obtained with a technique based on video imaging.

[1] Deltares | Delft Hydraulics, Coastal Structures and Waves Department, P.O. Box 177, 2600 MH Delft, The Netherlands. Pieter.vanGeer@deltares.nl, Marien.Boers@deltares.nl, Marcel.vanGent@deltares.nl

3. Measurement techniques

During each experiment wave heights, currents and the position of the dune crest were measured. At various moments in time also transects were measured to monitor the morphologic development. This was done with a wheel profiler. To increase the resolution of the bathymetry measurements in longshore direction we applied a new technique based on video-images, comparable to the use of video-images in the field for derivation of a beach bathymetry (Wijnberg *et al,* 2008). With intensity differences of succeeding images it is possible to identify the location of the water line. This can be seen as one contour line of the bathymetry. Analysis of video-images of the bathymetry with a rising or falling water level thus provides a representation of the bathymetry. Dune crest measurements were based on a comparable method. Determination of the dune crest position however was done with the use of cross shore color gradients. Figure 2 shows the measured dune crest retreat in time for a dune segment during one of the experiments. These video measurement techniques will be a substantial part of the paper.

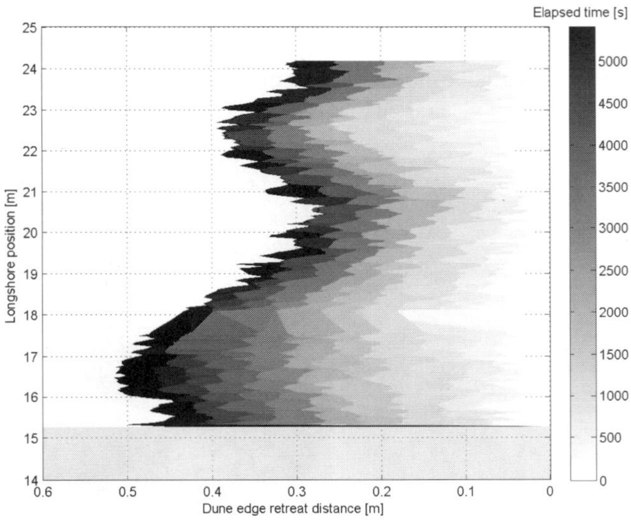

Figure 2. Representation of the measured dune crest retreat in time for a dune segment during one of the experiments.

4. Results on dunes and dikes

We obtained a valuable dataset with wave heights, currents, cross–shore profiles, bathymetry measurements, dune crest measurements, hand – held surface scans, sediment grain sizes and photos and videos. High resolution bathymetry and dune crest measurements were obtained with the use of video images. The data show an increased erosion at the location of a connection between a dune and a dike. Also the erosion in a breach is substantially larger than for situations without a breach.

References

Van Gent, M., Van Thiel de Vries, J., Coeveld, M., De Vroeg, H., Van de Graaff, J. 2008. Large-scale dune erosion tests to study the influence of wave periods. *Coastal Engineering Journal.* 55: 1041-1051.

Van Gent, M. 2008. Large – scale tests to analyse the influence of collapsed dune revetments on dune erosion. *Proceedings of the 31st International Conference on Coastal Engineering.* In press.

Wijnberg, K, Uunk, L., Morelissen, R., Cohen, A, Hulscher, S. 2008. Beach response derived from automated bathymetry measurements. *Proceedings of the 31st International Conference on Coastal Engineering.* In press.

Coastal Dynamics 2009
Paper No. 85

HISTORICAL MORPHOLOGICAL DEVELOPMENT OF THE EASTERN SCHELDT TIDAL BASIN (THE NETHERLANDS)

M. Eelkema[1], Z.B. Wang[2] and M.J.F. Stive[3]

1. Introduction

The Eastern Scheldt basin in the southwestern part of the Netherlands is an elongated tidal basin with a length of approximately 50 km and an average width of 9 km (Figure 1). The Eastern Scheldt is the original estuary of the Scheldt River, but lost this function during the 14th and 15th century, when the Western Scheldt became the primary distributary. The (former) islands around the Eastern Scheldt are the reclaimed salt marshes and tidal divides between the Eastern Scheldt and its neighboring estuaries, the Western Scheldt to the south and the Grevelingen to the north. The Eastern Scheldt basin and outer delta have shown impressive changes in both shape and size over the last six centuries. Especially the configuration of the tidal basin has been dramatically changed by human influence as well as extreme events (Figure 2). These changes have not only had their impact on the system of flats and gullies within the basin, but also on the shoals and channels of the outer delta.

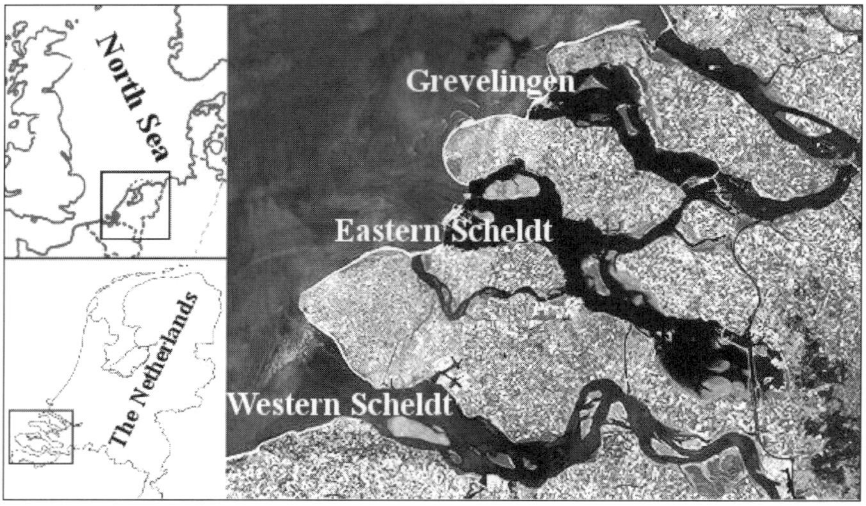

Figure 1: The Eastern Scheldt and its neighboring basins

2. Research methods

This research aims to describe the evolution of the basin from the end of the Middle Ages until the 20th century. The research focuses on the development of the basin as a whole, as well as certain elements in particular such as its outer delta and one of its tidal divides. We also investigated whether certain observed behavior of these elements can be linked to processes elsewhere in the basin. Van den berg (1986) already put forward several hypotheses on linked behavior within the Eastern Scheldt. To test these hypotheses, simplified basin geometries for different periods are reconstructed according to known records as well as general relations describing basin geometry. These reconstructions are then used to study the hydrodynamics inside the basin using a two dimensional numerical flow model

[1] Dept of Civil Eng & Geosc, Delft Univ of Techn., Stevinweg 1, 2628 CN Delft, NL, m.eelkema@tudelft.nl
[2] Deltares, P.O. Box 177, 2600 MH Delft, NL, wang@wldelft.nl
[3] Dept of Civil Eng & Geosc, Delft Univ of Techn., Stevinweg 1, 2628 CN Delft, NL, m.j.f.stive@tudelft.nl

(Delft3D). This approach is somewhat similar to the method applied by Van der Spek (1997), who studied the evolution of the tidal propagation in the Western Scheldt estuary. The reconstructions are highly simplified, consisting of no more than one main channel with shallow area on both sides, much like the geometries used by Friedrichs and Aubrey (1987) in their investigation on tidal deformation. The sole purpose of these simulations is to assess the differences in propagation speed and amplification of the tidal wave between these periods. For the evolution of the tidal divide this means assessing whether different geometries produce different locations of a minimum in average velocities.

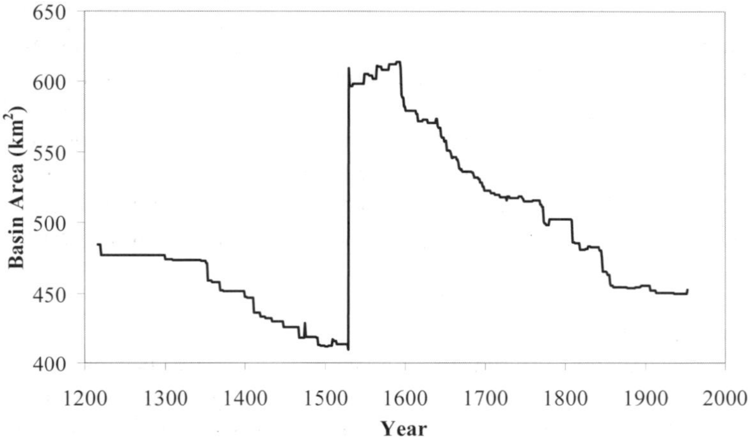

Figure 2: Evolution of basin area

3. Results

The results of this research are twofold. A description is made of almost five hundred years of basin evolution including the ebb-delta, based on records on land reclamations and historical charts and observations. Next to this, simplified models of the basin's geometry and tidal hydrodynamics give insight into the causes of changes within the basin as well as the nature of the overall sediment transport (import vs. export). From this analysis it appears that the inundations in the 16th century have caused a continuous export of sediment and scour of the main channels. Sediment balances from the 19th and 20th century show this process lasted well into the 20th century. Human influence has not turned this process around until the 1980's, when the tide inside the Eastern Scheldt was tamed by compartmentalization of the basin and the construction of the Storm-Surge Barrier. Not only does this research explain some of the morphological changes inside the Eastern Scheldt basin, but it also gives more general insight into the mechanisms behind macro-scale evolution of tidal basins and the stability of tidal divides.

Acknowledgements

This research was carried out as part of the Ecoshape: BuildingWithNature program, project ZW3.2. Part of the funding was provided by the Dr. Ir. Cornelis Lely Foundation.

References

Friedrichs, C. T. and D. G. Aubrey, 1988. Non-linear tidal distortion in shallow well-mixed estuaries: a synthesis. *Estuarine, costal and shelf science*, 27: 521-545.

Van den Berg, J. H., 1986. *Aspects of Sediment- and Morphodynamics of Subtidal Deposits of the Oosterschelde (the Netherlands).* Rijkswaterstaat Communications, no. 43/1986, The Hague.

Van der Spek, A. J. F., 1997. Tidal asymmetry and long-term evolution of Holocene tidal basins in the Netherlands: simulation of paleo-tides in the Schelde estuary. *Marine Geology*, 141: 71-90.

Coastal Dynamics 2009
Paper No. 86

FORECASTING AND UNCERTAINTY ESTIMATION OF NEARSHORE BATHYMETRIC EVOLUTION OVER ANNUAL TIMESCALES

Jose M. Horrillo-Caraballo[1], Vanesa Magar[1] and Dominic E. Reeve[1]

1. Summary

A new forecasting technique is presented and applied to a set of historic bathimetric surveys of sandbanks and channels off Great Yarmouth, UK, which is an extension of the method proposed by Reeve et al. (2008). The method combines three procedures. The first one consists of an Empirical Orthogonal Function (EOF) analysis to identify principal bathymetric patterns and their dynamics. The second is to use a jack-knife resampling technique to produce an ensemble of bathymetric realisations from 1846 to 1998. Finally, an Auto Regression (AR) model is used to find an ensemble of forecasts for 2006. Two AR techniques were used, both giving comparable results, indicating the robustness of both methods. Also, forecasted bathymetries agreed well with measurements of the 2006 bathymetry.

2. Methodology and Results

The region off the coast of East Anglia was chosen as a test case, because in this region several sandbanks and channels may be identified (see Figure 1a, from Horrillo-Caraballo, 2005). While most features are slowly varying, the region covered by Cross Sands and Barley Pickle has been observed to be very dynamic, with the sandbanks moving in and out of the depicted domain during the period of study, which extended from 1848 to 2006.

The objective of this study was to analyse the morphodynamics of the whole area, to develop a forecasting technique and find an estimate of the uncertainty of the forecast. The dynamics were analysed using EOFs, which were first applied to coastal studies by Winant et al. (1975). This is a data-driven technique used to find patterns of behaviour in observations. The method consists in solving an eigenvalue problem with the data covariance matrix as the kernel. Each eigenvalue is equivalent to a portion of resolved variance, and may be ordered in decreasing order. Here it was found that the first three EOFs captured 98% of the variance, while the first eight EOFs captured 99.9%, i.e. almost all the variance of the original dataset. A measured bathymetry and its corresponding reconstruction, for year 2000, using the first eight EOFs, is shown in Figures 1. b) and c), respectively. These figures show the EOF analysis successfully resolves the positions of the sandbanks and channels, in most of the domain, being less appropriate in the most dynamic region of Cross Sands and Barley Pickle.

The next step consisted on developing a resampling technique to produce an ensemble of realisations, which here was the jack-knife technique (Quenouille, 1949). The method consists of removing one of the measurements, here a bathymetric dataset, and interpolating over the missing data to produce a new realisation. This method leads to an ensemble of EOFs. This ensemble was then forecasted from 1999 to 2006 using two AR methods. The first one follows Mares and Mares (2003), where the best fit model is constrained to have small order, and the order is found for each EOF before the forecast. The second one, however, assumes that the AR order is the same for all EOFs and that it has no constraints; the order is, moreover, found by using both measured and forecasted bathymetries up to 2006, and by setting it to the value that gives the smallest error for the 2006

[1]University of Plymouth, School of Engineering, Centre for Coastal Dynamics and Engineering (C-CoDE), Reynolds Building, Drake Circus, PL4 8AA, Plymouth, UK. Jose.Horrillo-Caraballo@plymouth.ac.uk, Vanesa.Magar@plymouth.ac.uk, Dominic.Reeve@plymouth.ac.uk

forecasted and measured bathymetries. Both AR methods yield similar results, showing that in forecasting applications it is not necessary to constrain the order of the AR model to be small, and that both approaches are valid. Finally, it was assumed the mean of the ensemble forecasts was the best future prediction for that particular year, and the ensemble variance is used to make an estimate of the uncertainty in the prediction. Again, the forecasts predict well the slowly varying features of the test case under consideration when compared to measurements for that year.

3. Conclusions

A method to make ensemble predictions for yearly scales has been developed, which includes a statistical technique to analyse patterns of behaviour in observations, a resampling method to produce an ensemble of realisations based on the observed dataset, and two forecasting techniques based on an AR model. Uncertainty of the forecasts was assessed using the ensemble variance and estimate errors between the forecasted and measured bathymetries for a year when measurements are available. The work is ongoing and we expect to be able to present an analysis of the performance of the method for predicting sandbank changes over the order of a decade as well as the robustness of the predictions to the length of the dataset.

Acknowledgements

The authors acknowledge the support from the UK Engineering and Physical Sciences Research Council under Grant Number EP/F020511/1.The bathymetric data was kindly provided by International Port Holdings and the Great Yarmouth Port Authority.

References

Horrillo-Caraballo, J.M., 2005. *Investigation of the long-term evolution of the offshore sandbanks at Great Yarmouth, U.K.* PhD Thesis, The University of Nottingham, England, 197 pp.

Mares, C. and Mares, I., 2003. Improvement of Long-Range forecasting by EEOF extrapolation using an AR-MEM model. *Weather and Forecasting,* 18 (2): 311-324.

Quenouille, M., 1949. Approximation tests of correlation in time series, *Journal of the Royal Statistical Society, Series B*, 11: 18-84.

Reeve, D.E., Horrillo-Caraballo, J.M. and Magar, V., 2008. Statistical Analysis and Forecasts of Long-term Sandbank Evolution at Great Yarmouth, UK. *Estuarine, Coastal and Shelf Science*, 79 (3): 387-399.

Winant, C.D. Inman, D. L. and Nordstrom, C.E., 1975. Description of seasonal beach changes using empirical eigenfunctions, *Journal of Geophysical Research*, 80(15): 1979-1986.

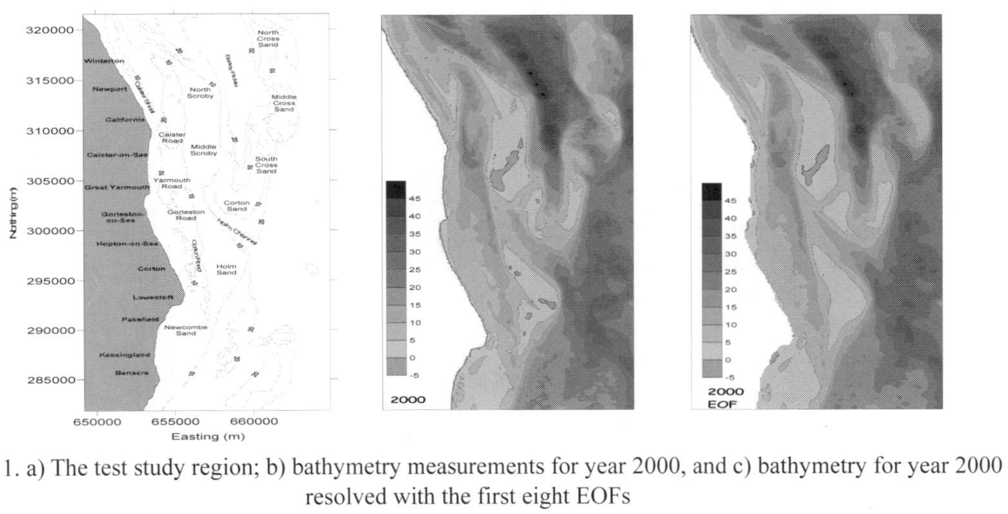

Figure 1. a) The test study region; b) bathymetry measurements for year 2000, and c) bathymetry for year 2000 resolved with the first eight EOFs

Coastal Dynamics 2009
Paper No. 87

INTERACTIONS BETWEEN SAND BANKS – THE GREAT YARMOUTH BANKS, UK

Karen Thurston[1], Chris Vincent[2] and Tony Dolphin[3]

1. Introduction

The Great Yarmouth sand banks are a group of nearshore, shore-parallel banks off the east coast of the UK, in the southern North Sea (figure 1). They are in close proximity to two busy ports (Great Yarmouth and Lowestoft) and host a wind farm (Scroby Sands). The group is highly dynamic, yet the mechanisms driving bank evolution are poorly understood.

Tidal currents along the Norfolk coast are shore-parallel and rectilinear, with predominantly southward residual currents (figure 1). The mean spring tidal range is 3.2 ms^{-1} at Winterton, and 2.5 ms^{-1} at Lowestoft. Tidal currents are strong, with spring tides of around 1.5 ms^{-1} (Dolphin et al., 2007). Park and Vincent (2007) found that tidal currents and waves act together to transport sediment in the area, with waves suspending material that is then carried along by the currents.

Dolphin et al. (2007) identify radical, cyclical reconfigurations of Newcombe Sand, the southernmost bank in the Great Yarmouth group, and Park and Vincent (2007) found similar patterns for Scroby Sands. Dolphin et al. (2007) hypothesize that these reconfigurations could be attributed to changes in the local hydrodynamics, driven by bank movement further north in the group.

In this paper, we develop this hypothesis, expanding the study area to encompass additional banks within the group and examine long-term bathymetric datasets for evidence of bank-bank interactions.

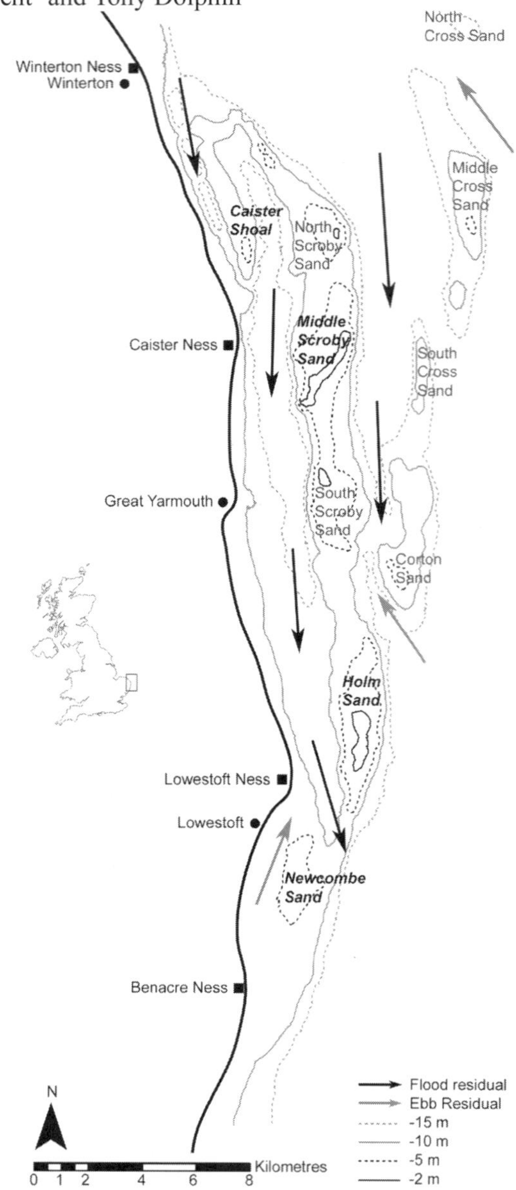

Figure 1. Great Yarmouth sand banks, Norfolk coast, UK. Bathymetry digitized from 1992 Admiralty Chart by ABP Mer. Tidal residual currents based on Robinson (1966)

2. Method

Since each bank's morphology both influences and is influenced by the local hydrodynamics, we investigate the possible ways that changes in tidal flow paths caused by one bank in the group could affect the morphology of a 'downstream' bank,

[1]School of Environmental Sciences, University of East Anglia, Norwich. NR4 7NQ. k.thurston@uea.ac.uk
[2]Address as above. c.vincent@uea.ac.uk
[3]Address as above. tony.dolphin@uea.ac.uk

resulting in cascading bank-tide-bank interactions throughout the group.

We have constructed a unique bathymetric dataset for the Great Yarmouth banks, comprising digitized UK Hydrographic Office fair sheet (survey) data from 1824 to 2002. The use of fair sheets instead of navigational charts offers improvements in temporal and spatial resolution and accuracy (Dolphin et al., 2007; Van der Wal and Pye, 2003). Charts are an amalgam of the 'most recent' surveys and frequently there are areas that were not surveyed between subsequent editions of charts, resulting in the impression that banks had not changed; using fair sheets allows us to select areas and define time periods where bathymetric changes are known. This dataset is used to study the long-term (decadal to centennial scale) evolution of individual banks within the group – Holm Sand, Middle Scroby Sand and Caister Shoal (figure 1, shown in bold type) – and look for evidence of interactions between the banks.

Time-series of bank volume, area, elevation and shape are being constructed and compared to the same parameters from Newcombe and Scroby Sands. We expect to identify time-lagged correlations in the behaviour of banks within the group. The TELEMAC suite of hydrodynamic (tidal current and waves) and sediment transport models will be used to further investigate the plausibility of scenarios suggested by correlations between banks. A number of historical bank configurations, at times when the survey data indicate that morphology was robustly known and the timing of subsequent morphological changes well constrained, will be modelled to investigate the bank-tide-bank feedback hypothesis.

3. Discussion

The detailed examination of the survey data from 1824 to 2002 will help to identify, initially qualitatively, the existence of time-lagged correlations in the behaviour of these sand banks. The hydrodynamic modelling of a variety of historical configurations will identify semi-quantitatively whether the tidal current streams are likely to have been able to drive the observed changes. Morphodynamic modelling is not yet reliable enough to allow the complete, or even partial, evolution of sand banks on the scale known to have occurred around the Great Yarmouth banks, but changes to the tidal currents (both magnitude and direction), coupled with the addition of realistic wave-stirring, will indicate areas of erosion and accretion possibly associated with changes in the flood- and ebb-dominant channels

The Great Yarmouth sand banks are highly mobile, have well-documented historical records and are ideally suited to an investigation of the processes by which sand banks evolve.

Acknowledgements

We are grateful to NERC for funding the PhD studentship of Karen Thurston.

References

Dolphin, T. J., Vincent, C. E., Coughlan, C. and Rees, J. M., 2007. Variability in sandbank behavior at decadal and annual time-scales and implications for adjacent beaches. *Journal of Coastal Research*, SI 50 (Proceedings of the 9th International Coastal Symposium): 731-737.

Park, H-B. and Vincent, C. E., 2007. Evolution of Scroby Sands in the East Anglian coast, UK. *Journal of Coastal Research*, SI 50 (Proceedings of the 9th International Coastal Symposium): 868-873.

Reeve, D., Li, B. and Thurston, N., 2001. Eigenfunction analysis of decadal fluctuations in sandbank morphology at Gt Yarmouth. *Journal of Coastal Research*, 17(2): 371-382.

Robinson, A. H. W., 1966. Residual currents in relation to shoreline evolution of the East Anglian coast. *Marine Geology*, 4(1): 57-63.

Van der Wal, D. and Pye, K., 2003. The use of historical bathymetric charts in a GIS to assess morphological change in estuaries. *The Geographical Journal*, 169(1): 21-31.

Coastal Dynamics 2009
Paper No. 88

SEDIMENT TRANSPORT MODULE FOR A B-GRID COASTAL SHELF OCEAN MODEL

Laurent O. Amoudry[1], Alejandro J. Souza[2] and Jason T. Holt3

1. Introduction

Sediment transport and morphodynamic models are essential to the coastal engineering and policy making communities. Increasingly complex models have been developed in the past decades and sediment transport is now included in coastal three-dimensional hydrodynamic models, which tend to be mostly using C-grids. In contrast to C-grid models, B-grid models such as the Proudman Oceanographic Laboratory Coastal Ocean Modelling System (POLCOMS) (e.g., Holt and James, 2001) help prevent numerical dispersion of frontal velocity features. We will discuss here the implementation of suspended sediment transport, bed load transport and morphodynamical changes within such a model.

2. The model

The physical model is based on a three-dimensional baroclinic model and solves the incompressible, hydrostatic, Boussinesq shallow water equations of motions. The governing equations are formulated in spherical polar coordinates and discretized on a B-grid in the horizontal, while S or σ coordinates are used in the vertical. The model uses a forward time, centered space, time splitting method following which the barotropic mode is solved at a CFL limited time step and the baroclinic mode at a larger time step. A full description of the model is given in Holt and James (2001).

The suspended sediment transport model solves an advection-diffusion equation for the suspended sediment concentration which follows that of Holt and James (1999). Exchange with the bed due to vertical settling and erosion is included via an additional source/sink term. Deposition is taken to be continuous and erosion is modelled following a linear dependence to the excess shear stress. The advective terms are computed using a Piecewise Parabolic Method while the eddy viscosity and diffusivity are obtained by coupling the model with GOTM (see www.gotm.net). The bed load transport rate is calculated following the approach of Soulsby and Damgaard (2005) which is modified to account for the influence of the bed slope. Finally, the sediment bed is represented by a layered structure. Only the topmost layer, the thickness of which is limited, interacts with the flow and layer properties (thickness, porosity, age) are updated so that sediment mass is conserved.

3. Model validation

We use the trench migration experiment of van Rijn (1987) as a benchmark validation test for the model. The setup consists of a 30 meter long, and 0.5 m wide straight channel in which a trench is created in the sediment bed of well-sorted sand of median diameter 160 μm. The flow is steady with a depth-averaged velocity of 0.51m/s. The channel is discretized in a 300 x 8 points mesh horizontally and 20 levels vertically. In the numerical simulations, a k-ε turbulence closure (Canuto et al., 2001) is used and the bed shear stress is calculated assuming a logarithmic near-bed velocity profile. Velocity and concentration profiles are first obtained without bathymetry changes and the trench migration is observed as a second step.

The top panel of figure 1 presents the model-data comparison of the flow velocity and concentration profiles at five location along the trench. The bottom panel of figure 1 shows the

---

1 Proudman Oceanographic Laboratory, 6 Bownlow Street, Liverpool, L3 5DA, UK. laou@pol.ac.uk
2 Proudman Oceanographic Laboratory, 6 Bownlow Street, Liverpool, L3 5DA, UK. ajso@pol.ac.uk
3 Proudman Oceanographic Laboratory, 6 Bownlow Street, Liverpool, L3 5DA, UK. jholt@pol.ac.uk

morphological changes associated with the migrating trench. In spite of under-prediction of the flow velocity in the excavated trench region, and a small over-prediction of the trench migration and final depth, we find that there is a rather good model-data agreement.

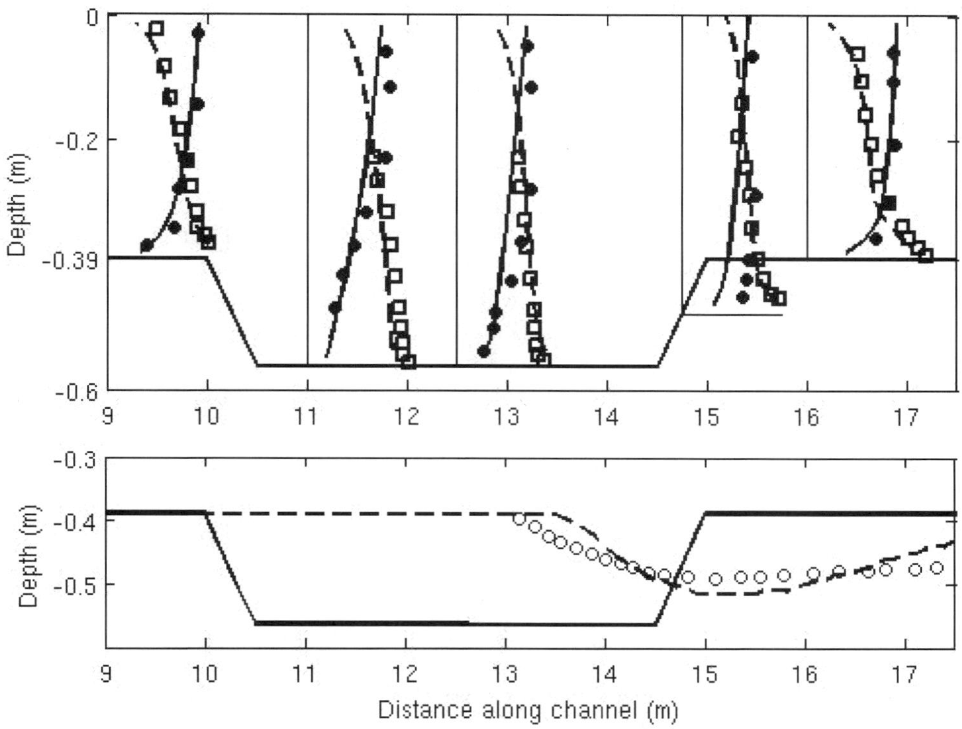

Figure 1. Comparison of velocity and concentration profiles (top) and trench migration (bottom). Top panel: measured (solid circles) and modelled (solid lines) velocities, measured (squares) and modelled (dashed lines) concentrations using a logarithmic scale. Bottom panel: initial (solid line), final measured (circles) and final modelled (dashed line) profiles.

References

Canuto V. M., A. Howard, Y. Cheng and M. S. Dubovikov, 2001. Ocean turbulence. Part I: One-point closure model – Momentum and heat vertical diffusivities, *J. Phys. Oceanography*, 31: 1413-1426.

Holt, J. T. and I. D. James, 1999. A simulation of the southern North Sea in comparison with measurements from the North Sea Project Part 2 Suspended Particulate Matter, *Continental Shelf Res.* 19: 1617-1642

Holt, J. T. and I. D. James, 2001. An S coordinate density evolving model of the northwest European continental shelf 1, Model description and density structure. *J. Geophys. Res.* 106 (C7): 14015-14034.

Soulsby, R.L. and Damgaard, J.S., 2005. Bedload sediment transport in coastal waters. *Coastal Eng.*52(8): 673-689.

van Rijn, L. C., 1987. *Mathematical Modelling of Morphological Processes in the Case of Suspended Sediment Transport*. Delft Hydraulics.

Coastal Dynamics 2009
Paper No. 89

MODELING THE EFFECT OF WAVE SHAPE ON SEDIMENT TRANSPORT

Tom J.J. van den Berg[1] and B.G. Ruessink[1]

1. Introduction

The local non-linearity of surface gravity waves propagating across a natural beach is important to sediment transport. Velocity skewness, such as found beneath shoaling waves, generally gives rise to a net transport of sediment in the direction of wave advance, except for fine sand and short wave periods, when unsteadiness in the sand flux during the wave cycle can lead to transport against the wave advance. Laboratory experiments involving acceleration-skewed waves (Watanabe and Sato, 2004), typical of the inner surf zone, have produced a net transport in the direction of maximum acceleration, with the magnitude of the transport increasing with the degree of acceleration skewness. Little is known about the net transport beneath waves that possess both velocity and acceleration skewness. In the present study, a 1DV single-phase model based on Reynolds-averaged Navier-Stokes and advection-diffusion equations (Ribberink et al., 2008) is used to explore the effect of wave non-linearity, velocity and acceleration skewness, and sediment size on net transport rates.

2. Methodology

We parameterize the free-stream orbital motion u_∞ of waves with combined velocity and acceleration skewness as

$$u_\infty(t) = U_w f\left[\sin(\omega t) + \frac{r\sin\Phi}{1+\sqrt{1-r^2}}\right] / \left[1 - r\cos(\omega t + \Phi)\right], \tag{1}$$

where t is time, U_w is the velocity amplitude, $\omega = 2\pi/T$ with T is the wave period, r is a non-linearity measure, Φ is a phase, and f is a normalization factor such that U_w equals $\sqrt{2}$ times the standard deviation of $u_\infty(t)$. For $0 < r < 1$, $u_\infty(t)$ has velocity skewness with $\Phi = -\pi/2$, acceleration skewness with $\Phi = 0$, and both velocity and acceleration skewness with $-\pi/2 < \Phi < 0$.

Time series of $u_\infty(t)$ with variable r [0 – 0.85] and Φ [$-\pi/2$ – 0] with $U_w = 1.2$ m/s and $T = 7$s were used as input in the Reynolds-averaged 1DV Navier-Stokes equations with k-ε turbulence closure. Combined with the advection-diffusion equation for sediment concentration, our model numerically simulates the time-dependent horizontal velocities, sediment concentrations, and suspended sediment fluxes by waves in flat sea-bed conditions ('sheet flow').

Prior to all simulations, we tested the model on existing sediment transport data measured in wave tunnels under full-scale velocity-skewed (Ribberink et al., 2008) and acceleration-skewed (Watanabe and Sato, 2004) data. We found the model to reproduce the net transport rates accurately (skill ≈ 0.8), including trends in net transport rates related to grain size, wave period, and acceleration skewness.

3. Results

Figs. 1a-c represent the modeled depth-integrated and wave-averaged flux q as a function of r and Φ using a median grain size d_{50} = 120, 200 and 300 μm. The dependence of q on r and Φ differs remarkably between the 3 grain sizes. For d_{50} = 300 μm (Fig. 1c) we see that q increases with increasing r for all Φ, especially for waves dominated by velocity skewness ($\Phi < -\pi/4$). For d_{50} = 200 μm (Fig. 1b), q is still positive ('onshore') for all r and Φ but is now largest for acceleration-skewed waves. Finally, for d_{50} = 120 μm (Fig. 1a), q is negative ('offshore) for all r, except for acceleration-skewed waves ($\Phi > -\pi/8$), where q is even larger than for d_{50} = 200 and 300 μm.

[1] Dept of Physical Geography, Faculty of Geosciences, Utrecht University, P.O. Box 80.115, 3508 TC Utrecht, the Netherlands. g.ruessink@geo.uu.nl

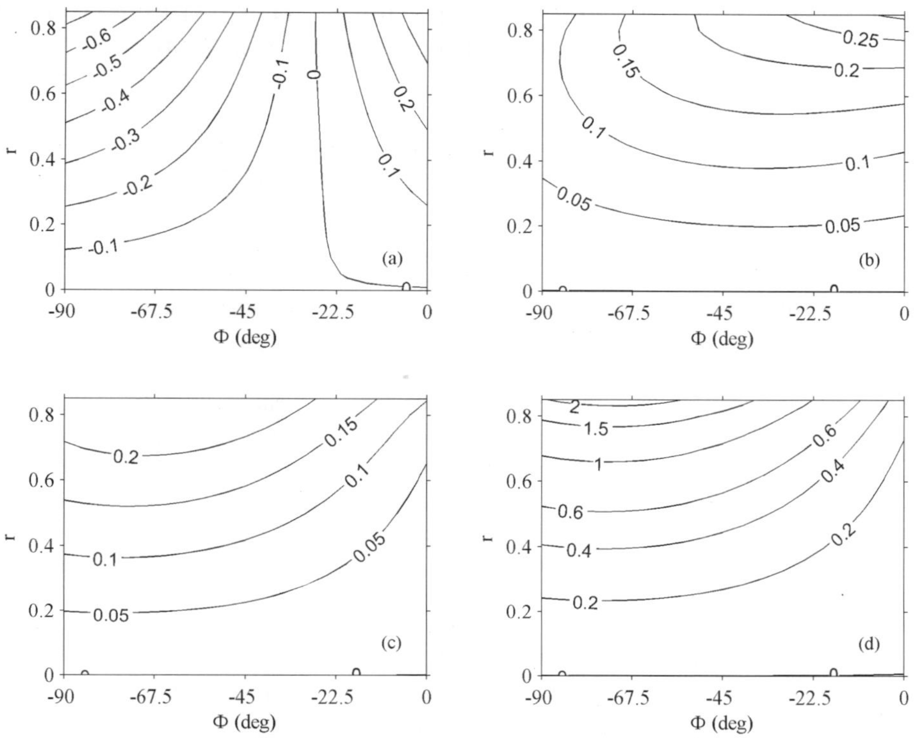

Figure 1. Contour plots of depth-integrated and wave-averaged sediment transport q (unit: kg/m/s) for d_{50} = (a) 120, (b) 200 and (c) 300 μm ($T = 7$ s, $U_w = 1.2$ m/s) as a function of the non-linearity measure r and phase Φ. (d) shows the parameter P, see Eq. (2). P is identical for all 3 grain sizes.

To assess the importance of unsteadiness to q, we computed the parameter P (Figure 1d),

$$P = \left(\int_0^T \Gamma\{\tau\} \mid \tau \mid^{3/2} \mathrm{d}t \Big/ \int_0^T \Gamma\{-\tau\} \mid \tau \mid^{3/2} \mathrm{d}t \right) - 1, \tag{2}$$

where Γ is the heaviside function ($\Gamma\{\tau\} = 1$, when $\tau > 0$ and 0 otherwise) and τ is the bed shear stress. If the dependence of q and P on r and Φ are similar, then the intra-wave depth-averaged flux responds (almost) instantaneously to the time-varying bed shear stress, and unsteady effects are thus small. For d_{50} = 300 μm this is indeed the case, compare Fig. 1d to 1c. For smaller d_{50} the functional dependence of q and P on r and Φ start to differ increasingly, in other words, unsteady effects become increasingly important. For waves dominated by velocity skewness ($\Phi < -\pi/4$), sand stirred during the positive flow phase remains suspended well into the negative flow phase, thereby reducing q and ultimately leading to negative q. For waves dominated by acceleration skewness, sand stirred during the positive flow phase has largely settled before flow reversal. However, sand stirred during the negative flow phase persists into the positive flow phase, thereby enhancing the net positive q. Our simulations suggest that unsteady effects may be more relevant to the prediction of wave-induced sediment transports than previously envisaged. We are currently working on extending our simulations to a wider U_w and T range.

References

Ribberink, J.S., Van der Werf, J.J., O'Donoghue, T. and Hassan, W.N.M., 2008. Sand motion induced by oscillatory flows: sheet flow and vortex ripples. *Journal of Turbulence*, 9(20): 1-32.

Watanabe, A. and Sato, S., 2004. A sheet-flow transport rate formula for asymmetric, forward-leaning waves and currents. *Proc. 29th Int. Conf. on Coastal Eng.*, Lisbon, pp. 1703-1714.

Coastal Dynamics 2009
Paper No. 90

NUMERICAL SIMULATION FOR VERTICAL SORTING OF GRADED PARTICLES IN SHEETFLOW BY TWO-PHASE TURBULENT FLOW MODEL

Eiji Harada[1] and Hitoshi Gotoh[2]

1. Introduction

As the armor coat formation process occurring in graded sediment influences bed load transportation significantly, understanding of this behavior is indispensable for a beach transformation prediction. In this study, the solid/liquid two-phase flow model, in which a distinct-element-method type solid-phase model is installed, is applied to examine coastal sediment transport mechanism from a viewpoint in sand particle scale. A numerical simulation for a vertical grading process of sediment in oscillatory flow is carried out. Although several previous works of this kind of simulation have been conducted, one-way type models in which interaction between solid and liquid phases is not considered had been mostly applied. Simulation in consideration of interaction between solid and liquid phases is carried out herein and to examine the difference in two-way type and one-way type models gives in a reproduction of grading process in a sediment transport process in sheetflow regime.

2. Numerical Model

In this study, process of unsteady graded sediment transport under oscillatory flow condition in closed-duct is tracked by two dimensional movable bed simulator (Gotoh and Sakai, 1997). In the movable bed simulator, interaction between particles is described by a distinct element method, and motion of an individual sand particle is evaluated by the translational and rotational momentum equations. The model constants are set in each model so that relation of a bottom shear stress and a sediment transport rate observed in previous experiments is reproduced in the transportation of uniform sand under unidirectional flow. An unsteady k-ε turbulence model is used for a velocity field, and wall surface boundary condition is given by the law of the wall, here roughness surface assumes mean particle diameter. An oscillatory flow condition has period T=5.0s, Shields number of ψ=0.3 and it is confirmed that the appearance domain where a sheetflow excels in this flow condition, separately.

3. Results

A snapshot of a grading process of the one-way type model and the two-way type model and number density distribution of particle size d_l are shown in Fig.1. Three particle size mixture: d_l=0.4cm, d_2=0.85cm, d_3=1.0cm are used, and specific density of particle is 1.318. Period boundary condition is set in the x-axis direction and the Lagrangian mass-less marker particles are arranged to track a flow pattern. From decrease of number density of smallest particle size d_l around the neighborhood of exchange layer, development of a sorting in an oscillatory flow is confirmed in both models clearly. In addition, a prompted sorting is shown in the two-way type model in comparison with a the one-way type model. The distribution of vertical component velocity by Eulerian observation in x/d_3=30.0 is shown in Fig.2. In the one-way type model, velocity in the vertical direction is quite small, and time-dependent change of velocity is small, too. On the other hand, in the two-way type model, significant velocity with large fluctuation in time is shown. And this will contribute to activation of motion of

[1]Dept of Urban and Environmental Eng., Kyoto University, Katsura Campus, Nishikyo-ku, Kyoto 615-8540, Japan. harada@mbox.kudpc.kyoto-u.ac.jp
[2]Dept of Urban and Environmental Eng., Kyoto University, Katsura Campus, Nishikyo-ku, Kyoto 615-8540, Japan. gotoh@mbox.kudpc.kyoto-u.ac.jp

particles in vertical direction greatly. This should be the reason why the two-way type model promotes a sediment sorting. If presentation at the conference will be accepted, internal structures in a development process of a sorting will be shown in detail.

References

Gotoh, H. and Sakai, T., 1997. Numerical Simulation of Sheetflow as Granular Material, *Jour. of Waterway, Port, Coastal, and Osean Engrg., ASCE*, Vol.123, No.6, pp.329-336.

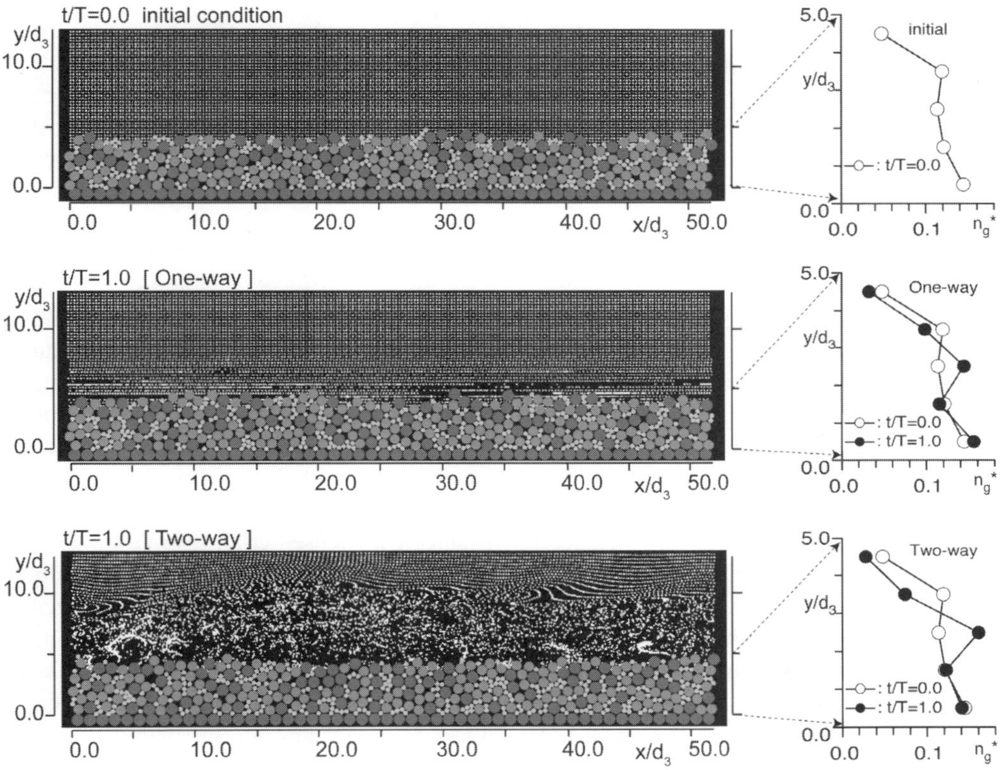

Figure 1. Snapshots and distribution of number density of d_3 at t/T=1.0

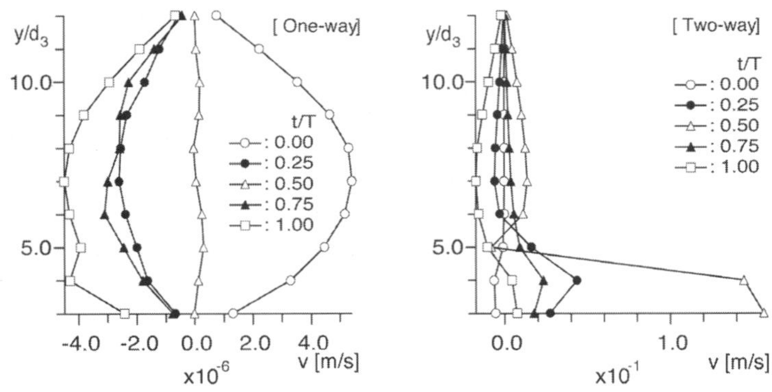

Figure 2. Velocity distribution of vertical component observed in x/d_3=30.0

Coastal Dynamics 2009
Paper No. 91

THE ROLE OF THE DEPTH-AVERAGED CONCENTRATION IN COASTAL MORPHODYNAMICS

Daniel Calvete[1], Albert Falques[1], Huib E. de Swart[2], Nicholas Dodd[3],
Francesca Ribas[1], Roland Garnier[1]

1. Introduction

There are many beaches where the irregularities in its topography along the coast are small at the scales comparable to the surf zone width. Their cross-shore profile can be assumed to be approximately uniform along the coast and hence, alongshore gradients can be safely ignored. They are called 2D beaches. On the contrary, other beaches exhibit very clear morphological patterns in plan-view at those lengthscales. Their hydrodynamics and their morphodynamics is profoundly influenced by these patterns so that the assumption of alongshore uniformity is by no means applicable any more. These beaches are called 3D beaches. These large scale patterns can range from quite irregular to fairly periodic in space along the coast. In the latter case they are referred to as rhythmic morphological patterns.

The most common rhythmic patterns are crescentic bars, shore-transverse bars, shore-oblique bars, finger bars or beach cusps are distinct morphological pattern pertaining to 3D beaches (Carter, 1988; Short, 1999). On the continental shelf, shoreface-connected ridges, tidal sands bancs and sand waves are well known rhythmic morphological patterns. A very plausible explanation is that their origin lies in positive feedback mechanisms between the bathymetry and the hydrodynamics which render unstable the 2D morphology. These inherent instabilities are associated to self-organization processes in the coupling between flow and morphology that eventually lead to those patterns. In recent years there has been much research in this line, see for instance Christensen et al. (1994), Caballeria et al. (2002), Damgaard et al. (2002), Reniers et al. (2004), Garnier et al. (2005), Vis-Star et at. (2008).

In this contribution a general discussion is presented on the development of self-organized coastal morphodynamic patterns which are due to the joint action of gradients in the depth-integrated concentration (also called potential stirring) and the flow perturbations produced by the bedforms. This is done in the context of a depth-averaged shallow water model. In spite of its remarkable simplicity, this analysis has proven to be a powerful tool to get insight into the underlaying feedback mechanisms between the morphology and the hydrodynamics.

2. The sediment stirring function and the bottom evolution equation

On many of these studies on coastal morphodynamic stability the advective part of the sediment transport (bedload, suspended load or total load transport) is parameterized to be directed towards the depth-averaged current:

$$\vec{q} = \alpha \vec{v}$$

where α depends nonlinearly on the hydrodynamics variables and sediment properties. Moreover, the hydrodynamics is modeled by the depth-averaged shallow water equations and the evolution of the bottom can be derived just considering the sediment conservation, i.e.

$$\frac{\partial h}{\partial t} + \vec{\nabla} \cdot \vec{q} = 0$$

[1]Dept Fisica Aplicada, Universitat Politecnica de Catalunya, Spain. calvete@fa.upc.edu
[2]Institute for Marine and Atmospheric Research, Utrecht University, The Netherlands. h.e.deswart@uu.nl
[3]School of Civil Engineering, University of Nottingham, United Kingdom, Nick.dodd@nottingham.ac.uk

Combing sediment and mass conservation equation yields to the bottom evolution equation (BEE):

$$\frac{\partial h}{\partial t} = -D\vec{v} \cdot \vec{\nabla} C \quad \text{where } C = \frac{\alpha}{D}$$

Then, C can be interpreted as the depth averaged concentration and some authors have named as potential stirring function (Falques et al, 2000). This equation gives the time evolution of the bed level at any particular location as a function of the water depth, D, the depth averaged current, v, and the depth averaged concentration, C, potential stirring gradient. In a morphodynamic model the BEE can be use instead on the sediment conservation, together with the hydrodynamic equations. However our aim is to present a powerful use of the BEE to interpret physically under which conditions accretion or erosion will occur and to understand the feedbacks between topography and water motions.

If by whatever reason the current and the distribution of the depth averaged concentration are known (for instance, from field observations, from numerical simulations or just qualitatively from physical reasons), the morphodynamic effect of such a current can be easily inferred. Indeed, according to BEE, $\vec{v} \cdot \vec{\nabla} C > 0$, will imply $\partial h / \partial t < 0$ and vice versa. In other words, as is sketched in the figure, any current running with the gradient in potential stirring will produce erosion and any current running against this gradient will cause accretion.

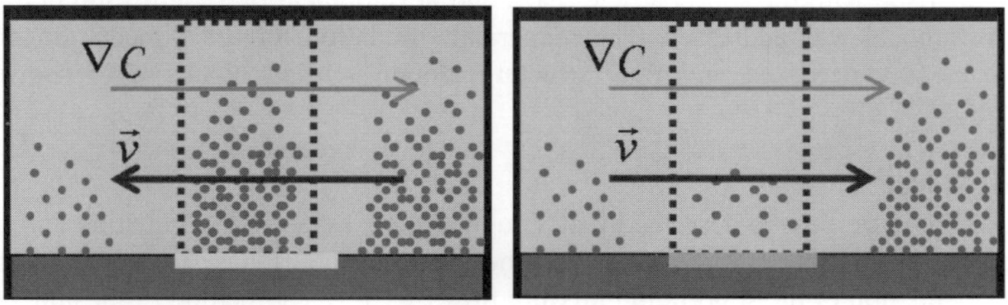

Figure. Sketch of two circustances under deposition (right) or erosion (left) occurs.

As an illustration, the formation of shoreface-connected ridges, nearshore bedforms and beach cusps will be discussed.

References

Caballeria, M., Coco, G., Falques, A., Huntley, D.A., 2002. *Self-organization mechanisms for the formation of nearshore crescentic and transverse sand bars.* J. Fluid Mech. 465, 379–410.

Carter, R.W.G., 1988. *Coastal Environments.* Academic Press.

Christensen, E., Deigaard, R., Fredsoe, J., 1994. *Sea bed stability on a long straight coast.* In: Coastal Eng. 1994, vol. 4. Am. Soc. of Civ. Eng., pp. 1865–1879.

Damgaard, J., Dodd, N., Hall, L., Chesher, T., 2002. *Morphodynamic modelling of rip channel growth.* Coastal Eng. 45, 199–221.

Falques, A., Coco, G., Huntley, D.A., 2000. *A mechanism for the generation of wave-driven rhythmic patterns in the surf zone.* J. Geophys. Res. 105 (C10), 24071–24088.

Garnier, R., Calvete, D., Falques, A., Dodd, N., 2008. *Modelling the formation and the long-term behavior of rip channel systems from the deformation of a longshore bar.* J. Geophys. Res. 113(C7), C07053, doi: 10.1029/2007JC004632

Reniers, A.J.H.M., Roelvink, J.A., Thornton, E.B., 2004. *Morphodynamic modeling of an embayed beach under wave group forcing.* J. Geophys. Res. 109 (C01030), doi:10.1029/2002JC001586.

Short, A. D., 1999. *Handbook of Beach and Shoreface Morphodynamics.* Willey, Chichester.

Vis-Star, N.C., de Swart, H.E., Calvete, D., 2008. *Patch behaviour and predictability properties of modelled finite-amplitude sand ridges on the inner shelf.* Nonlin. Process. Geophys. 15(6), 943-955

Coastal Dynamics 2009
Paper No. 92

A SEDIMENTATION PROCESS OF A GROUP OF BLOCKS BY DNS

Eiji Harada[1], Hitoshi Gotoh[2] and Naoki Tsuruta[3]

1. Introduction

For mound construction in relatively deep water, such as large-scale landfill or an artificial upwelling current mound, huge amount of sand and blocks are dumped into water from a sea surface to save a construction cost. If a precise prediction of behavior of sand and group of blocks dumped into water is enabled, a cost of construction can be reduced significantly. As for an effective tool which can evaluate solid and liquid phases interaction, solid/liquid two-phase flow model have been used in previous numerical simulations. However, computational grid in the flow field is larger than that of solid particles in the most of previous model, therefore, interaction between solid and liquid phases has been evaluated roughly by drag coefficient. Therefore flow structures around particles such as wake had not been resolved in detail, and an experimental result of a sedimentation process was not always reproduced well. In this study, a considerable fine computational grid in the flow field is applied in comparison with the solid particle scale, and development of a high-resolution solid/liquid two-phase flow model was addressed. The numerical simulation for a sedimentation process of a group of blocks is carried out and flow structure around settling blocks were shown in detail computationally.

2. Numerical Model

The solid/liquid mixture flow is described with coupling the Navier-Stokes equation and the Distinct Element Method. In this model, the motion of blocks in the liquid phase is evaluated as following procedure: 1) evaluation of the solid/liquid mixture flow by CIP-CUP method (Yabe and Wang, 1991), 2) calculation of flow force acting on solid body, 3) tracking of solid body motion by the DEM. For verification of basic performance of model, numerical simulation for flow structure around the column in the uniform flow is performed and confirmed that numerical results shows good agreement with previous experimental results. And numerical simulation for group of blocks behavior in the water is performed and flow structure around blocks is shown in detail. In addition, each block consists of two elements (refer to Fig.2) and motion of each block element is modified as showing rigid body by using the passively moving solid model proposed by Koshizuka *et al.* (1998) after the calculation of DEM.

3. Results

It is reported that three dimensional flow structure is shown significantly over particle Reynolds number Re_p=200. In this study, model is developed for two dimension, so simulation is performed under the Re_p=200. Fig.1 shows the calculation domain and the relation between Re_p and coefficient C_D, Strouhal number St. In the case in which the diameter of the column is 8 times of the calculation grid, good performance of model have been confirmed. Fig.2 shows snapshots of sedimentation process of a group of blocks. The complicated flow structure around blocks is confirmed in detail. In the full paper, sedimentation process of blocks with dynamic posture, like a repetition of tumbling

[1]Dept of Urban and Environmental Eng., Kyoto University, Katsura Campus, Nishikyo-ku, Kyoto 615-8540, Japan. harada@mbox.kudpc.kyoto-u.ac.jp
[2]Dept of Urban and Environmental Eng., Kyoto University, Katsura Campus, Nishikyo-ku, Kyoto 615-8540, Japan. gotoh@mbox.kudpc.kyoto-u.ac.jp
[3]Undergraduate School of Global Eng., Kyoto University, Katsura Campus, Nishikyo-ku, Kyoto 615-8540, Japan. tsuruta@t05.mbox.media.kyoto-u.ac.jp

depending on the initial condition will be also shown.

Acknowledgements

We thank REFEC (*Research Foundation for the Electrotechnology of Chubu*) for supporting a part of present study.

References

Koshizuka, S., Nobe, A. and Oka, Y., 1998. Numerical analysis of breaking waves using the moving particle sem i-implicit method, *Int.J.Numer.Meth.Fluids*, Vol.26, pp.751-769.

Yabe, T. and Wang, P. Y., 1991. Unified Numerical Procedure for Compressible and Incompressible Fluid, *J. Phys. Soc. Japan*, Vol. 60, No. 7, pp. 2105-2108.

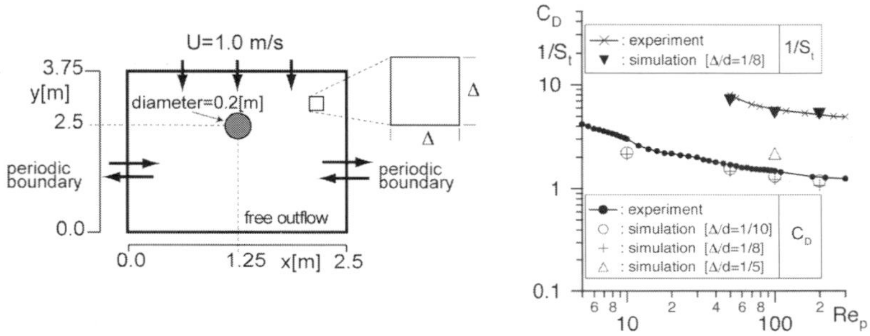

Figure 1. Particle Reynolds number vs. coefficient of drag force & Strouhal number

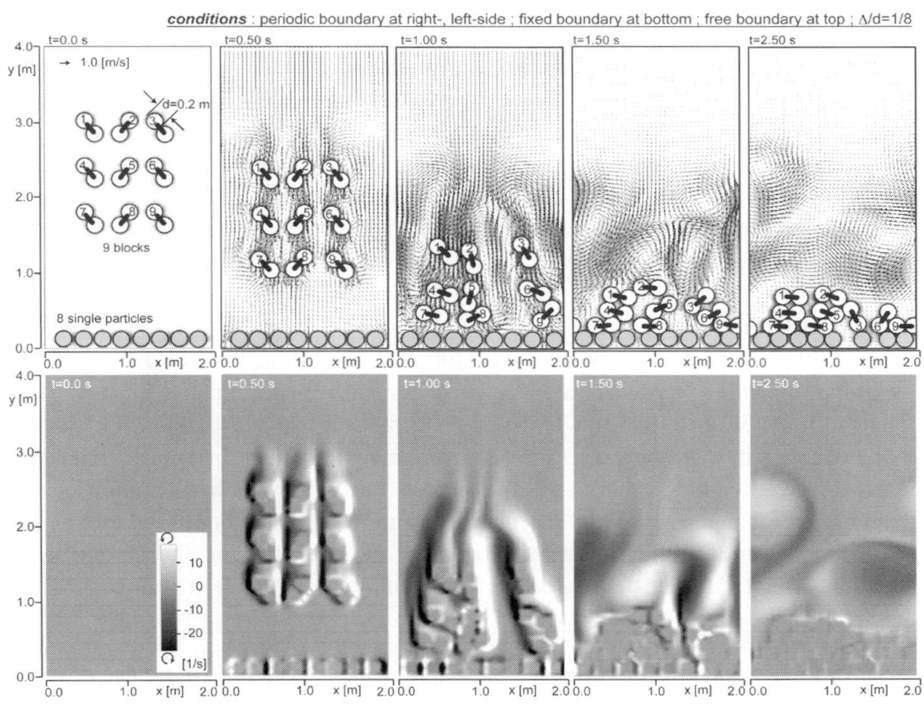

Figure 2. Sedimentation process of blocks (upper: velocity vector, lower: vorticity)

Coastal Dynamics 2009
Paper No. 93

NUMERICAL SIMULATION OF RAPID EROSION OF SEISHO COAST TRIGGERED BY STORM WAVES DURING TYPHOON 0709

Masumi Serizawa[1], Takaaki Uda[2], Kenji Suzuki[3], Shogo Maruyama[3], Hiroyuki Takano[3]
Toshiro San-nami[1] and Toshinori Ishikawa[2]

1. Introduction

With coastal development, public facilities such as highways have often been constructed along the coast, because of the ease of acquiring the land for construction. In recent years, sandy beaches have been rapidly disappearing owing to the decrease in sediment supplied from rivers, resulting in enhancing the vulnerability of those facilities. The Seisho Bypass has been built since the 1960s along the Seisho coast facing Sagami Bay. On September 7, 2007, this highway was severely damaged by erosion due to storm waves associated with Typhoon 0709, and the highway was forced to close for repairs. The seawall protecting the highway collapsed over a length of 1.1 km near the Ohiso IC. This study aims at predicting rapid erosion of this coast due to storm waves associated with T0709.

2. Beach Changes

Long-term shoreline changes were investigated using aerial photographs. The longshore and cross-shore scales of the aerial photographs were distorted so that the aspect ratio was 1/5, and images were produced (**Fig. 1**). In 1947 (a), the shoreline smoothly extended with an approximately 80-m-wide foreshore. In 1999 (b), the Seisho Bypass had already been constructed along the coastline, and the Ninomiya fishing port was built. In 2006 (c), the shoreline east of the port retreated. On September 13, 2007(d), the foreshore disappeared in extensive area, whereas on the east side of the fishing port, the foreshore was wider and the shoreline had retreated on the west side compared with its location in 2006. By November 2007 (e), the foreshore had widened over an extensive area, but the sandy beach in front of the damaged area did not recover. **Figure 2** shows the shoreline changes. The shoreline markedly retreated until September 13 2007, but it had partly recovered by November 21, 2007. The change in longitudinal profiles was also investigated before and after the typhoon. **Figure 3** shows the profile change along transect X=3.2 km. The foreshore was eroded until October 2007 after the storm, and sand accumulated between 4 and 9 m depth. Then, shoreward sand transport occurred under the normal wave conditions resulting in the formation of a berm.

3. Numerical Simulation

Fukuhama *et al.* (2008, ICCE) developed a model for predicting short-term changes in the foreshore slope and shoreline position due to cross-shore sand transport in response to changes in the wave intensity. They formulated their concept under the assumption that the equilibrium slope varies in response to the changes in wave height. If the equilibrium slope of sand becomes gentler in response to the increase in wave height, offshore sand transport is generated. We applied their concept to the analysis of 3-D beach changes caused by storm waves and normal waves after storm. Assuming the parallel contours as the initial condition using expanded coordinate system, and the breaker height of 3 m (T=9s) with $\alpha_b=10^\circ$ incident counterclockwise, numerical simulation was carried out. In the

[1] Coastal Engineering Laboratory Co., Ltd., 301, 1-22 Wakaba, Shinjuku, Tokyo 160-0011, Japan. coastseri@nifty.com
[2] Public Works Research Center, 1-6-4 Taito, Taito, Tokyo 110-0016, Japan. uda@pwrc.or.jp
[3] Keihin Office of River, Kanto Regional Development Bureau, Ministry of Land, Infrastructure and Transport, 2-18-1 Tsurumi Chuo, Tsurumi-ku, Yokohama City, Kanagawa Prefecture 230-0051, Japan.

calculation, the initial foreshore slope is 1/8 in the zone shallower than -4 m, and 1/30 offshore of - 4 m before the wave attack, and during storm waves the foreshore slope was assumed to be 1/30 which is the same slope in the offshore zone. After the storm event, the incidence of normal waves with the breaker height of 0.6 m assumed to continue for 83 days. The depth of closure is 9 m and the berm height is 3 m. **Figure 4** shows the results. The initial beach was eroded rapidly due to seaward sand transport and sand deposited in the offshore zone, but due to the wave incidence of normal waves after storm, beach was recovered except the damaged area of Seisho Bypass.

4. Conclusions

Applying the Fukuhama *et al.*'s concept, a model for predicting rapid erosion of the Seisho coast due to seaward sand transport by storm waves and the reformation of a berm due to shoreward sand transport after storm was developed. Predicted results are in good agreement with the measured beach changes.

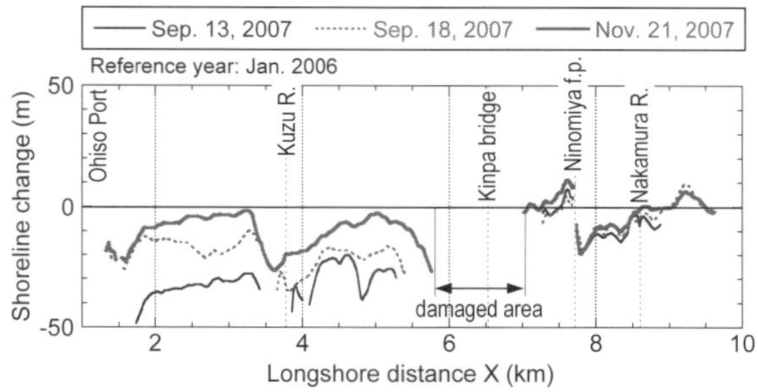

Figure 2. Short-term shoreline changes before and after typhoon

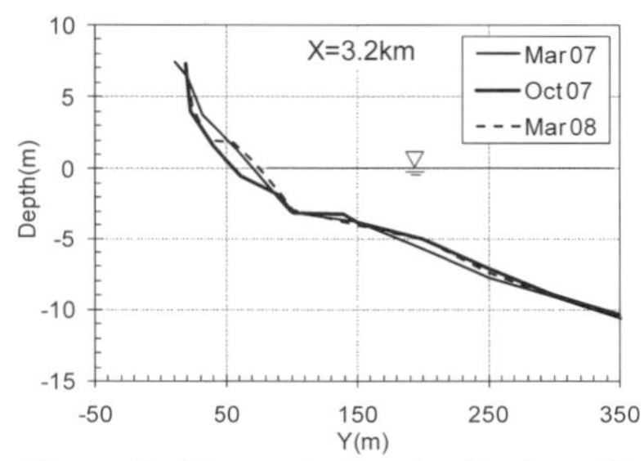

Figure 3. Change in longitudinal profiles

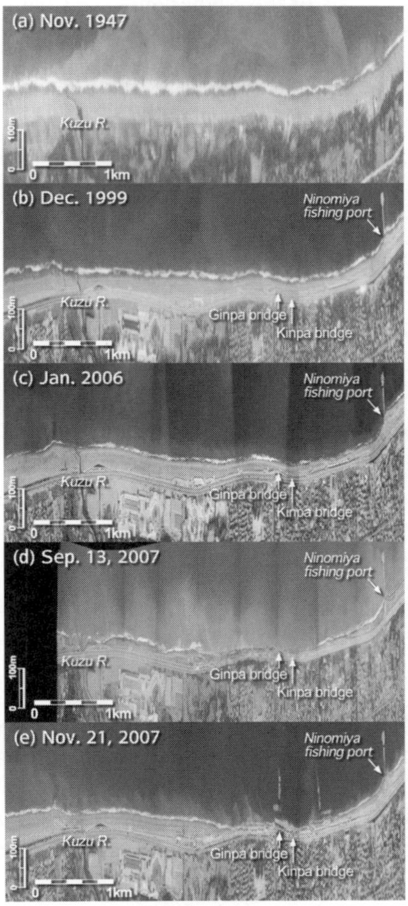

Figure 1. Comparison of coastal changes in damaged area using distorted aerial photographs

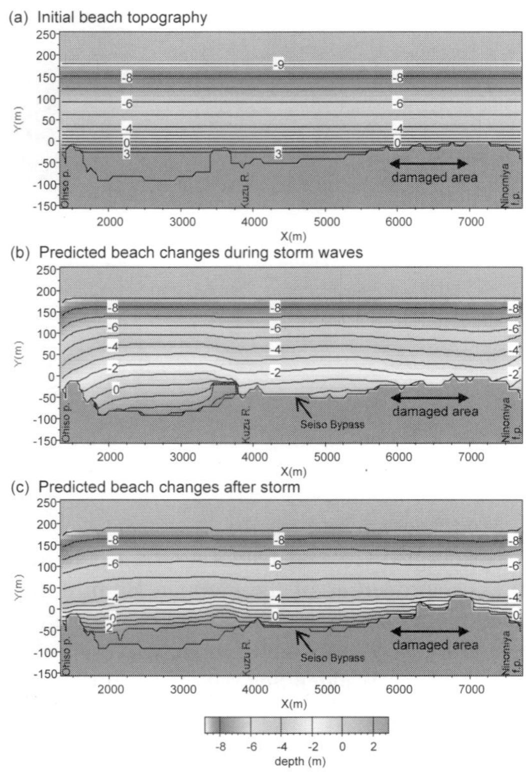

Figure 4. Predicted beach changes

Coastal Dynamics 2009
Paper No. 94

HYDRO- AND SEDIMENT- DYNAMICS IN THE GERMAN BIGHT: A FOCUS ON OBSERVATIONS AND NUMERICAL MODELLING

Joanna Staneva[1]

1. Introduction

This work deals with analysis of hydrographic observations and results of numerical simulations. The data base includes ADCP observations, continuous measurements on data stations and satellite data originating from the Medium Resolution Imaging Spectrometer (MERIS) onboard the ESA satellite ENVISAT with a spatial resolution of 300 m. Numerical simulations use nested models with horizontal resolutions ranging from 1 km in the German Bight to 200 m in the East Frisian Wadden Sea coupled with a suspended matter transport model. Modern satellite observations have now a comparable horizontal resolution with high-resolution numerical model of the entire area of the East Frisian Wadden Sea allowing to describe and validate new and so far unknown patterns of sediment distribution. The two data sets are consistent and reveal an oscillatory behaviour of sediment pools to the north of the back-barrier basins and clear propagation patterns of tidally driven suspended particulate matter outflow into the North Sea. The good agreement between observations and simulations is convincing evidence that the model simulates the basic dynamics and sediment transport processes, which motivates its further use in hindcasting, as well as in the initial steps towards forecasting circulation and sediment dynamics in the coastal zone.

The German Bight (Figure 1a), which is bordered by the coasts of the Netherlands, Germany and Denmark, is situated in the south eastern corner of the North Sea. The East Frisian Wadden Sea is on e of the shallowest areas of the German Bight, being characterised by a series of barrier islands, each 5-10 km long in the east-west direction and 2-3 km wide (Figure 1b).

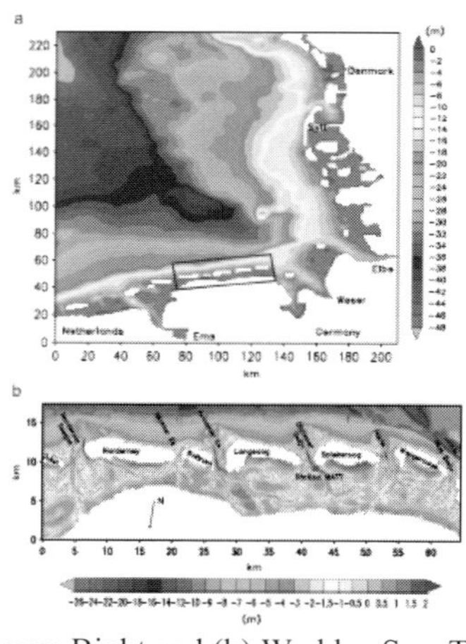

Figure 1. Topography of the (a) German Bight and (b) Wadden Sea. The -1.5 m isobath plotted in the Wadden Sea map contours approximately the dry area under low water.

2. Model set-up and results

A nested-grid General Estuarine Transport Model (GETM) is used to study the circulation and the physical properties of the German Bight during the last decade. The nested-grid model consists of three model configurations: a coarse-resolution North Sea-Baltic Sea (~3 nm) outer model, a fine-resolution (~ 900 m) inner model covering the German Bight and a very fine-resolution (~ 200 m) model for the Wadden Sea region resolving the barrier islands and the tidal flats. The model system is

[1]Institute for Coastal Research, GKSS Research Centre, Max-Planck-Straße 1,
D-21502 Geesthacht, Germany, e-mail: Joanna.Staneva@gkss.de

forced by: 1) the atmospheric fluxes estimated by the bulk formulation using 6-hourly ECMWF re-analyses data from 1986 to 2008, 2) river run-off and 3) time varying lateral boundary conditions of sea surface elevations as well as temperature and salinity.

The importance of prescribing correct fresh water forcing becomes obvious from Fig. 8, which describes the horizontal surface salinity distribution for two particular cases during high water and low water on 05.03.2005. The largest gradients are simulated in the areas of river mouths (Elbe, Weser and Ems). Protrusions of diluted waters are better seen during low water both deep in the German Bight, as well as over the tidal flats (water from Weser River into the tidal basins, or in case with eastward wind from Ems River into the tidal basins).

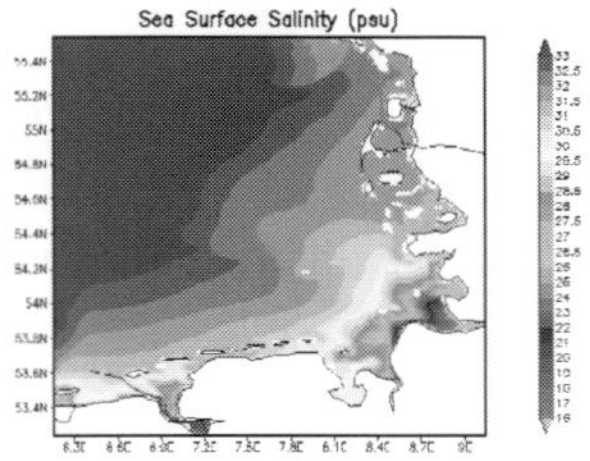

Figure 2: Surface salinity (psu) in the German Bight at 05.03.2005

The general pattern of Wadden Sea circulation (Figure 3) is characterized by extreme velocities in the tidal inlets, westward currents in front of the barrier islands during ebb, and eastward currents during flood.

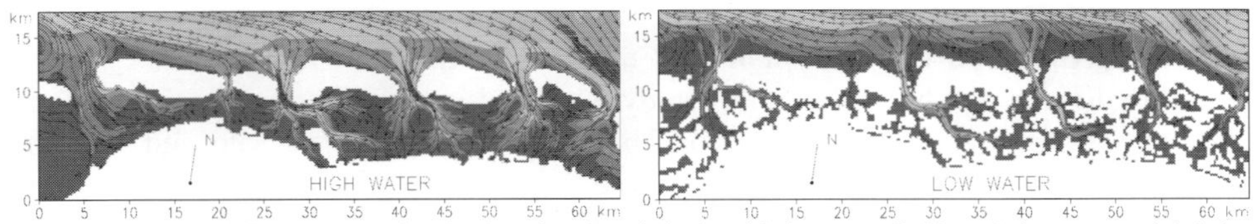

Figure 3: Vertically integrated velocity simulated during high water (left) and low water (right)

Bellow we provide a short demonstration of consistence between the two sediment data types analysed in the present study (MERIS observations and numerical simulations). We are using a cloud-free MERIS scene for image of the 3 Jun 2004 (Figure 4). For this period the wind velocity was low for at least the last 12 hours (3.5 m/s) and had an east/north-east direction. The influence of the wind is considered negligible at such low wind velocities. The chosen scene is a high water scene ensuring a water coverage of the whole area. The high water situation together with the above-mentioned low Secchi depths also ensures that MERIS does not see any sea bottom shining through the water column, except perhaps at locations very close to the coasts. The corresponding simulated SPM distribution agrees perfectly with the observations.

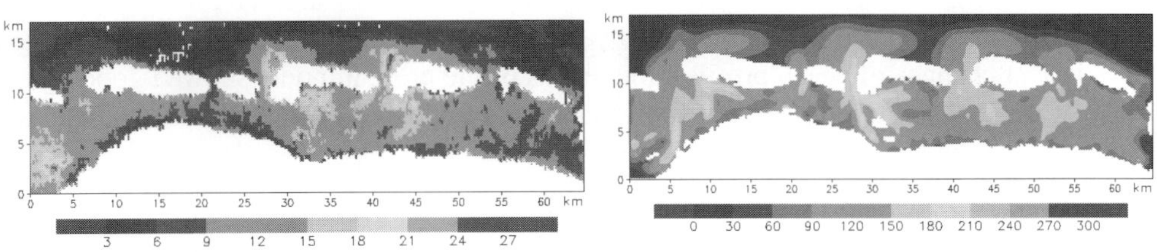

Figure 4. Surface sediment distribution during high water from MERIS observations potted in the model grid (left) and numerical simulations (right)

Coastal Dynamics 2009
Paper No. 95

NUMERICAL CALCULATION OF THE SEDIMENT TRANSPORTS ALONG THE LAGOONAL COAST ON MAJURO ATOLL

Daisaku Sato[1] and Hiromune Yokoki[2]

1. Introduction

Atoll islands are low-lying and consist of the biological organisms such as corals and foraminifers (Yokoki et al, 2004). These islands have high vulnerability to the coastal erosion and its acceleration due to the projected sea-level rise. For the sustainable use of the islands, implementation of the island conservation plan is indispensable. In this study, toward the development of the base model for the future land change simulation, we improved the sediment transport model, which was constructed by Sato and Yokoki (2008), by considering the interaction between waves and land changes. The accuracy of the calculated results was confirmed by the subsequent changes in beach profile data.

2. Location and Geomorphology of Majuro atoll

Majuro atoll is located at 171°22'E, 7°6'N and is about 40km wide in longitude and 9.7km wide in latitude (Fig 1(a)). Rita and Uliga districts, located in the eastern side of the atoll, are civilized area. Laura islet, located in the western side of the atoll, is traditional area. In the northern part, the atoll has a large passage. In this study, we focused on the sediment changes in Laura islet because we collected the beach profile data, which were measured by the SOPAC (1998) in 1997 and by the authors in 2007. Figure 1(b) shows the location of measurement lines in the Laura islet.

3. Calculation Method and Verification of the Calculated Sediment Changes

Calculations are divided into two parts in this study. The first part is the calculation of wave field and the second part is the calculation of the longshore sediment transports. Sato and Yokoki (2008) computed these two parts separately, then the extreme large erosion and accretion rate were calculated where the shoreline is complicated. In order to improve the model for long-term estimation, we added the interaction between waves and morphological changes in the longshore sediment transport model by iterating these two calculation parts mutually every month. Wave fields in Majuro atoll were calculated by the Simulating WAves Neashore (SWAN) model (Booij et al, 1999). And the longshore sediment transports were calculated by the CERC equation (Komar and Inman, 1970). Calculation domain was 45km wide in longitude and 25km wide in latitude and the rectangular grids of 32m width were set in it.

To confirm the accuracy of the calculated sediment changes, they were compared with the sediment volume change calculated by the differences of the beach profile data between 1997 and 2007 at each measurement line (Fig 2). For the comparison, our previous results that were calculated without interaction between waves and sediment transports (Sato and Yokoki, 2008) are also plotted in the figure. While the high accretion or erosion were computed in Sato and Yokoki (2008) where the shoreline is complicated, the sediment changes by the present model became smoother. In comparison with the sediment changes of beach profiles, the results of the present model showed good agreement. Consequently, the longshore sediment transport model was improved satisfactory to estimate the land change, due to including shoreline change feedbacks.

[1]Dept. of Information and System Sciences, Graduate School of Science and Engineering, Ibaraki University, 4-12-1 Nakanarusawa, Hitachi, Ibaraki 316-8511, Japan. 06nd308s@hcs.ibaraki.ac.jp
[2]Center for Water Environment Studies, Ibaraki University, 4-12-1 Nakanarusawa, Hitachi, Ibaraki 316-8511, Japan. yokoki@mx.ibaraki.ac.jp

4. Conclusion

In this study, the longshore sediment transport model in Majuro atoll was improved by considering the interaction between waves and land changes, regarding to construct a simple model for future sediment change estimation. The calculated results showed a good agreement with the measured sediment changes during 10 years. The present model has been verified to be a good estimation model for the future land change in Majuro atoll.

Acknowledgements

This study was supported by Research Fellowship from the Japan Society for the Promotion of the Science for Young Scientists, and partly by Global Environment Research Fund (Ba-085), Ministry of the Environment, Japan. The authors considerably appreciate it.

References

Booij, N., Ris, R. C. and Holthuijsen, L., 1999. A third generation wave model for coastal regions. Part I. Model description and validation, *J. Geophys. Res.*, 104, C4, 7649–7666.

Komar, P. D. and Inman, D. L., 1970. Longshore sand transport on beaches, *J. Geophys. Res.*, 30, pp. 5914-5927.

Sato D. and Yokoki H., 2008. Field investigation and numerical simulation toward the control of longshore sediment transport on Laura islet in Majuro atoll, *Annual Journal of Coastal Engineering, JSCE*, VOL. 55 (2), 1401-1405. (In Japanese)

SOPAC, 1998. Re-survey of Dud and Laura Beach Profiles Majuro atoll, Republic of the Marshall Islands, *SOPAC Preliminary Report*, No.93, APPENDIX1 Beach Profile Data.

Yokoki, H., Yamano, H., Kayanne, H., Sato, D., Minami, Y., Ando, S., Shimazaki, H., Yamaguchi, T., Chikamori, M., Ishoda, A. and Takagi, H., 2004. Comparison between longshore sediment transport due to waves and long-term shoreline change in Majuro atoll, Marshall Islands, *Global Environmental Research*, Vol. 9, 21–26.

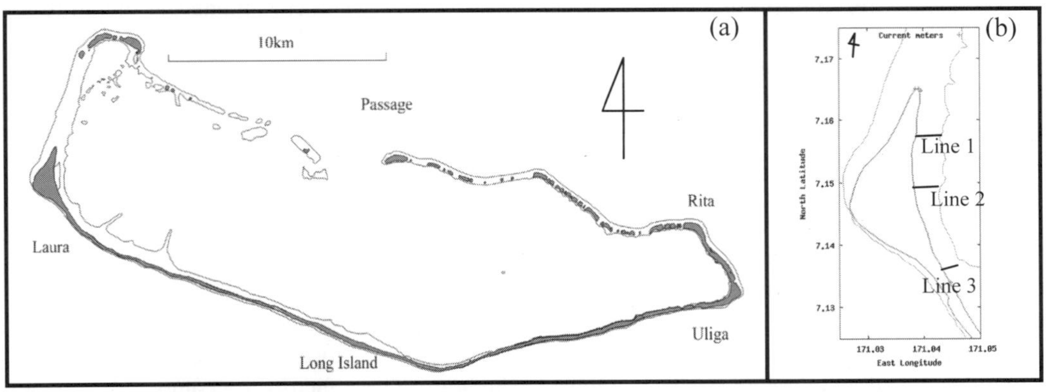

Figure 1. (a)Majuro atoll, Marshall Islands. (b) Expansion of the Laura islet. The lines in the figure showed the measuring lines of beach profiles in 1997 and 2007.

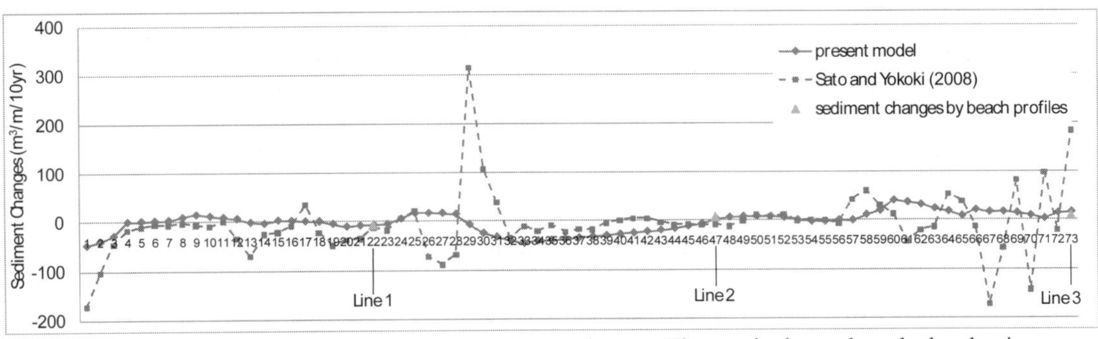

Figure 2. Comparison of the calculated sediment changes. The x-axis shows the calculated point along the lagoonal coast in Laura islet.

Coastal Dynamics 2009
Paper No. 96

EMPIRICAL RELATIONSHIP BETWEEN INLET CROSS-SECTION AND TIDAL PRISM

Marcel J.F. Stive[1], Nghiem T. Lam[2] and Tran T. Tung[3]

1. Introduction

Although other engineers had considered the relationship between tidal prism and inlet cross-sectional area before, it is O'Brien who is usually credited for deriving the familiar relationship

$$A = aP^m \tag{1}$$

where A is the cross-sectional area (relative to mean sea level) and P is the spring tidal prism. The coefficients a and m vary from entrance to entrance. However O'Brien (1969) showed that for 28 US entrances, $a=4.69\ 10^{-4}$ and $m=0.85$ are best-fit values applicable to all entrances when P is measured in cubic meters (m^3) and A in square meters (m^2).

This contribution will first compare some theoretical elaborations both qualitatively and quantitatively, confront these with US, Vietnam and Dutch empirical data and discuss their pros and cons for practical applications. Based on schematic and advanced process-based studies (Lam, 2009 and Tung, 2009), our most important contribution is that we are able to explain the scatter in the cross-sectional area and prism relationships that exist in literature.

2. Empirical relation between inlet cross-section and tidal prism

Many authors have attempted to apply Equation 1 to groups of entrances from the same or similar regions and have suggested, sometimes based on a theoretical derivation, many different values for the exponent m, resulting in many different values for a. However, scrutinizing the literature for the US and the Dutch entrances reveals that very good results are obtained for $m=1$ and best-fitting a. We will address theoretical approaches in the paper.

In summary, we observe that for a typical type of entrances, i.e. near-natural entrances, micro to meso tidal energy, medium wave energy, no fresh water discharge, and good sediment availability the exponent $m=1$ seems appropriate and the a-values are fairly close.

3. The scatter in the A-P relationship explained

Although there seems to be a convergence in the parameter values of the cross-sectional inlet area and tidal prism relationship, we note that individual data for different inlets display a deviation from the best fit relationship, both leading to larger and to smaller values. Based on schematic and advanced process-based studies (Lam, 2009 and Tung, 2009), we are able to explain the scatter in the cross-sectional area and prism relationships that exist in literature. Three factors are competing: the strength of the tidal prism, the strength of the littoral drift and the strength of the river discharge. Their qualitative influence is indicated in Figure 1. This figure displays the Escoffier relationship for a particular inlet at the central coast of Vietnam, including the effect of a river discharge (Lam, 2009). In the figure we have indicated our conceptual interpretation of the relative effect of each of the three factors. Although the Escoffier relationship is derived for this particular inlet, our conceptual interpretation is generally valid. We will underpin these concepts with the results from process-based modeling (Tung, 2009).

Acknowledgements

The second and third authors acknowledge the financial support from the Royal Netherland Embassy

[1]Dept of Civil Eng & Geosc, Delft Univ of Techn., Stevinweg 1, 2628 CN Delft, NL, m.j.f.stive@tudelft.nl
[2]Dept of Civil Eng & Geosc, Delft Univ of Techn., Stevinweg 1, 2628 CN Delft, NL, tlnghiem@yahoo.com
[3]Dept of Civil Eng & Geosc, Delft Univ of Techn., Stevinweg 1, 2628 CN Delft, NL, t.tranthanh@tudelft.nl

in Hanoi, Vietnam, under the Project Upgrading of the training capacity in coastal engineering between Delft University of Technology, the Netherlands and Water Resources University, Hanoi, Vietnam.

References

Lam, N.T., 2009. Hydrodynamics and morphodynamicsof a seasonally forced tidal inlet system, PhD thesis.

O'Brien, M.P., 1969. Equilibrium flow areas of inlets on sandy coasts. J Waterway Port Coast Ocean Eng 95 (l): 43—52

Tung, T.T., 2009. Morphodynamics and stability of tidal inlets in the central coast of Vietnam. PhD thesis.

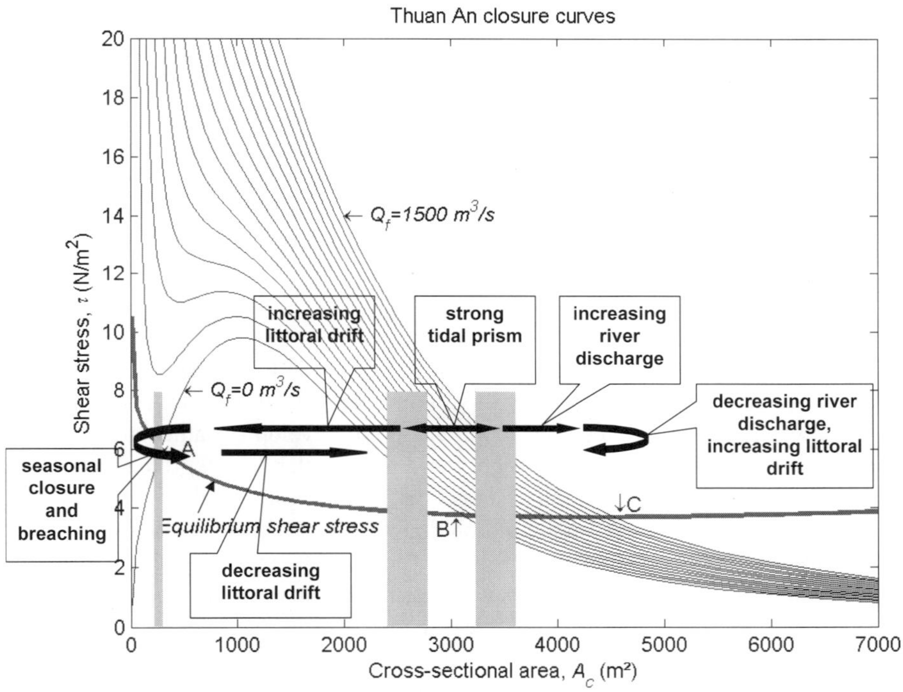

Figure 1. Escoffier relationship between inlet cross-sectional area and shear-stress magnitude. Point A is the instable equilibrium point, points B and C are the stable equilibrium points for no discharge and maximum discharge respectively. Whereas most empirical relations are assuming that point B is the "ideal" equilibrium point, we indicate the processes that force towards a different dynamic equilibrium.

Coastal Dynamics 2009
Paper No. 97

SEASONAL CLOSURE OF CHILAW INLET IN SRI LANKA: PHYSICAL PROCESSES AND MATHEMATICAL MODELING

Magnus Larson[1], Nalin Wikramanayake[2], and Hans Hanson[1]

1. Introduction

In many parts of the world inlets are seasonally closed due to sediment transport. Typical features of such inlets are that they occur in micro-tidal, wave-dominated coastal environments where the seasonal stream flow and wave climate variations are large. The Chilaw estuary is located on the west coast of Sri Lanka about 120 km north of the capital Colombo. The Chilaw lagoon system includes the lagoon, the 5-km long entrance channel, which connects the lagoon with the estuary and ocean, and the mouth of the Deduru River. A sandstone reef stretches for about 5.5 km in a southerly direction along the coast from the inlet. Chilaw inlet is located in a micro-tidal area and every year during the south-west monsoon, approximately from May to September, a spit is formed across the inlet that causes a closure. The main industry in Chilaw is fishing and the lagoon and the entrance channel are used for anchorage by a large fleet of fishing boats. When the inlet closes the access to the ocean becomes restricted which leads to large negative consequences for the Chilaw fishing industry. The inlet is typically opened artificially in order to gain access to the ocean for the fishing fleet. Another problem related to the closure of the inlet is the deterioration of the water quality in the lagoon and estuary.

2. Objectives and Procedure

The main objectives of the study were: (1) to gain an increased understanding about seasonally closed tidal inlets in Sri Lanka through the study of Chilaw Inlet; (2) to measure and analyze the spit formation at Chilaw during the south-west monsoon; and (3) to develop a mathematical model of the inlet morphological evolution, including the spit growth. First, available background information and data were compiled, focusing on Sri Lankan inlets in general and Chilaw Inlet in particular. Additional field data were then collected on the morphological change in connection with inlet closure at Chilaw Inlet during the southwest monsoon of 2006. The physical processes that control sediment transport and morphological change during the southwest monsoon were observed and investigated. The measurement campaign resulted in morphological maps covering the inlet closure caused by the spit growth. Finally, a mathematical model was developed of the inlet morphological evolution and spit growth. Model simulation results were compared with the measured evolution of the spit.

3. Selected Results

The field measurements started on May 20 and continued until July 30 2006, comprising a total of nine field days during which the beach topography was surveyed. Seven of the nine field days proved to have sufficient data for further detailed analysis. Fig. 1 shows the measured topographies, clearly illustrating how the spit is developing and finally closing the inlet. Towards the end of the study period a new channel was artificially dug to allow for passage through the inlet. The average spit growth rate during July was 13 m/day and the average volume growth rate was 360 m^3/day. Similar observations of the topographic evolution was made by Lanka Hydraulic Institute, who carried out field measurements during the southwest monsoon in 1996.

A mathematical model was developed to describe the evolution of the main morphological elements and associated sediment pathways at the inlet during the southwest monsoon. Fig. 2 shows

[1]Department of Water Resources Engineering, Lund University, P.O. Box 118, S-221 00 Lund, Sweden. Email: Magnus.Larson@tvrl.lth.se
[2]Department of Civil Engineering, Open University, Box 21, Nawala, Nugegoda, Sri Lanka.

the schematized sediment transport pathways and morphological elements at Chilaw inlet. The rate Q_{IN} is the sediment transported into to the area from beaches south of the sandstone reef by northerly directed longshore sediment transport. Along the reef sediment is added to the longshore sediment transport by erosion of the material deposited on top of the reef (Q_D). The sediment outflow from the reef beach (Q_R) is divided into two transport components at the tip of the reef. One is caused by diffraction behind the tip and results in a sand transport in an easterly direction across the inlet, which initiates the spit growth. The second sediment pathway is to the north-east, and the sediment is here assumed to be transported to an ebb shoal. The ebb shoal connects the sediment pathways from the reef beach, spit, and the down-drift side of the inlet. Sediment is transported to the ebb shoal either from Q_R or from the spit (Q_C), when the inlet current is strong and erodes the tip of the spit. Sediment leaves the ebb shoal through sediment bypassing (Q_{OUT}) or through Q_E, which is the wave-related cross-shore transport. Various sediment formulas were developed to compute the transport between the morphological elements. Fig. 3 shows the modeled and measured spit growth.

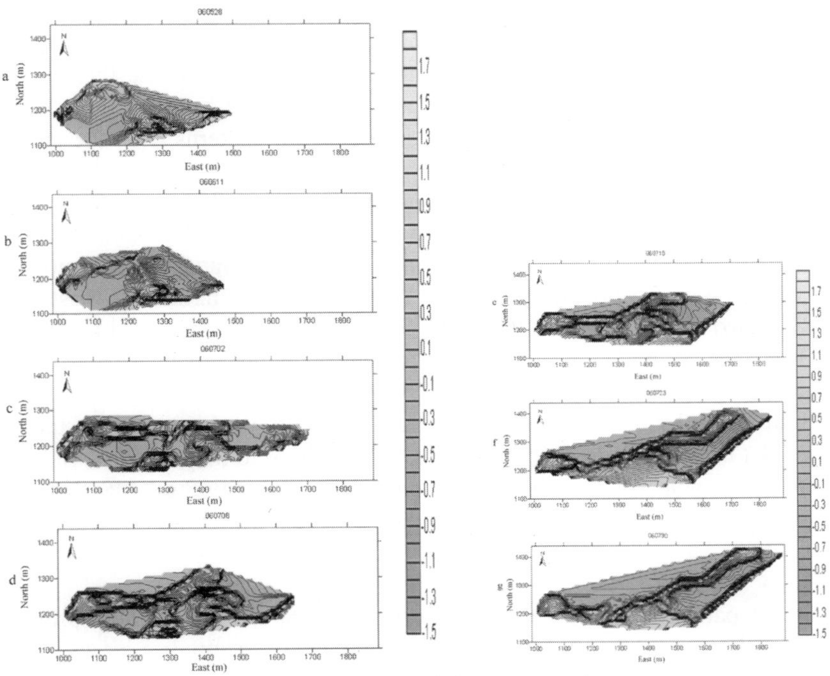

Figure 1. Measured inlet topography

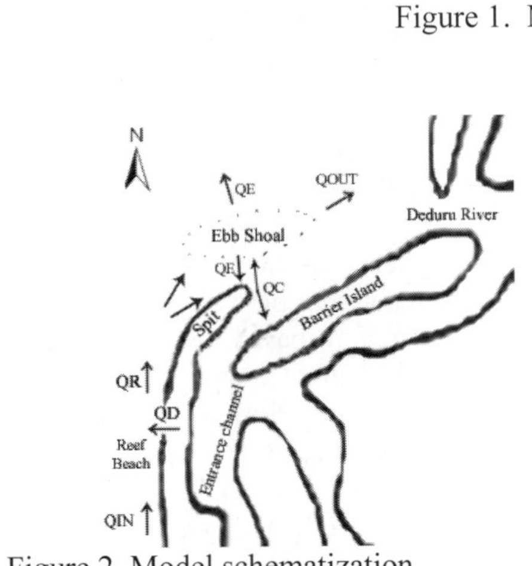

Figure 2. Model schematization

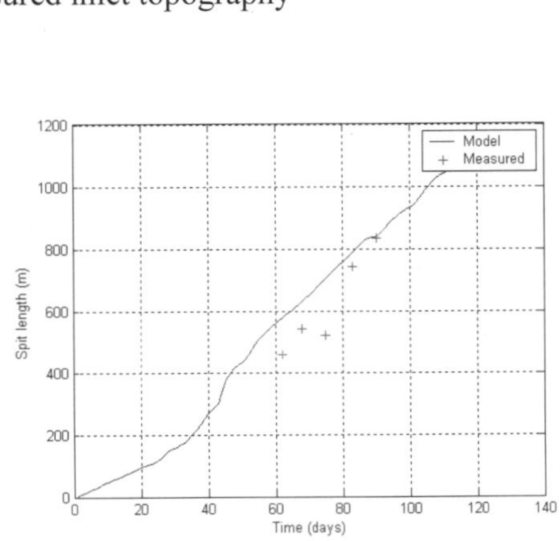

Figure 3. Modeled and measured spit growth

Coastal Dynamics 2009
Paper No. 98

INFLUENCES OF CHANNEL DREDGING ON FLOW AND SEDIMENTATION PATTERNS AT MICROTIDAL INLETS, WEST-CENTRAL FLORIDA, USA

Tanya Beck[1] and Ping Wang[1]

1. Introduction

The West-central Florida coast has 29 barrier islands and 30 tidal inlets with diverse morphologies ranging from tide dominated to wave dominated, despite the micro-tidal (range <1 m) and low-wave energy (typically wave height <0.4 m) setting. Cold fronts passing every 1 to 2 weeks during winter are the dominant mechanism driving the regional net southward longshore sand transport. The summer is typically calm, punctuated by high waves generated by distant tropical storms. There was no direct landfall of tropical storms during the study period. This study examines four inlets in two multi-inlet systems (Fig. 1). Johns Pass and New Pass have large ebb tidal deltas, whereas Blind Pass is an artificially stabilized migratory inlet with no significant delta. These three inlets are dredged regularly every 4-9 years, whereas Big Sarasota Pass has never been dredged. The goal of this study is to investigate the influences of channel dredging on the flow patterns over the ebb tidal delta and sediment bypassing.

Time-series aerial photos and bathymetric surveys from the 1920s were analyzed to assess the pathways of sand bypassing and morphology change at the inlets. The Coastal Modeling System (CMS), an integrated system computing wave, current, sediment transport, and morphology change, was applied. The CMS is developed by the U.S. Army Corps of Engineers' Coastal Inlets Research Program (CIRP). Two models, one for the Johns Pass-Blind Pass system and one for the New Pass-Big Sarasota Pass system, were constructed with recent bathymetry data. Various alternatives based on maintenance dredging projects were tested. The models were driven by measured tide from the study areas. Both upward looking and side-looking ADPs were deployed to measure current profiles throughout the inlet channel, and a ship-mount ADCP was used to map the flow field at Big Sarasota Pass during one tidal cycle. These data serve to calibrate and verify the CMS flow calculation. The modeled morphology changes are compared with data collected from time series surveys.

2. Summary

Different sediment pathways and associated morphology features are identified at the updrift and downdrift sides of the ebb-tidal delta. At the updrift side, the nearshore bar serves as a sand-transport pathway from the updrift beach to the delta. At the downdrift side, migration and attachment of swash bars play a significant role in sand bypassing. These pathways control the shoreline variation in the vicinity of the inlets. Dredging at Johns Pass and New Pass seem to correlate to a short-term trend of onshore migration of swash bars and subsequent attachment to the shoreline. This agrees with FitzGerald's (1988) ebb tidal delta breaching model for sediment bypassing, except that the breaching is artificially initiated. Dredging at Blind Pass has largely eliminated the development of the ebb delta and the potential for natural sediment bypassing.

The CMS reproduced water-level fluctuations accurately; however, peak flow velocity, especially that of ebbing spring tides, was under-predicted (Fig. 2). Comparison between measured and calculated velocity was improved by adjusting local bottom friction based on current-profile measurements. Intensified ebb flow was predicted in the dredged channel along with an overall decrease of flow velocity over the entire ebb delta (Fig. 3). The reduced ebb flow over the delta after the channel dredging may relate to the onshore migration of swash bars, as depicted from the aerial

[1]Dept. of Geology, University of South Florida, 4202 E. Fowler Ave, Tampa, FL 33620, USA.
Tbeck1@cas.usf.edu

photographs. The infilling of the dredged channel was predicted by medium-term simulations of morphologic evolution. Also, the erosional and accretionary trend in the nearshore zone, updrift and downdrift of the inlets, was reproduced.

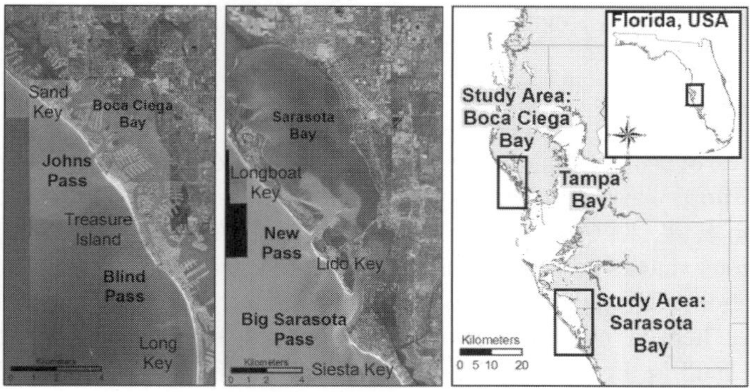

Fig. 1. Area map of the two inlet systems. Johns Pass and Blind Pass serve Boca Ciega Bay; New Pass and Big Sarasota Pass serve Sarasota Bay.

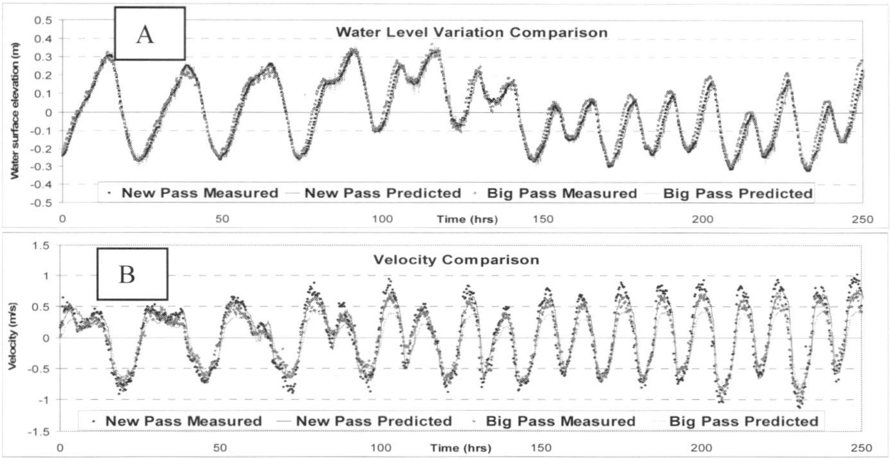

Fig. 2. A) Measured and calculated water surface variations for New Pass and Big Sarasota Pass. B) Measured and. calculated current velocities for New Pass and Big Sarasota Pass.

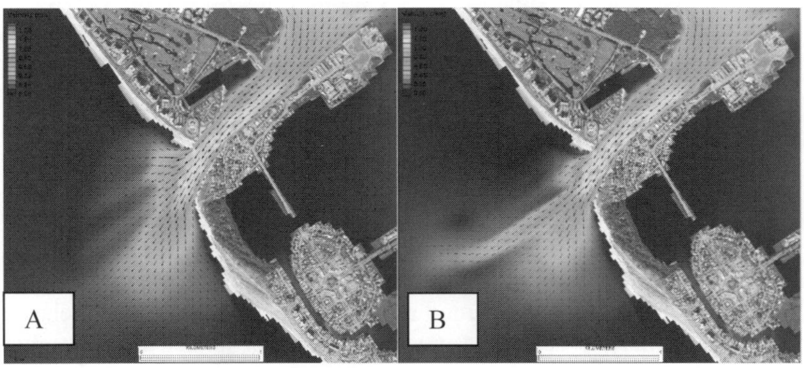

Fig. 3. Calculated flow field through New Pass during an ebb tide. A) Flow field over the 2005 bathymetry, 2.5 years after the dredging in 2003; B) Flow field over the 2003 dredged bathymetry.

Acknowledgements

This study is funded by U.S. Army Corps of Engineers' Coastal Inlets Research Program (CIRP) and by Pinellas and Sarasota Counties, Florida.

References

Fitzgerald, D.M., 1988. Shoreline erosional-depositional processes associated with tidal inlets. In D.G. Aubrey and L. Weishar, eds., *Hydrodynamics and Sediment Dynamics of Tidal Inlets*, New York. p. 412-439.

Coastal Dynamics 2009
Paper No. 100

2D HORIZONTAL PROPAGATION OF IRREGULAR WAVE ON SOFT MUD

Ali Oveisy[1], Kevin Hall[2]

1. Introduction

Attenuation of propagating water waves over mud bottom has been observed and investigated over paste decades. In recent years more efforts have been aimed in the numerical simulation of wave propagation over muddy bed (e.g. Zhao et al., 2006; Winterwerp et al., 2007; Soltanpour et al., 2008; Oveisy et al. 2008). Most of the previous studies have not considered irregular wave propagation or they just limited to crossshore propagation of irregular wave on muddy coast. Since the waves in real field condition have irregular nature, here a numerical model is proposed for irregular wave propagation and interaction with fluid mud in the 2D horizontal plan based on time dependent mild slope equation. The model includes wave shoaling, refraction, reflection, and dissipation due to both the wave breaking and wave-mud interaction

2. Wave Propagation Model

The governing equations for wave propagation are the time dependent mild slope equations (Horikawa, 1988):

$$\frac{\partial Q_x}{\partial t} + c^2 \frac{\partial}{\partial x}(\xi) + f_D \cdot Q_x = 0 \tag{1}$$

$$\frac{\partial Q_y}{\partial t} + c^2 \frac{\partial}{\partial y}(\xi) + f_D \cdot Q_y = 0 \tag{2}$$

$$\frac{\partial \xi}{\partial t} + \frac{1}{n}\{\frac{\partial (nQ_x)}{\partial x} + \frac{\partial (nQ_y)}{\partial y}\} = 0 \tag{3}$$

where Q_x and Q_y are flow rates, ξ is the water surface elevation, c is the wave celerity and n is the wave group velocity over wave celerity. The total wave energy dissipation coefficient f_D includes energy loss caused by wave-mud interaction (f_{DM})before the breaking line and both energy loss due to wave breaking and wave-mud interaction within the surf zone ($f_D = f_{DB} + f_{DM}$). The governing equations are discretized and a staggered mesh scheme is employed, details of the wave propagation model can be found in Oveisy et al. 2008, Soltanpour et al., 2008.

3. Wave mud interaction

Following Tsuruya et al. (1987), the fluid system is divided into N layers and the linearized Navier-Stokes equations, neglecting the convective accelerations, and the continuity equation are solved by assuming periodic solutions for the horizontal and vertical components of orbital velocity and dynamic pressure in each sub-layer. The calculation is repeated at each grid point to obtain the local attenuation coefficient, k. Considering the complexity of mud rheological behavior, the visco-elastic model has been adopted in the present study to develop a predictive behavior model for wave-mud interaction (Shibayama and An, 1993).

4. Irregular Wave Transformation

Using spectral approach the irregular waves discretized to number of regular wave components, and the wave model is run individually for each component Considering the incident wave spectrum $S_0(f_i, \theta_i)$ in the space of σ, θ and the exponential decay of wave height due to wave mud interaction,

---

[1]Dept of Civil Eng., Queens University, Ellis Hall, Kingston, Ontario, Canada,. a_oveisy@ce.queensu.ca
[2]Dept of Civil Eng., Queens University, Ellis Hall, Kingston, Ontario, Canada,. hallk@ce.queensu.ca

$$S_0(f_i, \theta_i) = \frac{1}{\Delta\sigma \cdot \Delta\theta} \cdot \frac{1}{2} a_{ix}^{\ 2} \tag{4}$$

$$a_{ix} = a_{i0} e^{-k_i(x-x_0)} \tag{5}$$

The attenuated wave spectra can be drived as,

$$S(f_i, \theta_i) = S_0(f_i, \theta_i) e^{-2k_i(x-x_0)} \tag{6}$$

where a_{i0} is the amplitude of the i-th wave component at point $x=x_0$, a_{ix} is the amplitude of the i-th wave component at distant x from point x_0, and σ_i, θ_i are corresponding frequency and direction.

5. Model Performance

The model has been applied for prediction of wave propagation over a mud berm off Dauphin Island in Alabama, an underwater mound composed of sediment dredged from the Mobile Ship Channel deposited in 1988 (Fig. 1). The design dimensions of the berm were 2,750 m alongshore, 300 m wide at the toe, and 6 m high with 6.5 m mean depth of water above the berm crest (McLellan, Pope, and Burke 1990). The placed material is highly graded, with a median size of about 15 μm. In this simulation the offshore wave was used as incident wave height. Comparison of simulation result against measurement for inshore station is shown in Fig. 2.

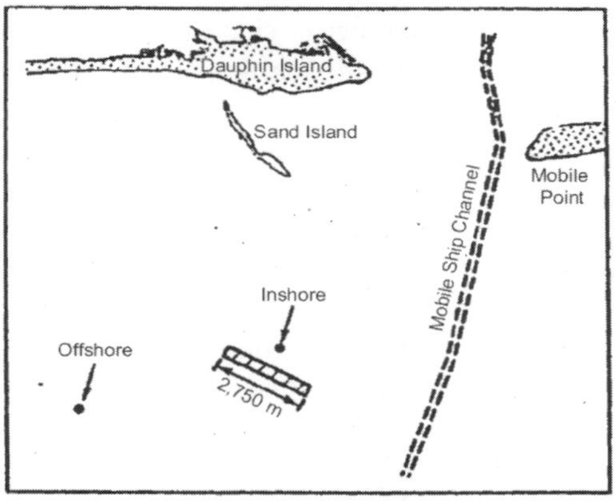

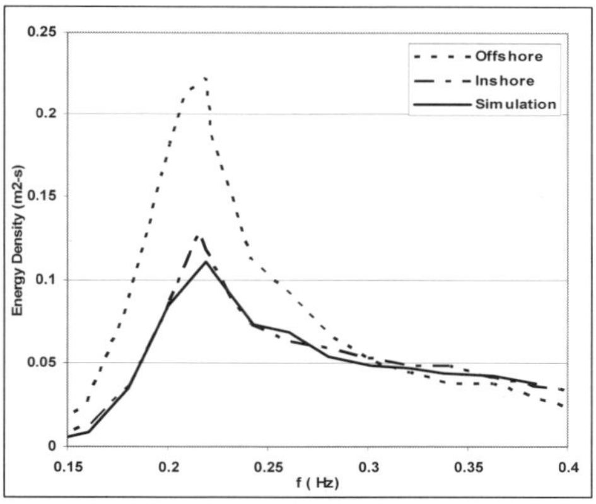

Figure 1. Location of the berm and measurement stations.. Figure 2. Comparison of measurement and simulation.

References

Horikawa, K., 1988. *Nearshore Dynamics and Coastal Processes*. University of Tokyo Press. Part III, chapter 2.

McLellan, T.N., Pope, M.K., Burke, C.E., 1990. Benefits of nearshore placement. In: Proceedings, Third Annual Technical Conference. Florida Shore and Beach Preservation Association, Tallahassee, FL, pp. 339–353.

Oveisy, A. Hall, K., Soltanpour, M., Shibayama, T., 2008. A two-dimensional horizontal wave propagation and mud mass transport model, *Continental Shelf Research*.

Shibayama, T., An, N. N., 1993. A visco-elastic-plastic model for wave-mud interaction. Coastal Engineering in Japan, Vol. 36, No. 1, pp. 67-89.

Soltanpour, M., Shibaiama, T. and Masuya, Y., 2007. Irregular wave attenuation and mud mass transport, *Coastal Engineering Journal* 49 (2), pp. 127–147.

Soltanpour, M., Oveisy, A., Shibayama T., 2008. Numerical modeling of wave transformation on muddy coasts. *Coastal Engineering Journal*, 50(2).

Tsuruya, H., Nakano, S., and Takahama, J., 1987. Interaction between surface waves and a multi-layered mud bed. Rep. Port and Harbor Res. Inst., Ministry of Transport, Japan, Vol. 26, No. 5, pp. 141-142.

Winterwerp J.C., de Graaff R.F., Groeneweg J., Luyendijk A., 2007. Modeling of wave damping at Guyana mud coast, *Coastal Engineering*, 54, 249–261.

Zhao, Z.D., Lian, J.J., Shi, J.Z., 2006. Interactions among waves, current, and mud: numerical and laboratory studies. *Advanced Water Resource* 29 (11), pp. 1731–1744.

Coastal Dynamics 2009
Paper No. 102

DYNAMIC THERMAL RESPONSE OF URBAN HEAT RELEASE INTO COASTAL AREA - HEAT ISLAND REDUCTION IN OSAKA REGION, JAPAN-

Nobuhito Mori[1], Yasuaki Sachi[2], Takaaki Shigematsu[2] and Masaki Nakao[2]

1. Introduction

A heat island phenomenon becomes a serious problem at urban area located mid-latitude. For example, the temperature at Osaka and Tokyo has been increased 2 degree the last one hundred years and it is two times faster than the climate change. Therefore, various measures are proposed to reduce the temperature increase during the summer time. Primary measures are reduction of artificial heat and albedo of land area. The other hand, the thermal capacity of water is several times lager than air, and the coastal water therefore has capability to store the urban heat owing to the heat capacity difference. Moreover, the stratified coastal water can be regarded as an unused energy in summer. Assessments of such large-scale thermal environmental change require estimation of heat transfer, thermal diffusion of ocean and atmosphere, and interactions with urban climate.

The main purpose of this study is estimation of the thermal environmental impact of urban heat release into Osaka bay, Japan. The three dimensional ocean model including thermal mixing and diffusion is used to predict thermal environmental mitigation from an ocean side of view. First, the total amount of urban heat is estimated along Osaka bay. Second, the thermal impact of urban heat release is discussed based on the numerical results.

2. Numerical Model

The urban heat database considering the monthly and hourly variations at Osaka area was made based on emission factors with GIS database. The total amount of heat as a function of distance from coastal line (denotes d) can be estimated using the database. The maximally possible heat capacitance to be released into the ocean can be estimated integrating along the coastal area using the database. The numerical model based on the Regional Ocean Modeling System so-called ROMS has been modified and developed to simulate released thermal behavior in Osaka bay. The bathymetry is used by Japan Coast Guard data and the computational domain covers latitude of 34.0 to 34.85 degrees and east longitude of 134.5 to 135.65 degrees. Main tidal components are given at the open boundaries and gives good agreement with the observed tidal elevation. The three different turbulent models are used for vertical mixing. The freshwater inflows are considered for four main rivers toward to Osaka bay. The high-order advection diffusion scheme is used for the momentum, density and temperature. The heat fluxes at ocean surface layer are considered for shortwave and longwave radiation using bulk formula based on Craig-Banner scheme.

3. Results and Discussion

The summer season is selected for the target in this study. The urban heat from artificial cooling system in waterfront area is estimated from the database as mentioned above section. The target urban areas are changed 500 m to 5000 m from the coastal line (d=500-5000m). It is assumed that the urban heat can be collected using the sewage and it can be released into the coastal zone as thermal effluent such as an electric power plant, and the urban heat therefore is converted to thermal effluent flowing which is 7.0 degrees warmer than the environmental coastal water. It is also assumed that the thermal effluent of urban heat discharged as a free jet from the bottom at waterfront to the bay.

First, the mixing of coastal water was verified by three different turbulent models. Figure 1 shows the velocity vectors at the sea surface by the numerical model with Mellor-Yamada level 2.5 closure

[1] Disaster Prevention Research Institute, Kyoto University, Gokasho, Uji, Kyoto 611-0011, Japan.
[2] Graduate School of Eng., Osaka City University, 3-3-138 Sugimoto, Sumiyoshi-ku, Osaka 558-8585, Japan.

model (MY) (a) and radar observed data (b), respectively. The MY model gives good agreement with the observed velocity profile than k-ε and KPP models. The stability functions based on the Richardson number are also verified to check the validity of the scheme beforehand. Using the developed ocean model, the environmental assessment to release thermal effluent into Osaka bay can be simulated. Fig.2 shows horizontal distribution of velocity and water temperature difference between thermal effluent release case and control run in the case of d=5000m. The maximally increased sea surface temperature near the release point is 2.5 and 1.3 degrees C at neap and spring tide, respectively. The released thermal effluent diffused following tidal current 10-15 km offshore from the coastal line. Fig.3 shows area where sea surface temperature is increased more than 1.0 degrees due to the urban heat release as a function of d. In the case of d=5000 m, the influenced area due to thermal effluent depends on the tide but is up to about 10 km^2 maximumly. However, the moderate conditions (d<3000 m) show that the sea surface temperature increased area is negligible and the water temperature is only increased 0.07 degrees C in average. These results indicate that the Osaka bay can be used thermal storage at the summer time.

The detail of the results will be presented at the conference.

References

Shchepetkin, A. F. and J. C. McWilliams, 2005. The regional oceanic modeling system(ROMS), *Ocean Modeling*, 9, 347-404.

Umlauf, L. and Burchard, H., 2003. A generic length-scale equation for geophysical turbulence model. *Journal of Marine Research*, 61(2): 235-265.

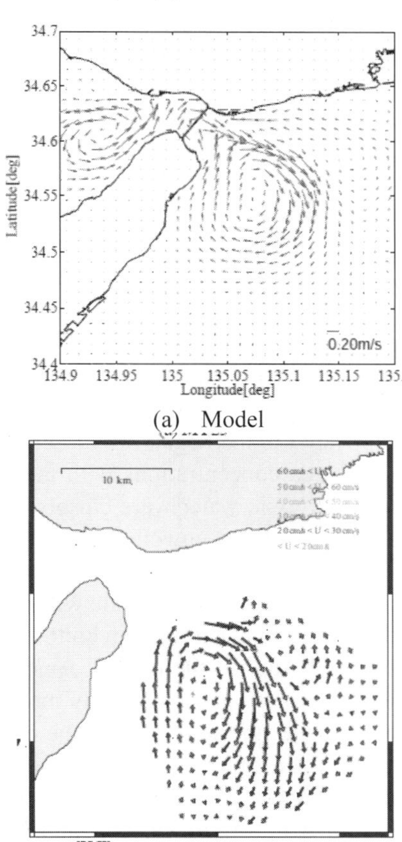

(a) Model

(b) Observed by VHF Rader

Figure 1. Comparison of sea surface velocity near the Akashi channel.

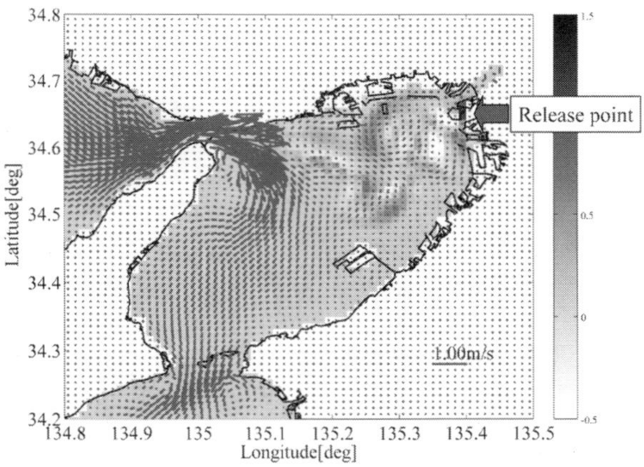

Figure 2. Water temperature change due to the urban heat release (contour: urban heat release case minus control run, vectors: surface velocity; neap tide, case d=5000m, August)

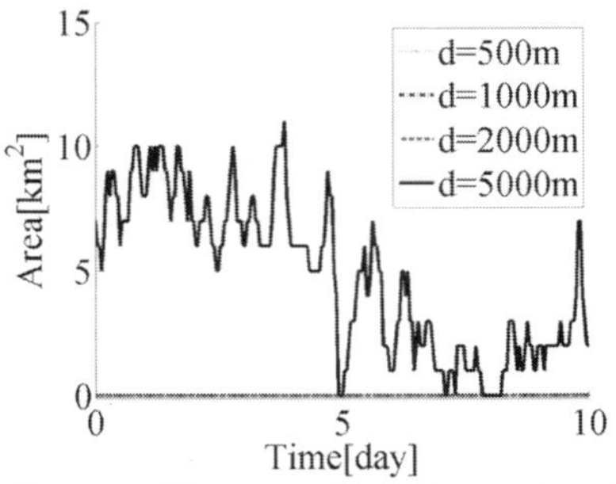

Figure 3. Time variation of area where SST increased more than 1.0 degrees than control run

Coastal Dynamics 2009
Paper No. 103

Temporal and spatial variations of hypoxic water mass in Tokyo bay, Japan

Hiroshi YAGI, Tanuspong, POKAVANICH, Susumu YASUI, Kazuo NADAOKA, Yasuyuki NAKAGAWA, Ryuichi ARIJI, Shoichi MATSUZAKA, Nobuaki SUZUKI and Kazunobu MOROHOSHI

1. Introduction

A new runway of Tokyo International Airport (Haneda), which is one of the largest ongoing infrastructural projects in Japan, is under construction at the Tama River Estuary on the west coast of Tokyo Bay. It is important to clarify the possible impacts of the new runway construction on the surrounding environments, because the largest estuarine type tidal flat in Tokyo bay, which is a valuable habitat for juvenile fishes, benthos and plants, is remained in this area. For investigating the environmental impacts, hypoxic water mass is one of the key factors. In the preset study, the temporal and spatial variations of hypoxic water mass in Tokyo bay, which is closely related to the water environment in the Tama River Estuary, were investigated based on field surveys and numerical simulations.

2. Outline of field surveys and numerical simulations

To understand the temporal and spatial variations of hypoxic water mass in Tokyo bay, the field surveys were performed once a week or two weeks intervals from July to September in 2007. The observation stations were placed along the bay-longitudinal and bay-transverse lines off Haneda as shown in Fig.1. Vertical profiles of dissolved oxygen concentration, temperature, salinity, turbidity and Chlorophyll-a were measured by using a CTD type multi-parameter water quality sensor (AAQ1183, JFE Arec LTD) at every observation stations. To measure the continuous velocity and water quality variations, the mooring system was also deployed off Haneda (Sta.12 as shown in Fig1) from 23 Aug. to 21 Sept. Furthermore, 3D hydrodynamic model (Delft 3D) was applied to simulate the hydrodynamic conditions for the field survey period.

3. Results and discussions

From the field observation results, it was revealed that the hypoxic water mass showed the large variations both in space and time, i.e., i) hypoxic water extended to the offshore region in the bottom layer in early August (14 Aug.), ii) the bottom hypoxic water mass was lifted up to the middle layer in mid August (23 Aug.), iii) the middle layer hypoxic water disappeared in the beginning of September (4 Sept.) as shown in Fig.2. The comparison of the dissolved oxygen concentrations with measured salinities and temperatures indicated that these large variations of hypoxic water were closely related to the movement of Tokyo bay water induced by the wind in the bay and the upwelling phenomenon occurred around the bay mouth. Figure 3 shows the numerical simulation results of water temperatures and residual currents along the longitudinal section. In the early August, the warm water was accumulated inside the bay and the weak outward flow was induced in the bottom layer corresponding to the continuous and strong south wind. In the mid August, the cold oceanic water mass intruded into the lower layer by the upwelling phenomenon occurred around the bay mouth. In the beginning of September, the wind direction changed from the south to the north and the oceanic water intruded into the middle layer of Tokyo bay. These results suggest that the large variations of measured hypoxic water mass (as shown in Fig.2) were induced by the wind driven currents and the upwelling around the bay mouth. In particular, the combination of wind driven currents and middle layer intrusion of oceanic water after the upwelling phenomenon is important for the displacement and disappearance of middle layer hypoxic water.

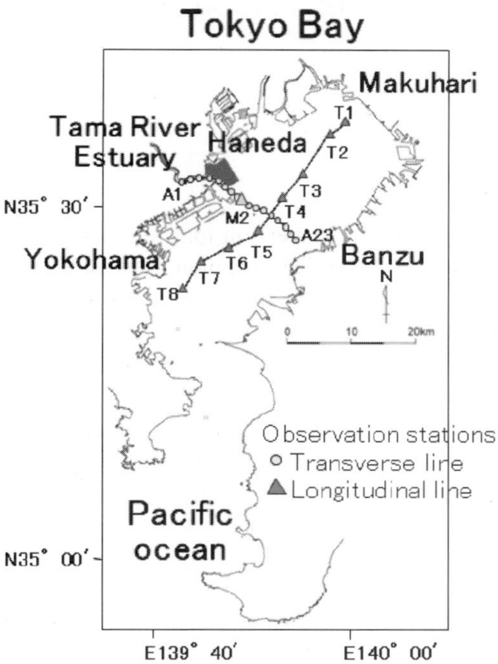

Fig.1 Locations of study area and observation stations

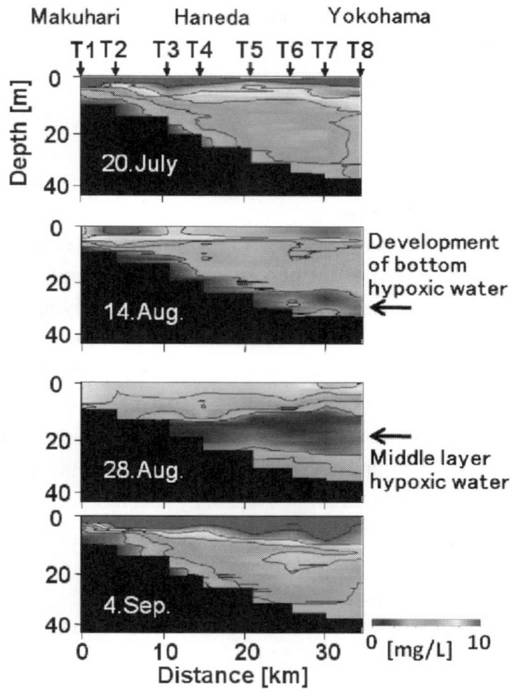

Fig.2 Longitudinal distributions of dissolved oxygen concentrations

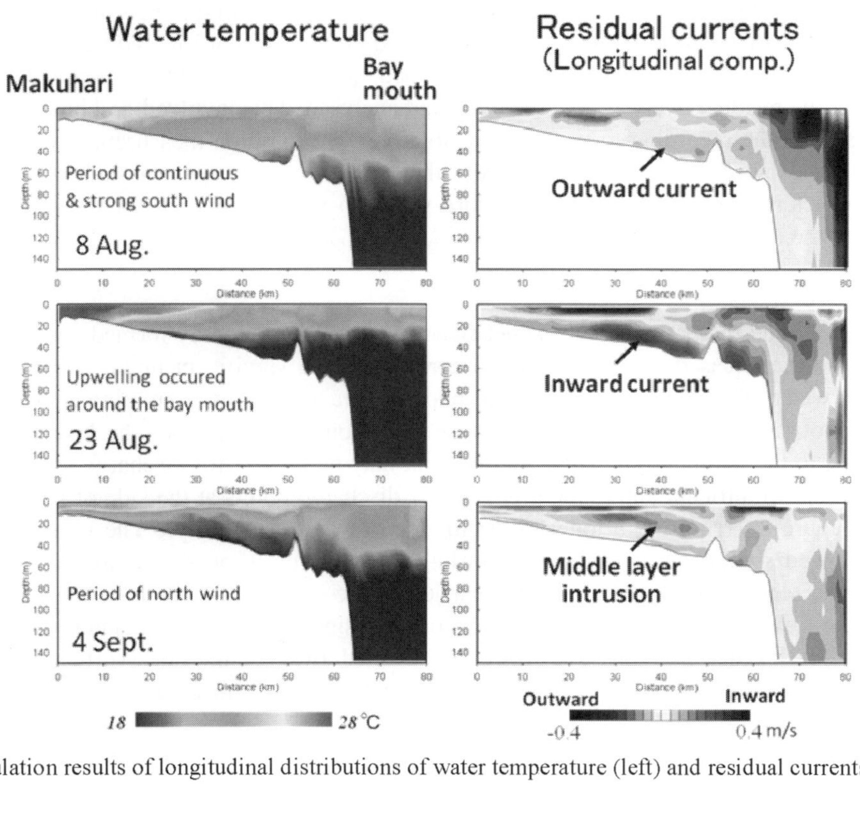

Fig 3. Simulation results of longitudinal distributions of water temperature (left) and residual currents (right).

Coastal Dynamics 2009
Paper No. 104

POLLUTANT DISPERSION AROUND THE PERIMETER OF A COASTAL HEADLAND

Fu Ee Tang[1]

1. Introduction

Point and nonpoint source pollutants often enter coastal waters. Its dispersion maybe complicated by the existence of numerous coastal geographical features such as headlands, promontories and islands. The water depths in coastal areas are typically put into the context of the shallow water flows. The shallow water layer has been studied extensively in previous research work, including Wolanski, et al (1984), Ingram and Chu (1987) and Chen and Jirka (1991, 1995 and 1997). When coastal flow passes a headland on the coastal line, a recirculation zone will be formed as the flow separates at the headland toe. Effective pollutant dispersion at the recirculation zone will be crucial to maintain the water quality of the region. If pollutant is discharged at different locations around the perimeter of a headland, the potential of pollutant trapping around the headland and its generated recirculation zone may change accordingly. This study investigates the characteristics of pollutant dispersion in the region around the headland perimeter.

2. Experimental Methodology

The experimental investigation was carried out in a purpose-built flume, located at the Manchester Tidal flow Facility (MTF) to study the pollutant dispersion around the perimeter of the headland and its generated recirculation zone in a coastal area with a rough, sloping bed in shallow steady flow and tidal flow. The averaged Reynolds number for steady flow was 6.35×10^3, and for tidal flow up to 2.3×10^5. Novel experimental methodology was used, utilising the Planar Laser-induced Fluorescence (PLIF) technique with fluorescent dye released at several points around the perimeter of the headland to perform flow visualization (Figure 1). The visualised flow was recorded, and the flow sequence and visual characteristics were scrutinised and described qualitatively. After a period of dye release, dispersion of the residual dye was also recorded and analysed. This involved grabbing and processing raw images from the recording to produce quantitative concentration measurement. For steady flow cases, dye decay profiles were produced from the grabbed raw images to study the dispersion or flushing characteristics. A calibration procedure was also used to account for the absorption of the laser light by the released dye.

3. Results

Dye released around the perimeter of the headland in steady flow rapidly formed a steady dye plume that channeled the released dye into the main current without significant initial entrainment into the surrounding water. This may indicate that in real life situations, initial entrainment of the released pollutant is not significant. Dye released at discharge points at the toe of the headland was easily dispersed due to the main current, some entrainment may be present for discharge points at the lee of the headland and significant entrainment is seen for discharge points at the side of the headland near the shore (Figure 2), requiring a longer time to disperse the residual dye. The produced dye decay profiles followed similar decay patterns according to discharge location (Figure 3). From the main decay profiles, secondary analysis was also carried out to discern the trend for the dye decay. For tidal flow, the flow patterns for discharge points on opposing sides of the headland were symmetrical and mirrored each other. Residual dye was easily flushed away by the tide. Vortical movement was seen for discharge points near the shore due to the recirculatory currents.

[1]Dept of Civil and Construction Eng., Curtin University of Technology Sarawak Campus, CDT 250, 98009 Miri, Sarawak, Malaysia. tang.fu.ee@curtin.edu.my

Acknowledgements
This study was supported by a grant (No. GR/L/34570) from the UK Engineering and Physical Science Research Council. The PhD study of the author was supported by a UK Overseas Research Studentship.

References

Chen, D., and Jirka, G.H. 1991 "Pollutant mixing in wake flows behind islands in shallow water" *Proc. of the Int. Symp. on Environmental Hydraulics,* Hong Kong, 1991

Chen, D., and Jirka, G. H. (1995) "Experimental study of plane turbulent wakes in a shallow water layer" *Fluid Dynamics Research,* Vol. 16

Chen, D., and Jirka, G.H. (1997) "Absolute and convective instabilities of plane turbulent wakes in shallow-water layer." *J. of Fluid Mechanics*, Vol 338,157-172

Chen, D., and Jirka, G.H. (1999) "LIF study of plane jet bounded in shallow water layer" *Journal of Hydraulic Engineering,* Vol. 125, No. 8

Chen, D., Tang, F.E., and Chen, C. (2003) "PLIF study of pollutant trapping and 3-D plume structures in recirculation zones around a headland: steady and tidal currents." *Proc. of the Int. Symp. on Shallow Flows*, ASCE, Holland, 2003

Chen, D., Tang, F.E., and Chen, C. 2005. Pollutant trapping at a coastal headland. *ASCE Journal of Waterway, Port, Coastal and Ocean Engineering, 131(3).*

Ingram, R. G., and Chu, V. H. 1987. Flow around islands in Rupert Bay: An investigation of the bottom friction effect, *Journal of Geophysical Research*, 92(C13), 14521-14533.

Wolanski, E., Imberger, J., and Heron, M.L. 1984. Island wakes in shallow coastal waters *Journal of Geophysical Research,* 89(C6)

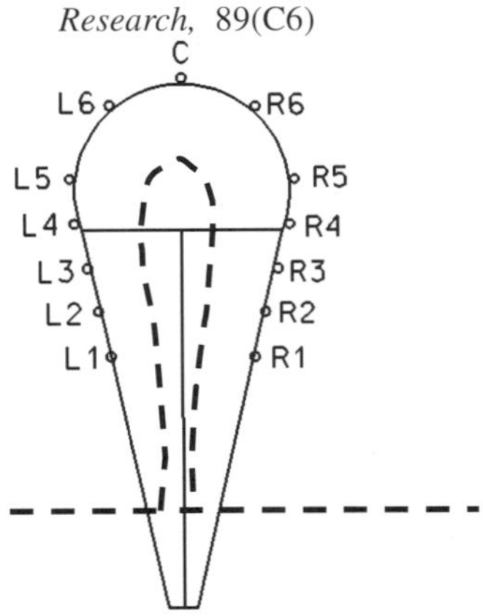

(Above) Figure 1.Location of discharge points around peri meter of headland. Dotted line indicates approximate wate r level

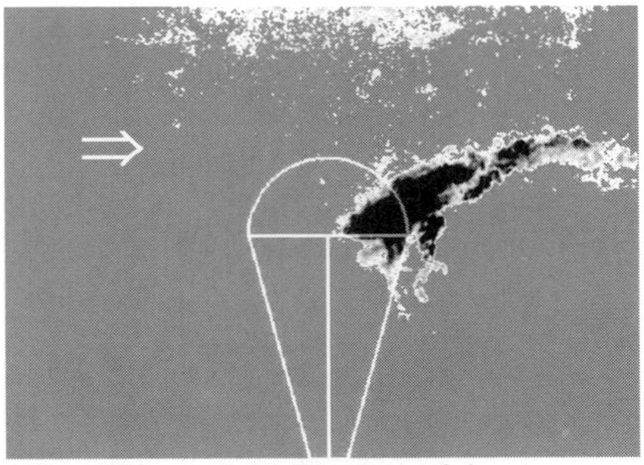

(Above) Figure 2. PLIF image of dye concentration distribution for developed dye plumes for dye discharge at R1, steady flow, arrow indicate flow direction.

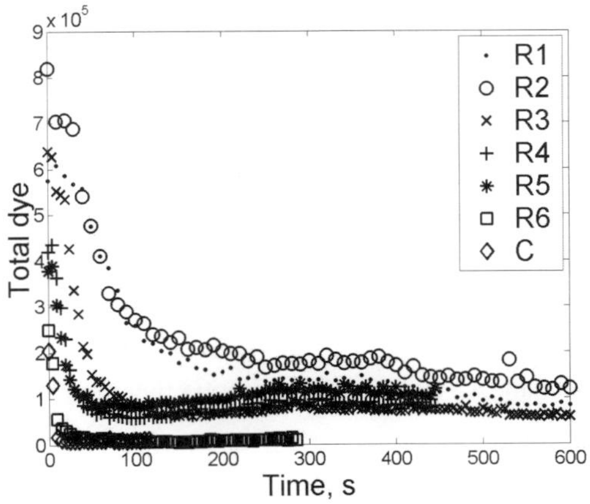

(left) Figure 3. Dye decay profiles for flushing of residual dye after dye discharge was stopped in steady flow, with absorption analysis incorporated. Units of dye are in milligram (mg) multiplied by a factor of 9.185×10^{-8} for the laser sheet thickness of 2mm.

Coastal Dynamics 2009
Paper No. 105

FATE OF POLLUTANTS IN THE BAY OF MARSEILLE-ASSESSMENT OF SEDIMENT DYNAMICS

Michalis I. Vousdoukas[1], Jean Louis Gonzalez[2], Samuel Meulé[3], Christel Pinazo[4], Didier Sauzade[5] and Romaric Verney[6]

1. Introduction

The METROC project is focused on the origin and fate of pollutants inside the Bay of Marseille and since fine sediments are the main pollutants' carriers; one of the project's main objectives is to investigate the sediment dynamics in the area. The Bay of Marseille is characterized by a mostly rocky steeply sloped shoreline and the presence of several small islands, which make it rather protected, with milder hydrodynamic conditions compared to the Gulf of Lions overall (Dufois et al., 2008). Few kilometers after the Bay's western limit lays an important industrial zone, with the largest commercial port of France and the Rhone River delta (see also Figure 1b), a major sediment source for the area.

2. Methods

The approach used in the project is described below: Bed cover information (i.e. grain size measurements, water content, concentration, carbonates and organic matter content) is obtained from sediment samples collected at ~60 locations on the study area (Figure 1b); in the meantime, cores with undisturbed sediment/water interface are collected and used to estimate the critical shear stress of erosion in unidirectional flume tests (Le Hir et al., 2006). Autonomous benthic stations (see also Figure 1a) are deployed for periods of few months to provide in-situ measurements of wave height, current velocities, turbidity and bed altimetry (ADCP/CTD/Turbidity/ALTUS). Turbidity and other measurements collected during independent field campaigns are supplementary to the above data. Finally, satellite images and hydrodynamic models are validated with the above data and are used to provide information for the wider area.

3. Results

The area is dominated by the Mistral (NW) and SE winds, with the latter being responsible for most of the occurring intense wave conditions. The bed is covered by sediments ranging from clay to medium sand and *Posidonia oceanica* patches. The estimated critical bed shear stress values from the erodibility tests (ranging between 0.15-0.6 N/m^2) were lower than the theoretical ones, obtained following a Shields approach (Shields, 1936).

The project is currently active and four frame station deployments have been already concluded, showing that sediment suspension is mainly dominated by wave conditions, rather than currents, while intense SW waves (H_s>2 m) have been found to be related with most of the observed erosion

[1] IFREMER, Département Biogéochimie et Ecotoxicologie. B.P.330 Zone Portuaire de Brégaillon, 83507 La Seyne/mer cedex, mvousdoukas@ualg.pt

[2] IFREMER, Département Biogéochimie et Ecotoxicologie. B.P.330 Zone Portuaire de Brégaillon, 83507 La Seyne/mer cedex, Jean.Louis.Gonzalez@ifremer.fr

[3] CEREGE, Universite de Provence, Europôle Méditerranéen de l'Arbois - BP 80, F-13545 Aix en Provence cedex 04, meule@cerege.fr

[4] LOPB, Universite de la Mediterranee, Station Marine d' Endoume, Chemin de la Batterie des Lions, 13007, Marseille, France, christel.pinazo@univmed.fr

[5] IFREMER, B.P.330 Zone Portuaire de Brégaillon, 83507 La Seyne/mer cedex, Didier.Sauzade@ifremer.fr

[6] IFREMER, Dyneco-Physed BP70 29280 PLOUZANE Cedex, Romaric.Verney@ifremer.fr

events (e.g. see Figure 2). On the other hand, the data imply that the Frioul Islands appear to divide the study area in two 'isolated' cells (Northern Bay and the Bay of Prado), in which material is exchanged between areas through the ephemeral longshore transport regimes. Areas covered with *Posidonia oceanica* showed accretion even under extreme conditions behaving like sediment sinks, while satellite images and current circulation model results since 2000 show that the Rhone River plume can only episodically contribute to the sediment budget of the Bay of Marseille. A hydrodynamic model has been validated with the frame stations' data and after being run for a 1-year period (results expected soon), will complete the overall description of the sediments sources, sinks and pathways in the study area.

References

Dufois, F., Garreau, P., Hir, P.L. and Forget, P., 2008. Wave- and current-induced bottom shear stress distribution in the Gulf of Lions. *Continental Shelf Research*, In Press.

Le Hir, P., Cann, P., Jestin, H. and Bassoullet, P., 2006. Instrumentation légère pour la mesure de l'érodabilité des sediments vaseux ou sablo-vaseux, *Actes IXèmes Journées Nationales Génie Côtier - Génie Civil,* Session VIII (Instrumentation et mesures), Brest, pp. 9-16.

Shields, A., 1936. Anwendung der Ähnlichkeitsmechanik und der Turbulenzforschung auf die Geschiebebewegung. Mitteilungen der Preußischen Versuchsanstalt für Wasser-, Erd- und Schiffbau, Heft 26, Berlin.

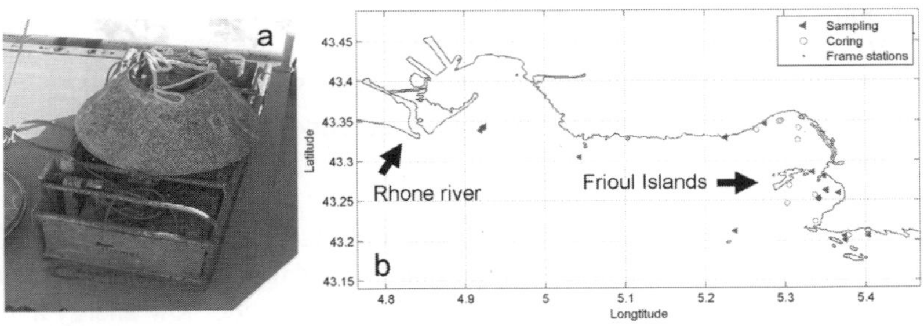

Figure 1. (a) "FRAME" autonomous benthic station directly after retrieval, (b) Map of the study area showing the frame stations deployment locations, as well as the areas were sediments or cores were collected

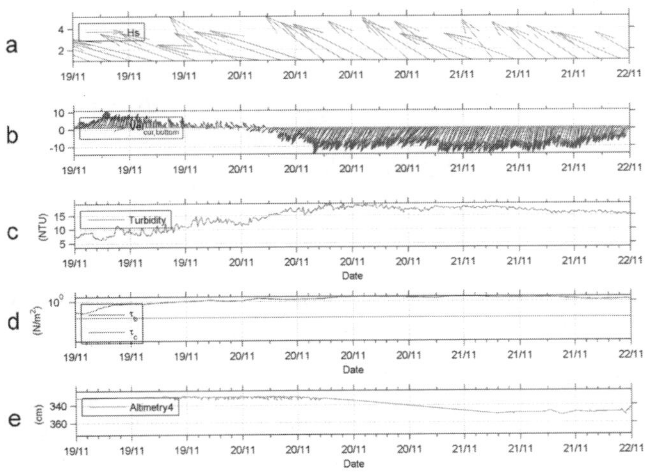

Figure 2. Example of an erosional event from the Fos area, under intense SE wave conditions. Key: a) Significant wave height; b) bottom current velocity; c) Turbidity; d) Critical (lower line) and instantaneous wave and current induced bed shear stress; e) Bed level

Coastal Dynamics 2009
Paper No. 106

NOVEL METHODOLOGY FOR ONE LINE MODEL CALIBRATION USING IMPOUNDMENT ON MIXED BEACH

Inés Martín-Grandes[1], Jason Hughes[2], David J. Simmonds[3], Andrew J. Chadwick[4] and Dominic E. Reeve[5]

1. Introduction

In the UK gravel and mixed (sand and shingle) beaches are common coastal features along the south coast and these have a great significance for the protection of coastal communities and environmental and agricultural sources (Mason et al., 2001). To date, only a few investigations (Nicholls et al., 1987; Chadwick, 1989) have attempted to estimate the LST (Longshore Sediment Transport) rate for coarse grain size beaches, using models derived from the CERC equation (SPM, 1984). This is one of the most familiar formulae in coastal engineering, which has been developed for sandy beaches. Tracers and traps experimental data, measured on shingle beaches in the South UK, have been used to calculate *K*, giving values of 7% and 9%, Nicholls et al. (1987) and Chadwick (1989) respectively, of those obtained for sandy beaches.

In this work an impoundment technique has been applied for the first time on a mixed beach to calibrate the CERC equation using field data. Volumetric estimations from beach profile surveys and wave measurements recorded at the field site, Milford-on-Sea (South UK), are being used to estimate the LST rate, thus enabling model calibration for future shoreline change scenarios.

2. Field Site

The field site, Milford-on-Sea, is located in Christchurch Bay (Hampshire, South UK). It shows gravel size above high water level, mixed sediment in inter-tidal beach, sand size below low water line, longshore bar exposed during spring tides and cusps formed under low energy and shore normal wave incidence. The site presents cliff eroding in the western side near Barton on Sea, natural beach in Hordle Cliff and coastal defence structures from Milford on Sea to Hurst Spit. Hordle Cliff is a mixed (gravel and sand) beach dominated by coarse size sediment acting as the main mechanism to protect the cliff.

3. Methodology

An impoundment technique has been performed during October-November 2007 in Hordle Cliff to develop a novel methodology to calculate the longshore transport rate for a mixed beach. A temporary groyne of 40m length and built up with Geotextile bags of 1m x1m x1m filled with native beach material was deployed from berm to low water line to act as a barrier for the sediments movement. Beach profiles were surveyed daily with a DGPS for the two months, at 10m intervals for 150m either side of the groyne, in order to estimate beach volume changes and relate those to the wave conditions also recorded at the field site. These experimental measurements are used to obtain the longshore transport rate by considering the Energy Flux Method (SPM 1984) and applying the CERC equation

[1]School of Eng.,University of Plymouth, Drake Circus, Plymouth, PL4 8AA, UK. ines.martin@plymouth.ac.uk
[2]School of Eng.,University of Plymouth, Drake Circus, Plymouth, PL4 8AA, UK. J.Hughes@plymouth.ac.uk
[3]School of Eng.,University of Plymouth, Drake Circus, Plymouth, PL4 8AA, UK. D.Simmonds@plymouth.ac.uk
[4]Dept of Civil and Environmental.Eng.,University of the West Indies, St Augustine Campus, Trinidad. Andrew.Chadwick@sta.uwi.edu
[5]School of Eng.,University of Plymouth, Drake Circus, Plymouth, PL4 8AA, UK.dominic.reeve@plymouth.ac.uk

for the longshore sediment transport rate calibration.

4. Modelling

A one-line modelling framework coupled with a wave computation model (Li, 1994) is used to predict the shoreline response to the groyne being placed at the site. This approach allows the beach-wave interaction to be captured in the simulations. The model of the shoreline movement includes both cross-shore and longshore sediment transport. In the one-line model theory, a line is chosen at a specified elevation, and the location is evolved with time, using an equation that relates the rate of change of location of the line to the sediment transport rate. The method of lines with an adaptive time step technique is used to solve this equation. The wave model is driven by offshore wave data, and is solved to provide the wave height, bearing and water depth at the breaking line. This information is then used to calculate the sediment transport rate variation along the section of coastline, using the formula by Hanson et al. (2006).

Experimental data is used to calibrate the constants in the formula for the sediment transport rate. Comparisons with contour maps produced from beach surveys taken during the groyne experiment, show that the simulation approach used can model the shoreline deformations that occur due to the presence of a groyne. The model, thus calibrated, will then be used to assess future evolution of this shoreline over the remainder of the observation period. Results will be presented comparing the model output and the observed shoreline response from field surveys.

Acknowledgements

This research is funded by the Faculty of Technology of the University of Plymouth and the EPSRC (Engineering and Physical Sciences Research Council) under the Grant EP/C005392/1.

References

Chadwick, A.J., 1989. Field measurements and numerical model verification of coastal shingle transport. *Advances in water modelling and measurements.* BHRA, The Fluid Engineering Centre, UK. Chapter 27, 381-402.

Hanson, H., Larson, M., Kraus, N.C. & Gravens, M.B., 2006. Shoreline response to detached breakwaters and tidal current: comparison of numerical and physical models. *Proceedings of the 30th International Conference on Coastal Engineering.*

Li, B., 1994. An evolution equation for water waves. *Coastal Engineering*, 23, 227-242.

Mason, T. and Coates, T. T., 2001. Sediment transport processes on mixed beaches: a review for shoreline management. *Journal of Coastal Research,* 17(3): 645-657.

Nicholls, R. J. and Wright, P., 1991. Longshore transport of pebbles: experimental estimates of K. *Proceedings of Coastal Sediments'91,*ASCE, 920-933.

US Army Corps of Engineers, 1984. *Shore Protection Manual.* Coastal Engineering Research Center.

Van Wellen, E., Chadwick, A. J. and Mason, T., 2000. A review and assessment of longshore sediment transport equations for coarse-grained beaches. *Coastal Engineering,* 40: 243-275.

Coastal Dynamics 2009
Paper No. 107

A Two-Dimensional Statistical Model for Morphodynamic Seabed Modelling and Prediction

Anna Bakare[1], Jeremy Morley[2], Richard Simons[3] and Serge Guillas[4]

1. Introduction

Data-driven statistical techniques have previously been used to analyse and characterise natural morphodynamic processes within the spatial and temporal domain of a bathymetric dataset. The use of such methods as a predictive tool have been subject to the limitation that past morphodynamic characteristics may not necessarily continue into the future due to stochastic effects (Larson *et al*, 2003). However there is the potential to mitigate this by deriving an ensemble average against which the likelihood of a specific occurrence can be calculated. Another limitation to existing techniques is that they are implicitly zero-dimensional (0D) point-based methods that have been adapted and applied at multiple dimensions. The extent to which such methods are able fully to capture spatial morphodynamic behaviour is still the subject of investigation (Rattan *et al*, 2005). This paper describes progress in the development of a new spatial and temporal statistical technique applicable to morphological problems with and without the inclusion of anthropogenic influence. The method employs an implicit 2D statistical analysis that predicts future morphological states based on the evolution of morphological behaviour through time. It has the potential to be extended to account for multiple variables in the morphological analysis and prediction. So far, it has been applied to a series of idealised morphological case studies and to data from a study area off the East Anglian coast.

2. Methodology

A statistical spatial regression method (Lai & Schumaker, 2007) previously applied to ozone modelling applications has been adapted and developed for the present study. The method involves deriving morphological spatial behaviour functions for each time step that minimise the error but maximise the spatial relationships between the available data points. Regression analysis is then applied to evaluate the evolution of the behaviour functions through time to predict future behaviour and the corresponding morphology. The spatial regression model is denoted by equation 1:

$$Y = \int_D g(s)X(s)ds + \varepsilon$$

where g(s) is the function space and is what is evaluated across the domain through time, ε is a real random variable that satisfies $E\varepsilon = 0$ (Cardot *et al*, 2003).

3. Idealised Morphological Scenarios and Morphological Application

Spatial regression has not previously been applied to morphological modelling. Therefore a set of idealised morphological scenarios were created to investigate its capabilities and validity. The idealised cases were designed to assess the model's ability to account for changes in bed levels, structure and form of three-dimensional (3D) features, feature movement, periodic behaviours and morphological response to stochastic occurrences. The experiments were also used to assess the sensitivity of the model to behavioural noise and to evaluate the influence of model parameters on the analysis and prediction outputs.

After the initial idealised scenarios the method was applied to real data from a study area off the East Anglian coast that is characterised by dynamic banks and channels. The analysis focused on bathymetric data for sixteen approximately decadal datasets from the study area. These were used to predict the subsequent morphological state which was then validated against the observed bathymetry.

[1]Dept of Civ, Env and Geom Eng., UCL, Gower St, London, WC1E 6BT, UK. a.bakare@cege.ucl.ac.uk
[2]Dept of Civ, Env and Geom Eng., UCL, Gower St, London, WC1E 6BT, UK. jmorley@cege.ucl.ac.uk
[3]Dept of Civ, Env and Geom Eng., UCL, Gower St, London, WC1E 6BT, UK. r.simons@ucl.ac.uk
[4]Dept of Statistical Science, UCL, Gower St, London, WC1E 6BT, UK. serge@stats.ucl.ac.uk

4. Results and Discussion

The results for the idealised scenarios demonstrate that the method is able to identify the main morphological features and predict future states with reasonable accuracy. This assessment is based on the difference between the predicted and actual height values being significantly less than the natural height variability between successive datasets. In particular the model is able to capture and predict feature migration including periodic behaviour and response to stochastic events (table 1). It was observed that in deriving the prediction of future states the model also maintains the morphological structure and form of the features.

Scenario	Description	Prediction error (m)	Actual variability (m)
A	Single alongshore migrating feature	0.0339	0.0864
B	Multiple non-uniform alongshore migrating features	0.0174	0.1448
C	Single migrating feature with a temporal periodicity	0.0146	0.1934
D	Single migrating feature with stochastic influence	0.0179	0.0957
E	Single migrating feature with stochastic influence and noise	0.0416	0.0957

Table 1: Averaged prediction errors and averaged bed variation for scenarios which simulate morphological movement (A & B), temporal periodicity (C), response to stochastic events (D) and evolution with noise (E)

Assessing the applicability to real morphological data the results demonstrate that the methodology is able to pick up various characteristics of the developing bathymetry. This is deduced from an observation that the prediction error is less than the actual natural bed variation rate as described for the morphological scenarios. It is noted that the predicted morphological state is interpreted to be at the same temporal scale as the successive data inputs.

5. Conclusion and Further Development

The analysis undertaken and the results obtained suggest that the methodology is a valid approach for 2D morphological analysis and prediction, in particular where investigating the natural evolution of morphological states and the impact of anthropogenic influences. The method is able to capture different morphological behaviour at the observational time scale and predict future states based on the derived behaviour. An advantage over existing 0D statistical analysis techniques is the ability to assess spatial relationships between data values without applying prior assumptions. However the behaviour being investigated and predicted needs to be present in the input datasets.

The discussion presented so far has focussed primarily on a single characteristic variable. However, there is the potential for extending the methodology to account for multiple variables. Ongoing work is being carried out to investigate the statistical correlation between bathymetric and forcing characteristics to derive the influence of environmental forcing on the morphological calculations within the spatial regression model. This will be included in the prediction calculations.

Acknowledgements

The authors would like to acknowledge the information provided by Ming-Jun Lai of University Georgia (USA), ABPMER for the provision of data and EPSRC for funding this research project.

References

Cardot, H., Ferraty, F. and Sarda, P., 2003. Spline Estimators for the Functional Linear Model. *Statistica Sinica*, 13: 571-591.

Lai, M. J. and Schumaker, L. L., 2007. *Spline Functions on Triangulations*. Cambridge University Press

Larson, M., Capobianco, M., Jansen, H., Różyński, G., Southgate, H. N., Stive, M., Wijnberg, K. M. and Hulscher, S., 2003. Analysis and Modelling of Field Data on Coastal Morphological Evolution over Yearly and Decadal Time Scales. Part 1: Background and Linear Techniques. *Journal of Coastal Research,* 19: 760-775.

Rattan, S. S. P., Ruessink, B. G. & Hsieh, W. W., 2005. Non-Linear Complex Principal Component Analysis of Nearshore Bathymetry. *Nonlinear Processes in Geophysics*, 12: 661-670.

Coastal Dynamics 2009
Paper No. 108

SPATIALLY-VARYING MORPHODYNAMICS OVER A SHORE-PARALLEL ELONGATE SHOAL, SOUTH-CENTRAL LOUISIANA, U.S.A.

Daijiro Kobashi[1,5], Gregory W. Stone[1], Syed M. Khalil[3], Richard E. Condrey[2], John W. Fleeger[4], Mark Grippo[4], Carey Gelpi[2]

Muddy coasts along the Louisiana inner shelf have offshore sand bodies that were formed during the final phase of the Mississippi River delta cycle. These sand bodies, including Ship Shoal, a shore-parallel elongate sand body (Figure 1), have been identified as a viable sand resource for restoration of Louisiana's rapidly deteriorating beach and barrier islands (Khalil et al., 2007). Through field deployments in spring 2006 and winter 2008 and numerical model analysis, spatially-varying unique bottom boundary layer dynamics for heterogeneous sediments have been quantified. This paper synthesizes these unique dynamics and their implications for benthic biological habitats on the shoal. Little documentation of such work exists in the scientific and engineering literature.

Results obtained from bottom boundary layer instruments comprised of high resolution current profilers, pressure sensors, and turbidity sensors, revealed two contrasting morpho-sedimentary dynamics in the shoal for fluid mud and sand. During spring 2006, the instruments deployed on the deeper eastern flank of the shoal, captured up to 40 cm in thickness of fluid mud; the fluid mud further underwent re-suspension, vertical mixing, hindered settling and re-distribution during a winter storm (Figure 2 left). Whereas during winter 2008, instruments were deployed on the shallower middle shoal revealing exposed bottom sediments comprised primarily of fine sand. Thus, resolving sediment dynamics was accomplished by used conventional dynamical solutions (Figure 2 right).

Results obtained from numerical modeling of spectral waves and three dimensional currents, demonstrated spatially-varying hydrodynamics over the shoal. The results were associated with hydrodynamic forcing and water depth. The magnitude of these hydrodynamic parameters has an east-west trend following changes in bathymetry (Table 1). Sediment re-suspension intensity (RI) defined as wave shear stress minus critical shear stress, had the highest values on the western flank, decreased gradually toward the east, and became lowest on the eastern flank (Table 1). During moderate and weak storm conditions, the magnitude of RI has a conspicuous difference between the western flank and eastern flank. The results were consistent with *in-situ* data, leading to the conclusion that the deeper eastern shoal favored fluid mud accumulation during winter storms.

The above spatially-varying dynamics further suggest a strong interplay between spatially-varying habitats of benthic organisms on the shoal whose species are likely influenced by granulometry and composition of bottom sediments. Our collaborative benthic biological study unveiled a spatially-varying distribution of benthic organisms (benthic micro-algae (BMA)) on the shoal (Dubois et al, in preparation; Grippo et al. submitted): high on the western flank and low on the eastern flank, suggesting the importance of water depth and likely of sediment heterogeneity. More detailed results will be included in the full manuscript and presentation.

---

[1]Coastal Studies Institute and Dept. of Oceanography and Coastal Sciences, Louisiana State University, 331 Howe-Russell Geo-science Complex, Baton Rouge, LA 70803, USA. dkobas1@lsu.edu

[2]Department of Oceanography and Coastal Sciences, 1002-Y, School of Coasts and Environment, Baton Rouge, 70803, USA.

[3]Louisiana Department of Natural Resources, 617 North Third Street, LaSalle Building, Baton Rouge, LA 70802, USA

[4]Dept. of Biological Science, Louisiana State University, Life Science Bldg, Baton Rouge, LA, 70803, USA

[5]Present address: Office of Ocean and Atmospheric Research, National Oceanic and Atmospheric Administration, Silver Spring, MD, USA

References

Dubois, S., Gelpi, C., Condrey, R., Grippi, M., and Fleeger, J., Diversity and composition of macrobe nthic community
associated with sandy shoal of the Louisiana continental shelf. In preparation.

Grippo, M., Fleeger, J.W., Condrey, R. and Carman, K.C. High Benthic Microalgal Biomass Found on Ship Shoal, North-central Gulf of Mexico. Submitted

Khalil, S.M., Finkl, C., Andrew, J.,Knotts, C.P., 2007. Restoration-quality sand from Ship Shoal, Louisiana: Geotechnical investigation for sand on a drowned barrier island. Proceeding of *Coastal Sediments '07*. New Orleans, Louisiana, 685-698.

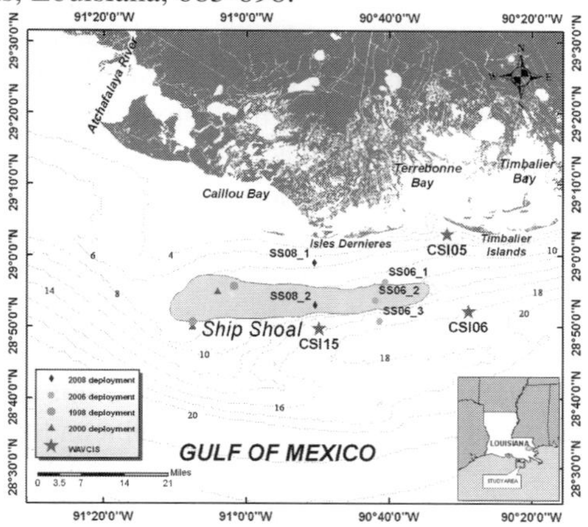

Figure 1 Map of study area

Table 1. Summary of hydrodynamic parameters over the shoal

Parameters	Ship Shoal			
	West	Middle	East	Offshore
Depth	Shallower ◀		▶	Deeper
H_S	Low ◀		▶	High
Current speed	High ◀		▶	Low
ΔH_S(N-S)	High ◀		▶	Low
RI	High ◀		▶	Low
BMA mass	High ◀		▶	Low

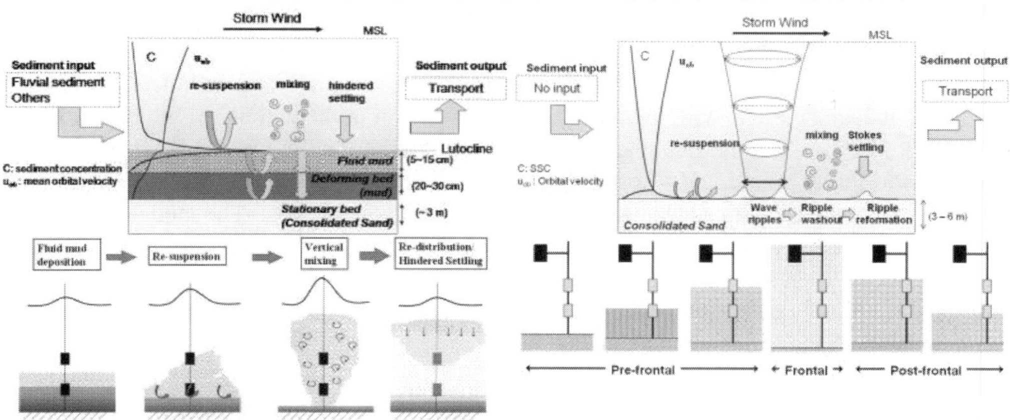

Figure 2 Ship Shoal sediment exchange during fluid mud regime (left) and sand regime (right)

Coastal Dynamics 2009
Paper No. 109

MODELING SEDIMENT SUPPLY AND DELTA MORPHOLOGY ON THE DAMMED ELWHA RIVER, WASHINGTON STATE, USA

Guy Gelfenbaum[1], Andrew Stevens[2], Edwin Elias[3], and Jonathan Warrick[4]

1. Introduction

The geomorphology of a river delta is determined by a balance among river sediment supply, relative sea level, and coastal processes, including waves and tides. Changes in any one of these forcings will alter the distribution of sediment and resulting delta morphology. A common anthropogenic influence on deltas arises from reducing sediment supply by the damming of coastal rivers. Dams within coastal watersheds can reduce sediment supply to the coast, resulting in shoreline erosion as well as modification of delta and nearshore habitats. Well-documented examples of shoreline modifications resulting from reduced sediment supply on a dammed river system include the Ebro Delta in Spain (Jimenez, et al., 1997) and the Nile Delta in Egypt (Frihy et al., 2008).

To study the effects of changing sediment supply on delta morphology in a mixed grain size sediment system we examine the Elwha River delta in northern Washington State, USA. The Elwha is ideal because 1) two dams significantly reduce sediment supply to the delta (Randle et al., 1996), 2) there is a wide range of sediment sizes on the beaches of the delta and trapped behind the dams, and 3) both dams will be removed in the near future. The goal of our work is to develop a process-based hydrodynamic and sediment transport model that can predict short- and long-term delta morphological response to changes in sediment supply in a mixed sediment system.

2. Setting

The Elwha River drains approximately 800 km^2 of the Olympic Mountains of Washington State, and empties into the Strait of Juan de Fuca, the entranceway to Puget Sound. Early in the last century two dams were constructed on the Elwha River for hydroelectric power. Elwha Dam was built 8 km from the river mouth in 1913 and Glines Canyon Dam was built 21 km from the mouth in 1927 (Duda et al., 2008). Field surveys estimate that the reservoirs behind the two dams trapped 13.5 million m^3 of sediment, composed of 52% silt/clay, 35% sand, and 13% gravel/cobble (Gilbert and Link, 1995).

The Elwha River delta is a nearly symmetrical protrusion into the Strait of Juan de Fuca. Beaches are mixed sand and gravel. Shoreline erosion rates on the delta have increased from 0.3 m/yr during 1936-1990 to 1.7 m/yr during 1990-2006 to 2.8 m/y during 2004-2006 suggesting a greater impact from the reduced sediment supply to the delta (Warrick et al., submitted).

3. Approach

We have developed a spatially explicit, process-based hydrodynamic and sediment transport model for the Strait of Juan de Fuca, with increased spatial resolution in the vicinity of the Elwha River delta. The model is forced with tidal boundary conditions where the Pacific Ocean meets the Strait and at the entrances to Puget Sound and Georgia Basin. Waves are modeled using SWAN and include both swell entering from the Pacific Ocean and locally-generated wind waves. The model is calibrated and verified against water levels at tide stations throughout the Strait, and current and wave data collected at two sites, one on each side of the delta during the spring of 2005 and the winter of 2006.

The model will be scaled up to simulate long-term morphological responses to changing Elwha

[1] U.S. Geological Survey, 345 Middlefield Rd., Menlo Park, CA 94025, USA. ggelfenbaum@usgs.gov
[2] U.S. Geological Survey, 345 Middlefield Rd., Menlo Park, CA 94025, USA. astevens@usgs.gov
[3] Deltares, 345 Middlefield Rd., Menlo Park, CA 94025, USA. eelias@usgs.gov
[4] U.S. Geological Survey, 400 Natural Bridges Dr., Santa Cruz, CA 95060, USA. jwarrick@usgs.gov

River sediment supply using a morphological tide and a representative wave climatology (deVriend et al., 1993). The model is operated in 3D to capture buoyant river-plume dynamics and dispersal of fine-grained sediments, and in 2D depth-averaged mode to simulate wave and tidal processes driving littoral transport of coarse-grained sediments.

4. Initial Results

Field measurements indicate that most wave energy (about 90%) reaching the Elwha River delta originates in the Pacific Ocean. Despite significant decay after propogating through the Strait, wave energy is sufficient to move gravel-sized sediment on the beaches. When storm waves coincide with high tides, beach erosion and gravel overwash occur. Model simulations suggest that instantaneous tidal currents are greater on the east side of the delta, while residual currents are greater on the west side. In addition, large-scale residual eddies form on both sides of the delta, suggesting that spatial heterogeneity in tidal flow could impact fine-grained sediment dispersal (Figure 1).

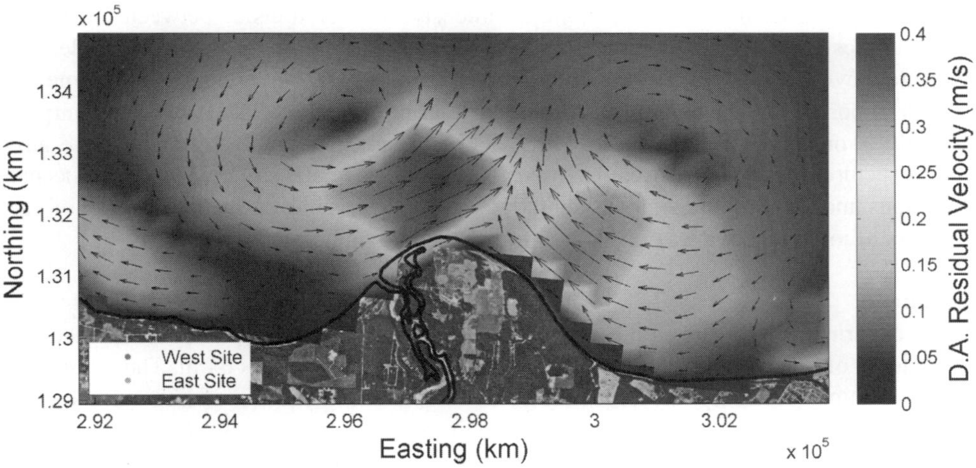

Figure 1. Modeled depth-averaged residual tidal currents around the Elwha delta showing large-scale eddies.

Acknowledgements

This work is funded by the U.S. Geological Survey, Coastal and Marine Geology Program.

References

De Vriend, H.J., Capobianco, M., Chesher, T., De Swart, H.E., Latteux, B. and Stive, M.J.F., 1993. Approaches to long-term modelling of coastal morphology: a review. *Coastal Engineering*, v. 21, p. 225-269.

Duda, J.J., J.E. Freilich, and E.G. Schreiner, 2008. Baseline studies in the Elwha River ecosystem prior to dam removal: Introduction to the special issue. *Northwest Science*, v. 82, p. 1-12.

Frihy, O.E., S.M. Shereet, and M.M. El Banna, 2008. Pattern of beach erosion and scour depth along the Rosetta Promontory and their effect on the existing protection works, Nile Delta, Egypt. *Journal of Coastal Research*, v. 24, n. 4, p. 857-866.

Gilbert, J.D. and R.A. Link, 1995. Alluvium distribution in Lake Mills, Glines Canyon Project, and Lake Aldwell, Elwha project, Washington: Boise, Idaho, Bureau of Reclamation, Elwha Technical Series PN-95-459, 61 pp.

Jimenez, J., A. Sanchez-Arcilla, H. I. Valdemoro, V. Gracia, and F. Nieto, 1997. Processes reshaping the Ebro delta, *Marine Geology*, v. 144, issues 1-3, p. 59-79.

Randle, T. J., C. A. Young, J. T. Melena, and E. M. Ouellette, 1996. Sediment analysis and modeling of the river erosion alternative. Elwha Technical Series PN-95-9, U.S. DOI, Bureau of Reclamation, Boise, ID.

Coastal Dynamics 2009
Paper No. 110

EXPERIMENTAL STUDY OF IRREGULAR WAVES TRANSFORMATION IN A COASTAL ZONE

Yana Saprykina[1] and Sergey Kuznetsov[1]

1. Introduction

Many specific problems in coastal zones require a precise description of wave field and knowledge of wave parameters. Recent investigations have shown that sediment suspension and transport processes as well as extreme wave loads on coastal structures are determined by individual waves or wave groups and can never be estimated correctly on the basis of time averaged parameters. For example, high order wave moments and their time-space variability determine the wave breaking, sand transport, infragravity waves and current generation (Symonds et al., 1982). During nonlinear transformation of waves in the coastal zone an intensive energy transfer between the first and the second harmonics exists, it is periodical in space and could cause formation of underwater bars (Boczar-Karakiewicz et al., 1987). So, variability of parameters of individual waves in time and space is extremely important for the dynamical processes description in the coastal zone. The purpose of the present study is on the base of laboratory and field experiments to investigate how the amplitude-frequency structure of individual irregular waves varies in space and time, what is the mechanism of such variations and whether the simple qualitative model is possible for the description of variability of parameters of individual waves during their transformation in coastal zone.

2. Experiments

Laboratory experiment was carried out in a laboratory flume of Institute of Hydroengineering of the Polish Academy of Sciences, Gdansk, in 2005. The length of a flume was 64 m. The duration of wave records was 4 min, with sampling frequency of 200 Hz. For registration of free surface elevation 7 resistance and 8 capacitance type wire gauges were used simultaneously. Gauges were placed at distances 4 m up to 47 m from the wave maker through each 3 m. Transformation of initial bichromatic regular waves over a horizontal and inclined bottom.

For analysis of irregular waves the data of field experiment “Shkorpilovtsy2007” (Bulgaria, Black sea, 2007) were used. Registration of the free surface elevation at 15 points of coastal zone was performed with 7 capacitance and 8 resistant type wire gauges. The measurements by all of the gauges were taken simultaneously at sampling frequency 5 Hz during one hour.

3. Methods

A description of amplitude-frequency contents of individual wind waves is not simple task because waves in coastal zone have a group structure and are irregular and nonlinear. Therefore we restrict our investigation by next main assumption: we will take into account only first and second nonlinear wave harmonics; harmonics of higher order are neglected.

By analogy with the theory of Stokes waves it is possible to define a time chronogram of free surface elevation ($w(t)$) on some distance from coast as a sum of multiple harmonics with slowly varying in time amplitudes:

$$w(t) = a_1(t)\cos(\omega\, t) + a_2(t)\cos(2\omega\, t + \phi) \quad (1)$$

where the instantaneous amplitudes $a_1(t)$ and $a_2(t)$ are the envelopes of waves of corresponding frequency bands. The frequency bands of the harmonics are defined visually by localization of wave

[1]P.P. Shirshov Institute of Oceanology of Russian Academy of Sciences, Nakhimovskii prospect, 36, Moscow, 117997,Russia. e-mail: saprykina@ocean.ru, kuznetsov@ocean.ru

spectrum minima and maxima. Further the envelopes of the first and the second harmonics are calculated by the following formula :

$$e_{1,2}(t) = \sqrt{\{L[w(t)]\}^2 + \{H[L[w(t)]]\}^2}\ , \tag{2}$$

where *H[ ]* - Hilbert transform, *L[ ]* - linear operator of filtration by required frequency band, *w(t)* - the measured free surface elevation.

The variations of the amplitude-frequency structure of waves during their transformation were studied by analysis of the evolution of wave spectra, spectra of the envelopes $e_{1,2}$ and function of coherence between envelopes. To investigate relation higher order a bispectral analysis was applied.

4. Results

The results obtained for laboratory waves were confirmed by analysis of experimental field data.

It was revealed that during the transformation of the waves due to nonlinear energy transfer between harmonics the amplitude-frequency structure of individual waves varies quasi-periodically in space, as well as in time. The phase shift between the first and second harmonics strongly depend on the main direction of energy transfer. Because the ratio between the amplitudes of the first and second harmonics of waves also varies in time, waves with small first harmonics can have large higher harmonics and there is the effect of "filling" of intervals between wave groups with large amplitudes of first harmonics by wave groups with large amplitudes of highest wave harmonics. The statistical reliability of "filling effect" is proved by significant values of correlation function between envelopes of first and highest harmonics.

It was shown that construction of the simple qualitative model describing variability of individual waves in a coastal zone based, for example, on generalization of the theory of Stokes waves is possible only at an initial stage of deformation of waves then the amplitude of the second nonlinear harmonic is proportional to a square the main. At approach of waves to a coast, the amplitude-frequency structure of individual waves varies chaotically and is not possible to describe waves on the base of any simple qualitative model. The main reason of this is the additional irregularity of waves due to backward nonlinear exchange of energy between first and highest harmonics of waves.

As main result the physical concept of formation and evolution of irregular waves in a coastal zone in which the basic mechanism are nonlinear processes is suggested.

Acknowledgements

This work has been developed within the financial support of Russian Foundation for Basic research, project 08-05-00648.

References

Boczar-Karakiewicz, B., Davidson-Arnott, R. 1987. Nearshore bar formation by non-linear process – a comparison of model results and field data. *Marine Geology*, 77: 287-304.

Symonds, G., Huntley, D., Bowen, A. 1982. Two dimensional surf-beat: long wave generation by a time-varying break point. *Journal Geophysical Research*, 87(C1): 492-498.

Coastal Dynamics 2009
Paper No. 111

SPRAY SIZE DISTRIBUTIONS AT FLIP-THROUGH WAVE OVERTOPPING

Yasunori Watanabe[1] and Shin-ichiro Ishizaki[2]

1. Introduction

When a breaking-wave crest approaches a wall, the forward wave trough rapidly rises to focus with the crest on the wall, resulting in violent overtopping with significant impact wave pressure -- a so-called flip-through overtopping (Cooker and Peregrine 1990). The sprays occurring in this flip-through mode diffuse at high velocity over 5 m/s disperse over wide regions, which often cases traffic disturbances, structural damages of facilities and human damages behind coasts. In coastal engineering, most attention has been on overtopping discharge and water volume behind structures in terms of structure design rather than spray dispersions, and therefore little is known about geometric and dynamic features of overtopping sprays.

In this study, backlight-imaging measurements were performed to acquire fundamental spatial distributions of spray sizes at flip-through overtopping. A novel technique has been used for detecting the spray edges on high-speed video images. The mechanisms to form sprays and the details of size-spectrum through the flip-through event are discussed on the basis of the experiments.

2. Experiments and Imaging Process

The experiments were performed in a wave flume (L: 24m, W: 0.6m, H: 1.0m) with a piston type wave maker with absorption function. A transparent acrylic vertical wall was installed at water depth of 0.1m on a uniform sloping bed (slope of 1/30). A square LED light array (100*100mm) was set on a rear of the wall, and shadows of the sprays passing through an area illuminated by the LED back-light were recoded by a high-speed video camera from the shore-side of the wall (See figure 1).

The size distributions of the sprays on the acquired video images were estimated by detecting edges of the spray shadows. There have been many edge detecting techniques: simple image binarization, Canny and Sobel detections. While these methods need to give any threshold for the image intensity or its gradient that often depends on optical environment and recording devices, no reasonable thresholding method for general backlight shadowgraph has been found. In this study, a level-set representation of an active contour model based on fitting energy algorithm (Chan and Vase 2001) was applied to the shadow images. The shadow edges are determined to minimize deviations from the mean image intensity over the regions in- and out-sides of the edge boundary in this method (See figure 2).

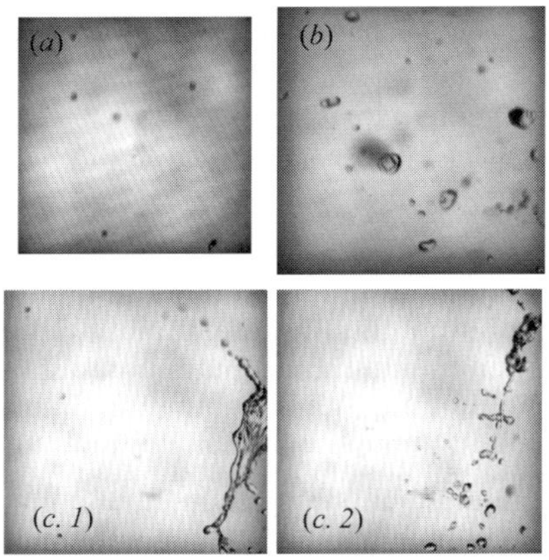

Figure 1 Spray backlight images: (a) initial fine s prays, (b) sprays dispersed due to collapse of the air pocket, (c) fragmenting sprays

[1]School of Engineering, Hokkaido University, North 13 West 8, Sapporo 060-8628, Japan, yasunori@eng.hokudai.ac.jp.
[2]School of Engineering, Hokkaido University, North 13 West 8, Sapporo 060-8628, Japan, ishizakix@eng.hokudai.ac.jp.

The estimated spray-sizes were statistically evaluated over 20 trials. The backlight and camera were vertically traversed at 50mm spacing from 0.35m to 0.95m above the still water level. Monochromatic waves with the incident wave height of 145mm and wave period of 1.8s were used in the experiments.

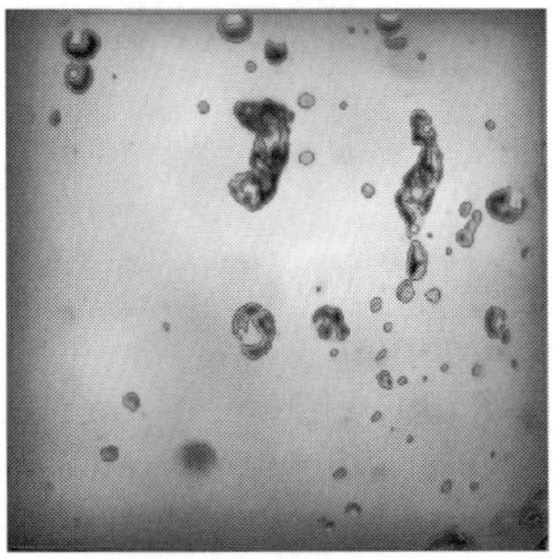

Figure 2. Detected edges of spray s hadow by the level-set method.

3. Results

Three different types of spray dispersions have been observed (See figure1): (a) fine sprays, of which the diameters are smaller than 0.8mm, rising at high velocity over 6 m/s at the inception of the flip-through event, (b) large numbers of sprays in moderate size range (about 1.6mm) radically diffuse when an air pocket formed by a wave face and wall is squashed, (c) relatively large sprays (about 2mm) fragmented from the overtopping jet, which are transported with the same velocity as the jet (1—2 m/s). It has been observed that large sprays re-fragment into multiple smaller sprays during the uprushing phase.

The spray number density exponentially decreases with vertical level, and fine sprays (smaller than 1mm) predominate at an early stage of the event. Then, the mean spray diameter gradually increases up to 1.5 – 2.0 mm until the spray falls at 0.55s after the inception of overtopping.

The maximum spray-size spectrum occurs at 0.7 — 0.8 mm at any measurement level (See figure 3). It is also found that another spectrum peak appears about 2 mm at relatively high measurement levels. These double spectrum peaks are possibly resulted from differences of the mechanisms to form spray, described earlier. The spectrum slope becomes steep with vertical level during radical spray dispersions and consecutive fragments.

References

Cooker, M.J. and Peregrine, D.H, 1990. A model for breaking wave impact pressure, Proce. 22nd Int. Conf. on Coastal Engineering,: 1473 – 1486.

Chan, T. and Vese, L., 2001. An active contour model without edges, IEEE Trans. on Image Processing, 10: 266 – 277.

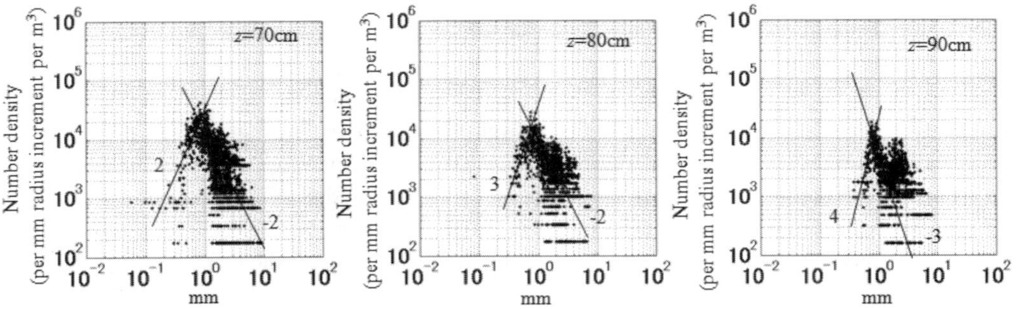

Figure 3 Spray-size spectra at 0.7 m (left), 0.8 m (middle) and 0.9 m (right) above the still water level.

Coastal Dynamics 2009
Paper No. 112

SPRAYS AND FINGER JETS EVOLUTION FROM SPLASHING JETS

Ayumi Saruwatari[1] and Yasunori Watanabe[2]

1. Introduction

A large number of sprays are generated and dispersed over wide regions in the surf zone. The sea sprays often damage coastal environment, structures and facilities behind coasts, and causes traffic disturbances on coastal roads and railways. Furthermore, a large amount of fine spray-aerosols are also known to be produced in surf zone, which enhances local temperature and moisture exchange between atmosphere and ocean. Fundamental properties of the spray dispersion need to be understood to estimate the above engineering and environmental issues.

A spray formation under splashing jets on a still water surface has been found computationally in Saruwatari & Watanabe (2008); that is, the front edge of the splashing jet is fragmented into finger-like jets, and the sprays are pinched off from the tips of the fingers via surface tension instability. Another spray formation at contacts between a droplet and receiving surface has been observed in experimental studies (e.g. Throddsen, 2002). In order to understand details of evolution of the sprays and fingers, in this study, distributions of the sprays at the jet splashing were measured by visual experiments. The evolutions of the size distributions and volume fractions of the finger jets and sprays, and the dispersion velocities were investigated.

2. Experimental Setup

A planar fluorescent water jet was obliquely splashed on a still water surface as a model of a wave-breaking jet. The fluorescent light was excited and emitted from the resulting sprays and secondary jet, projecting from the splashing point, by blue LED illumination, which was recorded by a 12bit high-sensitive CCD camera installed above the flume (see Fig. 1). The shapes of the secondary jet and sprays on the images were measured by a conventional image binarization. The binarization threshold was determined based on a fluorescent intensity induced from a reference water sheet at the measurement location in a calibration procedure. Thirty trials of measurements for six cases of different initial jet velocities and splash angles were preformed for statistically evaluating the results.

3. Results

Two different mechanisms to form the sprays were identified during the splashing process. (A) At the initial stage when the planar jet touches on the still water surface, fine spray jets, so-called "ejecta jet", are rapidly produced on the contacts between the initial planar jet and receiving surface (Fig. 1 (*b*)). The fine sprays are released from the ejecta jets at high velocity over 5m/s that is identical with the velocity of the moving contact, which defines the maximum dispersion velocity throughout the splashing event. (B) When the finger jets are stretched at the front edge of the secondary jet, toes of the fingers are fragmented into multiple sprays via surface tension instability (Fig. 1 (*c*)). These fragmented sprays are consecutively formed until the finger jet re-splashes on the forward water surface, whereas the ejecta jets are generated only for short duration before the secondary jet is formed. These fragmented sprays have much larger in size and slower in dispersion velocity than those of the ejecta-jet induced sprays.

Size distributions of the fingers and sprays were found to be approximated by log-normal distributions throughout the event (Fig. 2 (*a*1), (*a*2)). It is found that the finger width gradually increases as the fingers develop, and the predominant spray diameter is in a range of 2.0-2.5mm

[1]School of Eng., Hokkaido Univ. N13 W8, Sapporo, Hokkaido, 0608628, Japan. saruwata@eng.hokudai.ac.jp
[2]School of Eng., Hokkaido Univ. N13 W8, Sapporo, Hokkaido, 0608628, Japan. yasunori@eng.hokudai.ac.jp

The spray dispersion velocities can be described by following three representative velocities. (i) Maximum velocity subjected to the ejecta jet, representing the maximum spray dispersion velocity. (ii) Velocity of the moving contacts of the planar jet and still surface, which determines mean dispersion velocity. (iii) Initial projecting velocity of the finger jet, representing minimum spray dispersion velocity. A spray dispersion area in the splash event can be estimated using these representative velocities.

Spatial distributions of volume fraction of the finger jets and sprays have been statistically described. For the finger jet, it can be approximated by a cumulative distribution function of a Gaussian distribution (Fig. 2 (*b*1)) --- the volume fraction varies from 1.0 at the neck to 0 at the toe of jet. This distribution is consistent with the computed one presented in Saruwatari & Watanabe (2008). The volume fraction of sprays at the inception of the splashing process is well fitted by a Gaussian distribution (Fig. 2 (*b*2)), suggesting the initial spray dispersion can be described simply by a diffusion equation.

References

Saruwatari, A. and Watanabe, Y., 2008. Transition from splashing jets to fragmented fingers and sprays. *Proc. 31st ICCE*, printing.

Throddsen, S. T., 2002. The ejecta sheet generated by the impact of a drop. *J. Fluid Mech.*, 451, pp. 373-381.

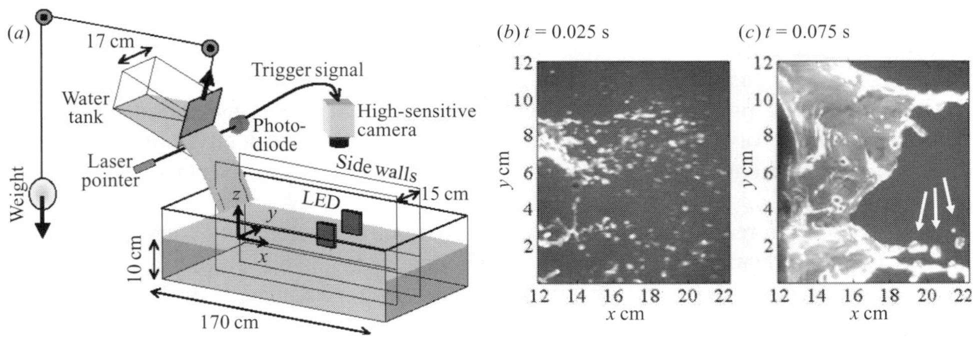

Figure 1. Experimental setup (*a*), and typical images of ejecta jets (*b*) and sprays fragmented from finger jets (*c*)

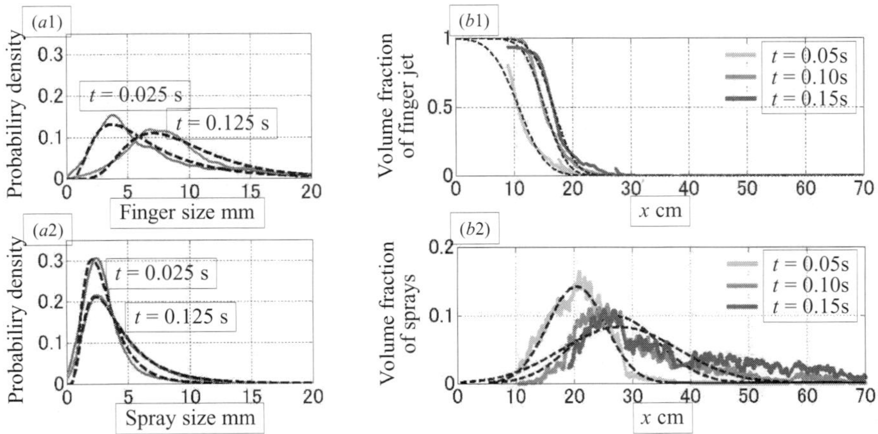

Figure 2. Size distributions of finger jets (*a*1) and sprays (*a*2). Distributions of void fractions for finger jets (*b*1) and sprays (*b*2) (solid lines: experimental results, broken lines: fitting curves)

Coastal Dynamics 2009
Paper No. 113

COMPARISON BETWEEN ANALYTICAL WATER WAVE THEORIES AND MEASUREMENTS MADE IN MEXICAN PORTS

Guillermo Cardoso-Landa[1]

1. Introduction

This paper presents four computer programs developed to calculate the physical characteristics of the progressive water waves using linear, Stokes second order, Stokes third order and cnoidal of Keulegan and Patterson theories: profile of water surface, celerity, wave length, horizontal and vertical orbital velocities, horizontal and vertical orbital accelerations, horizontal and vertical orbital displacements, and under water surface pressure; in some theories energy, potency, group velocity and mass transport. The Turbo-Pascal, version 6.0 programming language was used (likewise the rest of the programs).

LINEAR program requires the following data to be introduced: period, depth, height and phase angle of waves, as the depth 'z' of the particle to be analyzed. In the first part of the program, the Newton-Raphson method is used to determine the wave length. In its second part, the thirteen physical characteristics of waves, providing of equations with hyperbolic functions, are calculated: profile of water surface, celerity, wave length, group velocity, horizontal and vertical orbital velocity, horizontal and vertical orbital acceleration, horizontal and vertical displacements, under water surface pressure, energy and potency.

STOKES-2 program requires the following data to be introduced: period, depth, and height and phase angle of waves, as the depth 'z' of the particle to be analyzed. The eleven physical characteristics of waves, providing of equations with hyperbolic functions, are calculated: profile of water surface, celerity, wave length, horizontal and vertical orbital velocity, horizontal and vertical orbital acceleration, horizontal and vertical displacements, under water surface pressure, and mass transport.

STOKES-3 program requires the following data to be introduced: period, depth, height and phase angle of waves, as the depth 'z' of the particle to be analyzed. The eleven physical characteristics of waves, providing of equations with hyperbolic functions, are calculated: profile of water surface, celerity, wave length, horizontal and vertical orbital velocity, horizontal and vertical orbital acceleration, horizontal and vertical displacements, under water surface pressure, and mass transport.

CNOIDAL program requires the following data to be introduced: depth, height and phase angle of waves, as the depth 'z' of the particle to be analyzed. In the first part of the program, the Newton-Raphson method is used to determine the wave length. In its second part, the nine physical characteristics of waves, providing of equations with elliptic jacobians functions, are calculated: profile of water surface, celerity, wave length, horizontal and vertical orbital velocity, horizontal and vertical orbital acceleration, under water surface pressure and period.

2. Development

Four computer programs were designed in turbo-Pascal programming language to calculate their physical characteristics for twenty particles simultaneously. An equation was developed to relate directly the full elliptical integrals of first and second class with their modulus, necessary for the cnoidal computer program (equation 1).

---

[1]Dept of Civil Eng., Instituto Tecnológico de Chilpancingo, Av. J. F. Ruiz Massieu # 5, Col Villa Moderna, Chilpancingo, Gro., México, mail: gclanda@prodigy.net.mx

$$k^2 = m = \frac{E(k)\dfrac{H}{d}}{\left\{\left[\dfrac{H}{d}\left\{\dfrac{\{K(k)[K(k)-E(k)]\}}{[kK(k)]^2}\right\}\right]+2-\dfrac{H}{d}\right\}K(k)} \tag{1}$$

3. Results

Results after application these four computer programs validate the application region proposal by Littman and Struik (Figure 1) and the experimental works of Lé Mehauté, Divoky and Lyn in the Tetra Tech wave tank, of 32 meters of length, square across section of 1.219 meters and waves generator with a speed control in a rank from 1 to 12 seconds of period and with waves ranks 0.244 $m/s^2 \geq d/T^2 \geq 0.015\ m/s^2$. These results were compared with the calculus of the programs proposed in this work for the free surface profile, speeds and accelerations of the particles and it was observed that although none of these theories adjust precisely, the most accurate was the Keulegan and Patterson cnoidal theory.

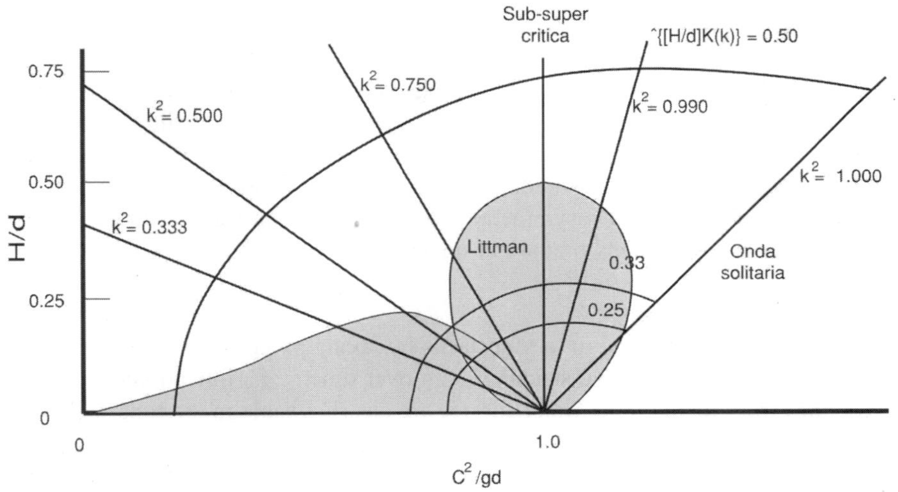

Figure 1 Approximates validity regions for existents profiles for cnoidal water waves (Littman) and Stokes water waves (Struik)

4. Conclusions

Importance of these four computer programs establish oneself not necessary use graphical solutions, and it is possible to compare results between different theoretical water wave models and direct measurements. Several measurements of water wave physical characteristics in some Mexican ports (Acapulco, Zihuatanejo, and Salina Cruz ports) were compared with the application of these computer programs and we present these results.

References

Cardoso Landa, G., 1999. Modelos teóricos del oleaje. *Master thesis*, U.N.A.M. Publisher.

Dean, Robert G., 1970. Relative validities of water wave theories. *Journal of the waterways and harbors division*. Vol. 96, No. WW1, 105-119.

Fuentes Mariles, O. A.; Sánchez Bibriesca, J. L. 1985. *Compendio de hidráulica marítima* (texto y apéndices). Instituto de Ingeniería, U.N.A.M Publisher.

Laitone, E. V., 1962. Limiting conditions for cnoidal and stokes waves, *Journal of Geophysical Research*.

Wiegel, R. L., 1960. A presentation of cnoidal wave theory for practical application, *Journal of Fluid Mechanics, Volume 7, No. 2.*

Coastal Dynamics 2009
Paper No. 114

REMOTE SENSING OF SWASH ZONE BOUNDARY CONDITIONS USING ARGUS

Hannah E. Power[1], Meg Palmsten[2], Rob A. Holman[3], Tom E. Baldock[4]

1. Introduction

One of the standard descriptions for breaking wave swash is the Shen and Meyer (1963) solution of the nonlinear shallow-water equations. While this solution effectively models the shoreline motion, it does not provide a realistic model for the internal hydrodynamics (e.g. Guard and Baldock, 2007). To address this problem, Guard and Baldock (2007) presented characteristic solutions to the nonlinear shallow-water equations that allow the value of the characteristic variable, α, on the seaward boundary of the swash to vary through time, such that $\alpha=2+kt$. Varying k changes the flux of mass and momentum into the swash zone, which in turn controls the subsequent swash hydrodynamics. Unique solutions occur for any given value of k, which can then be interpreted as a boundary condition; the Shen-Meyer solution is recovered for $k=0$. Higher values are suggested for usual swash conditions, as increases in k lead to solutions which have a more symmetric velocity field and a later flow reversal. Figure 1 shows an (x,t) plot of the horizontal velocity, u, for $k=0$ and $k=1$; the gradient of the locus of $u=0$ is quite different in each case. However, the appropriate value of k for different wave conditions and beach gradients is presently unknown. This paper outlines a method to measure k for multiple swash events using the ARGUS camera system. This is achieved by measuring the gradients of the locus of flow reversal for real swash, and comparing these to the predicted values from Guard and Baldock (2007). The ARGUS system provides a powerful tool to obtain the time of flow reversal across the full swash zone for a range of beach slopes and offshore wave heights, particularly under energetic conditions. If k can be measured, then it may also be possible to infer other important parameters, for example, the net asymmetry in the flow kinematics. Indeed, the complete hydrodynamic solution is given solely by the run-up amplitude and k.

2. Method

ARGUS video data can be converted in to timestacks by taking pixel intensity profiles along a cross-shore line from subsequent images and stacking them over time (e.g. Holland and Holman, 1993 and Fig. 2a). Streaklines from foam on the water surface are clearly visible in the timestacks, and the time of flow reversal is clearly indicated as the streaklines change direction. In addition to this, the slope of the streaklines is inversely proportional to swash surface velocity (Palmsten and Holman, 2006). Timestack images of ARGUS data from the swash zone were obtained for calm and storm conditions on a range of beach types, from dissipative to reflective. Each timestack image was used to make a difference timestack image (Fig. 2b), which was created by subtracting the pixel intensities of one timestack from itself with an offset of one row of pixels. The difference images show how much change has occurred at each cross shore location between each time step, with greater pixel intensities (i.e. whiter pixels) indicating greater change. Hence, dark zones represent no flow or flow reversal. For each swash event, the position of the start of the uprush and the maximum uprush were picked on the timestack image by hand, although this can be automated with an appropriate algorithm. To obtain the line along which flow reversal occurs, mean pixel intensities were calculated (from the difference timestack) along profile lines that fan out from around the maximum uprush point. The line with the lowest mean pixel intensity describes the locus of flow reversal in (x,t) space (Fig. 2c). The gradient of the line may then be used to back calculate a value of k for that swash event. The complete process

---

[1]Division of Civil Engineering, University of Queensland, St Lucia, Qld, 4072, Australia. h.power@uq.edu.au.
[2] Coastal Imaging Laboratory, Oregon State University, Corvallis, OR, USA. mpalmste@coas.oregonstate.edu.
[3] Coastal Imaging Laboratory, Oregon State University, Corvallis, OR, USA. holman@coas.oregonstate.edu.
[4]Division of Civil Engineering, University of Queensland, St Lucia, Qld, 4072, Australia. t.baldock@uq.edu.au.

is easily semi-automated in Matlab; the user only has to select the start of the uprush and the maximum uprush for each swash by clicking on the image. Figure 2c shows that the gradient of the locus of flow reversal corresponds to $0.8<k<1$, consistent with estimates for laboratory swash. The full paper will consider how k varies with offshore wave conditions and the beach slope.

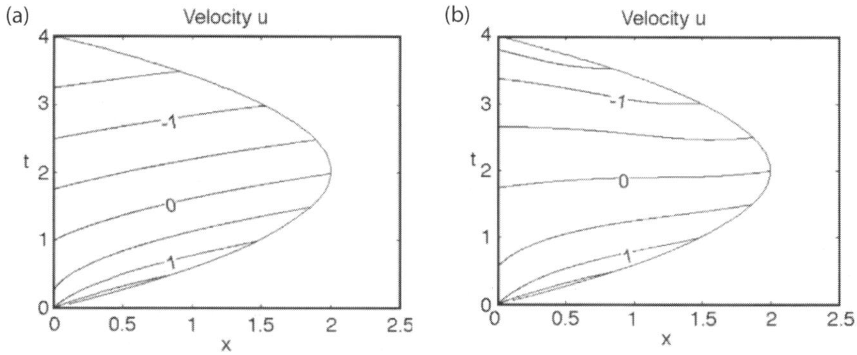

Figure 1. Locus of the position of horizontal flow reversal across the swash zone for (a) k=0 and (b) k=1 given by Guard and Baldock (2007).

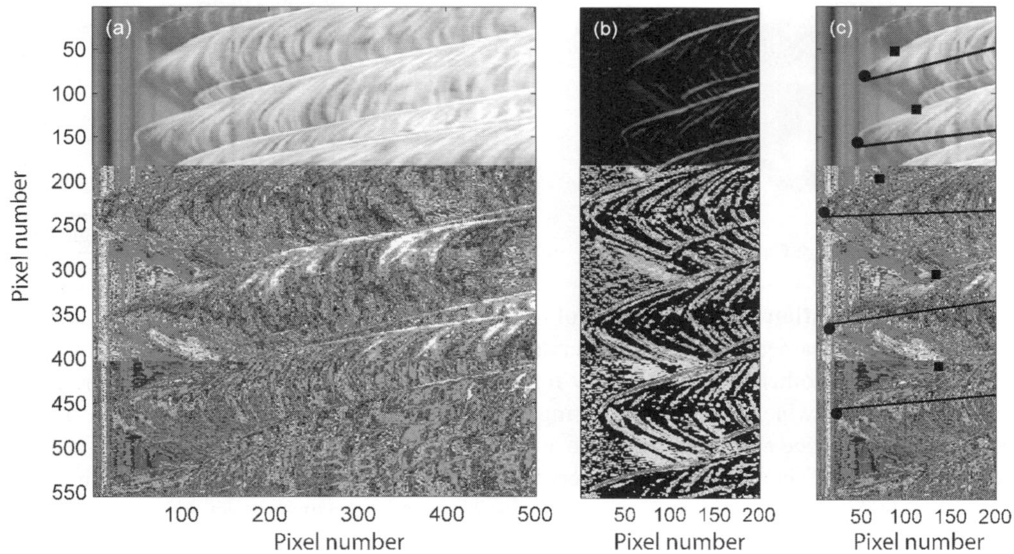

Figure 2. (a) ARGUS timestack collected from Black's Beach, CA, USA during NCEX showing the swash and inner surf zones ($?_o = 0.19$, $H_s = 1.13$ m, $T_p = 5.88$ s, $\beta = 0.027$), (b) difference timestack of the swash zone only, and (c) timestack of the swash zone with the start of the uprush (¦), the maximum uprush (?), and the automatically calculated flow reversal lines (?).

References

Guard, P.A. and Baldock, T.E., 2007. The influence of seaward boundary conditions on swash zone hydrodynamics. *Coastal Engineering*, 54(4): 321-331.

Holland, K.T. and Holman, R.A., 1993. The statistical distribution of swash maxima on natural beaches. *Journal of Geophysical Research*, 98(C6): 10,271-10,278.

Palmsten, M. and Holman, R.A., 2006. Optical measurements of swash kinematics. Proc. 30th Int. Conf. Coastal Engineering. ASCE, pp. 887-895.

Shen, M.C. and Meyer, R.E., 1963. Climb of a bore on a beach Part 3. Run-up. *Journal of Fluid Mechanics*, 16(1): 113-125.

Coastal Dynamics 2009
Paper No. 115

OBSERBATION OF NEARSHORE PROCESSES USING NETWORK CAMERA

Kojiro Suzuki[1] and Shinichi Yanagishima[2]

1. Introduction

Recently, network cameras are installed at a lot of beaches to see coastal areas. These network cameras are installed by municipal governments and private companies who inform wave conditions for surfers. If we can use these cameras, it is economical to observe coastal processes We developed the system to use these network cameras in order to observe coastal processes economically.

2. Network Camera

Figure.1 is Canon web camera which is used worldwide. Table 1 is the camera specification. The pixel resolution is VGA (640×480) and enough to analyze the coastal processes. Since Canon Inc. provides SDK (System Design Kit) for this camera, we can easily program the remote control code written by Visual C or Basic.

Figure 1. Network camera Canon VB-C50iR

Table 1. Specification of the network camera

Name	Canon VB-C50iR
Dimensions	W136×D130×H110(mm)
Pixels	1/4" CCD 340,000 Effective Pixels
Pan	+/-170° 1-90°/sec
Tile	-90°/10° 1-70°/sec
Zoom	26X Optical Zoom/12X Digital Zoom

3. Long Term Observation and Estimation of the trend of shoreline change

Using the network camera, long term observation is conducted in several beaches. Some cameras are installed by us and others are installed by municipal governments. Figure 2. shows the tidal flat around Haneda Airport which is captured by single camera. The camera is controlled every 5 minutes via internet from our office and can cover wide area.

Figure 4. shows the shoreline change observed at Hasaki beach (Figure.3). The shoreline position varies with the astronomical tide, wave setup, storm surge, sand deposit, erosion and so on. In order to estimate the trend of shoreline position movement due to sand deposit and erosion, we separate the effect of these terms using Akaike's Bayesian Information Criterion (ABIC) (Akaike, 1980) as expressed by equation 1 and 2. The detail of these equations will be explained in the conference.

The smooth line in the figure 4(a) shows the trend due to the sediment erosion. Shore line moved offshore after several typhoon hits.

$$\underbrace{x_n}_{\text{Shoreline}} = \underbrace{\sum_{m=1}^{M}(a_{2m-1}\cos\omega_m t_n + a_{2m}\sin\omega_m t_n)}_{\text{Astronomical tide}} + \underbrace{\sum_{k=0}^{K2} bss_k SS_{n-k}}_{\text{Storm Surge}} + \underbrace{\sum_{k=0}^{K3} bwa_k WA_{n-k}}_{\text{Wave Setup}} + \underbrace{d_n}_{\text{Trend due to sediment movement}} + \overset{\text{Noise}}{\varepsilon_n} \quad (1)$$

$$\mathrm{ABIC} = N\ln 2\pi + N + N\ln\hat{\sigma}^2 - K\ln v^2 + \sum_{i=1}^{2M+K}\ln S_{i,i}^2 \quad (2)$$

[1]Marine observation Group, PARI (Port & Airport Research Institute), 3-1-1 Nagase, Yokosuka, 239-0826, Japan. suzuki_k@pari.go.jp
[2]Coastal Sediments and Processes Research Group, PARI, yanagishima@pari.go.jp

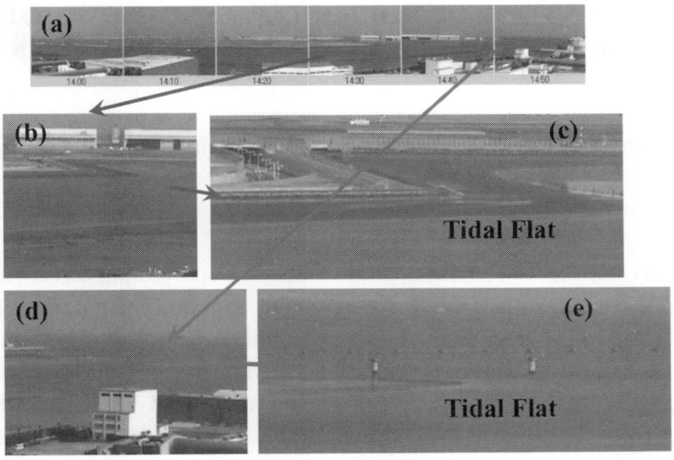

Figure 2. Tidal flat around Haneda Airport observed by single camera

Figure 3. Time exposure images at Hasaki beach, Oct.7.2007.

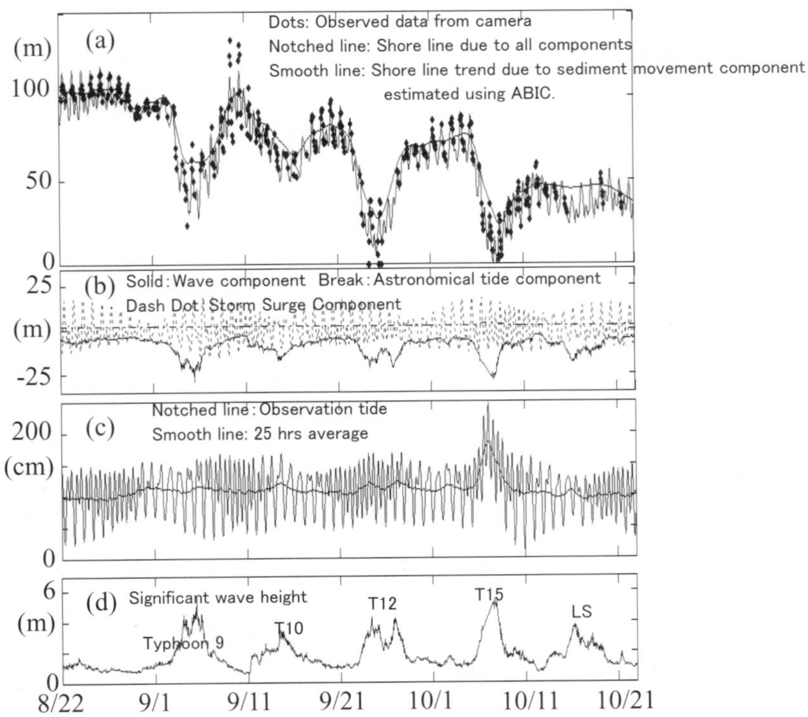

Figure 4. Estimation of shoreline change at hasaki beach in 2007 autumn.

References

Akaike, H., 1980. Likelihood and Bayes procedure, Bayesian Statistics, *University Press, Valencia, Spain,* pp.143-166.

Coastal Dynamics 2009
Paper No. 116

OBSERVATION OF MORPHOLOGY AND FLOW MOTION AT THE RIVER MOUTH OF TENRYU WITH X-BAND RADAR

Satoshi Takewaka[1], Yu Takahashi[2], Yoshimitsu Tajima[3] and Shinji Sato[4]

1. Introduction

X-Band radar observation has been conducted to observe morphology, flow and waves at the river mouth of Tenryu River, Japan. This imaging radar captures spatial distributions and temporal variation of water lines of the river channel and coast lines, and wave propagation in the shallow area. Breaching of the sand bar by a storm event and its recovery process, and behavior of the front formed at the river mouth by the river discharge are displayed in the study. The river flows into Enshu Coast which is suffering from severe erosion and enhancement of sediment supply from the catchment is planned to ease the degraded coast. The observation, in this context, aims to offer basic understanding of the behavior of the river mouth in morphological and hydrodynamic aspects.

2. Study Area and X-Band Radar Observation

The radar is installed on the roof of a sewage plant which locates on the right side bank of the river as shown in Figure 1. The measurement started in June 2007 and is still continuing (January 2009). Two typhoons traveled thorough the vicinity of the area in July and September of 2007. Averaged radar images over 17 min. are stored hourly and individual radar images are preserved for energetic periods. Features like, shore positions, breaker zone, front formed by the river discharge, etc., are interpreted from the averaged images. Procedures of the radar data analyses for waves and morphology are established in the previous works (Hasan and Takewaka, 2007; Elsayed and Takewaka, 2008).

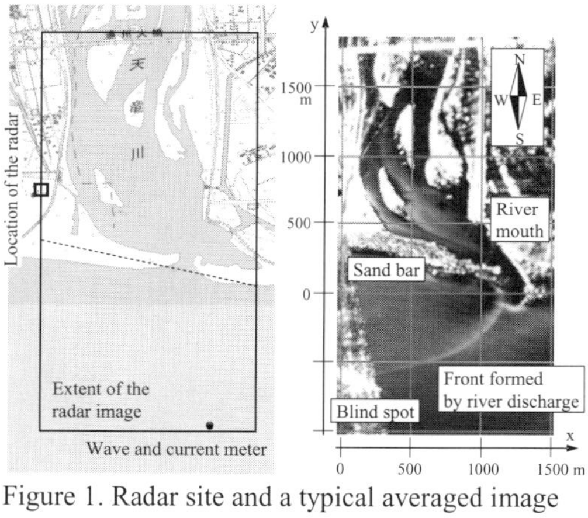

Figure 1. Radar site and a typical averaged image

3. Breaching of the sand bar and its recovery

Typhoon T4 approached the region in the middle of July 2007 accompanied with heavy rainfall, and the flood flow (maximum discharge ~ 9,000 m^3/s) breached the sand bank which was developed at the river mouth. Unfortunately, the radar failed to capture the detailed breaching process due to system trouble; however, amount of the eroded area of the sand bank is estimated from the intermittent averaged images. In the beginning of September 2007, another typhoon T9 traveled along the area and high waves attacked the sand bar, but, deposition was observed rather than erosion this time. The pre/post situations of these storms are shown in Figure 2. Figure 3 and 4 show relative variations of sand bar area, tip location of the sand bar and width of the river mouth, respectively.

Averaged images of Figure 2 indicate that the channel appeared in the sand bar by breaching was almost filled in the following 2 months, and the widened river mouth narrowed gradually. The image

---

[1] Dept. of Eng. System and Energy, University of Tsukuba, Ibaraki 305-8573, Japan. takewaka@kz.tsukuba.ac.jp
[2] Ditto.
[3] Dept of Civil Eng., University of Tokyo, Tokyo 113-8656, Japan. yoshitaji@civil.t.u-tokyo.ac.jp
[4] Ditto. sato@civil.t.u-tokyo.ac.jp.

captured in December 2007 shows that the configuration of the area is close to the pre-storm situation. Breached amount was approximately 100,000 m^2, and half of this was recovered after 5 months. Detailed analyses revealed that the sand was mainly supplied from the coastal side, which can be inferred from the eastern migration of the sand bar tip induced by the longshore sediment transport as shown in Figure 4.

4. Motion of the River Water Front

River discharge intrudes into the coastal water and forms a front which is captured in the averaged radar images as shown in Figure 1. It was confirmed that linear pattern captured in the radar image corresponds to a strip which appeared in satellite remote sensing image where surface color showed a sudden change. Also an in-situ towing measurement on water temperature, salinity and turbidity was conducted across the pattern. All of the measured items showed a sudden jump when the boat crossed the pattern, indicating that river water was intruding into the coastal water. The intensities of the pattern captured in the radar images increased at ebb tides. They also showed a dependence on the width of the river mouth; a clearer front was frequently observed when the river narrowed.

Acknowledgements

This research is supported by the Special Coordination Funds for promoting Science and Technology of Ministry of Education, Culture, Sports, Science and Technology and the River Fund, The Foundation of River & Watershed Environment Management.

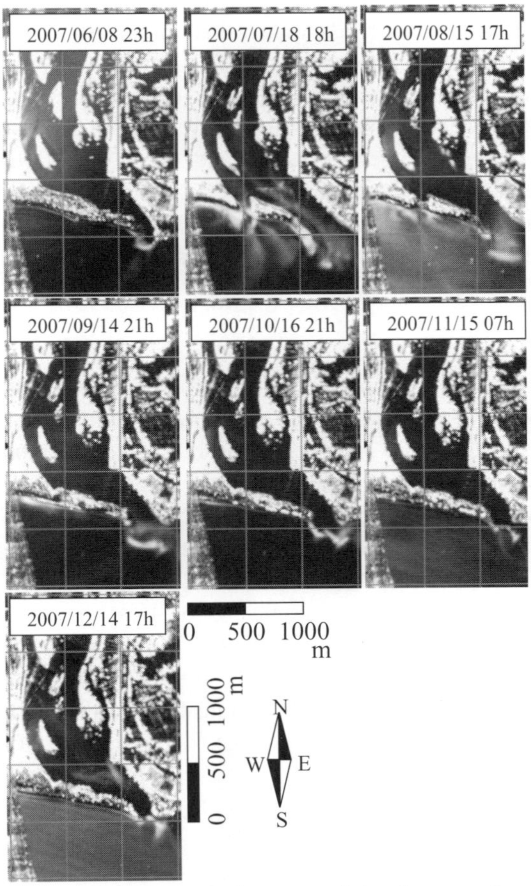

Figure 2. Morphological process of the river mouth. The images are captured at the same tide level.

References

Hasan, G. M. J. and Takewaka, S., 2007. Observation of a stormy wave filed with X-band radar and its linear aspect, *Coastal Engineering Journal*, 49, 149-171.

Elsayed, M. G. and Takewaka, S., 2008. Longshore migration of shoreline mega-cusps observed with X-band radar, *Coastal Engineering Journal*, 50, 247-276.

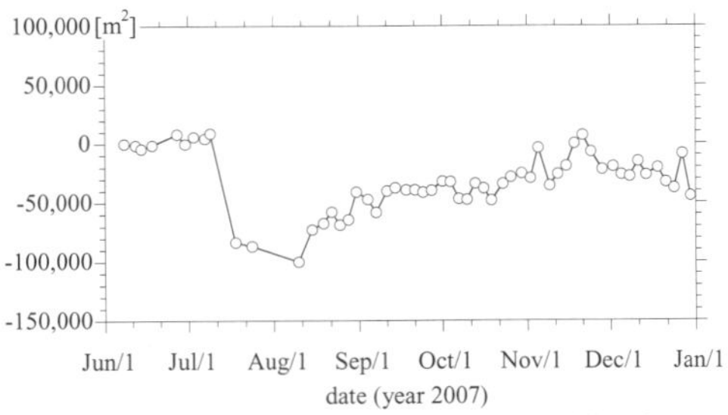

Figure 3. Relative variation of the sand bank area

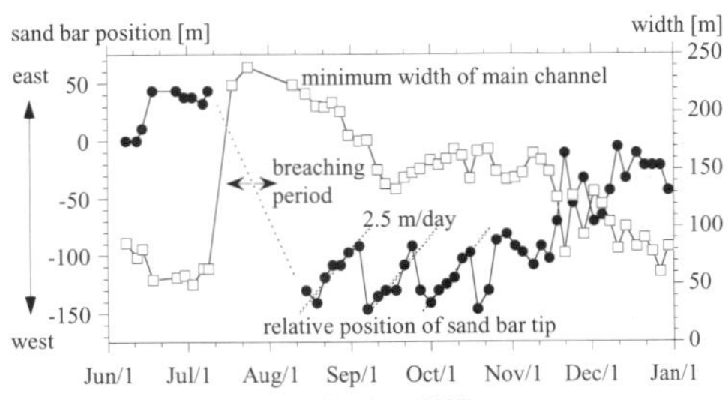

Figure 4. Variation of the relative sand bank tip location and width of the river mouth.

Coastal Dynamics 2009
Paper No. 117

SHORELINE IDENTIFICATION USING SATELLITE IMAGES

Gabriela García-Rubio[1], David Huntley[2], Kenneth Kingston[3], Luciana Esteves[4]

1. Introduction

Satellite images have helped to improve the understanding of coastal changes and nearshore dynamics in remote areas or areas with little or no direct measurements, providing a means of measuring shoreline location and, potentially, changes in shoreline position over time (Kumar *et al* 2007, Chu *et al* 2006, Lafon *et al* 2004, White and El Asmar 1999, De Wolf 1994). This study is focussed on the beaches at Progreso on Yucatan, in the SE part of Mexico between the Caribbean Sea and the Gulf of Mexico. The study area has typically low wave heights (Hs < 30 cm) and a small tidal range (around half a meter). The incident waves show two main directions North and East (Capurro *et al* 2006). The site is located within the Atlantic Hurricane Belt, with a hurricane season between June and November. The current study tests the detection limits of shoreline identification from satellite images by comparing shorelines from in situ data and satellite images taken within a few days of each other. Comparison between images taken nearly 20 months apart is used to assess the ability of satellite-derived shorelines to monitor medium and long term changes.

2. Methods

Two SPOT satellite colour images with pixel resolutions of 10 meters from September 20th 2008 and February 6th 2006 for Progreso beaches were georeferenced using 31 in situ Ground Control Points obtained during the fieldwork, giving an RMS error of less than a half pixel. Shoreline identification in the georeferenced images was made using an unsupervised classification, followed by visual examination and editing of the shoreline where obvious inaccuracies occurred. In situ waterline position was obtained on September 9th 2008 using a DGPS. Although there are 11 days between one of the satellite images and the field measurement, weather and wave conditions remained calm throughout the intervening period. Hence it is assumed that no significant changes occurred over that time. Ground proof of shoreline position obtained from the 2006 satellite image was assisted by 10 beach profiles which were measured on January 10th by Capurro *et al* (2006), only thirty one days before the 2006 image was taken.

3. Results

Comparisons between the measured and satellite-based shoreline positions for 2008 show good agreement, with variations generally of the order of one pixel (10m). In figure 1 we show a comparison between the 2008 in situ shoreline and the 2006 image. Figure 1a shows a region where the agreement between the two waterlines is within a pixel, suggesting little shoreline change over the two and a half year interval between the measurements. However there are some stretches on the beach where measureable horizontal differences occur. A clear example is to the west of the jetty at Progreso, shown in Figure 1b, where acretion of 24 to 32 m occurs. This difference cannot be explained by differences in tide level or methodological error. Beach profiles collected in January 2006, within a month of the image, suggest a typical slope at the water line of about 0.05 which, for an expected tide level difference of 23 cm leads to a possible horizontal difference of no more than 4.6 m. Furthermore, the accretion has a limited alongshore extent, with matching shorelines on either side. The magnitude of the observed change is consistent with

[1]School of Earth Ocean and Environmental Sciences (SEOES). University of Plymouth, (0) 1752 584715 Plymouth, United Kingdom, gabriela.garciarubio@plymouth.ac.uk,
[2]SEOES, University of Plymouth, (0)1752 584592 Plymouth, UK.D.Huntley@plymouth.ac.uk
[3]SEOES, University of Plymouth, (0)1752 584720 Plymouth, UK, K.Kingston@plymouth.ac.uk
[4] School of Geography, University of Plymouth, (0) 1752 585971, luciana.esteves@plymouth.ac.uk

measured beach profiles which show changes in shoreline position ranging between 0 and 7 m over a few months at specific beach points, corrected for the tide. Although we do not have a direct in situ confirmation of this accretion, it is likely that the satellite-derived shoreline has captured a real feature of the order of 30 m probably created by the changing direction of wave incidence and the influence of the jetty to the east.

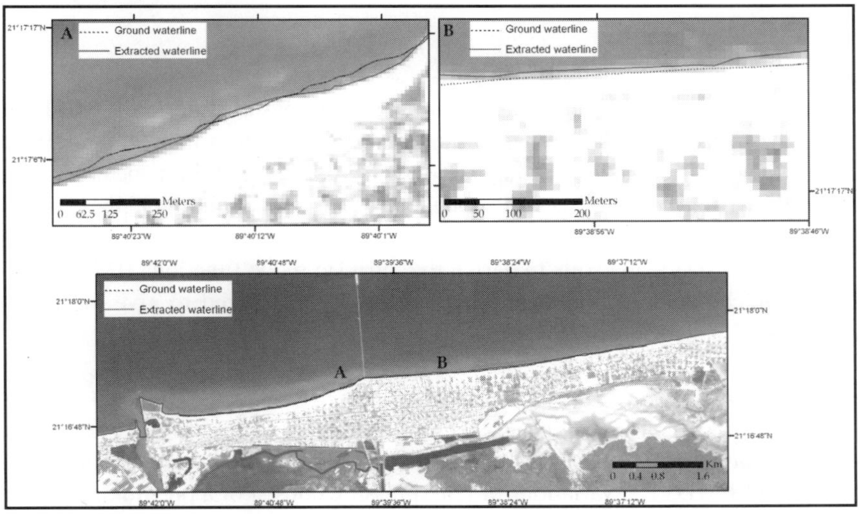

Figure 1. Spot image from February 6th in 2006 overlaying the shoreline extracted from the image and the waterline measured on September 9th 2008. 1A and 1B are close-ups of two regions indentified in the general image.

4. Conclusions

Shoreline identification from georefenced SPOT satellite images has been carried out using an unsupervised classification scheme, but we find that visual examination is also required in order to remove localized anomalies. However, with these methods, a comparison between an image-based shoreline and a shoreline measured less than a month earlier shows good agreement, with variations of the order of one pixel (10m). Comparison between the measured shoreline and an image taken more than 2 years earlier suggests that satellite images can be used to determine shoreline changes as small as 20-30 m.

References

Chu, Z.X., X.G. Sun, S.K. Zhai, K.H. Xu, 2006. Changing pattern of accretion/erosion of the modern Yellow River (Huanghe) subaerial delta, China:Based on remote sensing images. *Marine Geology*, 227:13-30.

Capurro Filograsso, L.,J. Herrera Silveira,J. Euan Ávila, I. Mariño Tapia, C. Enríquez Ortiz, V. Castillo, J. Acosta, 2006. *Coastal Evolution of the Western Littoral of the Yucatan Peninsula-Erosion and Water Quality*. Final report. CONACYT and World Bank funded research project.

De Wolf, P., D. Fransaer, J.Van Sieleghem, R.Houthuys, 1994. Morphological Trends of the Belgian Coast shown by 10 Years of Remote-Sensing based Surveying. *Coastal Zone Proceedings*. 245-257, New Orleans, US-LA.

Dinesh Kumar P. K., G. Gopinath, C. M. Laluraj, P. Seralathan, D. Mitra, 2007. Change Detection Studies of Sagar Island, India, using Indian Remote Sensing Satellite 1C Linear Imaging Self-Scan Sensor III Data. *Journal of Coastal Research,* 23(6):1498-1502.

Lafon V., D. De Melo Apoluceno, H. Dupuis, D. Michel, H. Howa, J. M. Froidefond, 2004. Morphodynamics of nearshore rhythmic sandbars in a mixed-energy environment (SW France): I. Mapping beach changes using visible satellite imagery. *Estuarine, Coastal and Shelf Science*, 61:289-299.

White K. and H. M. El Asmar, 1999. Monitoring changing position of coastlines using Thematic Mapper Imagery, an example from the Nile Delta. *Geomorphology,* 29:93-105.

Coastal Dynamics 2009
Paper No. 118

PRELIMINARY APPLICATION OF THE HHT TO ANALYZE THE CHARACTERISTICS OF NONLINEAR WAVE DECAY OVER A FLUIDIZED BED

Yung-Lung Chen[1] Shiaw-Yih Tzang[2] and Shan-Hwei Ou[3]

1. Introduction

To investigate the characteristics of nonlinear wave decay in coastal waters, the HHT method was applied to analyze the measurements from wave flume experiments. In principle, the HHT consists of two parts including the empirical mode decomposition (EMD) and the Hilbert spectral analysis (HSA). To overcome the drawback on the frequent appearance of "mode mixing", the Ensemble EMD (EEMD) purposed by Wu and Huang (2009) was adopted in present study to decompose the data of waves and pore pressures inside the soils to derive the intrinsic mode functions (IMF). After the HSA performance, the characteristics of nonlinear wave decay were analyzed to give the time-frequency distribution and instantaneous energy (*IE*).

2. Results and Discussions

Fig. 1(a) and Fig. 1(b) show typical nonlinear waves and wave-induced resonantly fluidized (RF) pore pressure response at a depth of 30 cm. Fig. 1(c) and Fig. 1(d) show their corresponding IMFs and the residue derived by the EEMD. The variations of pore pressure consist of a pre-fluidized (I), a transient (II) and a post-fluidized (III) stage (Tzang and Ou, 2006). Fig. 2 shows the results of Hilbert amplitude spectrum $S(f,t)$ derived by the HHT. It illustrates that waves decay on the peak frequency component in a fluidized response while the superharmonic component is slightly amplified. However, the results of the oscillating pore pressure derived by summing up IMF C_1 to C_6 in Fig. 1(d) (Chen et al., 2008) illustrate that the peak frequency component starts to be amplified from the transient stage and the superharmonic component exists exclusively in the transient stage as shown in Fig. 2(b).

Fig. 3 shows the excess pore pressures (or build-ups) ΔP derived by summing up IMF C_7 to C_{11} in Fig. 1(d) (Chen et al., 2008) and the instantaneous energy *IE* of waves. The instantaneous energy is defined as

$$IE(t) = \int_f S^2(f,t)df \quad (1)$$

It is clearly seen that waves decay significantly when excess pore pressures begin to increase in stage (II) and keep relatively steady after having attained to a maximum value. The result manifests that wave decay is mainly due to fluidized response. Thus, in order to quantitatively analyze, we define the wave-decay ratio (R_{wd}) as

$$R_{wd} = 1 - (\sqrt{\overline{IE_{III}} / \overline{IE_I}}) \times 100\% \quad (2)$$

Here $\overline{IE_I}$ and $\overline{IE_{III}}$ are the means of *IE* in stage (I) and (III). From the other four RF Tests of different wave conditions, the wave-decay ratio are found to be directly proportional to the magnitude of Ursell number (U_r) as shown in Fig. 4.

---

[1]Dept of Harbor and River Eng., National Taiwan Ocean University, 2, Pei-Ning Road, Keelung 202, Taiwan, R.O.C. d92520007@mail.ntou.edu.tw

[2]Dept of Harbor and River Eng., National Taiwan Ocean University, 2, Pei-Ning Road, Keelung 202, Taiwan, R.O.C. sytzang@mail.ntou.edu.tw

[3]Dept of Environmental Resources Management, Tajen University, 20, Weishin Road, Yanpu Shiang, Pingtung 907, Taiwan, R.O.C. oush@mail.tajen.edu.tw

Acknowledgements

This study is financially supported by the National Science Council, Taiwan, Republic of China under the grant No. NSC 97-2221-E-019-042.

References

Chen, Y.-L., Tzang, S.-Y. and Ou, S.-H., 2008. Application of the emd method to investigate pore pressure build-ups in a wave-fluidized sandbed. *Proc. 31st Int. Conf. on Coastal Eng.* (in press)

Tzang, S.-Y. and Ou, S.-H., 2006. Laboratory flume studies on monochromatic wave-fine sandy bed interactions: Part 1. Soil fluidization. *Coastal Engineering*, 53(11): 965-982.

Wu, Z. and Huang, N. E., 2009. Ensemble Empirical Mode Decomposition: a Noise-assisted Data Analysis Method. *Advances in Adaptive Data Analysis*, 1(1): 1-41.

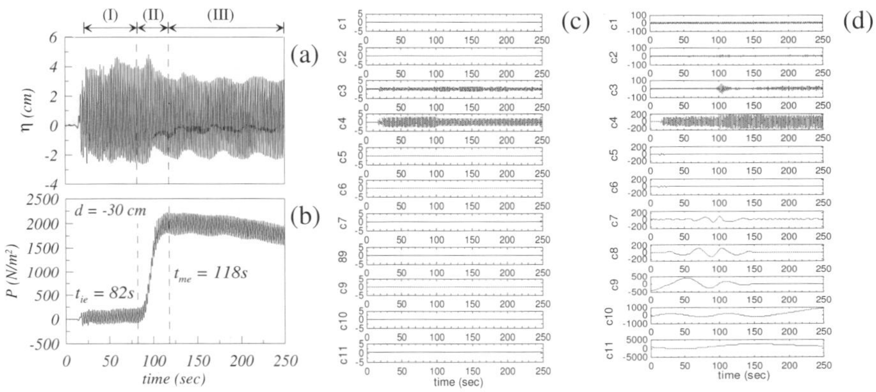

Figure 1. (a) Wave records, (b) wave-induced pore pressure responses taken at 30 cm below the bed surface, the IMFs of (c) waves and (d) pore pressures for fluidized response

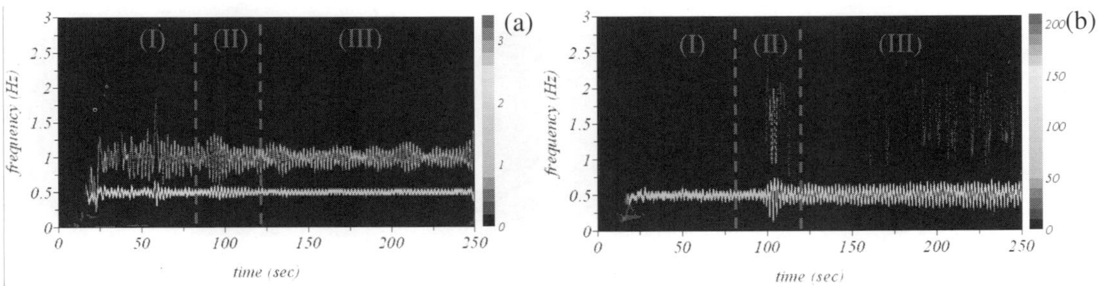

Figure 2. Hilbert amplitude spectrum of (a) waves and (b) wave-induced pore pressures

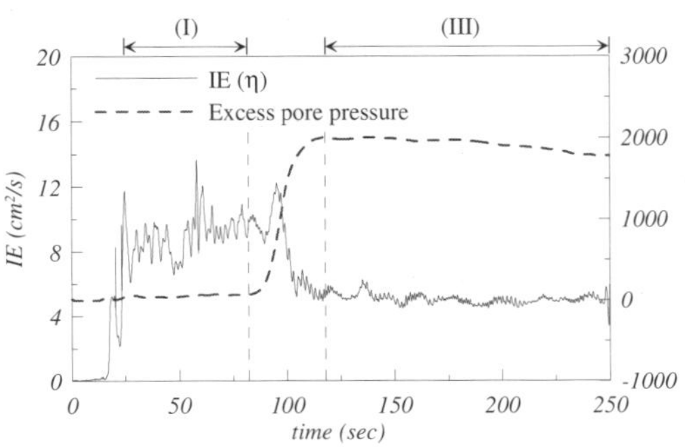

Figure 3. Instantaneous energy of waves and excess pore pressure

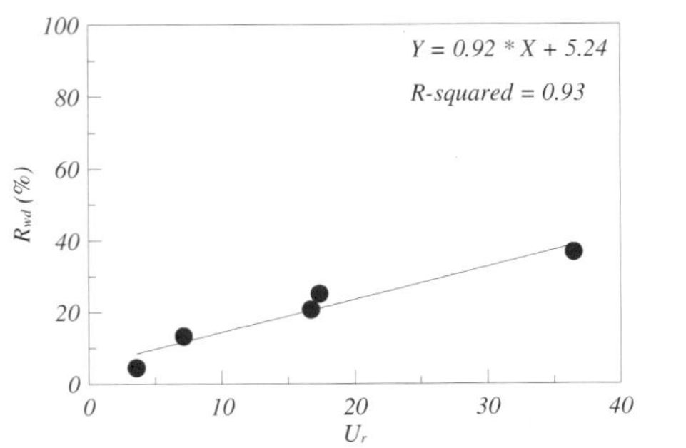

Figure 4. Wave decay rate (R_{wd}) versus the Ursell nu mber (U_r)

Coastal Dynamics 2009
Paper No. 119

MEASURING SUSPENDED SEDIMENT AND ITS WAVE AND TURBULENCE FORCING IN THE DEE ESTUARY

Rodolfo Bolanos[1] and Alejandro Souza[2]

1. Introduction

The Dee estuary (Fig. 1), located in the eastern Irish Sea, is 20 km long, 8 km wide at the mouth and it is characterized by a 10m tidal range. The Dee has undergone huge levels of sedimentation over the past two centuries, resulting in land reclamation and salt marsh colonization. The Dee estuary presents a mixture of sediments containing a range of non-cohesive and cohesive sediments and, therefore, the threshold of motion of the bed might be a complex process dependent on several conditions. The estuary has been the subject of several instrumentation deployments to measure and understand hydrodynamics, turbulence and sediment transport in the area. New data and methods have been used to explain turbulence dynamics in the Dee and its implications for the sediment transport.

2. Data and methods

The data presented in here comes from 2 deployments involving rigs such as STABLE (Sediment Transport and Boundary Layer Equipment) in the Hilbre channel, and mini-STABLE in the Welsh channel (Fig 1), both including instruments such as ADV, ADCP, ABS, CTD and LISST. The data cover a wide range of environmental conditions being useful to build an overall picture of the physical processes controlling the estuary. Special focus in this work is the link between the bottom stresses, turbulence and suspended sediment concentration in order to estimate proper sediment transport patterns.

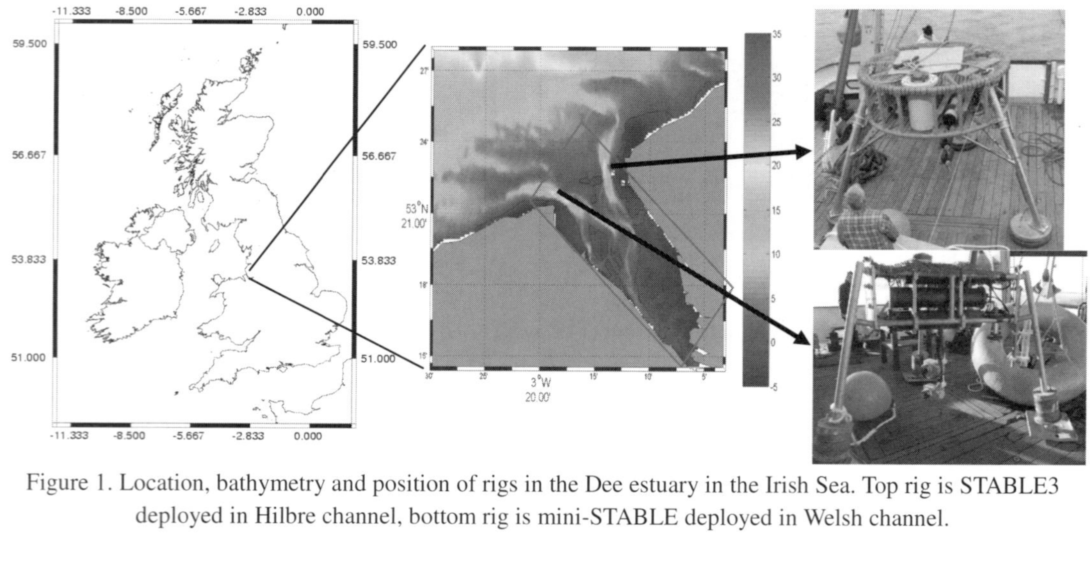

Figure 1. Location, bathymetry and position of rigs in the Dee estuary in the Irish Sea. Top rig is STABLE3 deployed in Hilbre channel, bottom rig is mini-STABLE deployed in Welsh channel.

[1]Proudman Oceanographic Laboratory, Joseph Proudman Building, 6 Brownlow Street Liverpool L3 5DA, UK. rbol@pol.ac.uk
[2]Proudman Oceanographic Laboratory, Joseph Proudman Building, 6 Brownlow Street Liverpool L3 5DA, UK. ajso@pol.ac.uk

The ADV data have been cleaned to remove spikes (Wahl, 2003) that can appear in this kind of instruments. Wave parameters have been obtained by applying the PUV method modified to include the effect of background currents. Bottom stresses have been estimated by the inertial dissipation method (Green, 1992) and compared with the covariance method (Souza and Howarth, 2005).

2. Results

The Dee is a macrotidal estuary and thus all processes are very tidal dominated. Bottom stresses and suspended sediment concentration follow the tidal theory (Souza et al., 2004), wave parameters (specially wave period) are very well modulated by tidal currents. The figure 2 (left) shows time series of water depth, current velocity and significant wave height in the Welsh channel, significant wave height reaches 2 m. The figure 2 (right) shows the distribution of peak sediment size at at about 1.5 m above the bed in the Hilbre channel and the current velocity, there is a clear decay of size when increasing velocity; this suggests that the suspended matter is highly controlled by aggregation-desegregation. The brake up of flocs occurs in the energetic phase of the tide and in agreement with high turbulence periods. The concentration near the bed is directly controlled by bottom stress with a quaterdiurnal tidal modulation. Wave effects in the Hilbre channel (deep areas) do not appear to have much effect in the amount of sediment in suspension near the bed.

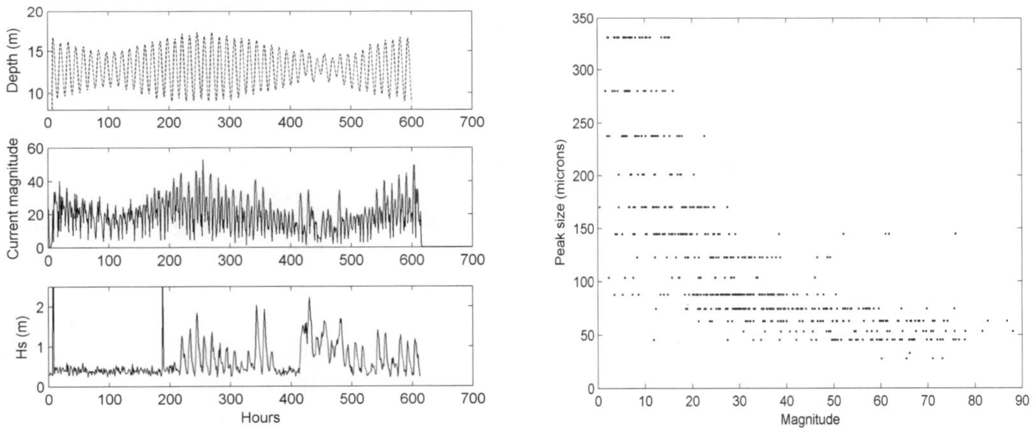

Figure 2. Time series of water depth, current velocity and significant wave height (left). Distribution of sediment in suspension peak size and current velocity (right).

Acknowledgements

Authors thank the support of the FORMOST (Field observation and modelling of the sediment triad) NERC project.

References

Wahl, T. L. 2003. Discussion of "Despiking acoustic doppler velocimeter data" by Derek G. Goring and Vladimir I. Nikora. *Journal of Hydraulic Engineering*, 129(6) 484-487.

Green, M. 1992. Spectral estimate of bed shear stress at subcritical Reynolds numbers in a tidal boundary layer, *Journal of Physical Oceanography*, 22, 903-917.

Souza, A. J. and Howarth, M.J., 2005. Estimates of Reynolds stress in a highly energetic shelf sea. *Ocean Dynamics*, 55, 490-498.

Souza, A.J., L.G. Alvarez and Tommy Dickey. 2004. Tidally induced turbulence and suspended sediment. *Geophys. Res. Lett.31*, L20309, doi:10.1029/2004GL021186.

Coastal Dynamics 2009
Paper No. 120

OSCILLATORY TURBULENT FLOW INSIDE AND AROUND POROUS MEDIA

Sota Nakajo[1], Takaaki Shigematsu[1], Naoya Sakashita[2], Gozo Tsujimoto[3], Kosei Takehara[4]

1. Introduction

The flows through porous media such as permeable coastal structures and gravel bed and so on have been treated as porous media flow in the past. Chiefly relationships between bulk flow velocity and pressure drop were measured. However, it was difficult to measure local velocity inside porous media. Recently, Nakajo et al. (2008) achieved to measure velocity distribution inside porous media directly using the refractive index matching method. They examined the relation between approach velocity and turbulent energy generated in porous media settled in unidirectional flow. However, developing and decay process of turbulence existed in unsteady flow. And the relationships of time-averaged velocity and turbulent energy obtained using steady flow experience have theoretical limitations for applying to unsteady flow. On the other hand, Hall et al. (1995) has shown the difficulty to apply Forchheimer pressure drop model, which concerned with turbulence, obtained using steady flow experience data to oscillatory unsteady flow. Consequently, it is necessary to measure local velocity for understanding this unsteadiness mechanism. In this research, we measured velocity distribution inside porous media settled in oscillatory unsteady flow by using image velocimetry and examined the effects of velocity amplitude and oscillation period to turbulent properties.

2. Experimental procedure

The experiment was conducted using a pipe channel showed in Figure 1. We achieved measurement of local velocity distribution inside porous media by using the refractive index matching method, which used iodide sodium solution and porous media model composed by silicon spheres to measure accurate tracer position. Oscillatory flow was generated by a disk which motion of velocity amplitude V_0 and oscillation period T was controlled by a computer ($V=V_0\sin(t/T)$). Experimental conditions are shown on the Reynolds number ($R_{ep}=V_0\phi D/\nu$) and the KC number ($KC=V_0T/D$) in Table 1, where ϕ is porosity, D is a diameter of silicon sphere and ν is a kinematic coefficient of viscosity. We used three high-speed cameras for capturing laser-illuminated tracer images, and its field of view was 0.45m widths and 0.1m heights. Image capturing was carried out at 1/250-second intervals during about 12 seconds. And a laser light sheet was slid at intervals of 0.007m to measure velocities at different sections. We used a PTV algorithm for measurement of each tracer motion and calculated local average velocities $\boldsymbol{v}_S$ and difference values $\boldsymbol{v}'$ from measured velocities. Furthermore we estimated turbulent energy K_S using $\boldsymbol{v}'$.

3. Primary results and conclusions

It is clear from the results of velocity fluctuation at x/D=-0.75 (inside porous media, Figure 2) that there are two phases, which have different flow regimes. One phase have significant velocity fluctuations and the other phase have weak fluctuations. This result gives us an implication that developing and decay process of turbulence existed in a period. There was a large difference between

---

[1]Dept of Urban Eng., Osaka City University, 3-3-138 Sugimoto, Sumiyoshi-ku, Osaka 558-8585, Japan. nakajo@hy2.civil.eng.osaka-cu.ac.jp / shige@urban.eng.osaka-cu.ac.jp
[2] Sakai City, Japan
[3] Dept of Civil Eng., Kobe City College of Technology, 8-3, Gakuen-Higashi-Machi, Nishi-ku, Kobe, 651-2194, Japan. tujimoto@kobe-kosen.ac.jp
[4] Dept of Civil Env. Eng., Kinki University, 3-4-1, Kowakae, Higashi-Osaka, Osaka, 577-8502, Japan. takehara@civileng.kindai.ac.jp

u_S and V at x/D=-4.5 (outside porous media) in phase of V<0, which means measurement point was wake region. According to a series of measured velocities transition, the wake region was enlarged with time and the largest wake region size increased in the KC number and the Reynolds number. Cross-sectional averaged turbulent energy $<<K_S>>$ varied with V although $<<K_S>>$ had almost constant value at each phase inside porous media (-2<x/D<2) (Figure 3). However, its transition outside porous media occurred later than that of one inside porous media. This phase difference between $<<K_S>>$ outside and inside porous media depended on the KC number and the Reynolds number and it can be concerned with developing process of wake region. Increase of Reynolds number induced the maximum and minimum value of internal (-2<x/D<2, -2<y/D<2, 0<z/D<0.5) average of turbulent energy $<K_S>_V$ in a period. On the other hand, increase of KC number induced decrease of the minimum value of it (Figure 4). It was found that the turbulent energy at the maximum velocity phase was larger than that as same one of unidirectional flow due to unsteadiness.

References

Hall, K. R., Smith, G. M., Turcke, D. J. 1995. *Coastal Engineering*, 24: 217-232.
Nakajo, S., Shigematsu, T., Tsujimoto, G., Takehara, K. 2008. *Proceedings of ICCE 2008*, in printing.

Table 1. Experimental conditions

V0 [m/s]	*T* [s]	*Rep*	*KC*
	2.0		2.8
2.8	6.0	212	8.4
	10.0		14
	1.4		2.9
4.1	4.0	310	8.2
	6.6		13.5
	1.0		2.8
5.5	3.0	416	8.3
	5.0		13.8

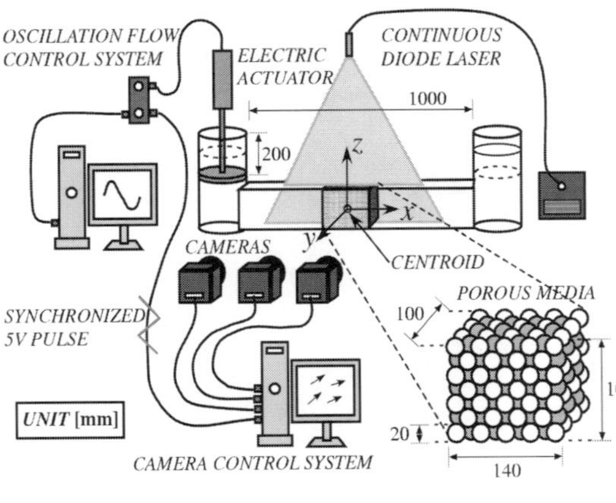

Figure 1. Experimental apparatus

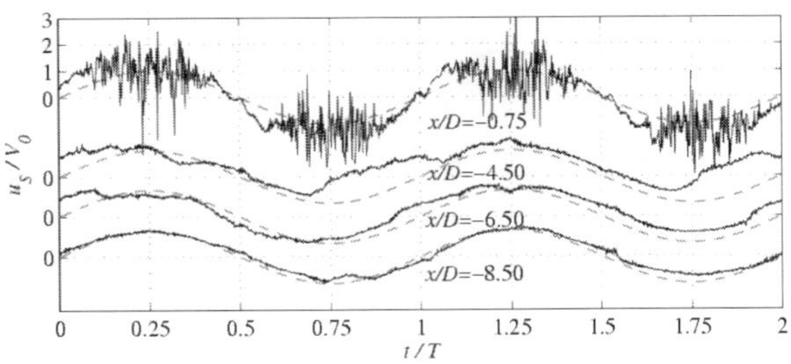

Figure 2. Fluctuations of local average velocity at fixed grid poi nts (R_{ep}=416, KC=8.3, y/D=0.0, z/D=0.0, broken line; V)

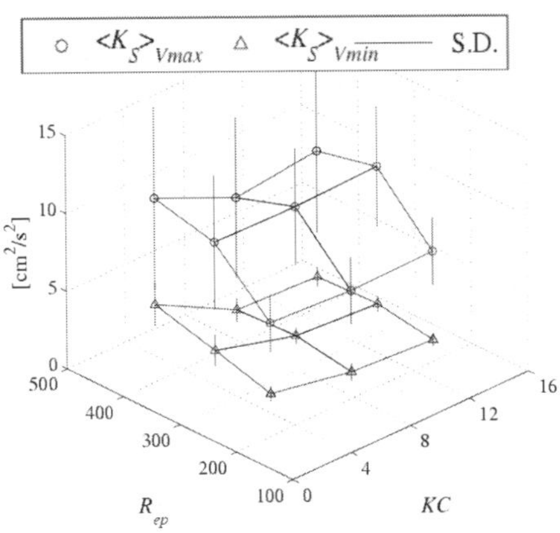

Figure 4. Relationships between spatial average of turbulent energy and oscillation properties

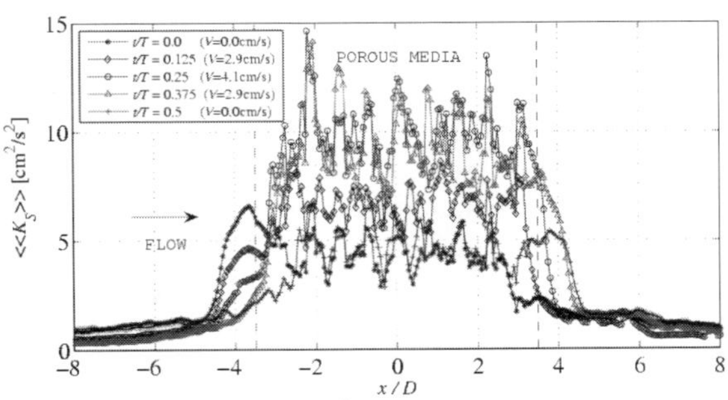

Figure 3. Phase variation of cross-sectional average turbulent energy (R_{ep}=310, KC=2.8)

Coastal Dynamics 2009
Paper No. 121

THE CHARACTERISTICS OF WIND-BLOWN SAND IN THE NAKATAJIMA DUNE

Junaidi[1,*], Naoto Wakae[1], Shigeru Kato[1], and Shin-ichi Aoki[1]

1. Introduction

The Nakatajima dune is approximately 1 km length (West to East) and 0.6 km width, located near Magome River mouth on the Enshu-Nada coast, Hamamatsu city of Japan. The Enshu-Nada coast is located both sides of the Tenryu River. The sand median diameter on this dune is around 0.29 mm. The dune field is being pushed eastward by the predominant westerly wind. Beach erosion and coastal forest developments have been decreased the sediment supply to the Nakatajima dune. Since the development of coastal forest in West side of the dune (1979), which is cutting–off the sediment supply from the shoreline, large topographic changes have occurred at the Nakatajima dune (Figure 1). The coastal forest has been covered with edge of sand dune at the Eastern end of the dune. Moreover, it has been shown that the edge of dune develops 6-7 m per year with more than 7 m height.

In this paper, we highlight the various properties of wind-blown sand within a coastal dune field in Hamamatsu city, Japan, such as, measurement of topographical changes of the dune, sand flux, direction, and airflow over the dune. In the measurements, we have found the sand flux across the field and it has been correlated with soil surface conditions such as topography, which is one of important aspect for coastal dune deformation.

2. Topographical Changes

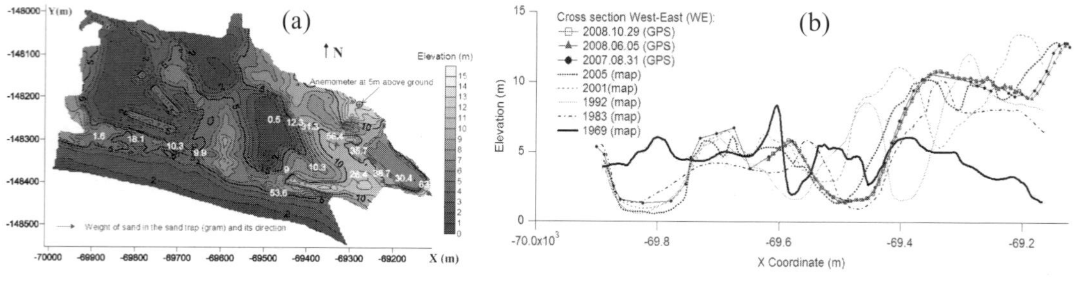

Figure 1. (a) Contour map of Nakatajima dune was generated with surfer 8 software, grids data were obtained from topographical survey on December 2008 using Global Positioning System (GPS) with Real time Kinematics (RTK). The arrow lines on the map showing the location of sand traps, direction of sand movement, and the weight of sand (gram) in the sand traps during sand trap experiment on January 13, 2008; (b) cross section West-East (WE) in the middle part of the dune. The cross section of 1969, 1983, 1992, 2001, and 2005 were obtained from topographical maps, and the cross section of 2007 and 2008 were obtained from the grids data of topographical survey using GPS with RTK.

Figure 1 clearly shows the Nakatajima dune affected by deflation and accumulation. By 1983, the Western part of Nakatajima dune decreased significantly in the elevation. In the middle part of the dune, where is an actively eroding region by wind, windward side and steep slip faces started to deform. Simultaneously, the windward side and the steep slip faces migrated eastward of the dune.

[1]Dept. of Arch. & Civil Eng., Toyohashi University of Technology, Hibarigaoka 1-1, Tempaku cho, Toyohashi 441-8580, Japan. (E-mail: junaidi@jughead.tutrp.tut.ac.jp; wakae@ jughead.tutrp.tut.ac.jp; s-kato@ tutrp.tut.ac.jp; aoki@ jughead.tutrp.tut.ac.jp).
* Corresponding author.

3. Wind-Blown Sand Transport

Following Sauermann (2003), the sand flux measured on the dune (Figure 1a and Figure 2) using traps consisting of a cylinder (diameter 5 cm, height 1 m) closed on the bottom and having vertical slot of 1 cm width. It was turned with its facing the wind and then its bottom was buried in the sand. In the Eastern part of Nakatajima dune, the sand transported into the traps was higher than the Western part. The sand fluxes on the central slice of the dune (T12, T13, T14, T15, T16, T9, and T10) were small in the foot of the dune and increased toward to dune crest (Figure 2). In the T11 (near beach), which was low-elevation and flat in topography, the sand flux was also high.

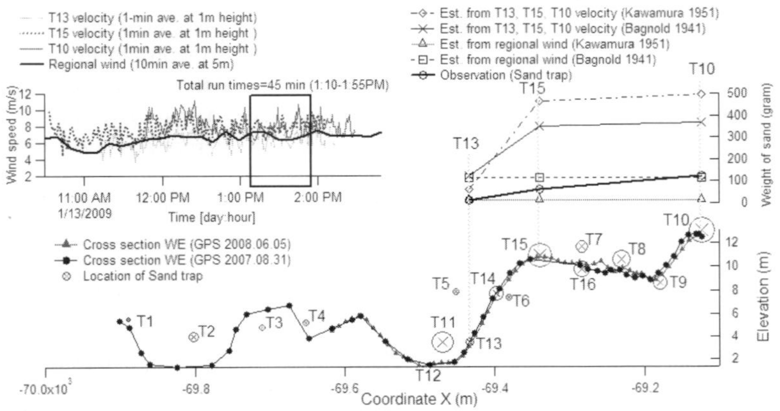

Figure 2. The cross section West-East (WE) in the middle part of the dune obtained from GPS survey in 2007 and 2008, and position of the traps (weight of sand in traps is scaled by size of circle). Upper left is 1-min averages wind speed were measured at 1 m height closed to the traps T13, T15, and T10, and regional wind of 10-min averages wind speed at 5m height mount in the Northern of the dune (wind direction is WNW during this study). Upper right is the weight of sand in the traps and estimation weight of sand in the traps by Bagnold (1941) and Kawamura (1951) formulae using wind data of T13, T15, T10, and regional wind.

4. Discussion

In the Nakatajima dune, sand availability is fundamental parameter for dune deformation due to the wind blown-sand. The source of the sand transported mostly from the dune itself and limited amount of sand come from the shoreline. The dune has a windward and steep slip face to maintaining dune stability and motion. The steep slip face/dune crest is an actively eroding region by wind. The sand was transported and accumulated to the lowland area/dune gap and back of dune crest. The windward and steep slip face migrated as predominant wind direction in years or decades.
Wind that encounters a dune, or hill, is expected to lose velocity as a result of pressure build up at the base of the obstruction (Lynch *et al.*, 2008). We also found that the wind velocity at dune's foot was lower than at the top of the dune/dune crest, and the sand flux was varied across the dune field due to the airflow pattern over dune.

References

Bagnold, R.A., 1941. *The Physics of Blown Sand and Desert Dunes.* Mineola, New York: Dover Publication, Inc., 265p.

Kawamura, R., 1951. Study on sand movement by wind (in Japanese). *Report of the Institute of Science and Technology, University of Tokyo*, 5 (3-4): 95-112.

Sauermann, G.; Andrade Jr., J.S., Maia, L.P.; Costa, U.M.S., Araujo, A.d., Hermann, H.J., 2003. Wind velocity and sand transport on a barchan dune. *Geomorphology*, 54: 245-255.

Lynch, K.; Jackson, D.W.T.; Cooper, J.A.G., 2008. Aeolian fetch distance and secondary airflow effects: the influence of micro-scale variables on meso-scale foredune development. *EarthSurface Processes and Landforms*, 33: 991-1005.

Coastal Dynamics 2009
Paper No. 122

A LARGE-SCALE LABORATORY EXPERIMENT OF RIP CURRENT CIRCULATIONS OVER A MOVEABLE BED: DRIFTER MEASUREMENTS

Bruno Castelle[1], Hervé Michallet[2], Adrien Lambert[2,3], Céline Berni[2], Frédéric Bouchette[3], Vincent Marieu[1], Philippe Bonneton[1], Olivier Kimmoun[4], Damien Sous[5], Rafael Almar[1]

1. Introduction

In recent years, there has been significant interest in rip currents systems (MacMahan et al., 2006), as they are important for beach safety and localised beach and dune erosion during storms (Thornton et al., 2007). Rip current circulations are driven by longshore variations of wave-induced radiation stress (Longuet-Higgins and Stewart, 1964). These gradients can be due to spatial variability of the incident wave field due to wave groups (Dalrymple, 1975), wave-current interactions (Dalrymple and Lozano, 1978), wave field interaction with lower-frequency waves such as edge waves (Symonds and Ranasinghe, 2000) or local topographic variations (Bowen, 1969). The latter is called 'topographically-controlled rip current', while the three other ones refer to 'transient rip currents' (Johnson and Pattiaratchi, 2004).

A substantial number of laboratory experiments, all performed with a fixed bed, have been undertaken to more easily grasp topographically-controlled rip current circulation information than on natural beaches. Eulerian measurements (Hamm, 1992; Haller and Dalrymple, 2001) and, more recently, lagrangian techniques (Kennedy and Thomas, 2004) have been used to investigate rip current systems. The latter, when a sufficient number of drifters are released during a sufficient duration, can be transformed into a horizontal mean circulation field. The same drifter deployment strategy was recently attempted in the field by MacMahan et al. (in press). In addition to the mean circulation information, the authors interestingly introduced new thoughts of rip current behaviors, suggesting that rip currents retain more floatsom material within the surf zone as opposed to transporting floatsom material offshore, as only ~10% of the drifters that entered a rip current exited the surf zone over the course of the experiment when caught by a pulsing jet. On the field, such a technique requires small tidal ranges, a large number of investigators on the field devoted to this task. Therefore, undertaking an extensive study comprising a large number of seabed morphologies and wave conditions is difficult on the field. This was one of the reasons that motivated a recent large-scale laboratory experiment that we describe below.

2. Experiment and expected results

Recently, during a 5-week period from the 10th of November to the 12th of December, a large scale laboratory experiment over a moveable bed has been undertaken. The tests were conducted in the multidirectional wave basin at the SOGREAH (LTF facility), France, extending 20 m and 30 m in the cross-shore and alongshore direction, respectively. Only shore-normal waves were used here to favor the formation of rip channel(s). The moveable bed consisted in fine sediment with d_{50} = 164 μm. Accurate bathymetric surveys have been undertaken every day. A strongly variable beach morphology was observed during the experiment, ranging from an alongshore-uniform geometry to a well-developed bar-rip morphology, favoring a large range of rip current systems. In addition to eurlerian measurements, drifters were tracked using captured images from a shore-mounted video camera.

[1]UMR EPOC-5805, Université de Bordeaux, Avenue des Facultés, 33405 Talence Cedex, France
b.castelle@epoc.u-bordeaux1.fr
[2]Laboratoire des Ecoulements Géophysiques et Industriels (UJF-INPG-CNRS), BP53, 38041, France
[3]UMR 5243, Université Montpellier 2, Place E. Bataillon, 34095 Montpellier cedex 5, France
[4]UMR 6594, Technopôle de Château-Gombert 49, rue Joliot Curie, 13384 Marseille Cedex 13, France
[5]LSEET, Université de Toulon et du Var, 83957 La Garde Cedex, France

Image coordinates were then rectified to still water level Cartesian coordinates. Figure 1 shows an example of a captured image, with drifters and a dye release when a well-developed rip channel was observed.

At the time of writing this abstract, data processing has is at its early beginning. Eulerian measurements, and particularly drifter velocities will be processed to assess the mean horizontal circulations and drifter retention during the experiment as a function of wave forcing and beach morphology. We therefore plan to communicate the results at *Coastal Dynamics 2009*.

Acknowledgements
The authors acknowledge financial support from the Project MODLIT (RELIEFS/INSU). The authors are also thankful to all the people involved in this experiment.

References

Bowen, A., 1969. Rip currents: 1. Theoretical investigations. *Journal of Geophysical Research*, 74, 5479-5490.

Dalrymple, R., 1975. A mechanism for rip current generation on an open coast. *Journal of Geophysical Research*, 80, 3485-3487.

Dalrymple, R. and Lozano, C., 1978. Wave current interaction model for rip currents. *Journal of Geophysical Research,* 83, 6063-6071.

Haller, M. and Dalrymple, R., 2001. Rip current instabilities. *Journal of Fluid Mechanics*, 433, 161-192.

Hamm, L., 1992. Directional nearshore wave propagation over a rip channel: an experiment. *Proceedings of the 23rd International Conference on Coastal* Engineering (Venice, Italia, ASCE), pp. 226-239.

Kennedy, A.B. and Thomas, D., 2004. Drifter measurements in a laboratory rip current. *Journal of Geophysical Research*, 109, C08005, doi: 10.1029/2003JC001927.

Longuet-Higgins, M.S. and Stewart, R.W., 1964. Radiation stress in water waves, a physical discussion with applications. *Deep Sea Research*, 11(4), 529-563.

MacMahan, J. H.; Thornton, E.B. and Reniers, A.J.H.M., 2006. Rip current review. *Coastal Engineering*, 53, 191-208.

MacMahan, J.H., Thornton, E.B., Stanton, T.P., Reniers, A.J.H.M., Brown, J.A., Brown, J.W., in press. Measurements of rip current cisculation, diffusion and dispersion. *Proceedings of the 31th International Conference on Coastal Engineering* (Hamburg, Germany, ASCE),

Thornton, E.B.; MacMahan, J.H. and Sallenger Jr., A.H., 2007. Rip currents, mega-cusps, and eroding dunes. *Marine Geology*, 1-4, 151-167.

Figure 1. Snapshot of the experiment showing the deployment of drifters in a rip current system. The rip current system is highlighted by the dye trajectory, about one minute after it was released in the feeder.

Coastal Dynamics 2009
Paper No. 123

MEASUREMENTS AND MODELLING OF SAND TRANSPORT UNDER FULL-SCALE SURFACE WAVES

Jolanthe J.L.M. Schretlen[1], Jan S. Ribberink[1] and Tom O'Donoghue[2]

1. Introduction

Existing models for the wave-related (cross-shore) sand transport are primarily based on data from oscillatory flow tunnel experiments. First experiments in the large wave flume of Hannover (e.g. Ribberink et al., 2000) indicated that, although large similarities exist between the transport processes in wave flumes and tunnels, the flow differences (e.g. residual streaming and drift under surface waves) may have a substantial effect on net transport rates. The new experiments described in the present paper were initiated to investigate more deeply the near-bed flow and wave related sand transport processes under regular velocity skewed non-breaking surface waves for various sediments in the sheet-flow regime. Besides providing more insight in sheet-flow dynamics, the data are used to further develop transport models in the Dutch/UK research project SANTOSS.

2. Experimental Set-up

The experiments were performed in the Große WellenKanal (GWK) in Hannover, Germany; a 280 m long, 5 m wide and 7 m deep wave flume. In the flume, a horizontal sand bed was placed with a 1:20 beach slope at the far end. The majority of the measurements were performed under regular wave conditions (height 0.7 to 1.6 m, period 5.0 to 9.1 s) in both sheet-flow and ripple regime. The tests are performed with two different types of sediment; medium (D_{50} = 0.25 mm) and fine sand (D_{50} = 0.15 mm), both well sorted. To capture the bed-surface processes under the waves, measurements were done on a range of scales; from flow velocities inside the sheet-flow layer up to bed level changes along the whole flume. Instruments used for the (detailed) flow velocity and concentration measurements include UVP, Vectrino, CCM, EMF, UHCM, TSS, pressure sensor, wave gauges and ABS. Net sand transport rates were then determined by applying the mass conservation principle to measured bed profiles. These profiles (4 over the width of the flume) were made with echo sounders before and after each test run.

3. Results

Figure 1 shows the results of measured net sand transport rates from the new experiments, compared to existing flume and oscillatory flow tunnel measurements. The new results of the medium sand coincide well with the existing results from former GWK measurements. This underlines the conclusion from Ribberink et al. (2000) that under full scale surface waves, the net transport rates are higher than for similar sand and wave conditions in oscillatory flow tunnels. In the new experiments, also fine sand was used under surface waves. From former oscillatory flow tunnel experiments it was found that with fine sediment, the net transport rates (after a slight increase) decrease and become negative (off-shore) with an increase of flow velocity ($<U^3>$). This phenomenon was explained by the domination of phase lead effects for fine sand conditions. However the new fine sand experiments under full scale surface waves show an increase of positive (on-shore) transport with increasing velocity ($<U^3>$). This indicates that other, on-shore directed, flow mechanisms (e.g. residual streaming or drift) are decisive for the net sand transport.

[1]Dept of Water Engineering & Management, University of Twente, P.O. Box 217, 7500 AE, Enschede, The Netherlands. J.L.M.Schretlen@utwente.nl, J.S.Ribberink@utwente.nl
[2]School of Engineering University of Aberdeen, King's College, Aberdeen, AB24 3UE, Scotland. T.ODonoghue@abdn.ac.uk

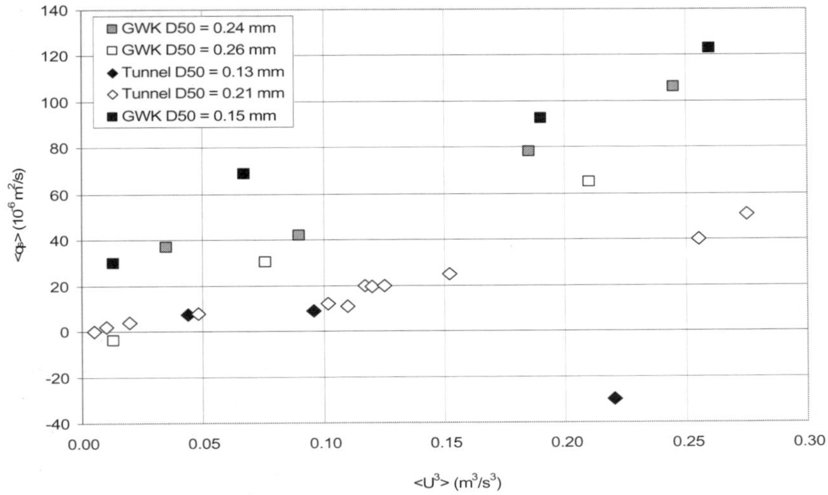

Figure 1. Measured net sand transport rates of medium and fine sand, in oscillatory flows and free surface waves

In the final paper, results of detailed flow velocity and concentrations measurements inside the wave boundary (and sheet-flow) layer are used to explain the variations in net sand transport results between oscillatory flows and full scale surface waves.

4. Sand Transport Modelling

The measured data are used for the development of a new, practical sand transport model for cross-shore sand transport. This model development will also be informed by insights from a detailed process-based (RANS) wave-current-sediment model (Uittenbogaard and Klopman, 2001). An important contribution of this model is the capability to determine and control the contribution of the flow differences between oscillatory flows and surface waves.

5. Conclusions and Future Work

The new measurements clearly show an influence of surface wave effects on the net sand transport, which seem to be especially important for finer sediment. Meticulous analysis of the detailed flow velocity measurements, combined with model simulations are to be done to identify the specific processes determining the net sand transport under these conditions.

Acknowledgements

The experiments in the GWK are performed in i) the framework of the Access programme of the Integrated Infrastructure Initiative Hydralab-III of the European Community's Sixth Framework Programme (contract no. 022441) and ii) SANTOSS, funded by the Dutch Technology Foundation STW, applied science division of NWO and the technology program of the Ministry of Economic Affairs (TCB 6586) and the UK's Engineering and Physical Research Council EPSRC (GR/T28089/01). The data-processing, data analysis and the transport model development are also mainly carried out within SANTOSS. Part of the data processing is done within the EU-project SANDS, a joint research activity (JRA) within Hydralab-III.

References

Ribberink, J.S., Dohmen-Janssen, C.M., Hanes, D., McLean, S. & Vincent, C., 2000. Near-bed sand transport mechanisms under waves, a large flume experiment. *Proc. 27th Int. Conf. Coast. Eng.;* 3263 – 3276, Australia.

Uittenbogaard, R.E. and Klopman, G. (2001). Numerical simulation of wave-current driven sediment transport. *Proc. Coastal Dynamics '01*, 568 – 577, Lund, Sweden.

Coastal Dynamics 2009
Paper No. 125

LARGE SCALE DATA OF SUSPENDED SEDIMENT IN THE SWASH ZONE

Jose M. Alsina[1], Agustín Sánchez-Arcilla[1], Iván Cáceres[1] and Joaquim Sospedra[1]

1. Introduction

Sediment transport prediction in the swash zone still represents an unsolved problem for coastal researchers and Engineers especially under low energetic or non-saturated conditions which might result in shoreline accretion and berm development. For such low energy conditions the energetic based models have shown their limitations in predicting sediment transport rates (Puleo *et al.*, 2000) due to the swash velocity asymmetry which displays a negative (off-shoreward) tendency. Moreover there is a range of different processes that have been argued to affect sediment transport in the swash zone and contribute to the uncertainties. They include acceleration effects, in-/ex-filtration, boundary layer conditions during the up-rush and backwash and turbulence due to surface wave breaking.

Some authors (Aagaard and Hughes, 2006) have shown that sediment concentration measurements might not correlate well with the locally measured hydrodynamics (velocity, instantaneous water surface), suggesting lag effects. They have shown improvements when considering turbulence fluctuations. The experimental results suggest that the enhanced turbulence due to wave breaking or bore collapse might play an important role in sediment transport in the swash zone. Recent laboratory results and numerical simulations highlight the importance of sediment advection from the bore collapse point to the swash zone limit (Alsina *et al.*, 2009).

To gain understanding about these issues a series of large scale experiments have been carried out at the CIEM wave flume of UPC (Catalonia University of Technology). These heavily instrumented tests have been designed to measure hydrodynamic conditions, sediment concentrations and bed level elevations in the swash zone under erosive and accretive conditions. The obtained results provide sufficient accuracy and resolution within the swash zone to allow a meaningful physical analysis and a comparison with numerical simulations.

The present paper focuses, thus, on analyzing the obtained hydro-morphodynamic data to assess the effects of bed and free-surface induced turbulence on the resulting water and sediment fluxes.

2. Experiment description

In the Large Wave Flume facility CIEM of the Maritime Engineering Laboratory (UPC) a series of test have been performed to characterize sediment transport conditions in the swash zone. The wave flume dimensions are 100m long, 5m deep and 3m wide. The swash zone consists of a beach with a slope of 1:15 made of mobile sand with a medium grain size, d_{50}, of 0.25 mm.

Two test series of irregular waves reproducing a *Jonswap* spectrum were performed during the experiments. The erosive test consists of 47 sets of irregular waves with 500 waves in each time series. The significant wave height and peak period were 0.53m and 4.14s respectively. A bar was developed during the erosive tests and, without any beach reshaping, the accretive test series were performed next. They comprised 35 sets of irregular accretive waves with a significant wave height of 0.32m and a peak period of 5.44s.

During all these tests the swash zone was instrumented with 7 Acoustic Doppler Velocimeters (ADV) organized in 3 arrays at three different cross-shore positions, all inside the inner surf, swash zone. In addition 7 optical backscatter sensors (OBS) were deployed organized as well in 3 arrays coincident in cross-shore location and vertical elevation with the ADV locations. Finally a series of 13

[1]Laboratorio de Ingeniería Marítima (LIM), Universidad Politécnica de Cataluña, Barcelona, 08034, Spain.
jose.alsina@upc.edu; agustin.arcilla@upc.edu; i.caceres@upc.edu; j.sospedra@upc.edu.

resistive wave gauges plus 11 Acoustic Displacement sensors were deployed in the wave flume to measure water surface elevation across the inner surf and swash zones.

3. Preliminary results

Measured suspended sediment concentrations and velocities at different cross-shore locations within the inner surf and swash zones have been analyzed using a number of techniques and quality controls. Figure 1 shows an example of registered sediment concentration and swash velocities corresponding to the erosive time series no.2 and two different cross-shore locations in the inner surf, swash zone. The solid blue line represents the velocity and concentration obtained by an ADV and OBS set located close to the run-down position but still in the inner surf zone. The dashed red lines correspond to velocity and concentration measured 2 meters on-shoreward, in the lower swash zone landward of the run-down position. It is apparent from figure 1 that suspended sediment concentration peaks show a lag with respect to velocity peaks (see e.g. velocity and concentration at time = 1135 s).

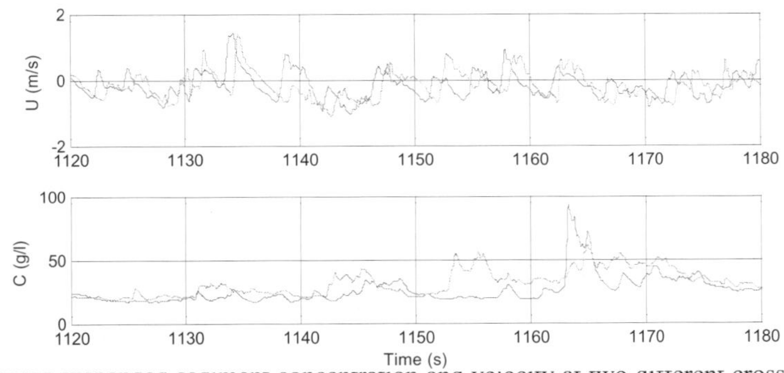

Figure 1.- Measured suspended sediment concentration and velocity at two different cross-shore locations in the swash zone. The solid blue line corresponds to the inner surf zone close to the run-down point while the dashed red line represents measurements in the lower swash zone. The instrument elevation above the bed level was 4 cm.

It can be observed that high suspended sediment concentrations are also associated to the backwash (see e.g. time = 1170s), preceding a peak of high concentration during the up-rush. This might be due to suspension during backwash and an offshore-wards drift but it could also be related to sediment advection from the wave collapse point which is located offshore-wards of the observation point. Measured velocity and concentration are being analyzed and turbulence fluctuations computed.

The final paper will discuss all these results, displaying all sediment turbulence correlations against the computed sediment transport rates from ADV, OBS and from measured bed levels. The role of the various possible forcing factors and the implications for practical calculations of swash zone fluxes will be also presented in the full paper.

Acknowledgements
The work here presented was done within the SANDS project, funded by the European Commission within the I3 project HYDRALAB III.

References

Aagaard, T. and Hughes, M.G., 2006. Sediment suspension and turbulence in the swash zone of dissipative beaches. *Marine Geology*, 228(1-4): 117-135.

Alsina, J.M., Falchetti, S. and Baldock, T.E., 2009. Measurements and modelling of the advection of suspended s ediment in the swash zone by solitary waves. *Coastal Engineering*, in press.

Puleo, J.A., Beach, R.A., Holman, R.A. and Allen, J.S., 2000. Swash zone sediment suspension and transport and the importance of bore-generated turbulence. *Journal of Geophysical Research*, 105(C7): 17021-17044.

Coastal Dynamics 2009
Paper No. 126

DEGREE OF EXPERIENCE FOR EXTREME WAVE STATISTICS

Toshikazu Kitano[1], Wataru Kioka[2] and Rinya Takahashi[3]

1. Aim

How far can we extrapolate the line which satisfies the relation between return period and return level of wave height? It is a most dominant factor of the *R*-year return wave height for the design of coastal structures. So we are used to see the graph of a set of sample wave heights based on a certain plotting position rule for the return periods, as seen in Figure 1. For that purpose, the confidence intervals of return levels are often examined in detail. It is, however, a relative discussion whether the interval is wide or narrow. Even when we have just a short record of extreme wave heights, there is no restriction to estimate a wave height of a longer return period, though it is accompanied with a very wide confidence interval. Engineers have been bothered with a problem: how much wide interval is no worth being considered. As a solution we introduce here newly the degree of experience for the estimation of occurrence intensity of extreme wave height.

2. New Index

In general, as the return period is given, then the corresponding return level will be estimated, and the statistical error should be taken into account for the return level. It should be a routine work in the design process of coastal structures. In this study, we change the points of view of the given object and the one for estimation. That is to say, the return period is regarded as a statistical variable, which carries the statistical variety around a given true value. Moreover, we deal with the occurrence rate λ in place of the return period R, because it is easily assumed that the rare occurrence numbers follows a Poisson distribution and the estimated rate does a Gamma distribution.

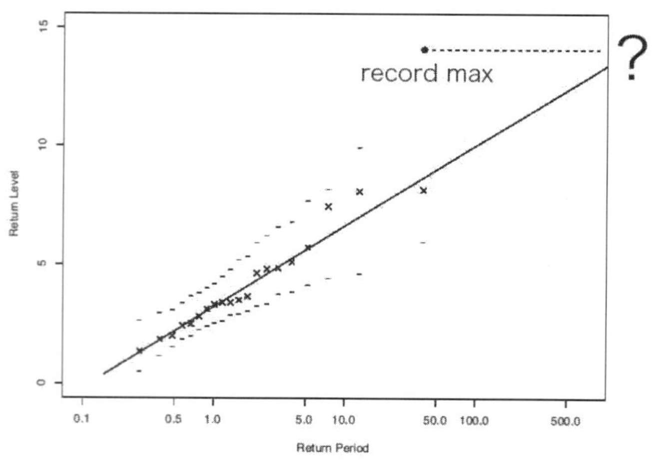

Figure 1. Return level vs return period

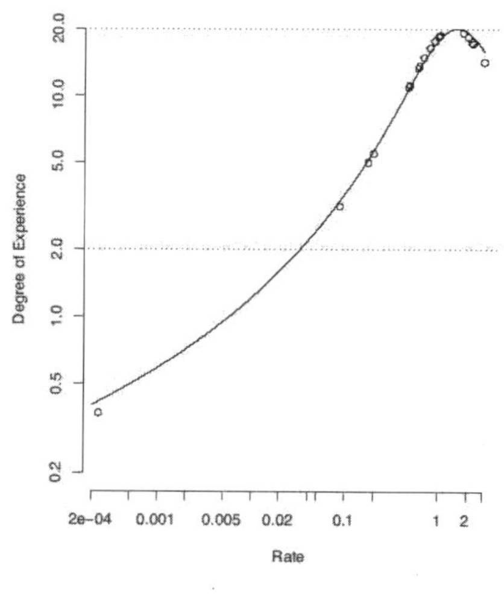

Fig. 2 Degree of experience for a sample

[1]Dept. of Civil Engineering, Nagoya Inst. of Tech., Gokishocho, Showaku, Nagoya, Japan. kitano@nitech.ac.jp
[2]Dept. of Civil Engineering, Nagoya Inst. of Tech., Gokishocho, Showaku, Nagoya, Japan. kioka@nitech.ac.jp
[3]Dept. of Marine Science, Kobe Univ., 5-1-1 Hukae, Higashinadaku, Kobe, Japan. r-taka@maritime.kobe-u.ac.jp

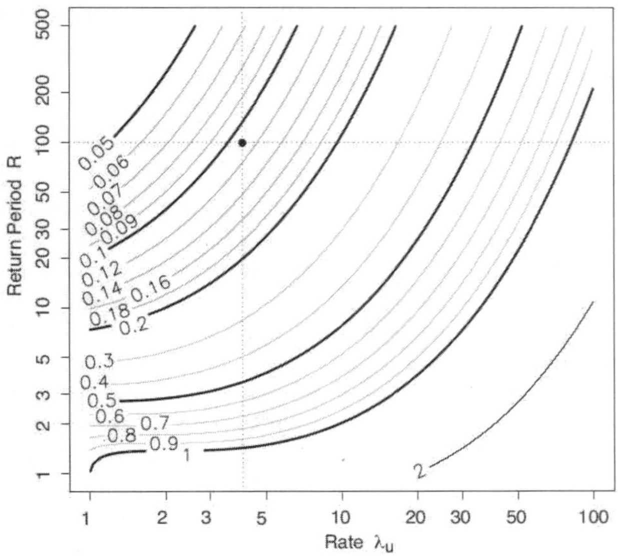

Figure 3. Theoretical calcuration of degree of experience

The coefficient of variation C will be obtained by the extreme value analysis:

$$C = \frac{\sqrt{V\left(\lambda_x \middle| \hat{\mu}, \hat{\sigma}, \hat{\xi}\right)}}{\lambda_x} \tag{1}$$

where $\hat{\mu}, \hat{\sigma}, \hat{\xi}$ are the location, scale and shape parameters of GEV distribution, and they affect the varance of occurence rate. Then, as an index of measuring the effective sample size for the estimation of rate, Kitano (2009) propose the degree of experience K, which is simply given by

$$K = C^{-2} \tag{2}$$

Fig. 2 shows the change of the degree of experience K for a sample set of annual maximum wave heights of 20 yrs. It should be noted that the maximum value of K is equivalent to the actual sample size. It is also seen that the value of K is diminishing as the rates becomes smaller. It will be revealed that the significance level for the value of K will be above 2.0.

3. Remarks

The values of K can be also obtained theoretically without a sample data. The total observation length N of extreme waves (in year) and the average number lu of available data over the adequate threshold (by year) are only required for the value of K for the target return period.
For example, a data whose total length N = 20 years and λu = 3.9 (i.e. sample size of extreme wave heights = 78) is significant, as seen in Fig.3 where the value of K = 2.2 (K/N = 0.11) for return period R =100 years.

References

Kitano, T. (2009): paper presented in Cooperative Research Report 224, Extreme Value Theory and Applications (6), The Institute of Statistical Mathematics (in Japanese).

Coastal Dynamics 2009
Paper No. 127

WAVE-INDUCED MORPHODYNAMIC RISKS. CHARACTERIZATION OF EXTREMES

Agustín Sánchez-Arcilla[1], José A. Jiménez[1] and Carlos Peña[2]

Coastal morphodynamics arise from the complex interaction between water and sediment fluxes plus the associated sea-bed morphology. The resulting coastal evolution takes place mainly under energetic climatic events since sediment transport is a highly non-linear function of water velocities. Based on this a driver-response framework can be applied by considering the probability distribution function of oceanographic drivers such as waves and mean water levels. This will serve to infer the distribution function of some illustrative coastal "responses". The probabilistic characterization of this response has been comparatively much less studied although it is of direct interest for coastal engineering and management decisions.

In this paper a methodology to estimate storm-induced morphodynamic risks will be presented. Following Mendoza and Jiménez (2008) the storm-induced coastal response is separately assessed in terms of erosion and inundation. Erosion is characterized by the eroded volume from a beach profile due to the impact of the storm which is obtained following the parametric approach developed by Mendoza and Jiménez (2006). Inundation is calculated by means of the run-up $Ru_{2\%}$ at the peak of the storm which is evaluated using the Stockdon *et al* (2006) method.

To estimate the resulting "hazards" (risk), we have followed an event-response approach (see e.g. Garrity et al., 2006) applied to both processes. To do this, we have built a 49 years long wave and water-level time series by using recorded and hind-casted data. As an example, Figure 1 shows the extreme distribution of the significant wave height *Hs* off the Ebro delta in the Spanish Mediterranean coast (see Prinos and Sánchez-Arcilla (2008) for a review on extreme value analyses). To apply the event approach for inundation and erosion, the storm event associated to a given return period was defined using the methodology presented by Sánchez-Arcilla et al. (2008).

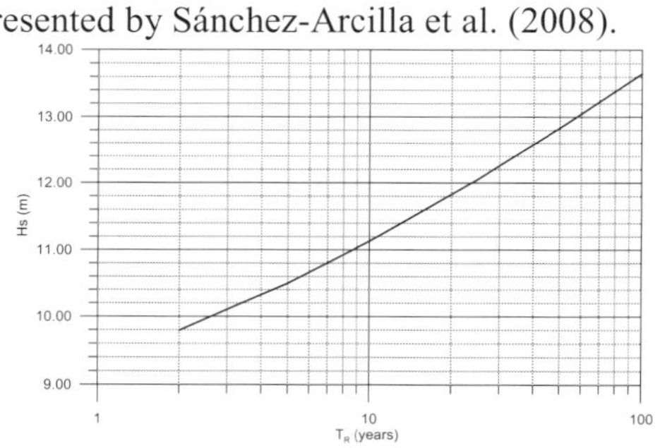

Figure 1. Extreme probability distribution function for the significant wave height *Hs* off the Ebro delta (Spanish Mediterranean coast).

From the built wave series, we have obtained several eroded-volume time series for typical profiles of the study area (the Catalan coast in this case). These time series are directly used to obtain an extreme distribution of eroded volumes (response approach) which is presented for a reflective beach profile in Figure 2. In the same figure we have also included the volumes eroded by wave events associated to a given return period (event approach). As it can be seen the eroded volumes associated to a T_R value significantly differ as T_R increases. In this case, the response approach predicts higher erosion volumes due to the nonlinearity of the transport formulation.

Figure 3 shows the same type of analysis for wave-induced run-up to estimate the part of the storm-induced risk associated to inundation. First, a $Ru_{2\%}$ time series was built from the wave data set

[1] Laboratori d'Enginyeria Marítima, Universitat Politècnica de Catalunya, Barcelona 08034, Spain. Agustin.arcilla@upc.edu, jose.jimenez@upc.edu
[2] Dirección de, Ministry of Environment and Rural and Marine Affairs. Plaza San Juan de la Cruz s/n, 28071 Madrid, Spain. cpena@mma.es

and it was directly used to obtain an extreme distribution of $Ru_{2\%}$ (response approach). In the same figure we have also included the $Ru_{2\%}$ induced by events associated to a given return period (event approach). As it can be seen $Ru_{2\%}$ values associated to given T_R differ for both approaches and, the pattern of variation changes from that observed for eroded volumes. This is because the dependence on wave variables is different for both processes.

In the final paper, the methodology will be presented in detail and applied to the selected Mediterranean beach profiles. The implications for estimating morphodynamic risks will be discussed based on the obtained results and the coastal damages produced during some recent storms.

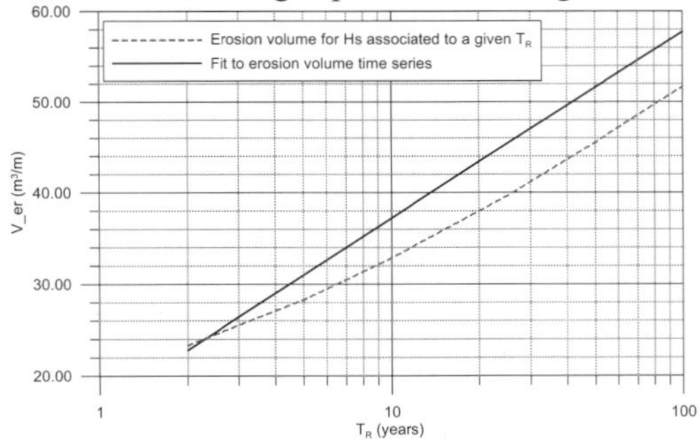

Figure 2. Extreme distribution function for eroded volumes in a reflective beach profile due to storm-induced cross-shore processes (solid line: response method, dashed line: event method).

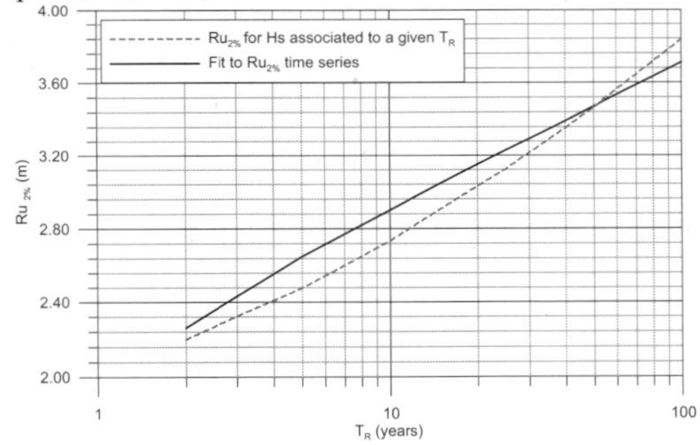

Figure 3. Extreme distribution function for wave-induced run up, $Ru_{2\%}$ (solid line: response method, dashed line: event method).

Acknowledgements

This work was partially funded by the Spanish Ministry of Environment (ComCat project) and the Spanish Ministry of Education (VuCoMA, CTM2008-05597/MAR and RIMA).

References

Garrity, N.J., Battalio, R., Hawkes, P.J. and Roupe, D. 2006. Evaluation of event and response approaches to estimate the 100-year coastal flood for Pacific coast sheltered waters. *Proc. 30th Int. Conf. on Coastal Engineering,* ASCE, 1651-1663.

Mendoza, E.T. and Jiménez, J.A. 2006. Storm-Induced Beach Erosion Potential on the Catalonian Coast. *Journal of Coastal Research*, SI 48, 81-88.

Mendoza, E.T. and Jiménez, J.A. 2008. Vulnerability assessment to coastal storms at a regional scale. *Proc. 31st ICCE,* ASCE, Hamburg, (in press).

Prinos, P. and Sánchez-Arcilla, A. (Eds) 2008. Analysis of riverine and coastal extremes. The FLOOD*site* approach. *Journal of Hydraulic Research*, 46.

Sánchez-Arcilla, A., Mendoza, E.T. Jiménez, J.A., Peña, C., Galofré, J. and Novoa, M. 2008. Beach erosion and storm parameters. Uncertainties for the Spanish Mediterranean. *Proc. 31st ICCE,* ASCE, Hamburg, (in press).

Stockdon, H.F.; Holman., R.A.; Howd, P.A. and Sallenger, A.H. 2006. Empirical parameterization of setup, swas h, and runup. Coastal Engineering, 53, 573–588.

Coastal Dynamics 2009
Paper No. 128

FORESHORE EROSION OF KASHIMA COAST BY 2006 ATUMN STORMS AND IT'S ALONGSHORE DYNAMICS

Elsayed M. Galal[1] and Satoshi Takewaka[2]

1. Introduction

Strong low-pressure systems traveled along Japanese Main Island in October 2006. High waves and storm surge attacked Kashima Coast resulting huge erosion over the area. This study analyzes the foreshore erosion caused by the storms using airborne laser data and compares the results with wave energies during the storm estimated with SWAN.

2. Study area and sea state

The focus of this research is to assess the impacts of the 2006 autumn storms on Kashima Coast, Ibaraki, Japan, which faces the Pacific Ocean. The northern part of the coast is 38 km long and bordered by Oarai Port at the north and Kashima port at south, while the southern part is 15 km long and bordered by Kashima Port at the north and Tone River at the south as shown in Fig. 1 (a). There are 33 headlands installed along the coast for shore protection, which are jetty-like structures.

From 5th to 9th, and from 24th to 26th of October 2006, strong low-pressure systems traveled along Japanese Main Island. High waves and storm surge caused by pressure fall and strong wind (maximum instantaneous wind speed 37.4 m/s was observed at Choshi which locates close to the Tone River mouth) attacked Kashima Coast resulting huge erosion over the area. Maximum offshore wave height measured by NOWPHAS Kashima station (water depth 24 m) was 7.0 m at 6th of October 2006. The tide level deviation from the astronomic tide observed at 6th of October 2006 was 0.94 m, which was the highest record measured by the Japanese Meteorological Agency at Choshi Fishery Port.

3. Erosion analysis

Airborne laser data measured on 23rd of October 2005 and 8th of November 2006 have been used and analyzed. Digital elevation models have been processed above the mean sea water level. Since the configuration of the coast is slightly concave, a longshore reference curve as a baseline and normal transects to the baseline with 10 m intervals has been set for the analyses. The cross-sectional change within the subaerial zone has been assembled along each transect from the elevation data of 2005 and 2006 as shown in Fig. 1(b). The results indicate that the amount of the erosion of the northern and southern parts decreased toward the south which may be raised due to the activation of longshore sediments transport along the coast, and this is assessed in the followings.

4. Longshore wave dynamics

The SWAN (Booij et al. 1999) wave model was applied to estimate wave condition along the coast during the storm event hours. The longshore component of wave power p_l, which is a proxy of the driving force for longshore sand transport, is estimated for the first storm from 6th to 8th of October 2006. The longshore distribution of p_l has been summed up over the two days as P_l. Fig. 1 (b) shows the longshore distributions of P_l and amount of erosion for the northern and southern parts of Kashima coast, and the locations of headlands indicated with squares. A positive P_l indicates that sediments should be transported toward the south, and a negative cross-sectional change indicates that subaerial

[1]Dept of Eng. Mechanics and Energy, University of Tsukuba, Tsukuba, Ibaraki, 305-8573, Japan. sayed@surface.kz.tsukuba.ac.jp.
[2]Dept of Eng. Mechanics and Energy, University of Tsukuba, Tsukuba, Ibaraki, 305-8573, Japan. takewaka@.kz.tsukuba.ac.jp.

zone was eroded. The distribution of the cross-sectional change indicates that the amount of erosion is less at the sections where the headlands are installed compared to untreated sections. Also, the results indicate that the amount of erosion and its distribution are in agreement with the magnitude and sign of a P_l along the coast, especially in the untreated areas which should be close to a response of an undisturbed beach.

Magnitude and longshore variability of run-up are important to understand beach processes such as beach/dune erosion and storm flooding. Run-up height R was estimated with

$$R/H = c\xi_o + d \tag{1}$$

in which c and d are constants, and ξ_o is the surf similarity parameter defined $\xi_o = \tan\beta/\sqrt{H/L}$. The foreshore beach slopes ($\tan\beta$) have been estimated from the airborne laser data measured in 2005 at each transect, which ranged from 0.035 to 0.109, while the offshore wave height H and wave length L have been estimated at water depth of approximately 12 m from SWAN computation. Run-up heights have been averaged over the period. Fig. 1 (c) shows the longshore distribution of the fluctuations of the run-ups and cross-sectional change which are defined as the deviations from the moving average. The results indicate that most of the high wave run-up values appear within the high eroded sections and vice versa.

Generally, the erosion pattern within the untreated areas shows a wavy fluctuation which is in accordance with the distribution of P_l and R, whereas the wave length of the fluctuation within protected areas are shorter which may be controlled by the headland locations.

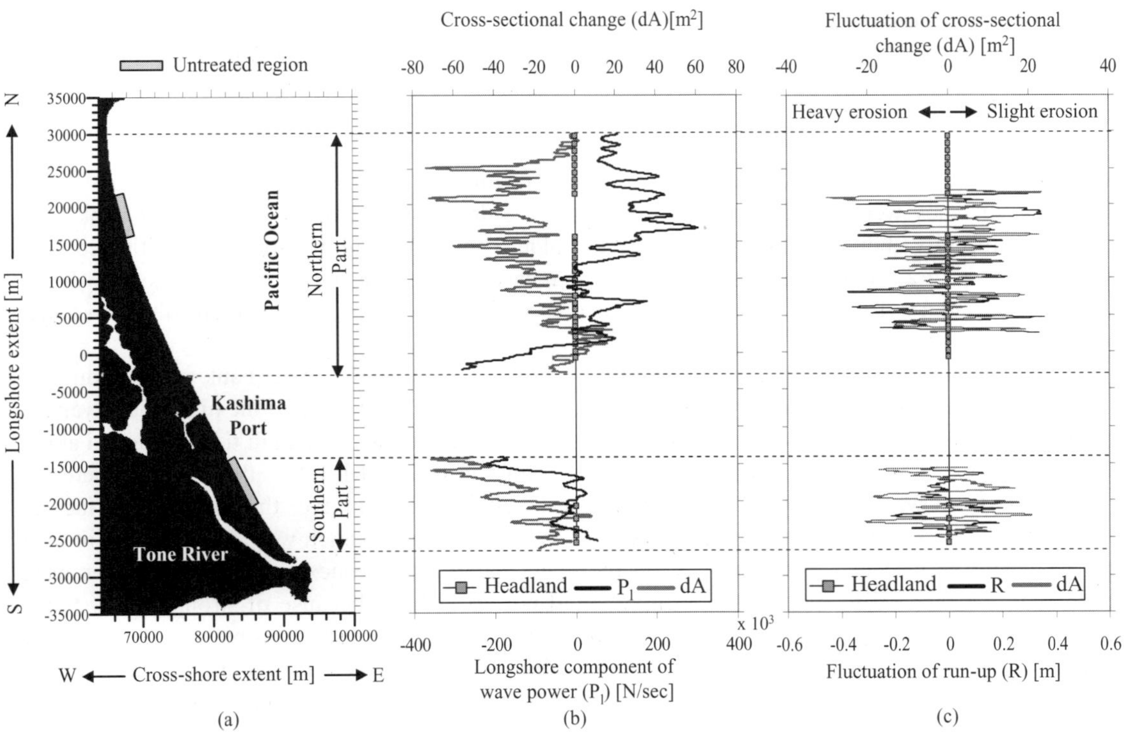

Figure 1. (a) Study area, longshore distributions of (b) cross-sectional change and longshore component of wave power, and (c) fluctuations of the cross-sectional change and wave run-ups

References

Booij, N., Ris, R. C. and Holthuijsen, L. H. 1999. A third-generation wave model for coastal regions, Part I: Model description and validation. *Journal of Geophysical Research*, 104 (C4): 7649-7666.

Coastal Dynamics 2009
Paper No. 129

FIELD SURVEY OF STORM SURGE DISASTER DUE TO CYCLONE SIDR IN BANGLADESH

Tomoya Shibayama [1], Yoshimitu Tajima[2], Taro Kakinuma[3], Hisamichi Nobuoka[4], Tomohiro Yasuda[5], Raquib Ahsan[6], Mizanur Rahman[7] and M. Shariful Islam [8]

1. Introduction

Field surveys were performed in the southwest of Bangladesh to learn lessons out of severe disasters due to Cyclone Sidr. Spatial distributions of inundation heights were measured around the most damaged area.

2. Numerical Prediction (Nobuoka, 2008)

Fig. 1 shows sea level change at 5 points due to wind setup and barometric pressure suction except wave setup. These results of sea level anomaly indicate that the height over 6 m was generated by the concentration of energy at river mouth and shallow depth. The calculated height at Kuakata was much different with that was obtained in the survey. The calculated wind field may be not good enough for evaluation of the height at Kuakata.

3. Field Measurements

Fig. 2 shows the survey route and distribution of measured storm surge heights. Basic method was to detect flood height and to measure distanced by using laser distance meters and to measure locations by using GPS.

(1) Southkhali

The measurement result of a cross section is shown in Fig. 3. Riverbanks were built around thirty years ago but they were broken due to erosion before Sidr attacked them. Though newer banks had been built recently about 150 m inside from these older banks, the high tide due to Sidr flowed over their top of about 2.5 m over the river water at the measurement time. These newer banks are shown in Fig. 3 around x = 151.6 m. The river depth decreases every year due to erosion, resulting in wider width of the river.

According to our hearing, the first wave of the high tide came to the newer bank at x is about 150 m, after which the second wave flowed over the newer bank. The third wave of the high tide advanced on the shelter. The wave period of these bores was around one minute. The high tide continued about fifteen minutes. Women and children took mainly refuge in the next level of the shelter, where the inundation depth was 0.59 m over the floor, whose height was 2.99 m over the basement surface of the shelter. Some water entered through windows, the lower edges of which were of 3.8 m height over the basement surface of the shelter. When the water came into the next level, the refugees were frightened and wanted to go outside but they could not open doors because of strong winds. Some traces were left when they forced themselves to open the doors. Thousand people evacuated on the shelter top, where they felt the space were not enough. Many people were flowed before they reached this shelter, including a child flowed about 3 km. Several children caught trees to be spared. In this area about 300 people were died.

In regard to evacuation, the following points are important: 1) education of disaster prevention, 2)

[1]Dept. of Civil Eng., Yokohama National University, 79-5 Tokiwadai, Hodogaya-ku, Yokohama, 240-8501, Japan. tomo@ynu.ac.jp
[2]University of Tokyo, [3] Kagoshima University, [4] Ibaraki University, [5] Disaster Prevention Research Institute, Kyoto University, [6-8] Bangladesh University of Engineering and Technology

certain communication channel of evacuation information, 3) construction of shelters considering adequate evacuation distance and enough capacity.

(2) Kuakata

Characteristic features observed in Kuakata are outlined as follows. Fig. 4 overviews observed features in Kuakata. Since the distance from the shoreline to the embankment was relatively long, L= 690m, and various plants were vegetated on the mild slope to the embankment, storm surge did not overtop the embankment. Before the storm, 1.5m-high sand dune was developed (dune A in Fig. 4) and restaurant house was built on the dune. Storm surge completely washed both house and the sand dune although the surface of the dune was fully vegetated. Estimated volume of the washed sand dune is about 8000m^3. A part of sand dunes (dune B in Fig. 4) was also washed away. The width and height of the washed sand dune were about 30m and 1m, respectively.

4. Concluding Remarks

Inundation heights along the Baleshwar River and the Burishwar River were relatively high compared to those observed on the coast of Kuakata although these sites are far from the coast. Embankments along the river had been eroded before the storm while dikes on the coast significantly functioned to reduce the damages of the coastal area behind. Many people witnessed that storm surges inundated with bore-like waves. Counter measured against storm surges should account for the physical mechanisms for the development of such bore-like waves and possible damages due to such waves. As observed in Kuakata and Somboniya, embankment showed significant roles to minimize the damage. Development of riverbanks especially around the river mouth is one of most essential countermeasures to be carried out in Bangladesh. Shelter functioned to save significant number of lives although the numbers of shelter still needs to be increased to match the number of residents.

References

Nobuoka, H., 2008. *Numerical simulation of storm surge of Cyclone Sidr*, unpublished report of Ibaraki University.

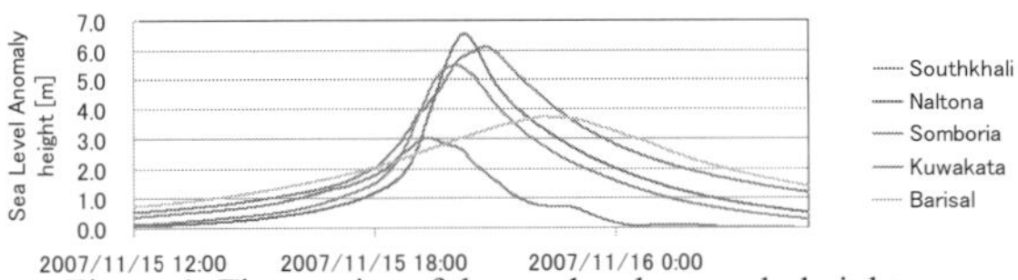

Figure 1. Time series of the sea level anomaly height

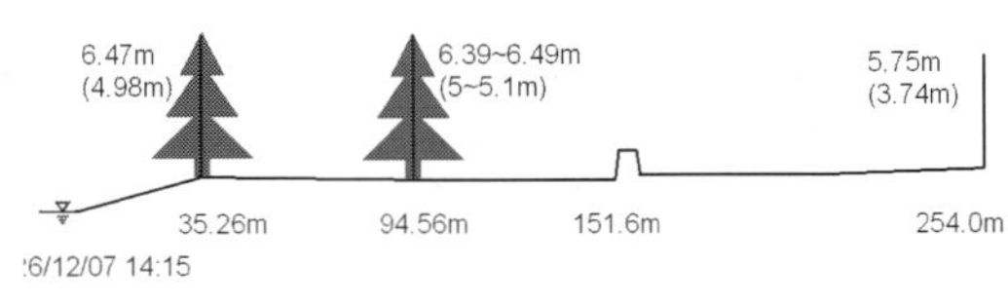

Figure 3. Measurement result of a cross section in Southkhali

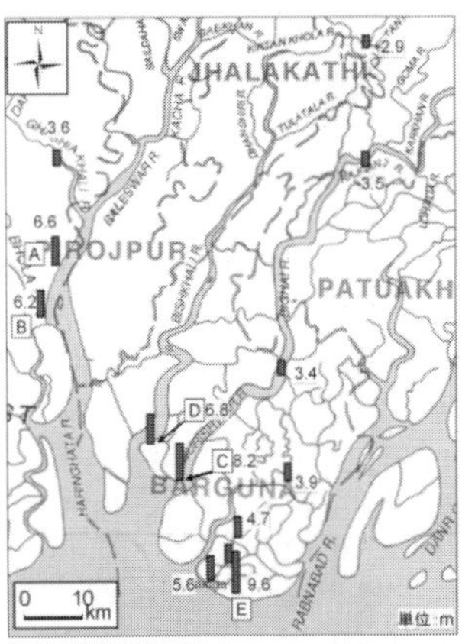

Figure 2. Survey route and measured storm surge height

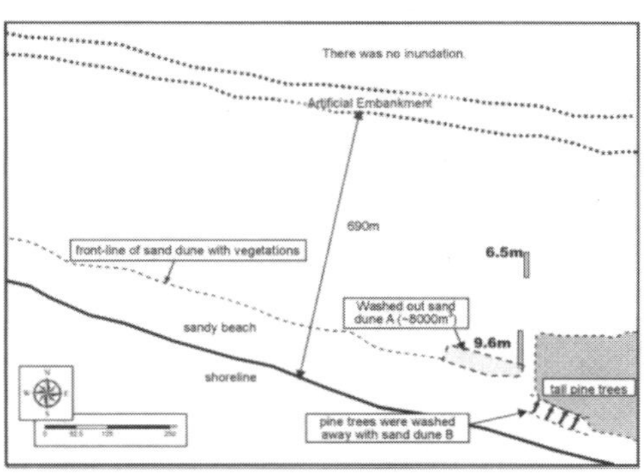

Figure 4. Overview of observed damages around Kuakata

Coastal Dynamics 2009
Paper No. 130

A FRENCH PROJECT TO UNDERSTAND AND PREDICT THE SHORT-TERM DYNAMICS OF SANDY COAST: ECORS

Thierry Garlan and Fabrice Ardhuin[1]

1. Introduction

ECORS is managed by the French Hydrographic Office (SHOM) and aims at demonstrating short-term predictions of the morphodynamics of sandy coasts. Such predictions are needed for naval amphibious operations, and the understanding of hydrodynamic and morphodynamic processes that will be generated by the project will certainly find many other applications, including coastal management. Over the years 2006 to 2011, the ECORS program include three main tasks: theoretical research, experiments and numerical modeling. The project include both field and laboratory experiments. Such a wide scope was made possible by the additional contributions from many research institutes and local authorities. The first step is to improve the quality of nearshore and offshore models over timescales ranging from one day to some months. Based on these research and campaigns, a coastal forecasting simulator will be created.

2. Studies on dynamics of sandy coasts

Planned investigations involve 147 sedimentologists and oceanographers from 62 research laboratories affiliated to INSU (French national institute for earth sciences and astronomy). The different subjects of these studies concern the definition of suitable methods for the study of rapid changes of morphology, the analysis of the impact of hydrodynamic factors (waves, currents, tides) and wind on the sediment transport, and the classification and magnitude of beach morphology parameters. Beach morphodynamics studies are the essential key to improve the knowledge of the beach system evolution. So the studies cover waves and currents effects, morphologic changes from detailed bottom evolution investigated at the time-scale of the wave morphology changes to high-resolution optical remote sensing, interactions between water and sediments, sediment properties, ... These efforts build upon a decade of previous field experiments at the same sites, and thus extend the existing database. Considering the variations of sandy beach morphologies, several experimental beaches have been selected for studying macrotidal and microtidal environments, with or without sand bars in the lower intertidal and/or in the subtidal domain. These beaches are all other the French coast; along the coasts of North Sea, English Channel, Bay of Biscay and Mediterranean sea.

3. ECORS experiments

A five-day deployment in Biscarosse (June 2007) was a pilot experiment in which a large variety of instruments and sampling techniques were tested. The wide range of offshore shore-normal waves conditions ($0.5<Hs<3m$), provided a persistent and intense rip current. A large array of sensors was deployed on a strongly alongshore non-uniform bar and rip system, as well as a drifter experiment.

The Six weeks ECORS international intensive field campaign took place in early 2008 at Truc Vert Beach (Aquitaine coast, France) during a wide range of offshore conditions, with offshore significant wave height ranging from 0.5 m to 8.2 m. Our goals were to obtain measurements of hydrodynamic processes, sediment characteristics and transport, and morphodynamic response of the beach on a large spectrum of scales under very energetic swell and meso- macro-tidal ranges. Thanks to the offered logistical support and an open invitation to anybody interested, 120 scientists, engineers and technicians from sixteen institutions and three continents were involved (Australia, France, Great Britain, New Zealand, The Netherlands, USA). These measurements combine on-site observations (current, pressure, turbidity, granularity of sediments), the use of video to estimate the bathymetry and

[1]SHOM, HOM/Recherche, 13 rue du Chatellier, BP 30316, 29603 Brest Cedex , France.
thierry.garlan@shom.fr

surface conditions, ... Frequent bathymetric surveys were conducted from hydrographic launches and personal water crafts, and provided continuous coverage with the daily beach topography surveys

During this experiment, surf and swash zones dynamics of a barred macro-tidal beach have been studied. A particular attention is paid to infragravity energy and groundwater dynamics, and their impact on sediment transport. Alongshore non-uniform rip channel was instrumented with point and profiling (horizontal and vertical) current-meters in particular in rip feeder currents and rip necks. Particle image velocimetry (PIV) techniques were used to calculate a 2D velocity field and to study wave interactions with rip currents in the surf zone.

During the experiment, Truc Vert Beach was exposed to 4 severe storms with significant wave height Hs > 4 m, comprising a 10-year return storm with Hs larger than 8 m during spring tidal ranges. The high-energy forcing resulted in highly-variable and dynamic double sandbar geometries and the collection of a high quality dataset. The combination of measurements of currents and topography evolution is expected to significantly improve our understanding of the hydro- and morphodynamics of sandy barred beaches under storm conditions.

4. Models

Between numerical models and field measurement, the laboratory experiments are used to test detailed hypothesis under different controlled conditions. Some laboratories involved in this project have developed experiments based on real beach parameters to reconstruct bedforms and to observe their movements. One of these has just been completed in a large 30*30 m square flume tank to compare field observations with laboratory results. On a larger scale, the on-going directional wave measurement off Truc Vert Beach has been combined with wave spectra from the ENVISAT satellite to provide the first quantitative estimation of the large scale swell dissipation. This knowledge has already been translated into new parameterizations for wave models, with a proven enhanced accuracy compared to pre-existing models (e.g. Ardhuin & al. 2008, http://preview.tinyurl.com /7bz6jj). Likewise, a similar effort is pursued to translate the more accurate offshore wave conditions into better coastal and surf zone wave and current fields, and to turn this hydrodynamic forcing into accurate sediment fluxes and morphological response.

Finally, the results of the project will be used to create a simulator which will combine geographic data from data base, software and numerical models to characterize nearshore environments and their short time evolution.

5. Conclusion

The environment has a determining influence on air, land and maritime force deployment operations, and the beaching boat in distress has nearly the same needs of knowledge. The bathymetric and topographical information available are in general limited to static information, which does not take into account the variability of the environment. Moreover, understanding the beach morphodynamics is still challenging because of the multitude of control parameters. To remedy the lack of validation data on sand transport and on coastal dynamics studies, campaigns and studies on physical and numerical models have been done by all the French laboratories involved in coastal dynamics. Some very interesting results are obtained and many others will arrived in the next few months and years.

Acknowledgements

The authors wish to thank all the teams involved in the ECORS Project. In particular, they are grateful to N. Sénéchal, F. Levoy, P. Bonneton and R. Certain They dedicate this work to Denis Michel who tragically passed away during one of the field experiment.

References

Ardhuin F., Collard F., Chapron B., Queffeulou P., Filipot J.F., Halon M. (2008). Spectral wave dissipation based on obervations : a global validation. Proceedings of the Chinese-german joint symposium on hydraulic and ocean engineering, Darmstadt, pp 391-400.

Coastal Dynamics 2009
Paper No. 131

ACOUSTICAL OBSERVATIONS OF VELOCITY AND SUSPENDED SEDIMENT STRUCTURES DUE TO OSCILLATORY FLOW OVER A RIPPLED BED

Peter D. Thorne[1] and David Hurther[2]

1. Introduction

Wave induced ripples are ubiquitous features in shallow coastal waters. The ripples are usually of the order of decimetres in length and centimetres in height. The ripples impact on the near bed hydrodynamics and the suspended sediment entrainment. For ripples of low slope, $?_r/?_r$ =0.1, where $?_r$ is the ripple height and $?_r$ is the ripple wavelength, the ripples enhance the roughness relative to a plane bed. If the ripples are relatively steep with $?_r/?_r \geq 0.1$, then the mixing close to the bed is considered to be dominated by a coherent process involving boundary layer separation on the lee-side of the ripple crest during each wave half-cycle near maximum flow velocity. Direct measurements of the hydrodynamics and sediment structures, within the wave boundary layer, above a rippled bed, at field scales are relatively rare and it is only recently that detailed observations of the processes have begun to become available. Here we describe some of the latest observations of simultaneous collocated flow and suspended sediment measurements, collected above a rippled bed, using novel acoustic profiling instrumentation.

2. Equipment and measurements

The measurements were made in the large scale flume facility of Campus Nord, UPC Barcelona, Spain, shown in figure 1. The flume is 100 m in length, 5m in depth and 3 m wide. The sand used in the experiments had a d_{50}=250 μm and was placed in the flume to a depth of 0.5m.

Figure 1. The Barcelona flume at Campus Nord UPC Barcelona, Spain.

The water depth for the experiments was between 1.6m-1.9 m above the sand bed. Regular waves were propagated along the flume with a constant period of T=6.5 s and heights between H=0.3m-

[1]Proudman Oceanographic laboratory, 6 Brownlow Street, Liverpool, L3 5DA, UK. pdt@pol.ac.uk
[2]Laboratoire des Ecoulements Geophysiques et Industriels (LEGI-CNRS), BP 53, 38041 Grenoble cedex 9, France. david.hurther@hmg.inpg.fr

0.55m. The instruments used to make the measurements were a multi-frequency acoustic backscatter system, ABS, two acoustic Doppler velocity profilers, ADVP, and a ripple profiler, ARP. The set up is shown in figure 2.

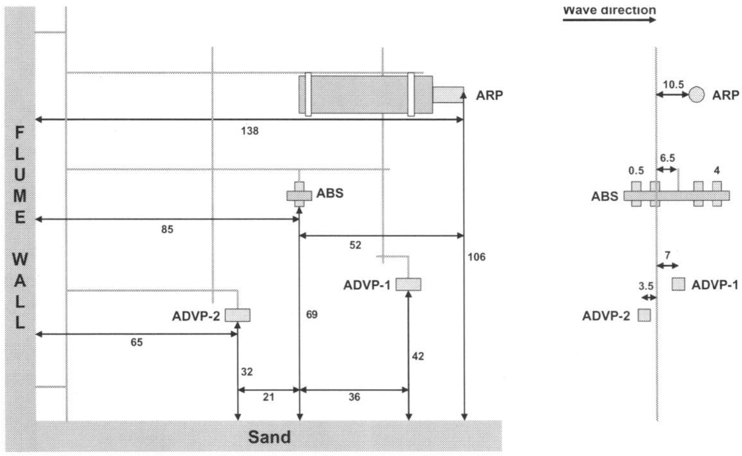

Figure 2. Schematic showing the layout of the instrumentation for the study.

Using the instrumentation, data were collected on the bedforms, the suspended sediments and the flow. A typical example of the results is show in figure 3. The figure shows patterns in the near bed flow and the suspended sediment at one location on a ripple with the phase of the wave. In the full paper, results of the type shown in figure 3 will be used to build up intra-ripple, intra-wave, measurements of the flow and suspended sediment structures. The aim of the work is to providing detailed descriptions of hydrodynamic and sediment processes, within the wave boundary layer, at field scales, under waves over a rippled bed.

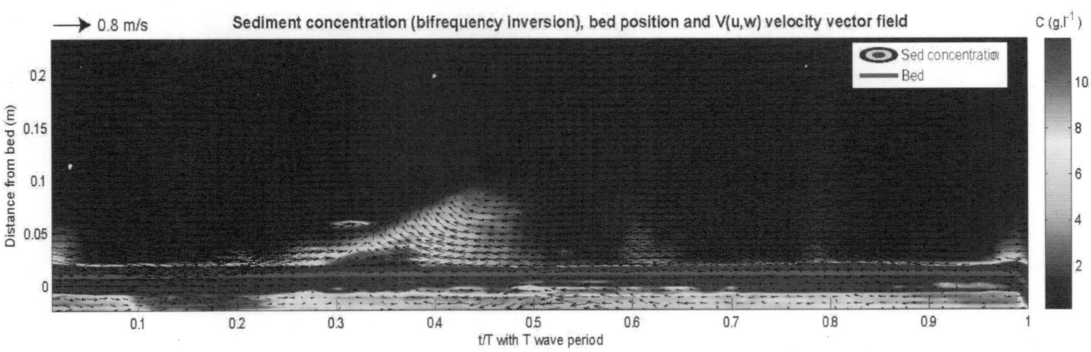

Figure 3. Example of simultaneously measured collocated flow and suspended sediments.

3. Acknowledgements

The authors would like to thank Dr Richard Cooke, Dr. Ivan Caceres and Francois-Xavier Chassagneux for helping to carry out the experiment and Professor Agustín Sánchez-Arcilla for providing financial support under the FP6 HydralabIII-SANDS project and an ICTS grant from the Spanish Ministerium of Education and Science.

Coastal Dynamics 2009
Paper No. 132

BEDLOAD SEDIMENT TRANSPORT BUDGET USING CT-SCANNING

Stephane Montreuil[1] and Bernard Long[2]

1. Introduction

Bedload sediment transport budget determination is a trivial problem specially because is not easy to measure the interaction between sediment dynamics and fluid hydraulics. To approach this problem, we use X-Ray measurements from a medical CT-Scanner associated with a hydraulic flume of 7 m long and 0.3 x 0.3 m section, filled with 50 mm Ottawa sand. This approach gives a 4D non intrusive density measurement (fig. 1).

The following experiment is set at two current velocities over two sediment grain sizes to obtain the different phases of ripple formation. Contrary to Montreuil et al. (2008), the CT-Scanner gantry stays on static position. The measurements are made using a constant time interval (Δt) of 30 seconds over a 50 minutes period. This protocol gives a Lagrangian representation of the movement of the sediment during the bedload process.

Figure 1. Scanning room with CT-Scanner, PIV and flume at Institut National de la Recherche Scientifique.

2. Measurement and results

The migration of internal structure of bedload density is measured on the central part of the flume on a 67 mm large by 200 mm height by 1 mm thick volume and for a 50 minutes period. Those measurements are made on the different ripples migrating phases from sinusoidal to rhomboidal. The sediment mass transport is calculated from 172 x 100 density profiles of 512 points and converted to voxel (volume element). The mass of sediment is computed for each voxel using sediment

[1]Institut national de la recherche scientifique, Université du Québec, 490 de la Couronne, Québec G1K 9A9, Canada. stephane.montreuil@ete.inrs.ca

[2]Institut national de la recherche scientifique, Université du Québec, 490 de la Couronne, Québec G1K 9A9, Canada. bernard.long@ete.inrs.ca

concentration. For this study, only voxel composing the bedload transport are treated. Those voxels of 0.39 x 0.39 x 1 mm, located between the frontier density point (FDP) and the base of lower bedload transport zone (LBTZ) (Montreuil and Long, 2007), are integrated to get a total mass of sediment transported (fig 2). The total sediment transported is 93.94 grams in a 67 mm section during 50 minutes,. The mass unit is used instead of volume to avoid error due to compaction fluctuation of bedload.

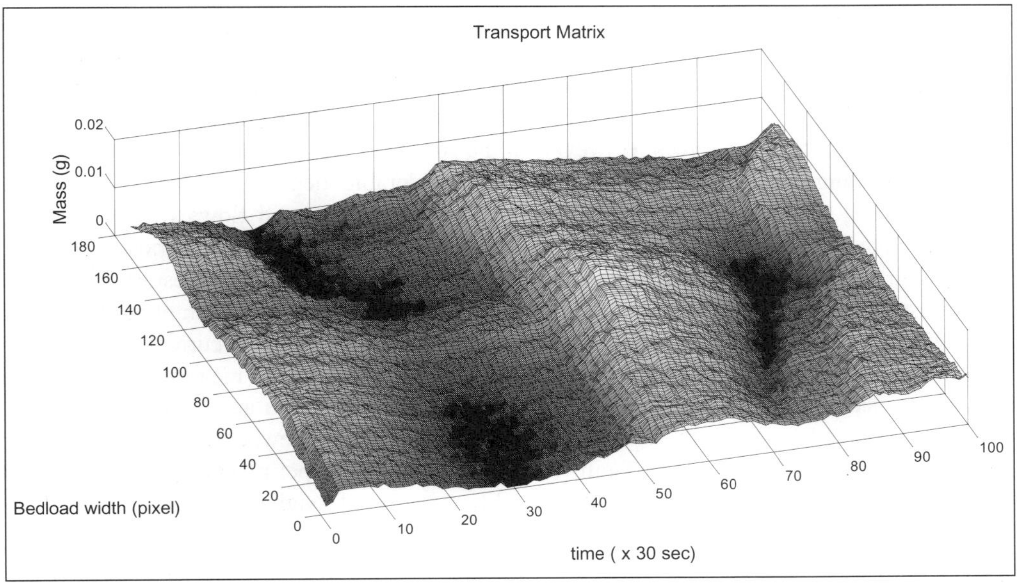

Figure 2. Mass of sediment transported in 50 minutes by step of 30 seconds. Examination window of 67 mm large (172 pixels)

3. Conclusion

CT-Scanning of a flume experiment opens a new perspective to investigate the 4D architectural evolution of bedload transport. The density variation obtained during migration under a stationary flow regime reveals hydraulic feedback between fluid and sand, influencing the shape and mass of sediment transport. This approach is used to compare the bedload sediment transport along sand ripples or sand dunes depending on grain sizes and flow velocities.

Acknowledgements

Special acknowledgment to Canadian found for innovation (CFI) and the GEOIDE project for there contribution and to J. W. Kamphuis for his collaboration to this project.

References

Montreuil S. and Long B. 2007, Flume Experiments under CAT-Scan to measure internal sedimentological parameters during sediment transport, Proceedings Coastal Sediments '07 Conference, ASCE, New Orleans, LA,1:124-136.

Montreuil S., Long B. and Kamphuis J. W., Surface density variations along a sand ripple under stationary flow measued by CT-Scanning, Proceedings International Conference on Coastal Engineering '08, CERC, Hamburg, Germany. In Press

Coastal Dynamics 2009
Paper No. 133

SHEET FLOW TRANSPORT FOR WAVES WITH ACCELERATION SKEWNESS

Dominic A. van der A[1], Tom O'Donoghue[2] and Jan S. Ribberink[3]

1. Introduction

When wave-generated velocities close to the seabed are high, sand is transported in a thin layer of high concentrations, generally a few mm to a few cm thick. Many studies have been conducted on this so-called 'sheet flow' transport since it is a major contributor to the yearly overall sand transport budget for coastal areas. The majority of these studies have focused on oscillatory flows induced by sinusoidal or skewed (Stokes) waves. However, in the surf zone waves are generally characterised by their forward-leaning 'sawtooth' shape, having a considerable degree of acceleration skewness. Previous laboratory experiments in which short period sawtooth-shaped flows were generated over sand beds produced non-zero transport rates, and the magnitude of the net transport increased with increasing degree of acceleration skewness (Watanabe and Sato, 2004). These results have prompted much interest amongst sand transport modellers, with many researchers attempting to incorporate "acceleration effects" into practical models (e.g. Nielsen, 2006). However, current understanding of acceleration effects is still poor, mainly because of the lack of detailed measurements. This paper reports on detailed measurements of sediment transport rates, velocities and concentrations for full-scale sawtooth-shaped flows, with varying degrees of acceleration skewness. The experiments were conducted in a large-scale laboratory facility. These mobile bed experiments complemented related research on sawtooth-shaped flows in the same facility using fixed rough beds (Van der A et al, 2008).

2. Experiments

The experiments were conducted in the Aberdeen Oscillatory Flow Tunnel (AOFT), a large tunnel facility capable of generating oscillatory flows with periods and amplitudes equivalent to those of full-scale waves. The AOFT has an overall length of 16m with a glass-sided, 10m long, rectangular test section, 0.75m high and 0.3m wide. The 0.25m deep sand bed in the test section alternately consisted of well-sorted 'fine' (d_{50} = 0.13mm), 'medium' (d_{50} = 0.27mm) and 'coarse' sands (d_{50} = 0.46mm). The piston was programmed to generate sawtooth shaped flows with specified period T, maximum velocity U and acceleration skewness parameter β, with

$$\beta = \frac{\dot{u}_{max}}{\dot{u}_{max} - \dot{u}_{min}} \qquad (1)$$

where $\dot{u}$ is the flow acceleration. Thus $\beta = 0.5$ corresponds to sinusoidal flow and $\beta > 0.5$ indicates (forward-leaning) acceleration skewness. The flow conditions consisted of fixed T-U combinations, with periods ranging from T = 5-9s and velocities U = 0.8-1.3m/s, for which β was varied in the range $0.55 < \beta < 0.8$. During all conditions the bed remained flat and sheet flow was the dominant transport regime. Sediment transport was measured by applying a mass conservation principle to the measured pre- and post-test bed profiles with the masses of sand collected from the two ends of the test section. Bed levels were measured using a Laser Bed Profiler (LBP) measuring bed level across the tunnel width every 2mm longitudinally. For selected conditions time-dependent concentrations in the sheet-flow layer were measured using Concentration Conductivity Meters (CCMs) and horizontal flow velocities above the sheet-flow layer were measured using Particle Image Velocimetry (PIV).

---

[1]School of Engineering, University of Aberdeen, Aberdeen AB24 3UE, Scotland. d.a.vandera@abdn.ac.uk
[2]School of Engineering, University of Aberdeen, Aberdeen AB24 3UE, Scotland. t.odonoghue@abdn.ac.uk
[3]Water Engineering and Management, Faculty of Engineering Technology, University of Twente, P.O. Box 217, 7500 AE Enschede, The Netherlands, J.S.Ribberink@utwente.nl

3. Results

Figure 1a presents example results from the net transport measurements. It shows net transport rates against acceleration skewness for the three different sand sizes and flow conditions with identical flow period T = 6s and peak velocity U ˜ 1.3m/s. For all conditions and all sand sizes the net transport rates are found to be positive (i.e. onshore directed) and consistently increase with increasing acceleration skewness. For example, for the coarse sand the net transport rate increases by a factor ~8 between $\beta = 0.58$ and $\beta = 0.73$. The high positive transport rates seen for the fine sand conditions are particularly interesting. The positive transport in the case of the fine sand under acceleration skewed (sawtooth) flow is in contrast to the negative net transport reported for fine sand sheet flow conditions under velocity-skewed flow, caused by unsteady phase lag effects (Dohmen-Janssen et al, 2002). This onshore transport is very important and could potentially lead to net onshore transport for real waves in the surf zone, where both velocity and acceleration skewness are present. Figure 1b shows example results of instantaneous erosion depth and sheet flow layer thickness for one of the coarse sand conditions. More sediment is eroded and entrained during the positive flow half cycle compared to the negative flow half cycle, which is due to the higher shear stress caused by higher flow acceleration, as was shown in the previous fixed bed experiments (Van der A et al, 2008). The moments of maximum pick up and entrainment around peak velocities create unsteady effects which can enhance the onshore transport and might well be responsible for the large transport rates found for the fine sand conditions.

Acknowledgements
This work is part of the SANTOSS project ('SANd Transport in OScillatory flows in the Sheet-flow regime') funded by the UK's EPSRC (GR/T28089/01) and STW in The Netherlands (TCB.6586).

References

Dohmen-Janssen, C.M., Kroekenstoel, D.F., Hassan, W.S., Ribberink, J.S., 2002. Phase lags in oscillatory sheet flow: experiments and bed load modeling, *Coastal Engineering,* 46: 61-87.

Nielsen, P., 2006. Sheet flow sediment transport under waves with acceleration skewness and boundary layer streaming. *Coastal Engineering*, 53: 749-758.

Van der A, D.A., O'Donoghue, T., Davies, A.G., Ribberink, J.S., 2008. Effects of acceleration skewness on rough bed oscillatory boundary layer flow. *Proc. 31st ICCE*, Hamburg, World Scientific, (in press)

Watanabe, A. and Sato, S., 2004. A sheet-flow transport rate formula for asymmetric forward-leaning waves and currents. *Proc. 29th ICCE,* Lisbon, World Scientific, 1703-1714.

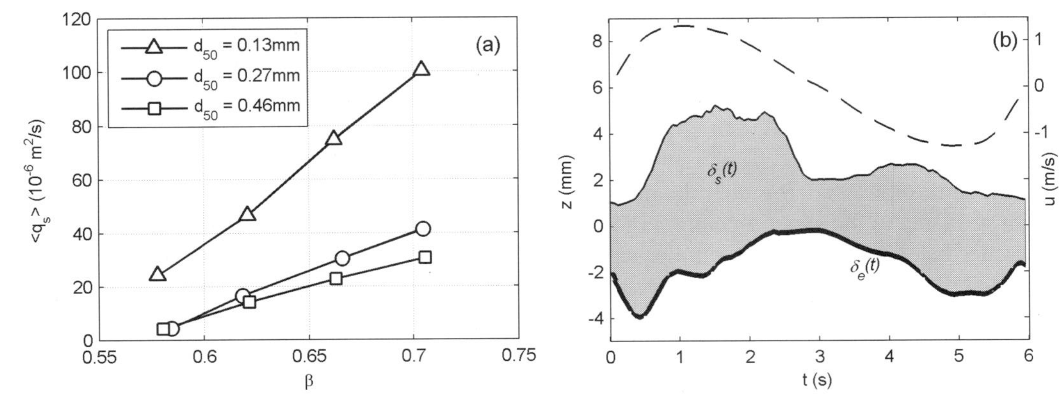

Figure 1. (a) Net transport rates against acceleration skewness for three sand sizes, all conditions have T = 6s and U ˜ 1.3m/s, and (b) time-dependent erosion depth (thick line) and sheet flow layer thickness (shaded area) for T = 6s, U = 1.3m/s, β = 0.73. The dashed line represents the free stream velocity time series.

Coastal Dynamics 2009
Paper No. 134

A PROCESS BASED APPROACH TO DERIVE PROBABILISTIC ESTIMATES OF COASTAL RECESSION DUE TO SEA LAVEL RISE

Roshanka Ranasinghe[12], David Callaghan[23] and Marcel J. F. Stive[2]

1. Introduction

Any rise in the mean sea level will result in the retreat of unprotected coastlines (Bruun, 1962). As global average sea levels are expected to rise at an unprecedented rate during the 21st century (Leuliette et al., 2004; IPCC, 2007), it is likely that coastal recession will also accelerate. The socio-economic impact of such accelerated coastal recession could be massive due to the unprecedented growth of coastal communities over the last 50 years or so which has led to $ billions worth of developments within the coastal zone. To avoid massive economic losses in the future, predictions of coastal recession due to sea level rise (SLR) need to be highly accurate; now more so than ever before.

The method most commonly used to estimate coastal recession due to SLR is the Bruun Rule. Essentially, the Bruun Rule predicts a landward and upward displacement of the cross-shore sea bed profile in response to a rise in the mean sea level and is expressed as:

$$R = LS/(B+h) \tag{1}$$

where, h = the maximum depth of sediment exchange, L = horizontal distance from the shoreline to depth h, B = berm/dune elevation for the eroded area, S = sea level rise, and R = coastal recession.

While the concept underpinning the Bruun Rule has been qualitatively verified (eg. Rosen, 1978, Dean, 1990), to date, it has not been have convincingly verified quantitatively. Futhermore, while the Bruun concept implies that SLR will result in the movement of sand from the berm/dune to the submerged nearshore profile, the actual physical processes by which this dune erosion will occur are not directly accounted for in this approach. Therefore, by necessity, the Bruun rule requires the specification of several input parameters that are associated with significant uncertainty (eg.. depth of closure DoC). Furthermore, the common practice of adopting a single value of coastal recession is also proving inadequate with the emergence of risk management style coastal planning frameworks which require probabilistic estimates of coastal recession (Cowell et al., 2006, Ranasinghe et al., 2007). Given these limitations associated with the Bruun rule, it is imprudent to depend on the Bruun rule for predictions of future coastal recession, particularly in view of the potentially massive socio-economic losses that may result from ill-informed planning decisions. The aim of the present study is to completely depart from the Bruun concept and develop an innovative process based approach for estimating coastal recession due to SLR within a probabilistic framework.

2. Method

The basic philosophy underpinning the method presented herein is that the net recession of the coastal berm/dune is due to the superimposition of higher return interval storms upon the gradually increasing mean water level due to SLR. In simpler terms, assume the MWL is zero at present and dune is at x = 0 (horizontal axis). Now say a 1 in 10 year storm occurs next month (MWL ~ 0) resulting in a dune retreat of say 10m. Now assume that the next 1 in 10 year storm occurs in 10 years when the MWL is elevated to 0+10 x (annual SLR rate ~ 0.3mm/yr). As beach recovery is an extremely slow process, it is very likely that the dune has not completely recovered to its original position. Say the recovery was

[1] UNESCO-IHE, PO Box 3015, 2601 DA Delft, The Netherlands. r.ranasinghe@unesco-ihe.org
[2] CiTG Technical University of Delft, PO Box 5048, 2600 GA Delft, The Netherlands. m.j.f.stive@tudelft.nl
[3] Division of Civil Engineering, University of Queensland, QLD 4072, Australia. d.callaghan@uq.edu.au

5m (seaward translation) from its eroded position. Then, due to the second storm, a further 10 m of dune retreat can be expected, which would result in a net dune retreat of 15m (5-2+5) compared to the present situation. As time progresses this process will recur, resulting in a gradual coastline retreat.

Callaghan et al. (2008) presented a Joint Probability Distribution Method (JPM) which incorporated a process based structural function to estimate storm erosion volumes within a probabilistic framework withoiut considering SLR. The method was developed and verified for Narrabeen beach, Sydney, Australia, using over 30 years of field data.
The JPM is implemented as follows:
1. Identify meteorologically independent storm events from offshore wave data
2. Fit extreme value distributions to wave height (H) and storm duration (D).
3. Fit the dependency distribution between wave height and storm duration.
4. Fit the wave period (T) and peak tidal anomaly R (using water level data) distributions.
5. Determine the empirical distribution for wave direction (θ).
6. Fit a non-homogenous Poisson distribution to the spacing between storms.
7. Perform a Monte Carlo simulation using the fitted distributions to simulate wave climate and derive a time series of storm erosion volumes.

In the present study, the JPM is extended to include SLR effects. The procedure adopted is as follows:
1. Generate a 110-year (1990-2100) long offshore wave climate using the JPM
2. Estimate the sea-level rise (S) (as per IPCC, 2007) at the time each storm occurs.
3. For each storm, estimate dune recession using the process based dune impact model
4. Obtain dune recession at 2100.
8. Repeat 1-7 many times until convergence is obtained
The resulting probabilistic estimate of coastal recession is shown in Fig. 1. Interestingly a simple Bruun Rule application to the same site returns a R/S of about 40 by the year 2100 which corresponds with the 1% probability R/S value in Figure 1.

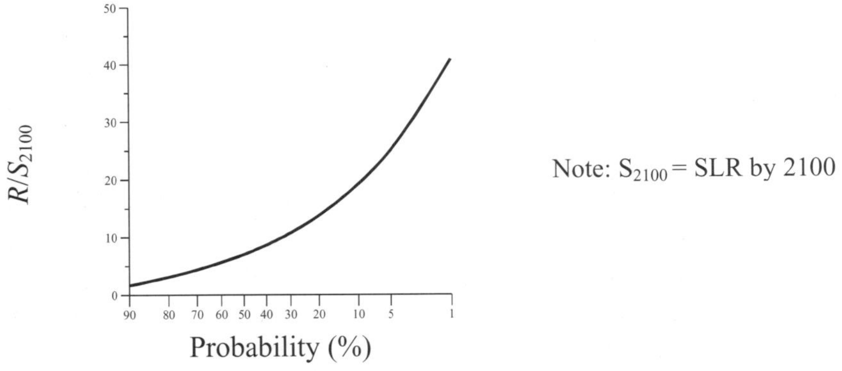

Figure 1. Probabilistic estimate of coastal recession by 2100 due to SLR at Narrabeen beach, Sydney, Australia.

3. Conclusions

A process based approach to derive probabilistic estimates of coastal recession due to SLR has been developed. The method requires as input long term water level and wave data which are now available via widespread tide gauges and global hind cast models (eg. WW3, ERA40) respectively. This method is proposed as a more appropriate and defensible alternative to the Bruun rule for the determination coastal recession due to SLR for planning purposes.

Coastal Dynamics 2009
Paper No. 135

IMPACT OF GLOBAL CLIMATE CHANGE ON WAVE CLIMATE

Nobuhito Mori[1], Tomohiro Yasuda[1], Ryota Iwashima[1], Tracey Tom[2] and Hajime Mase[1]

Introduction

The influence of global climate change due to green house effects will require impact assessment and adaptation strategies for future our society. The sea level rise is a static side issue of climate change and is important for human activity near the coastal zone. On the other hand, wave and storm surge are dynamic side of climate change and will be necessary to take into the consideration for coastal disaster prevention and reduction, if extreme weather events will become stronger than those in the present climate. This study analyzes and predicts the future ocean wave climate in comparison with those in present climate based on the climate model output.

Outline of computation and analysis

The research is conducted on the basis of climate model at Metrological Research Institute of Japan Metrological Agency (JMA-MRI; Kakushin, 2008). The JMA-MRI climate model is the atmospheric T959L60 single model with SST from coarse grid coupled run of atmosphere and ocean simulation, and is being computed for the three periods of 1979-2004 (present), 2015-2028 (near future) and 2075-2100 (future) following A1B scenario, respectively. The waves of the three periods are simulated using sea surface winds (U_{10}) of climate model by the SWAN model. The simulated U_{10} and wave height are analyzed to predict wind and wave climate change from present to future in global and regional area.

Results and discussion

There are regional dependencies of mean values of U_{10} and significant wave height. The mean wind speed at mid latitude westerlies and Antarctic circle are increased 0.3 - 0.8m/s in from the present to future climate, although their values at horse latitude are decreased as shown in Figure 1. The area of Japan belongs to corresponds slightly decreased region regarding U_{10}. Figure 2 shows the averaged significant wave height over the period both the past and future climate. There is significant increase of the averaged significant wave height near the Antarctic circle about 0.4m but is decreased at northern hemisphere. On the other hand, the maximum wave height due to tropical cyclones in future climate is increased in co mparison with the past climate (not shown). These results give the lower mean and higher maximum wave height at mid-latitude area, and the higher mean and maximum wave height at high-latitude. Both wind and wave climates likely depend on the region.
The detail of results and discussion will be presented at the conference.

References

Kakushin (2008) Innovative Program of Climate Change Projection for the 21st Century, Japan, http://www.kakushin21.jp

---

[1] Disaster Prevention Research Institute, Kyoto University, Gokasho, Uji, Kyoto 611-0011, Japan. Tel.+81(0)774-38-4146, Fax.+81(0)774-38-4321, email: mori@ceanwave.jp (NM)
[2] Surflegend, Tsujido Nishikaigan 3-1-1, Fujisawa, Kanagawa 251-0046, Japan.

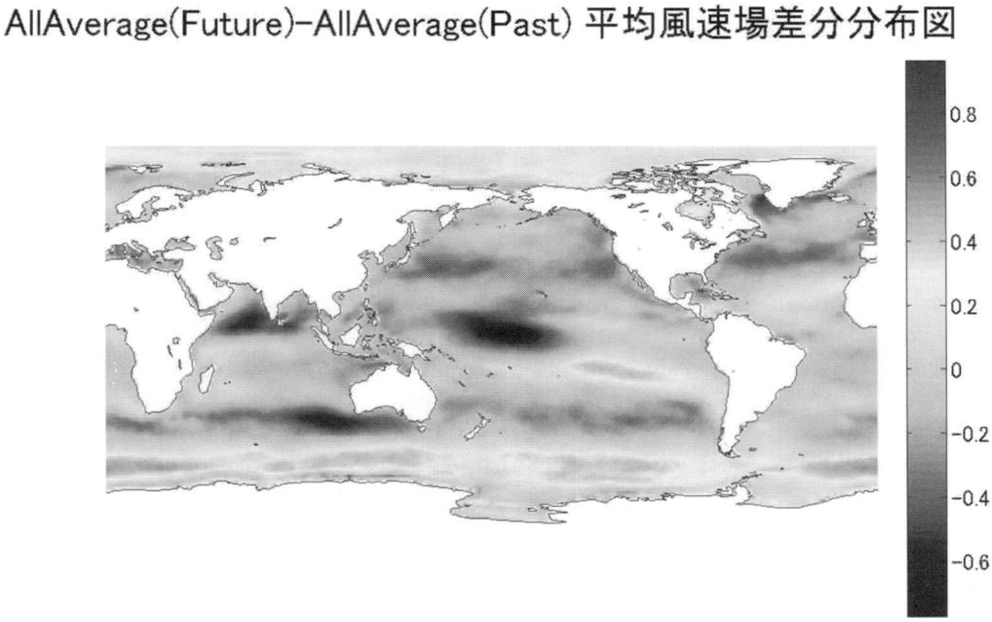

Figure 1. Difference of U10 between the future and present climate
(plus: increase future U10 than present climate)

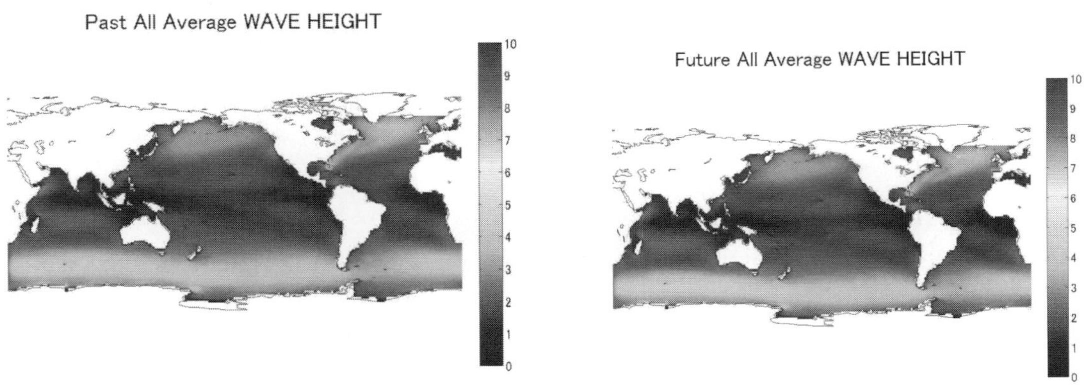

(a) Present climate (b) Future climate

Figure 2. Averaged H_s over the 25 years

Coastal Dynamics 2009
Paper No. 136

INFLENCES OF GLOBAL CLIMATE TO PEAK WAVE HIGHT IN TYPHOON SEASONS

Hajime Tanji[1] and Hirohide Kiri[2]

1. Introduction

Global warming will influence on the peak wave height in coastal areas. From observed wave data around Japan, Kawai said there is no remarkable trended in time series. The reason is actual climate is driven by not only long trend change but also variance. If the range of the variance is large, influence of long trend is difficult to be separated. In this paper, the authors will try to evaluate the variances as the component of warming. If data of target period is given and the target period can be separated into two periods, warm and cold, the variance between two periods can be used to evaluate the influences of global warming. This idea can be expressed that high wave will be occurred simultaneously. For the check of the hypothesis, this paper will search the simultaneous high wave events caused by typhoon and compare the index of global climate in each year.

2. Data and Method

For evaluation of high wave, 44 points data of yearly highest wave from 1978 to 2004 are selected from NOWPHAUS reports of Port and Airport Research Institute, Japan. Peak height and the month of each year are digitized. Monthly probability of each ports are calculated and matrix of monthly probability and ports are prepared.. Cluster analysis were done based on the matrix. Augusta and September were separated as a cluster. This means the effect of typhoon is the measure component in some ports. For the selection of the influence by typhoon, some ports which has the high probability in Augusts and September were selected. For these selected ports, based on the highest wave record in each year, rank of the year in each port were calculated. After that considering all selected ports, average rank of each year were calculated. Based on this average, rank of each year based on the risk of high wave was calculated.

Another data are indexes of climate conditions of each year. The following data are collected from 1978 to 2004. 1) Monthly average temperature bias of Augusts and September in Japan (*JPTAS*), 2) Average temperature bias of the Northern Hemisphere (*NGT*), 3) ocean surface temperature bias between 15N-EQ and 130E-150E (*NINOWST*) 4) ocean surface temperature bias between 5N-5S and 160E-150W (*NINO4*) 5) south oscillation index (*SI*), 6) monthly average temperature bias of Augusts and September of the world (*WAST*): Bias means the bias to the normal year values. These data are converted into rank between 1978 and 2004.

3. Selection of Waves Points

From the results of cluster analysis, typhoons have strong relation to peak wave height in Augusts and September. Figure 1 shows the probability that the highest wave in the year happens in Augusts or September. .There is a gap of probability at 0.4. Wave points that have probability of more than 0.4 are selected for the latter calculation. The number of selected points are 14 within 42.

4. Results and Conclusion

---

[1]Dept of Geotechnical and Hydraulic Engineering, National Institute for Rural Engineering, NARO, 2-1-6 Kannnondai , Tsukuba, Ibaraki 305-8609, Japan. tanji@affrc.go.jp

[2]Dept of Geotechnical and Hydraulic Engineering, National Institute for Rural Engineering, NARO, 2-1-6 Kannnondai , Tsukuba, Ibaraki 305-8609, Japan. kiri@affrc.go.jp

Figure 2 shows the cross plotting of order of selected parameters. Compare to *Wave*, linear relation can be found in *WAST* and *NGT*. If the table Spearman's rank coefficient values to *Wave* are as follows: *JPTAS* 0.22958327, *NGT* 0.62068053, *NINO4* 0.31479241, *NINOWST* 0.09903517, *SI* -0.17404581, *WAST* 0.69413919. From Figure 2 and these results, high wave event has strong relation to *NGT* and *WAST*.

If the wave point that are strongly influence by typhoons are selected, peak wave height is influence by *NGT* (Average temperature bias of the Northern Hemisphere) and *WAST* (monthly average temperature bias of Augusts and September of the world).

Acknowledgements

This research partly supported by the global warming research projects of ministry of environment and MAFF.

References

Nagai, N., Satomi, S. 2006. *NOWPHAS 2004*, Technical Note of The Port and Airport Research Institute, No.1118, 1-7. (in Japanese and other former notes)

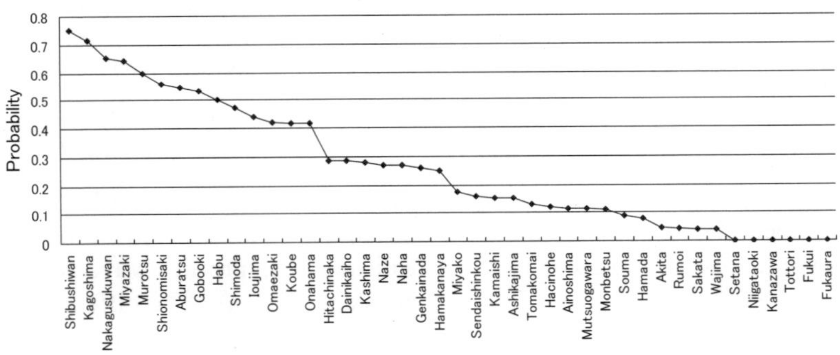

Figure 1. Highest wave probability in Augusts or September

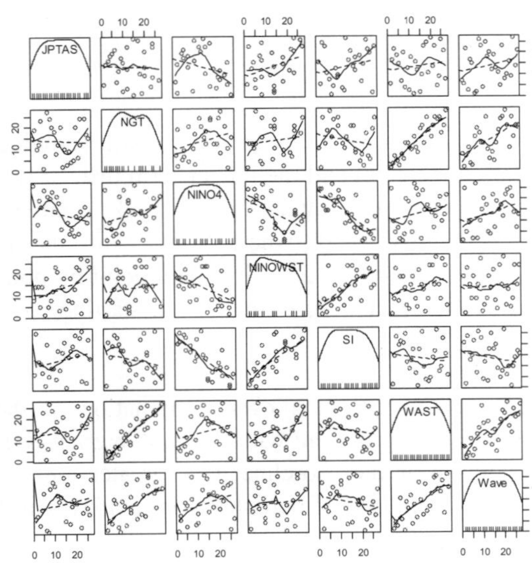

Figure 2.Scattered plot matrix of order of selected index

Coastal Dynamics 2009
Paper No. 137

FORMATION AND LONG-TERM EVOLUTION OF BEACH CUSPS WITH TRACKING PARTICLE MOVEMENT

Anurak Sriariyawat[1] and Nicholas Dodd[2]

1. Introduction

Beach cusps have interested scientists and engineers for decades. Previous research has attributed their development subharmonic edge waves (a hydrodynamic explanation stemming from linear shallow water theory) or to self-organisation theory (a morphodynamic explanation that utilises an abstracted model incorporating ballistic theory). Most recently Dodd et.al. (2008) combined nonlinear shallow water theory with an erodible bed in the swash, to produce a fully-coupled morphodynamical model, application of which over a relatively short time (< 500s) points to a hydrodynamic feedback mechanism that is enhanced by morphodynamic feedback being responsible for cusp formation. This study uses this model to examine the longer-term dynamics of beach cusps, in particular looking at their development in Fourier space, identifying their linear growth regime, looking at changes in cusp spacings and focusing on the circulation associated with cusp formation. To this end particle tracking is employed to follow the particle (water) paths and thereby to identify how water is recirculated within the cusp system,.

2. Beach cusp simulations

Simulations utilise either an initially plane 8 degree beach slope (i.e., slope = 0.14) or a beach profile obtained from a 1D bed profile simulation after a finite time, so as to avoid starting the simulation from an initial profile that is not representative of a realistic swash profile. Wave period (T) is varied (T = 3, 4, 5, 6, 7, and 8s), while wave height (H) is constant (0.25m). The different incoming wave periods and initial beach profiles induce the variations of cusp formation shown in the term of cusps spacing (?$_c$), swash excursion (S_e), and the reflection coefficient ($|R|$). As the predicted beach cusps are not completely regular, the growth rate (s), representing the initial evolution of beach cusps, must be determined using discrete Fourier analysis to find the dominant wavelength. The Fourier analysis also reveals the dominant wavelength associated with the observed cusps up to the final state. Example results are shown in Fig. 1.

3. Tracking particle movement

Tracking the water particle movement reveals the circulation pattern, showing both the short term paths (associated with the wave period, here 5s) and the longer-term drift of water from embayments to horns, from which location the water is recirculated rapidly back to embayments. Here we use simple movement equations ($x = U?t$ and $y = V?t$, where U and V are the depth-averaged current velocity in x and y direction respectively) to create a particle path point by point. Figure 2 illustrates the particle paths in different areas of one beach cusp where * is the starting point.

4. Discussion and conclusions

The resulting cusp spacing and swash excursion allows computation of the constant f of self-organisation theory. Results are roughly in agreement with those of Werner & Fink (1993) and Coco et.al. (1999). The water particle paths near the shoreline have greater cross-shore movement than those further offshore, and the gradual migration of particles from embayment to horn and back can

---

[1] Department of Civil Engineering, University of Nottingham, NG7 2RD, UK. evxas5@nottingham.ac.uk
[2] Department of Civil Engineering, University of Nottingham, NG7 2RD, UK. nick.dodd@nottingham.ac.uk

be seen. Further long-term evolution dynamics will be further discussed in the presentation.

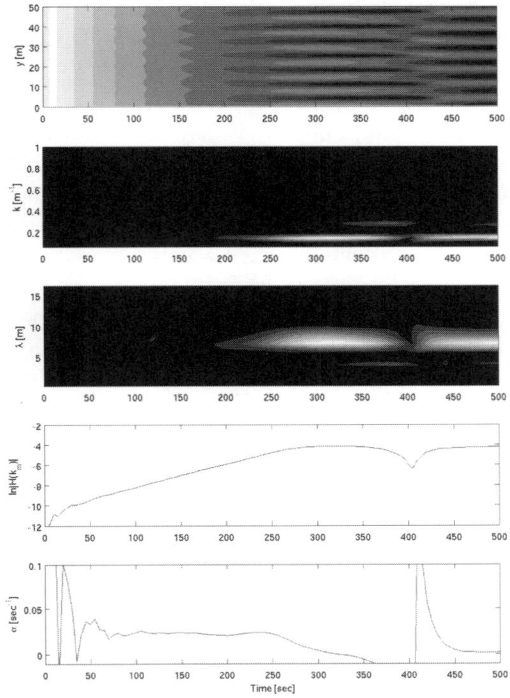

Figure 1. Linear growth rate computation using discrete Fourier analysis for T=5s.

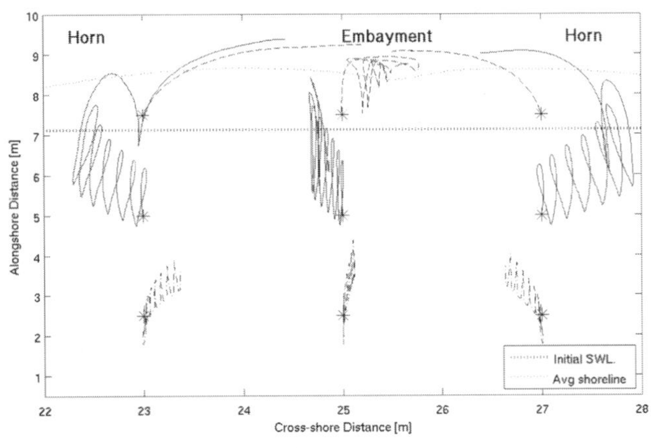

Figure 2. Particle movement along one beach cusp from t = 503-543 sec.

References

Coco, G., O'Hare, T. J., and Huntley, D. A. 1999. Beach cusps: a comparison of data and theories for their formation *J. of Coastal Research*, 15 (3): 741-749

Dodd, N., Stoker, A. M., Calvete, D., and Sriariyawat, A. 2008. On beach cusp formation. *J. of Fluid Mechanics*, 597: 145-169.

Werner, B. T. and Fink, T. M. 1993. Beach cusps as self-organized patterns *Science*, 260: 968-971

Coastal Dynamics 2009
Paper No. 138

MORPHOLOGY AND STRATIGRAPHY OF A DEGRADING DELTA

Hillen, M.M.[1], Geleynse, N.[2], Storms, J.E.A.[3], Walstra, D.J.R.[1,4], Stive, M.J.F.[1]

1. Introduction

Where a river debouches into a standing body of water, deposition leads to delta formation. This abstract presents the results of a numerical model study on sediment reworking in the deltaic environment, which links the fields of geology and hydraulic engineering. The study investigates the morphologic and stratigraphic behavior of deltas with a process-based coupled hydrodynamic-morphodynamic model (Delft3D) for a situation where waves rework the deltaic deposits.

The research of deltas is of interest to both hydraulic engineers and geologists, because of the presence of fluvial processes, the dynamic coastal environment and the presence of hydrocarbons. The research is conducted on different temporal and spatial scales. To study stratigraphic surfaces and sequences formed over long periods (10^3 to 10^6 years) large-scale models ($O(10^5)$) can be applied, whereas some problems can be modeled with a typical spatial scale up to $O(10^4$ m) over a period of seconds to years. This study investigates delta evolution from both a morphologic and stratigraphic perspective on meso-scale (temporal O(years), spatial O(kilometers)). This approach has already been successfully applied by Storms et al. (2007).

2. Model characteristics

The current study uses a 2D (depth-averaged) approach in the numerical model Delft3D, as described in detail by Lesser et al. (2004), which solves the unsteady shallow water equations. Two sediment fractions are present in the model; a sediment fraction with cohesive clay characteristics and a sediment fraction that represents non-cohesive fine sand. The sediment transport formulation of van Rijn (2007) is used to model both sediment fractions as suspended and bed load sediment transport. The 'Online' method (Roelvink, 2006) is applied for the morphodynamic update of the bed level with a morphological upscaling factor of 60. The wave climate is generated using the 3rd-generation SWAN model (Booij et al., 1999), which communicates with the flow field each simulated (hydrodynamic) hour.

The model setup represents an abandoned delta. The geometry and a simplified stratigraphy – using only two sediment fractions - of a delta developed in a gently sloped basin is used as the starting condition for wave reworking. For a situation with no active river discharge, the delta is subjected to waves for a period of 45 months (Figure 1).

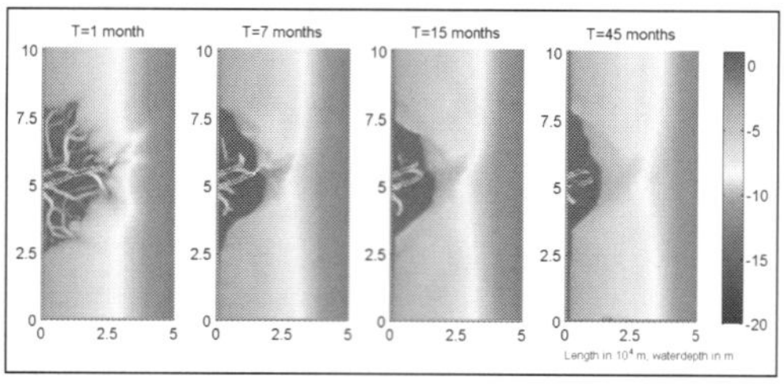

Figure 1. Different stages of delta degradation depicted by changes in the morphology of the original bird foot delta under influence of shore-normal waves (H_s = 1m, T_p = 5s) over a period of 45 months.

[1] Delft University of Technology, Faculty of Civil engineering and Geosciences, Section Hydraulic Engineering, PO Box 5048, 2600 GA, Delft, the Netherlands. MartenHillen@hotmail.com, M.J.F.Stive@TUDelft.nl
[2] Delft University of Technology, Department of Geotechnology, Section Geo-engineering, Delft, the Netherlands. N.Geleynse@TUDelft.nl
[3] Delft University of Technology, Department of Geotechnology, Section Applied Geology, Delft, the Netherlands. J.E.A.Storms@TUDelft.nl
[4] Deltares, PO Box 177, 2600 MH, Delft, the Netherlands. DirkJan.Walstra@Deltares.nl

3. Results

The model seems to accurately represent the process of delta degradation for the situation with no river discharge and a modest wave climate (shore-normal incident waves; H_s = 1m, T_p = 5s). Preliminary results show that channels are infilled by the clay fraction that is reworked from the retreating delta front. In addition, offshore transport of clay and alongshore deposition of sand occur (Figure 2).

The infill of channels by clay deposition starts at the ocean side, but over time nearly all channels fill in. Sediments are redistributed in alongshore direction by bed load transport and suspended sediment transport along the delta front, thereby leading to delta front retreat. On both lateral sides of the delta, sedimentation of sand occurs and flow circulation patterns are visible which transport the clay fraction offshore. These sediment dynamics are similar to what has been observed in the Atchafalaya Bay, Louisiana, where the sandy Wax and Atchafalaya deltas have been prograding under a modest wave climate and where suspended sediments have been transported offshore (Roberts et al., 2005). The stirring up of sediments by waves has a significant influence on the process of sediment reworking in the deltaic environment. This influence is has led to sedimentation on the lateral sides of the delta and the process of degradation of the delta slope.

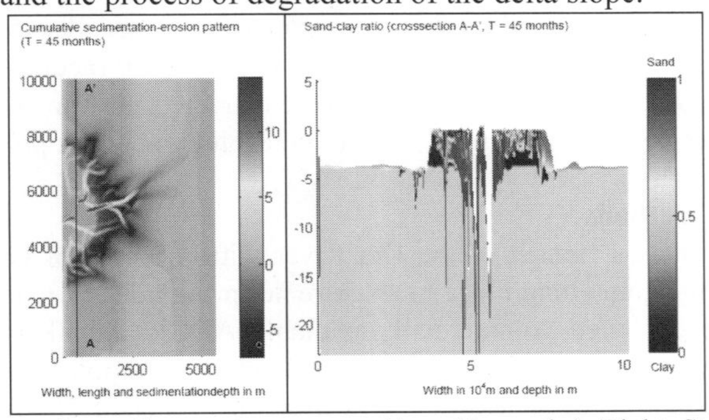

Figure 2. Left: Cumulative erosion-sedimentation patterns (Δt = 45 months), Right: Stratigraphy of lateral section.

4. Future work

Future research will include several wave climate scenarios. In addition, we will focus on the influence of river discharge and river sediment load during wave reworking to test the dependence of the delta deterioration on river characteristics. Subsequently, relative sea-level rise will be taken into account. Based on our preliminary results we expect our current work to confirm that the morphologic and stratigraphic behavior of deltas, for a situation where waves rework the deposits, can be investigated on meso-scale with the process-based model.

References

Booij, N., Ris, R.C. and Holthuijsen, L.H. (1999) A third-generation wave model for coastal regions. Part 1: Model description and validation. *Journal of Geophysical Research*, 104(C4), 7649–7666.

Lesser, G.R., Roelvink, J.A., van Kester, J.A.T.M. and Stelling, G.S. (2004) Development and validation of a three-dimensional morphological model. *Coastal Engineering*, 51(8-9), 883-915.

Roberts, H.H., Walker, N.D., Sheremet, A. and Stone, G.W. (2005) Effects of cold fronts on bayhead delta development: Atchafalaya Bay, Louisiana, USA. In: *High resolution morphodynamics and sedimentary evolution of estuaries*, FitzGerald, D.M. and Knight, J. (Eds.), Coastal Systems and Continental Margins series. *Springer, Dordrecht*, 269-298.

Roelvink, J.A. (2006) Coastal morphodynamic evolution techniques. *Coastal Engineering*, 53(2-3), 277-287.

Storms, J.E.A., Stive, M.J.F., Roelvink, J.A. and Walstra, D.J. (2007) Initial morphologic and stratigraphic delta evolution related to buoyant river plumes. *Proceedings of Coastal Sediments '07, New Orleans*, Kraus, N.C. and Rosati, J.D. (Eds), American Society of Civil Engineers, 736-748.

van Rijn, L.C. (2007) Unified view of sediment transport by currents and waves. II: Suspended transport. *Journal of Hydraulic Engineering*, 133(6), 668-689.

Coastal Dynamics 2009
Paper No. 139

LONG-TERM OBSERVATIONS OF FLOW AND SUSPENDED SEDIMENT IN A TIDALLY-DOMINATED ESTUARY

Mahdi Razaz[1], Kyoshi Kawanishi[2], and Yokoyama Tomoya[3]

1. Introduction

Coastal wetlands are under a great deal of pressure from forces of human population's intervention both along the estuary and in the river catchment. It is important to measure and assess erosion and deposition rate of suspended sediment in coastal zones to prevent shoaling in navigation channels or to estimate the amount of pollution carried by cohesive sediment flux. The latter involves precise *in situ* observations, although there are lots of factors involved in estuarine area that make it difficult to fulfill *in situ* measurements. Recently, devices based on the Doppler Effect have been developed to collect data with high quality and coherently throughout the water column. These instruments have been widely employed by scientist, e.g. Hill et al. (2003) used backscatter data to estimate resuspension rates in tidal waters, Fugate and Friedrichs (2003) used ADV (acoustic Doppler velocimeter) data to estimate turbulent kinetic energy (TKE) and the resulting Kolmogorov microscale. From ADCP (acoustic Doppler current profiler) backscatter data suspended sediment concentrations were extracted. Kawanisi (2004) studied structure of turbulent flow in the Ota River, Japan, using HRCP (high resolution current profiler) and ADV.

2. Study sites and field methods

Ota Floodway is the most-west branch of the Ota River with a length of nearly 9 km. It is a tidally-dominated river and the maximum tidal range in an extreme spring tide can reach 4 m (Figure 1).

In the present study we deployed commercially-available ADCP: AquaDopp which is developed by Nortek-AS. AquaDopps were fixed at the bottom of the floodway in 3 points A, B, and C that are 2.8, 4.9, and 7.9 km far from the mouth for at least 25 days over the winter of 2008. In point A measurements implemented from Nov. 27 to Dec. 25 in the right bank of the Ota Floodway, in point B from Nov. 17 to Dec. 15 in the left bank, and in point C from Nov. 13 to Dec. 8 at the center of the floodway. Settings of AquaDopps were similar for the all cases: 10 cm cell depth and sampling every 20 minutes.

3. Results and discussion

By processing data and applying necessary corrections, raw backscattered data can be translated into vertical profiles of velocity and suspended sediment concentration in different tidal phases. The advantage is that they are coherent either in depth or time. Vertical velocity profiles for each period were examined to study distortion from logarithmic distribution and distinguishing tidal straining strength. Tidal straining plays an important role in controlling the structure and intensity of turbulent stresses and TKE production on both ebb and flood phases of the tide in partially stratified estuaries (Simpson et al., 2005). Then, spatiotemporal changes of the velocity were plotted (Figure 2). As it is expected, difference between flood and ebb current speed is maximum near the mouth. However, it seems that unexpectedly flood currents

[1] Dept. of Civil & Environmental Engineering, Graduate School of Engineering, Hiroshima University, 1-4-1, Kagamiyama, Higashi-Hiroshima 739-8527, Japan. mahdirazaz@hiroshima-u.ac.jp
[2] kiyosi@hiroshima-u.ac.jp
[3] t-yokoyama@hiroshima-u.ac.jp

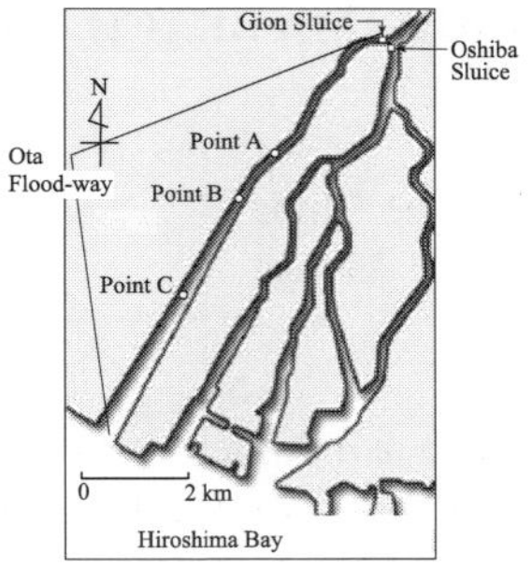

Figure 1. Schematic map of the observation points

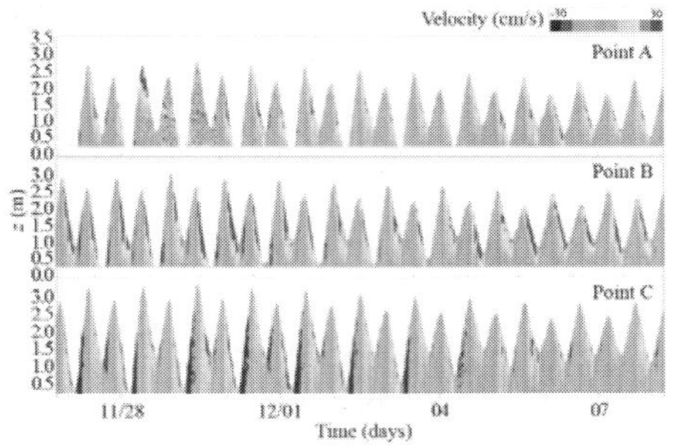

Figure 2. Spatiotemporal variations of stream-wise velocity

were faster in point B than the other sites. Moreover, in each tidal phase, highest and lowest current speed occurs before and after water slack which is a characteristic of tidally dominated rivers. Finally, the mechanism of sediment transport is studied by comparing spatiotemporal changes of the SSC (Figure 3) and velocity over the joint interval. In this figure, white areas below water surface denote very low SSC. As it is illustrated in Figure 4 which is the accumulated stream-wise sediment flux, most of the erosion activity occurred in the distance from the mouth to point B. Moreover, this figure reveals that most active part of the river from erosion/deposition point of view is in the range from point B to C. Yet, it can be translated as more activity in the left bank of the river. From point A to the Gion Sluice can be assumed as a closed boundary so we can conclude that a small part of eroded sediment is usually being deposited there. Since, we haven't measured SSC (suspended sediment concentration) at the Gion Sluice we cannot judge about erosion rates in this area, though it doesn't seem to be a very high value.

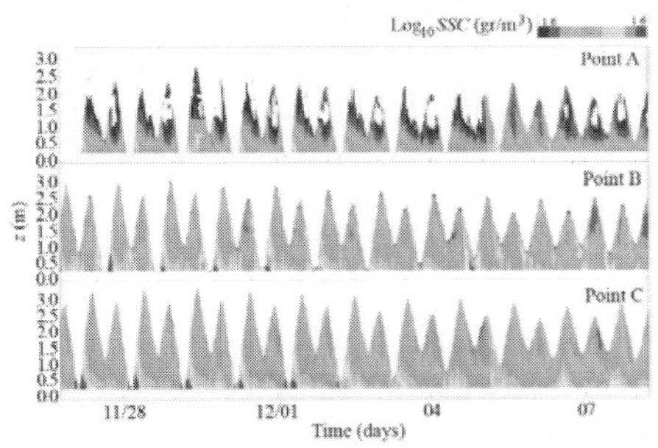

Figure 3. Spatiotemporal variations of SSC

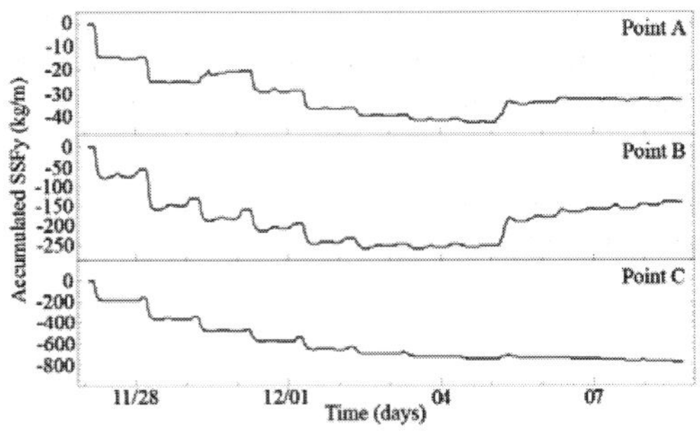

Figure 4. Accumulated Stream-wise SS flux

References

Fugate, D. C., and C. T. Friedrichs, 2003, Controls on suspended aggregate size in partially mixed estuaries, *Estuarine, Coastal and Shelf Science*, 58: 389-404.

Hill, D. C., S. E. Jones, and D. Prandle, 2003, Derivation of sediment resuspension rates from acoustic backscatter time-series in tidal waters, *J. Continental Shelf Research*, 23: 19-40.

Kawanisi, K., 2004, Structure of turbulent flow in a shallow tidal estuary, *J. of Hydraulic Engineering*, ASCE, 130(4): 360-370.

Simpson, J. H., E. Williams, L.H. Brasseur, and J. M. Brubaker, 2005, The impact of tidal straining on the cycle of turbulence in a partially stratified estuary, *J. Continental Shelf Research*, 25(1): 51-64.

Coastal Dynamics 2009
Paper No. 140

MIGRATION OF MULTIPLE SANDBARS ON THE CHIRIHAMA BEACH, JAPAN

Kazuhiro Hayakawa[1], Masatoshi Yuhi[2] and Hajime Ishida[3]

1. Introduction

Longshore bars play important roles in morphological processes in a number of sandy beaches. Recently, interannual behaviors of bars have been analyzed on the basis of the long-term survey records at several locations (e.g. Ruessink et al., 2003). Nevertheless, it is generally difficult to fully understand the temporal and spatial evolution processes due to the inherent complexity of the problem particularly for decadal scales. It is therefore important to continuously examine the characteristics of long-term variation of sandbars under various conditions. In this study, accordingly, the long-term variation of seabed profiles on the Chirihama Beach, Japan, is investigated using a set of field surveys. The focus is placed on the long-term characteristics of sandbar migrations.

2. Field Site

Chirihama Beach is located on the mid north coast of Japan on the Japan sea (Fig. 1). The coastline has a NNE-SSW general orientation. Typical sediment diameter is around 0.15mm. The beach has a driveway ('Nagisa Driveway') on the backshore (Fig. 1), where a number of tourists enjoy driving on the beach from spring to autumn. The tidal range at the coast is small (0.4m). Waves are calm in summer, while high waves have often been observed during winter season. Annual maximum significant wave height off the beach is in the range 5 to 8 m. The annual net longshore sediment transport is considered to be from the NNE to the SSW near the shoreline. It is known that large-scale multiple sandbars have evolved on the beach (Sunamura et al., 2007). However, the characteristics of temporal variations of these bars have not been examined yet in the existing studies.

3. Main Results

Bottom topographic survey has been carried out usually once a year on the Chirihama Beach by the Hokuriku Regional Development Bureau, Ministry of Land, Infrastructure and Transport, Japan. The location of the survey lines are shown in Figure 1.On the basis of the survey records obtained from 1998 to 2006, time variation of the cross-shore profiles has been examined (Fig. 2). The typical profile configurations vary from double barred to triple bar configuration. The movements of these bars are significant. Multiple bar systems along the beach clearly exhibits long-term cyclic behaviors; the bars migrate in net offshore direction repeatedly and 3 cycles are recognized in the figure. The offshore location and vertical elevation of the crest and trough of bars are then examined (Fig. 3). The geometric features such as bar height and bar length are also computed. Figure 4 shows the temporal variation of the cross-shore crest locations. The numbers near the circles indicate the bar heights, and the diameter of circle is proportional to them. It is recognized that the bars repeated periodic cycles consisting of generation, seaward migration and offshore decaying. The length of one cycle is approximately 8 years, and the interval between each cycle is approximately 3 to 4 years. When the offshore bar evolved most significantly, the height became as large as 3 to 4 m. The relation between the crest elevation and the trough elevations in the study area is plotted in Fig. 5. The slope of linear regression line is 1.53, which is slightly smaller than the value obtained by Keulegan (1945) for the west coasts of the United States (=1.69).

---

[1]Grad. Student, Kanazawa Univ., Kakuma-Machi, Kanazawa, 920-1192, Japan. haya1122@pref.ishikawa.lg.jp
[2]Faculty Env. Design, Kanazawa Univ., Kakuma-Machi, Kanazawa, 920-1192, Japan. yuhi@t.kanazawa-u.ac.jp
[3]Faculty Env. Design, Kanazawa Univ., Kakuma-Machi, Kanazawa 920-1192, Japan. hishida@t.kanazawa-u.ac.jp

4. Conclusion

Long-term migrations of sandbars on the Chirihama Beach, Japan, have been investigated using a set of field surveys collected over 9 years. The typical profile configurations are characterized by the presence of multiple bars. The movements of the bars are significant. For the time variation, it is shown that the net offshore periodic migrations of bars have been repeated. The length of one cycle is approximately 8 years, and the interval between each cycle is approximately 3 to 4 years. Further analyses are being made to investigate the relations between bar migration and retreat of shoreline.

References

Ruessink, B. G. et al., 2003. Intersite comparison of interannual nearshore bar behavior, *Journal of Geophysical Research*, 108(C8), 3249, doi:10.1029/2002JC001505.

Sunamura, T. and Takeda, I., 2007. Regional difference in the number of submarine longshore bars in Japan: An analysis based on breaking wave hypothesis, *Trans. Japanese Geomorphological Union*, 28(4), 381-398.

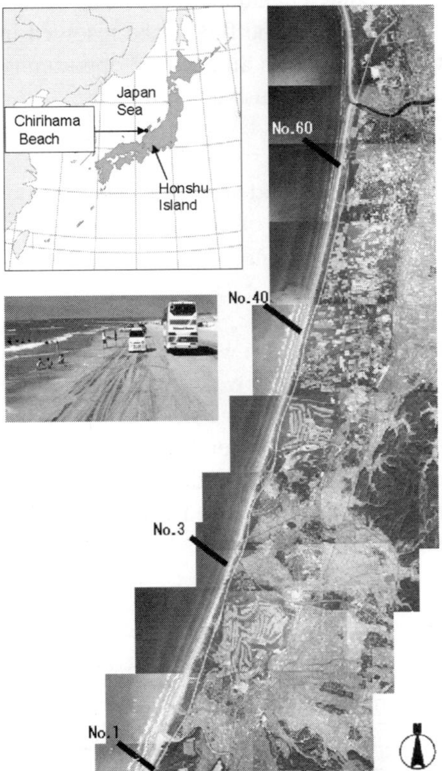

Figure 1. Field Site

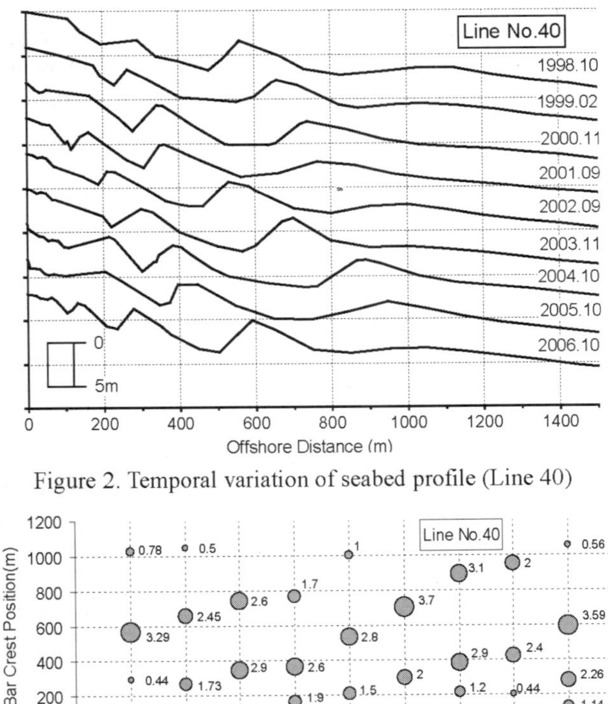

Figure 2. Temporal variation of seabed profile (Line 40)

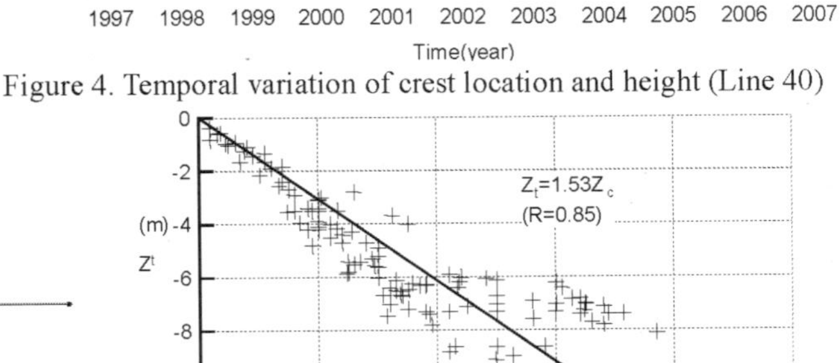

Figure 4. Temporal variation of crest location and height (Line 40)

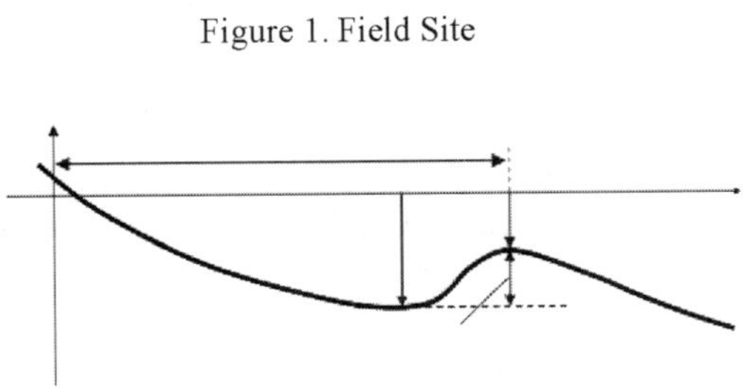

Figure 3. Definitions of bar dimensions

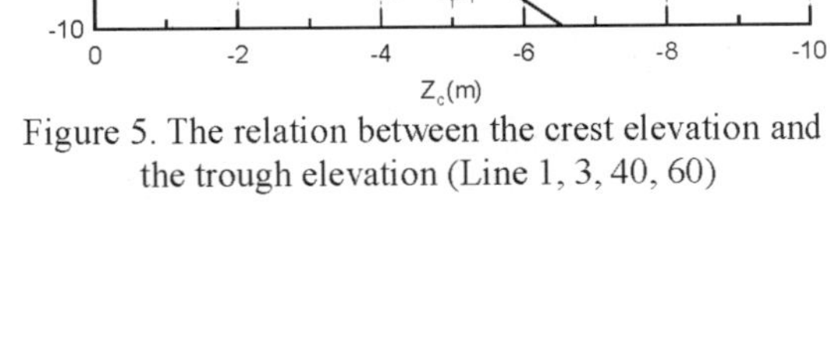

Figure 5. The relation between the crest elevation and the trough elevation (Line 1, 3, 40, 60)

Coastal Dynamics 2009
Paper No. 141

TURBULENCE AND SHEAR STRESS AROUND OFFSHORE WIND TURBINE PILE IN TIDAL CURRENTS USING PARTICLE IMAGE VELOCIMETRY

Dave McGovern[1], Suzana Ilic[1], Stuart McClelland[2], Andrew Folkard[1], Brendan Murphy[2]

1. Introduction

When a structure is placed in the marine environment it will alter the behavior of the flow around it; this can result in bed shear stress amplification, sediment transport, and consequently scour. Scour is estimated at a bridge pier or pile by laboratory derived empirical models and numerical models. Both models rely on an understanding of the scale effects and the fluid dynamics involved. Knowledge of such fluid dynamics in the case of tidal currents in the shallow sea environment around offshore structures, such as wind turbines is far from comprehensive. The technique of Particle Image Velocimetry (PIV) used for laboratory measurements allows large areas of the flow to be resolved in three-dimensions simultaneously and in a non-intrusive manner. This creates an exciting opportunity to map the flow field characteristics in greater detail than is possible with intrusive instruments.

2. Aims

To define in detail the spatio-temporal dynamics of the tidal current flow field around a scaled model of an offshore wind turbine pile. Furthermore, determine the effectiveness of collared and spiraled piles in altering the flow field to reduce bed shear stress.

3. Experimental Method

A series of tests were conducted over a rigid bed in a flume 16m long and 1.2m wide at the University of Hull's Deep Facility in the U.K. A pile 0.2m in diameter was placed in the exact mid point of the flume. Table one summarizes the experimental conditions for eight tests that correspond to approximately 1:20 scale with the prototype. To simulate the changes in current velocity (U) and water depth (H) throughout the tidal cycle, U and H are lowered and raised accordingly in the tests. Test P1B, P2E and P3F simulate a possible mid-tide scenario; tests P1C, P2G and P3H simulate a low-water scenario; while test P1D simulates a high-water scenario.

Table 1: Test conditions

Test	Current Velocity (U in m/s)	Water Depth (H in m)	Pile Reynolds Number (Re_D)	Froude Number (Fr)	Pile Type	Pile Diameter D (m)
P1A	0.22	0.28	3.24E+04	1.33E-01	smooth	0.2
P1B	0.33	0.25	4.85E+04	2.11E-01	smooth	0.2
P1C	0.23	0.1	3.38E+04	2.32E-01	smooth	0.2
P1D	0.15	0.4	2.21E+04	7.57E-02	smooth	0.2
P2E	0.33	0.25	4.85E+04	2.11E-01	spiral	0.2
P3F	0.33	0.25	4.85E+04	2.11E-01	ringed	0.2
P2G	0.23	0.1	3.38E+04	2.32E-01	spiral	0.2
P3H	0.23	0.1	3.38E+04	2.32E-01	ringed	0.2
Prototype	1	2-8	2.94E+06	1.43E-01	smooth	4

Extensive vertical PIV images ('slices') approximately 0.3 by 0.3m are captured along the centerline immediately behind the pile and up to 1.2m downstream for each test, providing a 2 minute time-

[1] Division of Geography, LEC, Lancaster University, Bailrigg, Lancaster, Lancashire U.K. d.mcgovern2@lancaster.ac.uk
[2]Department of Geography, University of Hull, Cottingham Road, Hull, U.K.

series of vector maps at a rate of 15Hz. Further slices are taken at three different positions off the centerline, parallel to it, and are also repeated downstream. The PIV slice set for each test gives a highly detailed time-averaged and instantaneous three-dimensional velocity field.

4. Results

Figure 1a and b show the 2 minute time-averaged velocity vectors and transverse velocity contours directly behind the smooth pile (figure 1a) and the collared pile (figure 1b). Bed shear stress amplification is calculated along the length of the slice. Figure 2a and b shows the amplification in bed shear stress related to both figure 1a and b respectively.

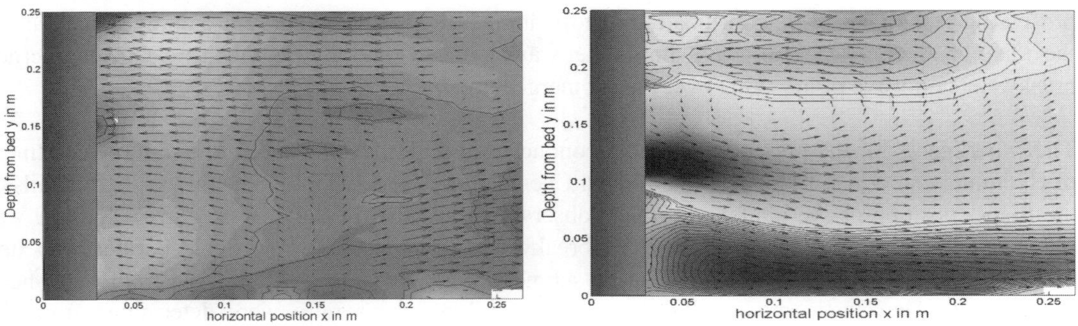

Figure 1a, b: 3D Velocity vector map for test P1B directly behind the smooth pile (1a, left); 3D Velocity vector map for test P2F taken directly behind the collared pile (1b, right). Pile is shown on the left hand side of 1a and b.

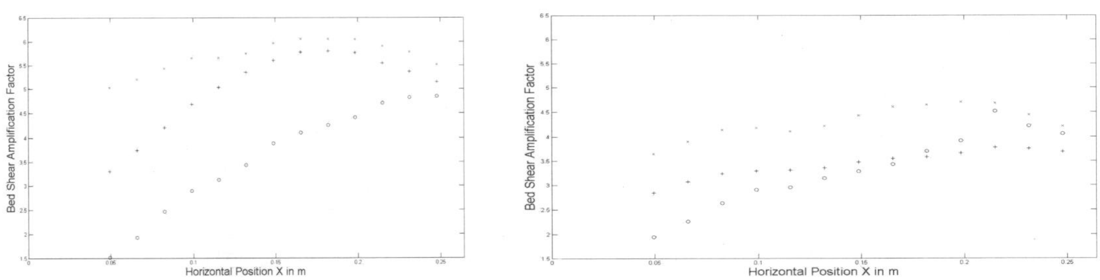

Figure 2a, b: Amplification of bed shear stress relating to the smooth pile (figure 1a) (2a, left) and the collared pile (figure 1b) (2b, right), where 'o' = centerline, '+' = 0.053m and '*' = 0.104m off the centerline.

Figure 1a shows a distinct stagnation point with reversal of flow behind the smooth pile. The strongest backflow is found near the surface, close to the pile, and the strongest stream wise flow is found beyond the stagnation point near the bed. Bed shear amplification shown in figure 2a shows the highest values on the centerline underneath the same area of this increase in velocity. Figure 1b shows a very different flow field behind the collared pile. There is a large vortical component near the surface with a horizontal axis that is dominant over the two minute period, and strong cross stream movements. Bed shear amplification (figure 2b) demonstrates that bed shear values are slightly lower at all positions. The rigid bed tests presented are supplemented by mobile bed tests involving full current reversal and measurements around the circumference of the pile. Calculation of turbulence and vorticity from the slices allows for a detailed spatial picture of the flow field.

Acknowledgements

This work is funded in full by the EPSRC.

Coastal Dynamics 2009
Paper No. 142

APPLICATION OF DIGITAL IMAGE PROCESS ON OBSERVING THE MOTIONS OF WAVE AND A SERIES OF PORO-ELASTIC SUBMERGED BREAKWATERS

Jian-Wu Lai[1], Yuan-Jyh Lan[2], Tai-Wen Hsu[3], Chia-Huan Ting[4] and Chen-Chen Chang[5]

The purpose of this study is to develop an image analysis system for laboratory observation based on a CCD camera. A high speed camera instead of intrusive wave gauges and linear displacement sensors is used to capture images for observing wave motions and structure deformation. Subsequent image processing techniques (e.g., coordinate transformation and edge detection) are employed to analysis the wave characteristics (e.g., wave amplitude, wave velocity) and the motions of elastic submerged breakwaters on wave propagating over the submerged breakwaters.

The experiments for the present study were conducted in a glass wall wave flume in the Department of Hydraulic and Ocean Engineering, National Cheng Kung University, Tainan, Taiwan. The details of the flume, position of the model and image observation system are shown in Figure 1. Sponge was selected to be the poro-elastic submerged breakwaters. The values of intrinsic permeability and turbulent drag coefficient for the poro-elastic submerged breakwaters were obtained from water-head experiments. The elastic modulus, shear modulus and poisson's ratio were determined from the standard test of Taiwan (CNS12352). The incident wave heights H_i are 1.5, 2.0, 2.5, 2.8 and 3.5cm, respectively. For each H_i, the incident wave period are determined by the appearance of Bragg resonant, where the ratios of bar spacing to wave length $2S/L$ are near 1.0 and 2.0, and non-resonant case that $2S/L$ is near 1.5.

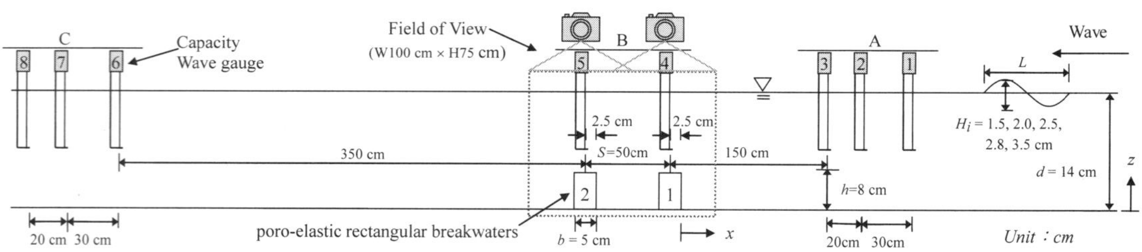

Figure 1. Schematic diagram of the experimental arrangement.

This image analysis system includes the following three procedures (Hsiao et al., 2008): First of all, the direct linear transformation (DLT) is used to calibrate the intrinsic and extrinsic camera model. After calibration, the image is transformed to physical coordinates parallel to the wave flume. Secondly, image reconstruction and merging are processed in order to carry out ortho-image. Thirdly, edge detection and data extraction are utilized to obtain time series products in order to obtain information of wave profile and structures deformation. The visualization outcome of image analysis system is demonstrated in Figure 2.

[1]Dept of Hydraulic & Ocean Eng., National Cheng Kung University, Tainan 701, Taiwan. laijw@mail.hyd.ncku.edu.tw
[2]Dept of Hydraulic & Ocean Eng., National Cheng Kung University, Tainan 701, Taiwan. yjlan@mail.ncku.edu.tw
[3]Dept of Hydraulic & Ocean Eng., National Cheng Kung University, Tainan 701, Taiwan. twhsu@mail.ncku.edu.tw
[4]Dept of Hydraulic & Ocean Eng., National Cheng Kung University, Tainan 701, Taiwan. mpl0204@gmail.com
[5]Dept of Hydraulic & Ocean Eng., National Cheng Kung University, Tainan 701, Taiwan. n8697423@mail.ncku.edu.tw

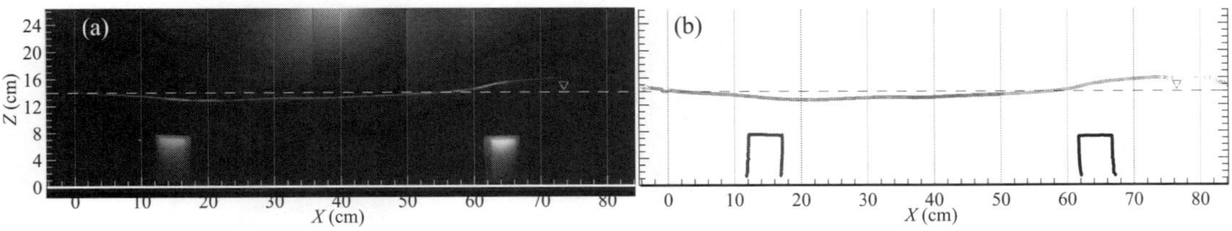

Figure 2. Outcome of image analysis system (a) Merged ortho-image (b) Extracted water and structures

The interaction between waves and a series of poro-elastic submerged breakwaters in Bragg resonant case and non-resonant case will be discussed and compared with the analytical solution proposed by Lan (2007). Figure 3(a) and Figure 3(b) shows one of the comparisons between experimental results and analytical solutions at right-top corner of first structure and left-top corner of second structure, respectively. Motions of water surface and poro-elastic structures are extracted from subsequent ortho-image. The amplitude, phase and frequency of water surface and poro-elastic submerged breakwaters are revealed. However, the motions of poro-elastic breakwaters in vertical direction are smaller than the image resolution where it is 0.04cm/pixel for the experimental image system. In this study, a polynomial regression for the gradient of image gray level will be employed to increase the resolution in the motions of poro-elastic structures.

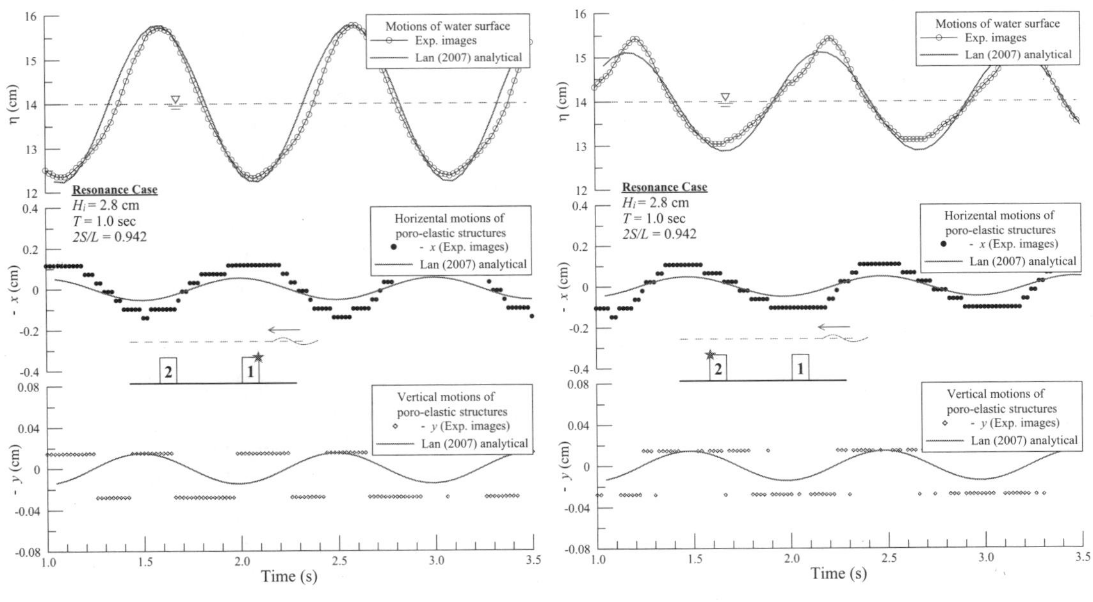

Figure 3. Motions of wave and poro-elastic submerged breakwaters (a) at right-top corner of 1st breakwater (b) at left-top corner of 2nd breakwater.

References

Lan, Y.J., 2007. Analysis Waves Interaction with a Series of Poro-Elastic Submerged Structures. *Proceedings of the 29th Ocean Engineering Conference in Taiwan, Republic of China*, 331-336. (in Chinese)

Hsiao Y.H., Lin Y.C., Chuang S.C. and Huang M.C., 2008. Data Analysis Procedure of Coastal Image Monitoring System at Hsi-tzu-wan. *Proceedings of the 30th Ocean Engineering Conference in Taiwan, Republic of China*, 781-786. (in Chinese)

Coastal Dynamics 2009
Paper No. 143

PIV Measurements of a Turbulent Round Jet in Wave Environment

Shih-Chun Hsiao[1], Tai-Wen Hsu[2], Jian-Feng Lin [3] and Kuang-An Chang[4]

Turbulent jets discharged into a wave environment have been an important problem to a wide spectrum of engineering practice, including hydraulic engineering, environmental engineering, coastal and ocean engineering, and the power industry. Primary examples are the outfalls located along a coastline such as the discharge of cooling water from a power plant cooling system into shallow water, and the treated/untreated discharge of waste water into the coastal water from municipality in a coastal region. It is of great concern to engineers, industry and government regulators that waste effluence disposed into a large water body needs to be effective in mixing and to maintain an acceptable level of concentration. In addition, as a wavy free surface exists near a turbulent jet, the behavior of the jet will change significantly.

The goal of the present study is to investigate the turbulent fields of a turbulent round jet under different wave conditions. The experiments were conducted in a glass-walled wave flume located at the Department of Hydraulic and Ocean Engineering, National Cheng Kung University. All statistical velocity fields were obtained from 1,320 instantaneous vector fields at the fixed phase for each test. In the post-processing, the mean flow and RMS (Root-Mean-Square) velocity fluctuation were extracted using the phase averaged method of all instantaneous velocity maps located at that spatial location.

To examine the properties of turbulence round jet under wave environments, the Particle Image Velocimetry (PIV) technique is employed. The mean velocities at several cross sections were measured from the jet centerline with and without waves. The self-similar behavior of the free jet is shown in Fig. 1(a), where symbols '*c*' indicates centerline value (due to the jet centerline is oscillating up and down when it under waves, the jet centerline velocity is defined as the location along the jet discharge direction where the maximum horizontal velocity of jet occurs at each vertical cross section.) and $< >$ indicates a phase averaged quantity. The plot covers data at five different cross section downstream of the nozzle at $x/D=[38.90, 46.32, 55.73, 61.73, 68.54]$. The jet half-width, $\eta_{1/2}$, is often used to describe the radial length scale of the round jet, and the spread rate of the jet. The jet half-width is defined as the location at z_a/x where $<u>/<u_c>=0.5$. The value of the $\eta_{1/2}$ for the free jet in Fig. 1 (a) is 0.101 which is in good agreement with the value of 0.094 reported by Hussein et al. (1994), 0.1 by Cowen et al. (2001) and 0.09 by Ryu et al., (2005). However, the normalized mean velocity profiles for the tip-top case of H = 3.0 cm (Fig. 1(b)) didn't collapse onto a single self-similar curve. We found that the jet spreading became wider as the wave height became larger. Fig. 2 (a) presents the RMS velocity fluctuation profiles of the free jet. The normalized mean centerline values averaging over five sections for the free jet was 0.282. This compared well with Wygnanski and Fiedler's (1969) value of 0.28 and Hussein et al.'s (1994) value of 0.28. As expected, the width of RMS velocity fluctuation of the jet increased when it under wave for the tip-top case (H = 3.0 cm) (Fig. 2 (b)). Fig. 3 (a) shows the RMS velocity fluctuation profiles in the near-exit region of the free jet and comparison with previous measurements of Hussein et al. (1994) and Mi et al. (2007). Fig. 3 (b) is the comparison of RMS velocity fluctuation profiles at the same section for all cases in wave

[1]Dept of Hydraulic & Ocean Eng., National Cheng Kung University, Tainan 701, Taiwan. schsiao@mail.ncku.edu.tw
[2]Dept of Hydraulic & Ocean Eng., National Cheng Kung University, Tainan 701, Taiwan. twhsu@mail.ncku.edu.tw
[3]Dept of Hydraulic & Ocean Eng., National Cheng Kung University, Tainan 701, Taiwan. n8895107@mail.ncku.edu.tw
[4]Dept of Civil Eng., Texas A&M Univ., College Station, Texas, 77843-3136, USA. kchang@tamu.edu

environment. Furthermore, the radial of $<u'^2>/<u_c>^2$ increased when the wave height increased. More experimental results related to the effects of wave nonlinearity and phase on the turbulent properties will be presented in the full paper.

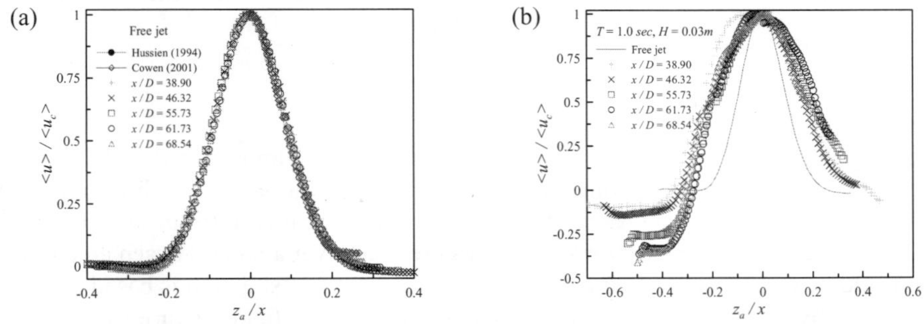

Fig. 1 Cross-sectional Jet mean velocity profiles of (a) Free jet and (b) $H = 3.0$ cm

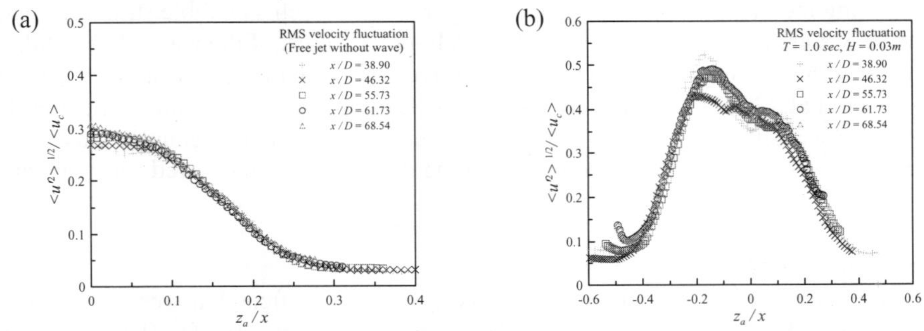

Fig. 2 RMS velocity fluctuation profiles of (a) Free jet and (b) $H = 3.0$ cm

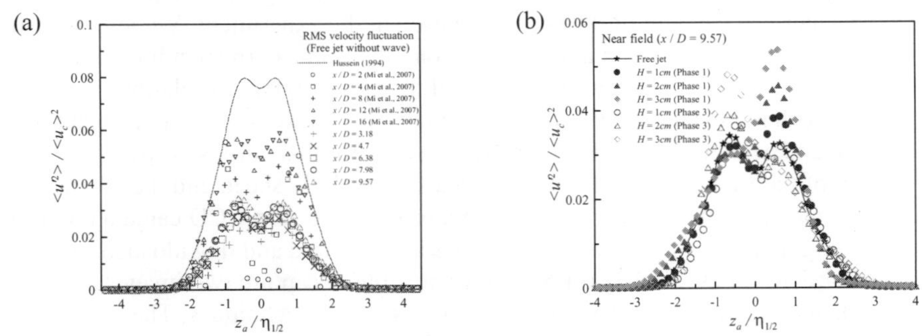

Fig. 3 (a) The RMS velocity fluctuation profiles in the near-exit region of the free jet. (b) The comparison of RMS velocity fluctuation profiles for all cases in wave environment (at $x/D = 9.57$)

References

Cowen, E.A., Chang, K.A. and Liao, Q., 2001. A single camera couple ptv-lif technique. *Experiments in Fluids*, 31: 63-73.

Hussein, H.J., Capp, S.P. and George, W.K., 1994. Velocity measurements in a high-Reynolds-number, momentum-conserving, axis-symmetric, turbulent jet. *Journal of Fluid Mechanics*, 258: 31-75.

Mi, J., Kalt, P., Nathan, G.J. and Wong, C.Y., 2007. PIV measurements of a turbulent jet issuing from round sharp-edged plate. *Experiments in Fluids*, 42: 625-637.

Ryu, Y., Chang, K.A. and Mori, N., 2005. Dispersion of neutrally buoyant horizontal round jet in wave environment. *Journal of Hydraulic Engineering*, 131: 1088-1097.

Wygnanski, I. and Fiedler, H., 1969. Some measurements in the self-preserving jet. *Journal of Fluid Mechanics*, 38: 577-612.

Coastal Dynamics 2009
Paper No. 144

DCIV DERIVED FLOW FIELDS FOR WAVES BREAKING OVER A BAR

Kessie Govender[1], Michael J. Alport[1], Gary P. Mocke[2], Hervé Michallet[3], Mathieu Mory[4] and Francois Smit[2]

1. Introduction

Natural beaches generally consist of one or more sand bars running almost parallel to the shore line. These features usually begin to occur coincident with the initial position of wave breaking marking the outer limit of the surf zone. The present investigation is an endeavor to fill in some of the gap in the understanding of the processes occurring in waves breaking over a natural beach and to provide a data set that would be useful for modeling purposes. The present investigation is based on the same bottom profile used by Mory (1994).The instantaneous velocity flow fields at various positions beyond the bar were measured using video camera based digital correlation image velocimetry (DCIV). By tracking the motion of neutrally buoyant particles, as well as bubble structures within the crest of the wave, we are able to measure velocities well into the crest of the wave. The validation of this measurement technique together with results for waves breaking on a plane slope beach can be found in Govender et al. (2002). The time averaged velocities, mass/velocity and momentum fluxes, and kinetic energies as well as turbulence intensities were then computed. A coupled wave roller and vertical turbulence model is further applied to simulating these detailed measured surf zone processes.

2. Experimental procedures

Experiments on regular two-dimensional waves breaking over an artificial bar were conducted in a glassed walled flume. The bar consists of a 1:20 up slope, followed by a 0.10 m flat section and then a 1:10 down sloping section followed by a flat section. Waves breaking on the bar, and with a period of 1.5s and a deep water wave height of 0.074 m were used in the experiment. Velocity measurements in these waves were restricted to horizontal positions along the flume corresponding to the down sloping part of the bar and the flat section shorewards of it. In the Digital-correlation-image-velocimetry (DCIV) technique the fluid velocity is estimated by tracking the displacement of a small region of the seeded and aerated flow. The experimental setup for this purpose is as follows: A longitudinal section of the flume was illuminated with a strobed light sheet, and the aeration/bubble structure and illuminated particles were imaged using a progressive scan CCD camera connected to a frame grabber residing in a PC. While the video images are captured and downloaded to the frame grabber using standard TV rates of 60Hz (or 50Hz) we are able to capture two images a few milliseconds(less than one frame time) apart using a novel strobing techniques. The instantaneous velocity field was then obtained by means of cross correlation techniques.

3. Results

The time averaged velocity vector field for positions beyond the break-point is shown in Figure 1. Figures 2 shows a comparison of our measurements of the undertow with those of Boers(2005) involving random waves breaking over similar bar structure. Whereas the LDA measurements are

[1] School of Physics, University of KwaZulu-Natal, Durban South Africa. govenderks@ukzn.ac.za , alport@ukzn.ac.za

[2]WorleyParsons Infrastructure & Environment, Al-Khuwair, Sultanate of Oman. gary.mocke@worleyparsons.com

[3]Laboratoire des Ecoulment Geophysiques et Industriels, Grenoble, France. herve.michallet@hmg.inpg.fr

[4] Ecole Nationale Supèrierure en Génie des Technologies Industrielles, Université de Pau et des Pays de l'Adour, France. mathieu.mory@univ-pau.fr

only available for positions below the trough level, the video DCIV results presented in the paper cover the entire water depth. This enables us to consider the conservation of mass, momentum and energy at various positions along the bar. The relative fluid density in the crest of the breaking waves is determined from the conservation of mass. The contribution of the different terms in the radiation stress is evaluated in the breaking zone. Turbulence intensities and turbulent kinetic energies are also presented. Numerical model predictions of several of these parameters obtained using a coupled roller and vertical turbulence model will also be compared with the measurements.

References

Boers, M., 2005. Surf zone turbulence, Ph.D thesis, Delft University of Technology.
Govender, K., Alport, M.J., Mocke, G., 2002. Video imaged surf zone wave and roller structures and flow fields. *J. Geophy Res.*, 107(C7), 3072, doi:10.1029/2000JC000755
Mory, M., 1994. Currents induced by breaking waves over a bar. Modelling of Coastal and Esturine Processes, Seminaire Franco Portugais de Modelisation en Hydraulique Maritime, 27-38.

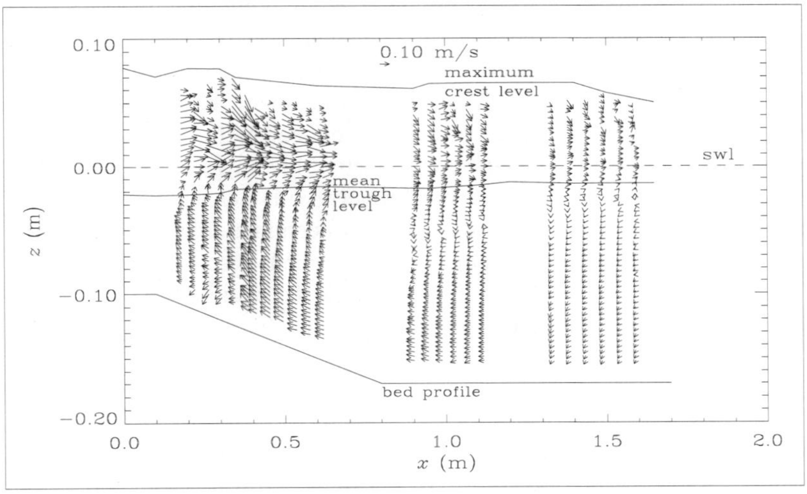

Figure 1. Time averaged vector field for positions beyond the break point

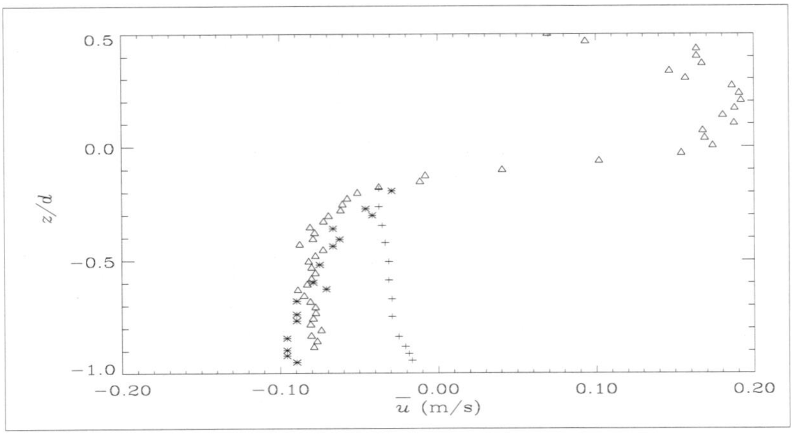

Figure 2. Undertow in the present experiment(diamonds) at x=0.37m compared with the LDA measur ements of Boers (2005)(Test1B-stars) (Test1C-pluses) for a similar position and setup

Coastal Dynamics 2009
Paper No. 145

IMPROVING BYPASS AND INCREASING NAVIGATION DEPTH: A VISION FOR HVIDE SANDE HARBOUR, DENMARK

Nicholas Grunnet[1], Ida Brøker[2], Erik Clausen[3] and Per Sørensen[4]

1. Introduction

A harbour on a coast blocks the littoral drift causing accretion and erosion at the up- and downdrift side, respectively. Eventually a significant part of the littoral drift may bypass the harbour. In this case it is important to minimize the deposition of the passing sediment in the harbour basin and the harbour mouth. Traditionally navigation channels are maintained, updrift sediment reservoirs are dredged and sand is bypassed mechanically or by repeated dredging and disposal. The sedimentation and depth in front of the harbour depends not only on the littoral drift but also on the layout of the protective breakwaters. The full paper presents an optimisation of breakwater layout in promoting bypass.

Hvide Sande harbour in Denmark is protected by shore-perpendicular breakwaters (see Figure 1) and experiences considerable sedimentation in its access channel. Facing the need to accommodate larger fishing vessels, the harbour wishes to increase the navigation depth in front of the harbour entrance while at the same time reducing the sedimentation in the access channel. An new scheme is proposed (see Figure 2) consisting of a combination of new protective breakwaters and an updrift capital dredging of the coastline: while the streamlined breakwaters will increase the bypass of sediment past the harbour mouth by increasing the flow velocity due to contraction, the retreat of the coastal profile will help maintain the required additional water depth needed. The present morphological study shows that the proposed scheme results in an increased equilibrium depth of the bypass bar in front of the harbour without increasing the downdrift erosion, and an improved natural bypass of the littoral drift.

2. The field site today

The West Coast of Denmark is exposed to the severe North Sea wave climate. Hvide Sande harbour is located in the central part of this very exposed coast where the net littoral drift is southward on the order of 1,000,000 m^3/year. Hvide Sande harbour is located at a tidal inlet on this coastline, on one of the narrow sandy barriers which divide the coastal lagoon from the sea. Sluices regulate the water exchange between the lagoon and the sea.

As a result of the updrift impoundment of sediment north of the harbour, the shoreline is advancing at a rate of 4-5 m/year. The downdrift erosion is effectively counteracted by the implementation of a combined shoreface and beach nourishment programme on the order of 300,000 m^3/year. Maintenance dredging on the order of 170,000 m^3/year is necessary to ensure a safe navigation depth presently on the order of 4.5-5.0 m.

3. Towards a new equilibrium

The main elements of the proposed scheme are the extension of the existing northern breakwater, kept to a minimum adding only 80 m to its present length, the construction of a new southern breakwater with a total length of approximately 700 m, and a capital dredging on the order of 1,200,000 m^3 over

[1]Coastal and Estuarine Dynamics, DHI Water Environment & Health, Agern Allé 5, DK-2970 Hørsholm, Denmark, ngr@dhigroup.com
[2]Coastal and Estuarine Dynamics, DHI Water Environment & Health, Agern Allé 5, DK-2970 Hørsholm, Denmark, ibh@dhigroup.com
[3]Hvide Sande Harbour, Fossanæsvej 22, Blåtårn, DK-6960 Hvide Sande, Denmark, ec@hvshavn.dk
[4]Danish Coastal Authority, Højbovej 1, DK-7620 Lemvig, Denmark, pso@kyst.dk

an alongshore stretch of approximately 1,600 m immediately north of the existing harbour. The dredging will result in an approximately 160 m retreat of the present shoreline. The dredged material is planned to serve as nourishment on the downdrift shoreline.

The study is supported by a comprehensive monitoring data set comprised of hourly wave, wind and water level measurements, regular beach profile data, and weekly soundings of the harbour entrance in its present configuration and makes use of a state-of-the-art morphological modelling tool.

Figure 1. Present breakwater configuration of Hvide Sande harbour

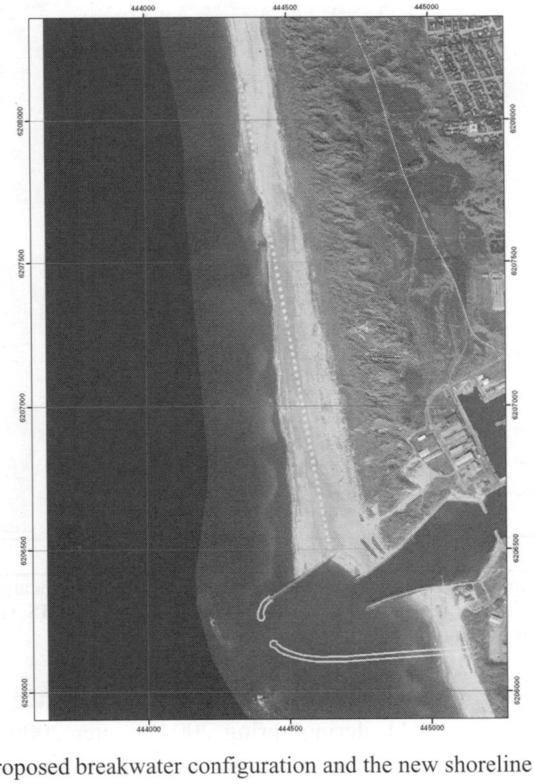

Figure 2. Proposed breakwater configuration and the new shoreline (dotted line)

Coastal Dynamics 2009
Paper No. 146

HYDRODYNAMIC RESPONSE OF A TRANSGRESSIVE SHOAL TO THE PROPOSED MINING FOR RESTORING ADJACENT BEACHES AND BARRIERS: SABINE BANK, OFF LOUISIANA-TEXAS COAST, UNITED STATES

Felix Jose[1], Gregory W. Stone[2], Daijiro Kobashi[2], Seyed M. SaidatMousavi[2] and Baozhu Liu[1]

[1] Coastal Studies Institute, Louisiana State University, Baton Rouge, LA-70803, USA
[2] Coastal Studies Institute and Department of Oceanography and Coastal Sciences, Louisiana State University, Baton Rouge, LA 70803, USA

Sabine Bank, a transgressive shoal located 30 km off Louisiana-Texas border, has been considered as one of the potentially viable resources for re-nourishment of the adjacent barrier islands and beaches due to the available sand volume and proximity to target restoration areas. The shallow shoal encompasses an area of 600 km^2 and is divided into two sections, viz., western and eastern Bank, by the Sabine Pass (see Figure 1). Fisheries survey from the region estimated that fish abundance was over two times greater on the east side of Sabine Pass compared to the west side. The study also demonstrated that the sandy areas of the shoal were not hypoxic at the time of the cruise, with a mean oxygen level of 4.4 mg O_2/L being recorded at the bottom. Not much has been reported on the bottom boundary layer dynamics and sediment transport from the shoal environment. A comprehensive field investigation, coupled with hydrodynamic modeling has been conducted for the Sabine Bank, as a first step before venturing for large scale targeted sand mining.

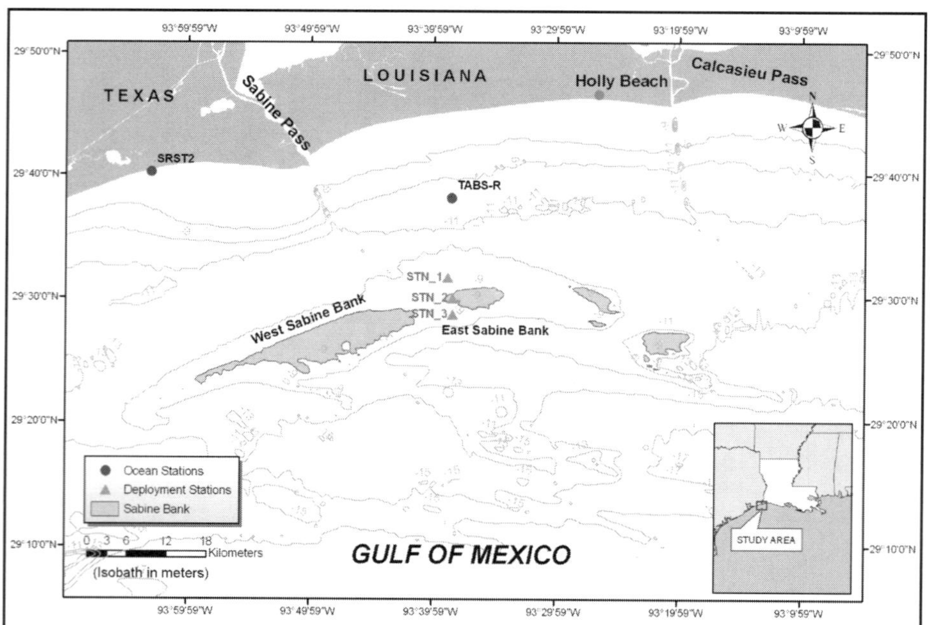

Figure 1. Location map. The yellow triangles show the deployment locations during 2004, 2006 and 2008. Additional met-ocean data were collected from TABS station and a met observation station near Sabine Pass (purple dots).

Three bottom boundary layer arrays were deployed on the crest of the shoal and nearshore and offshore flanks of the shoal (Figure 1), during spring 2004, winter 2006 and summer 2008. Time series observations of waves, currents, water level, near bottom suspended sediment concentration, temperature and salinity, were measured for 30-40 days each, during these deployments. Surficial sediments were collected during the deployment and retrieval cruises. Bottom boundary layer parameters, wave and current induced shear stress and shear velocities were computed. Sediment transport rate was quantified using Grant and Madsen Model.

The in situ observations showed that waves were weak and did not re-suspend sediment during fair weather conditions (see Figure 2). Except during calm conditions, currents were sufficiently strong to re-suspend sediments. Wave and bottom boundary layer interactions were strongly associated with the passage of cold fronts across the region. Strong southerly/southeasterly wind regimes also affected the wave and the bottom boundary layer interactions during the observation periods. During summer 2008, bottom boundary layer dynamics were significantly influenced by the high wind regime associated with high pressure system prevailed over the U.S. East Coast. The high wind regime, during fair weather, accompanied high wave height and relatively strong currents over the bank (Figure 2). Frequencies of this high wind regime were approximately every 2 to 3 weeks.

MIKE 21 Spectral wave model, developed by DHI Water and Environment, was implemented for the coast. The model was calibrated using offshore wave data from NDBC buoys and for the shoal, *in situ* observations from spring 2004, were used. Three case studies were conducted with case 1 corresponds to the pre-mining bathymetry and the other two with modified bathymetries in accordance with two realistic mining scenarios. It has been observed that alteration in significant wave height, peak period and dominant direction were insignificant with the modified bathymetries (Figure 3). Sediment re-suspension intensity (RI) was high over the inner shelf and shoal during severe and strong storms. During moderate storm conditions, the RI decreases from the shallowest western shoal to the deeper eastern shoal; the RI off the shoal was significantly lower than on the shoal. The modification in the RI values with targeted mining was minimal. It is suggested that small scale dredging should have minimal impacts on the hydrodynamics of the region. Detailed analysis of the wave-current regime for different met-ocean conditions as well as wave transformation over the shoal for different energetic conditions are discussed in the full paper.

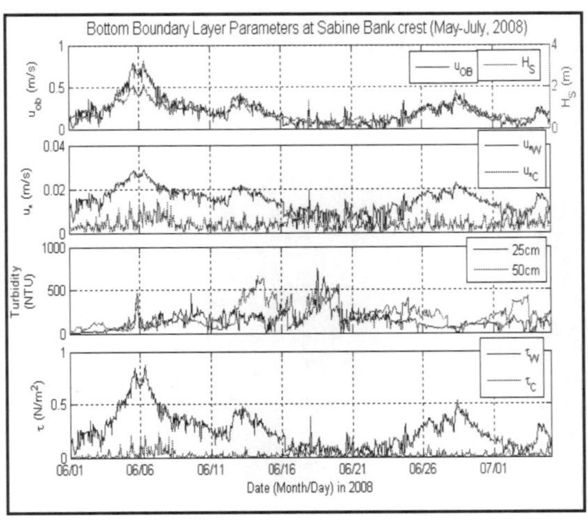

Figure 2. Time series of wave orbital velocity (left) and wave height (right), (b) shear velocity, (c) turbidity and (d) shear stress, measured at the crest of the shoal during the summer 2008 deployment

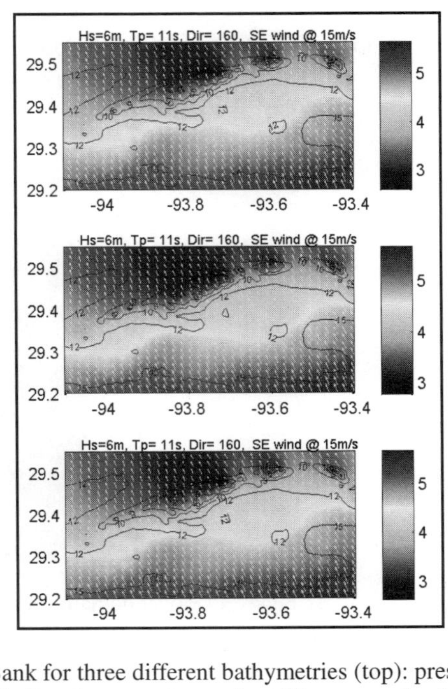

Figure 3. Significant wave height distribution over the Sabine Bank for three different bathymetries (top): present condition, (middle): Cumulative mining scenario, (bottom): Holly Beach restoration project. The area and volume for targeted mining were provided by U.S. Minerals Management Service (MMS)

Coastal Dynamics 2009
Paper No. 147

MODELING MORPHOLOGIC RESPONSES TO PROPOSED ENGINEERING MODIFICATIONS AT SEBASTIAN INLET, FL

Florian G. Brehin[1] and Gary A. Zarillo[2]

Numerical simulations of hydrodynamics and sediment transport were conducted for Sebastian Inlet, FL, using the Coastal Modeling System (CMS). The study is aimed at determining changes in morphology under the existing topography and configuration as compared with two proposed modification scenarios: 1) an extension of the south jetty and 2) dredging of the outer section of the ebb shoal. The morphology change modeling scheme is based on coupling the circulation and sediment transport model CMS-FLOW with predictions of the wave regime provided by CMS-WAVE under the Surface Water Modeling System (SMS) interface (Figure 1).

Runs were conducted for two 6-month periods within 2007, incorporating a variety of sea states and storm events. Availability of measured topographic data for winter 2007, June 2007 and November 2007 enabled verification of model performance. Model runs were driven by water surface elevation time series applied at the north and south interior boundaries and along ocean boundary of the model domain. These inputs were coupled with forcing from the CMS-WAVE model driven by hind cast wave data. Other model features included tagging both hard bottom/non-erodible cells in the inlet channel and observation cells along the shoreface to calculate littoral sand transport.

Analysis of existing conditions indicated that the predicted morphologic changes were in agreement with measured topographic data. Both measured and predicted morphologic data indicated sand deposition on the ebb shoal, on the attachment bar to the south of the inlet entrance, and in a dredged sand trap adjacent to the inlet throat section. Erosion was measured and predicted at the jetty tips and within the inlet throat. Model runs simulating dredging of the lower ebb shoal for beach fill sand indicated accretion in the borrow excavation and a decrease in sand bypassing around the inlet due to accumulation in the borrow cut section (Figure 2). Runs simulating the effects of the south jetty extension indicated accretion at the seaward end of the south jetty and erosion of the upper shoreface immediately south of the inlet.

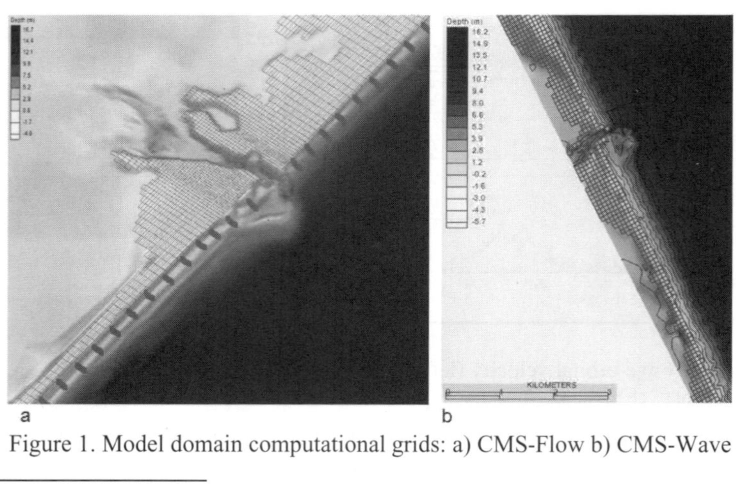

a b

Figure 1. Model domain computational grids: a) CMS-Flow b) CMS-Wave

[1] Dept of Marine and Environmental Systems, Florida Institute of Technology, Melbourne, USA fbrehin@fit.edu
[2] Dept of Marine and Environmental Systems, Florida Institute of Technology, Melbourne, USA zarillo@fit.edu

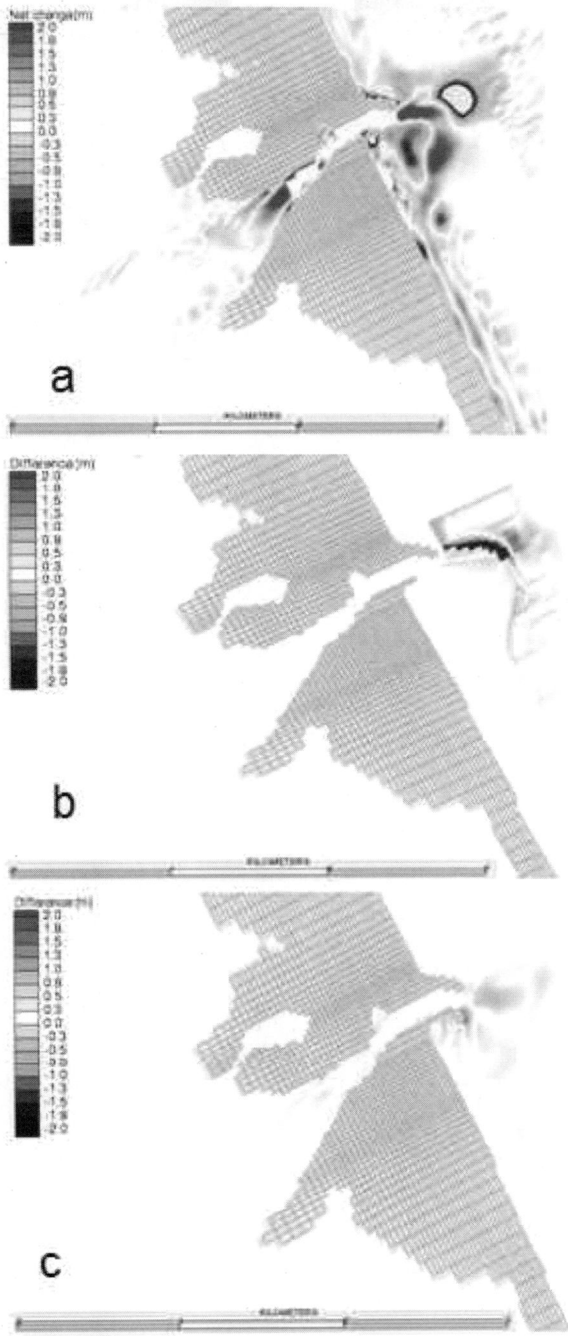

Figure 2. Example outputs from a 6- month model run (June to December 2007): a) net topography change for existing conditions, b) difference in net topographic change (dredging – existing), difference in net topographic change (south jetty extension – existing)